SWITCHMODE
POWER SUPPLY
HANDBOOK

Other McGraw-Hill Reference Books of Interest

Handbooks

CHRISTIANSEN • *Electronics Engineers' Handbook, 4/e*
COOMBS • *Electronic Instrument Handbook, 3/e*
COOMBS • *Printed Circuits Handbook, 4/e*
HARPER • *Electronic Packaging and Interconnection Handbook, 2/e*
JURAN AND GRYNA • *Juran's Quality Control Handbook*
JURGEN • *Automotive Electronics Handbook, 2/e*
JURGEN • *Digital Consumer Electronics Handbook*
OSA • *Handbook of Optics, 2/e*
RORABAUGH • *Digital Filter Designers' Handbook, 2/e*
SERGENT and HARPER • *Hybrid Microelectronics Handbook*
SHARRA • *Programmable Logic Handbook*
VALENTINE • *Motor Control Electronics Handbook*
WAYNANT • *Electro-Optics Handbook*
WILLIAMS and TAYLOR • *Electronic Filter Design Handbook*
ZOMAYA • *Parallel and Distributed Computing Handbook*

Other

BEST • *Phase-Locked Loops, 3/e*
DEVADAS • *Logic Synthesis*
GRAEME • *Optimizing Op Amp Performance*
GRAEME • *Photodiode Amplifiers Op Amp Solutions*
HECHT • *The Laser Guidebook*
JOHNSON • *LabVIEW Graphical Programming, 2/e*
JOHNSON • *LabVIEW Power Programming*
KELLER • *Mastering MathCAD*
KIELKOWSKI • *Inside SPICE, 2/e*
MASSOBRIO • *Bioelectronics*
PILLAGE • *Electronic Circuit and System Simulation Methods*
PRESSMAN • *Switching Power Supply Design, 2/e*
RORABAUGH • *DSP Primer*
SANDLER • *SMPS Simulation with SPICE*
SCLATOR • *McGraw-Hill Electronics Dictionary, 6/e*
SMITH • *Thin-Film Deposition*
TAYLOR • *Hands-on DSP*
VAN ZANT • *Microchip Fabrication, 3/e*

SWITCHMODE POWER SUPPLY HANDBOOK

Keith H. Billings, C.Eng., M.I.E.E.

President
DKB Power Inc.
Guelph, Ontario Canada

Second Edition

Boston, Massachusetts   Burr Ridge, Illinois
Dubuque, Iowa   Madison, Wisconsin   New York, New York
San Francisco, California   St. Louis, Missouri

Library of Congress Cataloging-in-Publication Data

Billings, Keith H.
    Switchmode power supply handbook / Keith Billings.—2nd ed.
        p.   cm.
    Rev. ed. of: Handbook of switchmode power supplies, c1989.
    Includes index.
    ISBN 0-07-006719-8
    1. Electronic apparatus and appliances—Power supply—Handbooks,
manuals, etc. I. Billings, Keith H. Handbook of switchmode power
supplies. II. Title.
    TK7881.15.B55   1999
    621.381'044—dc21   98-53827
                                                            CIP

McGraw-Hill

A Division of The McGraw·Hill Companies

        7 8 9 BKM BKM   0 9 8 7 6

ISBN 0-07-006719-8

*The sponsoring editor for this book was Stephen S. Chapman, the editing supervisor was Stephen
M. Smith, and the production supervisor was Modestine Cameron. It was set in Times Roman.*

CONTENTS

PART 2   DESIGN: THEORY AND PRACTICE

PART 4 SUPPLEMENTARY

PREFACE

During the thirty or so years that the author has specialized in the design of linear and switchmode power supplies, the desire for a general handbook on the subject has often been expressed. Here is the author's answer to this need, a practical, easy to read explanation of many of the techniques in common use today, together with some of the latest developments.

An easily assimilated, nonacademic treatment, using simplified theory and mathematical analysis, has been deliberately adopted. (The author craves the indulgence of his specialist colleagues for waiving the fully rigorous approach in the interests of simplicity.) As a result, the book should appeal to students, junior engineers, and interested nonspecialist users, as well as to the practicing professional power supply engineer. It covers the subject from simple system explanations (with typical specifications and performance parameters) to the final component, thermal, and circuit design and evaluation.

The author has drawn upon his own experience of the questions most often asked by students and junior engineers to address the subject in the most straightforward way, giving explicit design examples which do not assume any previous knowledge of the subject. In particular, the design of the wound components is covered very fully, since these are critical to the final performance, but tend to be rather poorly understood.

To simplify the design approach, considerable use has been made of nomograms, many of which have been developed by the author, originally for his own use. Some of the more academic supporting theory is covered in the chapter appendixes, and those who wish to go further should read these and the many excellent specialized books and papers mentioned in the references.

Over the last two decades, switchmode power supply design has developed from a somewhat neglected "black art" to a precise engineering science. The rapid advances in electronic component miniaturization and space exploration have led to an ever-increasing need for small, efficient power processing equipment. In recent years this need has caught and focused the attention of some of the world's most competent electronic engineers. As a result of intensive research and development, there have been many new innovations with a bewildering array of topologies.

As yet, there is no single "ideal" system that meets all needs. Each topology lays claim to various advantages and limitations, and the power supply designer's skill and experience is still needed to match the specification requirements to the most suitable topology to define the preferred technique for a particular application.

The modern switchmode power supply will often be a small part of a more complex processing system. Hence, as well as supplying the necessary voltages and currents for the user's equipment, it will often provide many other ancillary functions—for example, power good signals (showing when all outputs are within their specified limits), power failure warning signals (giving advanced warning of line failure), and overtemperature protection, which will shut the system down before damage can occur. Further, it may respond to an external TTL demand for power on or power off. Power limit and current limit circuitry will protect the supply and load from fault conditions. Overvoltage protection is often provided to protect sensitive loads from overvoltage conditions, and in some special applications, synchronization of the switching frequency to an external clock will be provided. Hence, the power supply designer must understand and meet many needs.

To utilize or specify a modern power processing system more effectively, the user should be familiar with the advantages and limitations of the many techniques available. With this information, the system engineer can specify the power supply requirements so that the most cost-effective and reliable system may be designed to meet these needs. Very often a small change in specification or rearrangement of the power distribution system will allow the power supply designer to produce a much more reliable and cost-effective solution to the user's needs. Hence, to produce the most reliable and cost-effective design, the development of the specification should be an interactive exercise between the power supply designer and the user.

Very often, power supply specifications have inflexible and often artificial boundaries and limitations. These unrealistic specifications usually result in overspecified requirements and hence an overdesigned supply. This in turn can entail high cost, high complexity, and lower reliability. The power supply user who takes the trouble to understand the limitations and advantages of modern switchmode techniques will be in a far better position to specify and obtain reliable and cost-effective solutions to power supply requirements.

The book is presented in four parts:

Part 1, "Functional Requirements Common to Most Direct-Off-Line Switchmode Power Supplies," covers, in simple terms, the requirements which tend to be common to any supply intended for operation direct from the ac line supply. It gives details of the various techniques in common use, highlighting their major advantages and limitations, together with typical applications.

Part 2, "Design, Theory and Practice," considers the selection of power components and transformer designs for many well-known converter circuits. It is primarily intended to assist practicing power supply engineers in developing conservatively rated prototypes with more speed and minimum effort. It provides examples, information, and design theory sufficient for a general understanding and the initial design of the more practical switchmode power supplies. However, to produce fully optimized designs, the reader will need to become conversant with the more specialized information presented in Part 3 and the many references.

Part 3, "Applied Design," deals with many of the more general engineering requirements of switchmode systems, such as transformer design, choke design, input filters, RFI control, snubber circuits, thermal design, and much more.

Part 4, "Supplementary," looks at the design of an active power factor correction system. The power distribution industry is becoming more concerned with the increasing level of harmonic content caused by noncorrected electronic equipment and in particular electronic ballasts for fluorescent lighting. Active power factor correction is still a relatively new addition to the power supply designer's tasks. It is difficult to display waveforms, and design power inductors, due to the dynamic behavior of the boost topology, with its low- and high-frequency requirements. This part should help remove some of the mystery regarding its subject.

In most switchmode power supplies, it is the wound components that mainly control the efficiency and performance. Switching devices will work efficiently only if leakage inductances are small and good coupling is provided between input and output windings. The designer has considerable control over the wound components, but it requires considerable knowledge and skill to overcome the many practical and engineering problems encountered in their design. The author has therefore concentrated on the wound components, and provided many worked examples. To develop a full working knowledge of this critical area, the reader should refer to the more rigorous transformer design information given in Part 3, and the many references.

No single work can do full justice to this vast and rapidly developing subject. It is hoped that this book will at least partly fill the need for a more general handbook on the subject.

ACKNOWLEDGMENTS

No man is an island. We progress not only by our own efforts, but also by utilizing the work of those around us and by building on the foundations of those who went before. The reference section is an attempt to acknowledge this, and I have no doubt that many more works should have been mentioned. I sincerely apologize for any omissions; it is often difficult to remember the original source.

I am grateful to the many who have contributed to this work, but worthy of special mention is my former engineering colleague and friend Rodney Weaving, who not only spent many hundreds of hours carefully checking the original manuscript and calculations, but also made many useful suggestions. Without his endorsement I do not think I would have had the courage to put this into print. I am also grateful to Betty Magee, who transcribed my dictation onto computer diskette for further word processing; Hammond Manufacturing Co., Ltd., of Guelph, Ontario, who generously provided illustrating equipment and encouragement; Gordon (Ed) Bloom, president of Bloom Associates, Inc., for his helpful suggestions; Unitrode and Lloyd H. Dixon, Jr., for permission to reproduce his work on "The Right-Half-Plane Zero"; the editors and staff of McGraw-Hill Publishing Company, who added the "professional touch"; and finally Line Donald of Douglas J. Donald (illustration and design), Guelph, Ontario, who produced the more than three hundred camera-ready illustrations and diagrams. Her professional skill has greatly enhanced this presentation.

UNITS, SYMBOLS, DIMENSIONS, AND ABBREVIATIONS USED IN THIS BOOK

Units, Symbols, and Dimensions

In general, the units and symbols used in this book conform to the International Standard (SI) System. However, to yield convenient solutions, the equations are often dimensionally modified to convenient multiples or submultiples. (The preferred dimensions are shown following each equation.)

The imperial system is used for thermal calculations, because most thermal information is still presented in this form. Dimensions are in inches (1 in = 25.4 mm) and temperatures are in degrees Celsius, except for radiant heat calculations, which use the absolute Kelvin temperature scale.

Some graphs and equations in the magnetics sections use CGS units where this is common practice. Many manufacturers still provide magnetic information in CGS units; for example, magnetic field strength is shown in oersted(s) rather than At/m. (1 At/m = 12.57×10^{-3} Oe.)

It is industry standard practice to show core loss in terms of milliwatts per gram, with "peak flux density $\hat{B}$" as a parameter. (Because these graphs were developed for conventional push-pull transformer applications, symmetrical flux density swing about zero is assumed.) Hence, loss graphs assume a peak-to-peak swing of $2 \times \hat{B}$. To prevent confusion, when nonsymmetrical flux excursions are considered in this book, the term "peak flux density $\hat{B}$" is used only to indicate peak values. The term "flux density swing ΔB" is used to indicate total peak-to-peak excursion.

Basic Entities

Unit symbol	Unit name	Quantity	Quantity symbol	Dimensions
kg	Kilogram	Mass	m	M
m	Meter	Length	l	L
s	Second	Time	t	T
A	Ampere	Electric current	I	Q/T^{-1}
K	Kelvin	Temperature	T	θ

Multiples and Submultiples of Units Are Limited to the Following Range

Symbol prefix	Prefix name	Power multiple
M	mega-	10^6
k	kilo-	10^3
m	milli-	10^{-3}
μ	micro-	10^{-6}
n	nano-	10^{-9}
p	pico-	10^{-12}

Symbols for Physical Quantities

Quantity	Quantity symbol	Unit name	Unit symbol	Formula
Electric				
Capacitance	C	farads	F	$S \cdot s$
Charge	Q	coulombs	C	$A \cdot s$
Current	I	amperes	A	V/Ω
Energy	U	joules	J	$W \cdot s$
Impedance	Z	ohms	Ω	—
Inductance, self-	L	henries	H	Wb/A
Potential difference	V	volts	V	Wb/s
Power, real (active)	P	watts	W	$VI \cos \theta$
power, apparent	S	voltamperes	VA	$V \cdot A$
Reactance	X	ohms	Ω	—
Resistance	R	ohms	Ω	V/A
Resistivity, volume	ρ	ohm-centimeters cubed	Ω-cm^3	$\dfrac{R \cdot A}{l}$
Magnetic				
Field strength	H	amperes per meter	A/m	—
Field strength (CGS)	H	oersteds	Oe	$4\pi 10^{-3}$ A/M
Flux	Φ	webers	Wb	$V \cdot s$
Flux density	B	teslas	T	Wb/m
Permeability	μ	henries per meter	H/m	$V \cdot s/A \cdot m$
Other				
Angular velocity	ω	radians per second	rad/s	$2\pi f$
Area	A	centimeters squared	cm^2	—
Frequency	f	hertz	Hz	cycles/s
Length	l	centimeters	cm	—
Skin thickness	Δ	millimeters	mm	—
Temperature	T	degrees Celsius	°C	—
Temperature, absolute	T	kelvins	K	—
Time	t	seconds	s	—
Winding height	φ	millimeters	mm	—

Symbols for Mathematical Variables Used in This Book

Variable	Parameter	Unit
A	area	cm^2
A	gain (without feedback)	dB
A'	gain (with feedback)	dB
A_c	minimum cross-sectional area of pole (transformer)	cm^2
A_{cp}	area of center pole (of core)	cm^2
ac	alternating current	A
A_e	effective area (of core)	cm^2
A_g	area of air gap (in core)	cm^2
A_L	inductance factor (inductance of a single turn)	nH
A_m	minimum area of core	cm^2
A_n	attenuation factor	——
A_p	area of center pole (of core)	cm^2
A_p'	area of primary winding	cm^2
AP	area product of core ($A_w \times A_e$)	cm^4
A_{pe}	effective area product ($A_{wb} \times A_e$)	cm^4
A_r	resistance factor (bobbin); also attenuation factor	——
A_w	winding window area (of core)	cm^2
A_{wb}	winding window area (of bobbin)	cm^2
A_{we}	effective area of copper in winding (total)	cm^2
A_{wp}	primary winding window area	cm^2
A_s	surface area	cm^2
A_x	area of copper (for a single wire)	cm^2
B	magnetic flux density	mT
$\hat{B}$	peak magnetic flux density	mT
β	feedback factor	——
ΔB	small change in B	mT
ΔB_{ac}	magnetic flux density swing (p–p)	mT
B_{dc}	steady-state magnetic flux density (due to H_{dc})	mT
B_{opt}	optimum flux density swing (for minimum loss)	mT
B_r	remanence flux density	mT
B_s	saturation flux density	mT
B_w	peak (working) value of flux density	mT
b_w	useful winding width (of bobbin)	mm
C	capacitance	μF
C_c	leakage (parasitic) capacitance	pF
cfm	cubic feet per minute (of air flow)	cfm
C_h	heat (storage) capacity (joules/in^3/°C)	Ws/°C)
C_k	interelectrode capacitance	pF
C_p	parasitic coupling capacitance	pF
D	duty ratio (t_{on}/t_p)	
d'	duty cycle (t_{on}/t_{off})	
D'	D' ($1 - D$) = "off" time	
dB	logarithmic ratio (voltage 20 $\log_{10} V_1/V_2$ or power 10 $\log_{10}$ P_1/P_2)	dB
dB_m	logarithmic power ratio with respect to 1 mW (10 $\log_{10}$ $P_1/1$ mW)	dB
DC	direct (nonvarying) current or voltage	A or V
di/dt	rate of change of current with respect to time	A/s
di_p/dt	rate of change of primary current with respect to time	A/s
di_s/dt	rate of change of secondary current with respect to time	A/s

Symbols for Mathematical Variables Used in This Book (cont.)

Variable	Parameter	Unit
dv/dt	rate of change of voltage with respect to time	V/s
d_w	wire diameter	mm
$\mathbf{e}$	emf, induced electromotive force (vector quantity)	V
e'	radiant emissivity of surface	
$|e|$	emf (magnitude of emf only)	V
E	electrical energy	J
f	frequency	Hz
F_1	layer factor (copper)	
F_r	ratio of ac/DC resistance (of winding)	
H	magnetic field strength	Oe
$\hat{H}$	peak value of effective magnetic field strength	Oe
h	conductor thickness (strip) or wire diameter	mm
H_{ac}	magnetic field strength swing, p–p	Oe
H_{dc}	magnetic field strength due to DC current	Oe
H_{opt}	optimum value of magnetic field strength	Oe
H_s	saturating value of magnetic field strength	Oe
ΔH	small change in magnetic field strength	Oe
I	current flow (DC)	A
i	rms current (ac)	A
$\hat{I}$	peak current	A
I_a	current density (in wire)	A/cm^2
I_{ave}	average value of current for a defined period	A
I_{cp}	peak collector current	A
I_{dc}	direct current (dependent variable)	A
I_e	effective input current	A
I_i	harmonic interference current	A
I_L	inductor or choke current (average)	A
i_L	ac inductor current	A
$I_{L(p-p)}$	ripple current p–p in choke or inductor	A
I_{max}	maximum value of current	A
I_{mean}	time-averaged current value	A
I_{min}	minimum value of current	A
I_P	primary current (in transformer)	A
I_s	secondary current (also snubber current)	A
ΔI	small change in current	A
I^2R	resistive power loss	W
j_{wc}	capacitive reactance, $1/2\,\pi fC$ (complex #)	Ω
j_{wl}	inductive reactance, $2\,\pi fL$ (complex #)	Ω
K'	copper utilization factor (topology factor)	
K_m	material constant	
K_p	primary area factor	
K_t	primary rms current factor	
K_u	packing factor (of wire)	%
K_{ub}	utilization factor of bobbin	
L	inductance (self-inductance of wound component)	H
l	length (length of magnetic path)	cm
l_e	effective path length	cm

Symbols for Mathematical Variables Used in This Book (cont.)

Variable	Parameter	Unit
l_g	total length of air gap (in core)	cm
L_{LP}	primary leakage inductance	μH
L_{Ls}	secondary leakage inductance	μH
L_{LT}	total (transformer) leakage inductance	μH
l_m	mean length of wire or magnetic path (or core)	cm
L_p	primary inductance	mH
L_s	secondary inductance	mH
M_{lt}	mean length per turn	cm
mmf	magnetomotive force (magnetic potential ampere-turns)	At
N	number of turns	
N_{fb}	number of turns of feedback winding	
N_{min}	minimum number of turns (to prevent core saturation)	
N_{mpp}	minimum primary turns for p-p operation	
N_p	primary turns (of transformer)	
N_s	secondary turns (of transformer)	
N_v	turns per volt (on transformer)	T/V
N_w	number of turns (or wires) per layer	
P	power	W
p	period (of time)	μs
P_c	power dissipated in core	W
P_f	power factor (ratio true power/VA)	—
P_{in}	input power	W
P_{id}	total internal dissipation	W
P_{out}	output power	W
P_{q1}	power dissipated in transistor Q1	W
P_q	heat energy (joules)	j
P_t	total internal dissipation	W
P_v/N	primary volts per turn	V/T
P_w	winding copper loss	W
Q	rate of heat flow (in watts by conduction or in $J/s/in^2$ by radiation)	W J/s
R	resistance	Ω
r	radius (or wire)	mm
R_{Cu}	DC resistance of wound component at specified temperature	Ω
R_e	effective DC resistance of transformer winding	Ω
R_{c-h}	thermal resistance, case to heat exchanger	°C/W
R_{h-a}	thermal resistance, heat exchanger to free air	°C/W
R_{ha}	thermal resistance of heat exchanger	°C/W
R_{j-c}	thermal resistance, junction to case	°C/W
rms	square root of the mean of the square of all the harmonic components	V or I
R_o	total thermal resistance	°C/W
R_s	effective resistance of prime source or network	Ω
R_{sf}	effective source resistance factor ($R_{sf} = R_s \times W_{out}$)	Ω
RT	temperature coefficient of resistance (copper = 0.00393 at 0°C)	$\Omega/\Omega/°C$
RT_{cm}	resistance of wire in Ω/cm at temp T, °C	Ω/cm
R_θ	thermal resistance (of heat-conducting path)	°C/W
$R_{\theta ja}$	thermal resistance, hot spot to free air	°C/W

Symbols for Mathematical Variables Used in This Book (cont.)

Variable	Parameter	Unit
R_{th}	thermal resistance	°C/W
R_w	effective resistance of wound component at frequency f	Ω
R_x	resistance factor of bobbin	
S_f	scaling factor	
T	temperature in degrees Celsius	°C
t	time	s
T_{amb}	ambient temperature (of air)	°C
T_c	temperature of copper (winding)	°C
t_d	time delay period	s
T_{ds}	temperature of surface (diode)	°C
t_f	fall time (time required for voltage or current decay)	μs
T_h	temperature of heat exchanger surface	°C
t_p	total period (of time), i.e., duration of single cycle	μs
t_{off}	nonconducting "off" time period	μs
t_{on}	conducting "on" time period	μs
ΔT	small change in temperature	°C
ΔT_a	small temperature rise (above ambient)	°C
Δt	small increment of time	μs
T_r	temperature rise (above ambient)	°C
VA	volt-ampere product (apparent power)	VA
V_c	transistor collector voltage	V
V_{cc}	supply line (voltage)	V
V_{ce}	voltage, collector to emitter	V
V_{ceo}	collector-to-emitter breakdown voltage (base open circuit)	V
V_{cer}	collector-to-emitter breakdown voltage (with specified base-to-emitter resistance)	V
V_{cex}	collector-to-emitter breakdown voltage (base reverse-biased)	V
V_e	effective volume of core	cm³
V_{fb}	feedback voltage	V
V_h	header voltage (voltage at input of regulator)	V
V_{hi}	harmonic interference voltage, rms	Vrms
V_{in}	input voltage	V
V_l	voltage across inductor	V
V_m	mean voltage	V
V_n	nominal (average normal) voltage	V
V/N	volts per turn	V/T
V_o	ripple voltage	V
V_{out}	output voltage	V
V_p	peak voltage or primary voltage	V
V_{p-p}	ripple voltage, peak–peak value	V
V_{ref}	reference voltage	V
V_{rms}	root mean square voltage	Vrms
V_{sat}	saturation voltage	V
W_{in}	true input power ($VI \cos \theta$, or VA × P_f, heating effect)	W
W_{out}	true output power ($VI \cos \theta$, or VA × P_f, heating effect)	W
W_j	heat dissipation at junction, J/s	W
X_c	capacitive reactance ($1/2\, \pi f C$)	Ω
X_L	inductive reactance ($2\, \pi f L$)	Ω

Symbols for Mathematical Variables Used in This Book (cont.)

Variable	Parameter	Unit
ρ	volume resistivity of copper (at $0°C = 1.588$ $\mu\Omega/cm^3$)	$\mu\Omega\text{-cm}^3$
ρ_{tc}	resistivity of copper at t_c °C ($R_{cu} = \dfrac{\rho_{tc} \cdot l}{A}$	$\mu\Omega\text{-cm}^3$
μ_0	magnetic field constant ($4\,\pi \times 10^{-7}$ H/m)	Vs/Am
μ_r	relative permeability (of core)	
μ_x	effective permeability (after gap is introduced)	
η	efficiency (power output/power input $\times$ 100%)	%
Δ	a small increment (change); also skin thickness, mm	mm
$\Delta\Phi$	a small change in total flux	Φ
φ	effective conductor height	mm
Φ	total magnetic flux, Wb	Vs
$\approx$	approximatly equal to	
α	proportional to	
ω	angular velocity ($\omega = 2\,\pi f$)	rad/s
0V	zero voltage reference line (often the common output)	V
$1-D$	1 − duty ratio (the "off" period)	s
π	physical constant (3.1416)	
$|x|$	magnitude of function (x) only	

ABBREVIATIONS

ac	alternating current
AIEE	American Institute of Electrical Engineers
AWG	American wire gauge
B/H	(curve) hysteresis loop of magnetic material.
CISPR	Comité International Spécial des Perturbations Radioélectriques
CSA	Canadian Standards Association
dB	decibels (logarithmic ratio of power or voltage)
DC	direct (nonvarying) current or voltage
DCCT	direct-current current transformers
e.g.	exemplia gratis
emf	electromotive force
EMI	electromagnetic interference
ESL	effective series inductance
ESR	effective series resistance
FCC	Federal Communications Commission
FET	field-effect transistor
HCR	heavily cold-reduced
HRC	high rupture capacity
IC	integrated circuit
IEC	International Electrotechnical Commission
IEEE	Institute of Electrical and Electronics Engineers
LC	(filter) a low-pass filter consisting of a series inductor and shunt capacitor
LED	light-emitting diode
LISN	line impedance stabilization network
mmf	magnetomotive force (magnetic potential, ampere-turns)

MLT	mean length (of wire) per turn
MOV	metal oxide varistor
MPP	molybdenum Permalloy powder
MTBF	mean time before/between failure(s)
NTC	negative temperature coefficient
OEM	original equipment manufacturer
"off"	nonconducting (nonworking) state of device (circuit)
"on"	conducting (working) state of device (circuit)
OVP	overvoltage protection (circuit)
PARD	periodic and random deviations (see glossary)
pcb	printed circuit board
PFC	power factor correction
PFS	power failure sense/signal
p–p	peak-to-peak value (ripple voltage/current)
PTFE	polytetrafluoroethylene
PVC	polyvinyl chloride
PWM	pulse-width modulation
RF	radio frequency
RFI	radio-frequency interference
rms	root mean square
RHP	right-half-plane (zero), a zero located in the right half of the complex s-plane
+s	positive remote sensing (terminal, line)
−s	negative remote sensing (terminal, line)
SCR	silicon controlled rectifier
SMPS	switchmode power supply
SOA	safe operating area
SR	saturable reactor (see glossary)
TTL	transistor-transistor logic
UL	Underwriters' Laboratories
UPS	uninterruptible power supply
UVP	undervoltage protection (circuit)
VA	volt amps (product; apparent power)
VDE	Verband Deutscher Elektrotechniker

SWITCHMODE
POWER SUPPLY
HANDBOOK

FUNCTIONS AND REQUIREMENTS COMMON TO MOST DIRECT-OFF-LINE SWITCHMODE POWER SUPPLIES

CHAPTER 1
COMMON REQUIREMENTS: AN OVERVIEW

1.1 INTRODUCTION

The "direct-off-line" switchmode supply is so called because it takes its power input directly from the ac power lines, without using the rather large low-frequency (60 to 50 Hz) isolation transformer normally found in linear power supplies.

Although the various switchmode conversion techniques are often very different in terms of circuit design, they have, over many years, developed very similar basic functional characteristics which have become generally accepted industry standards.

Further, the need to satisfy various national and international safety, electromagnetic compatibility, and line transient requirements has forced the adoption of relatively standard techniques for track and component spacing, noise filter design, and transient protection. The prudent designer will be familiar with all these agency needs before proceeding with a design. Many otherwise sound designs have failed as a result of their inability to satisfy safety agency standards.

Many of the requirements outlined in this section will be common to all switching supplies, irrespective of the design strategy or circuit. Although the functions tend to remain the same for all units, the circuit techniques used to obtain them may be quite different. There are many ways of meeting these needs, and there will usually be a best approach for a particular application.

The designer must also consider all the minor facets of the specification before deciding on a design strategy. Failure to consider at an early stage some very minor system requirement could completely negate a design approach—for example, power good and power failure indicators and signals, which require an auxiliary supply irrespective of the converter action, would completely negate a design approach which does not provide this auxiliary supply when the converter is inhibited! It can often prove to be very difficult to provide for some minor neglected need at the end of the design and development exercise.

The remainder of Chap. 1 gives an overview of the basic input and output functions most often required by the user or specified by national or international standards. They will assist in the checking or development of the initial specification, and all should be considered before moving to the design stage.

1.2 INPUT TRANSIENT VOLTAGE PROTECTION

Both artificial and naturally occurring electrical phenomena cause very large transient voltages on all but fully conditioned supply lines from time to time.

IEEE Standard 587–1980 shows the results of an investigation of this phenomenon at various locations. These are classified as low-stress class A, medium-stress class B, and high-stress class C locations. Most power supplies will be in low- and medium-risk locations, where stress levels may reach 6000 V at up to 3000 A.

Power supplies are often required to protect themselves and the end equipment from these stress conditions. To meet this need requires special protection devices. (See Part 1, Chap. 2.)

1.3 ELECTROMAGNETIC COMPATIBILITY

Input Filters

Switching power supplies are electrically noisy, and to meet the requirements of the various national and international RFI (radio-frequency interference) regulations for conducted-mode noise, a differential- and common-mode noise filter is normally fitted in series with the line inputs. The attenuation factor required from this noise filter depends on the power supply size, operating frequency, power supply design, application, and environment.

For domestic and office equipments, such as personal computers, VDUs, and so on, the more stringent regulations apply, and FCC class B or similar limits would normally be applied. For industrial applications, the less severe FCC class A or similar limits would apply. (See Part 1, Chap. 3.)

It is important to appreciate that it is very difficult to cure a badly designed supply by fitting filters. The need for minimum noise coupling must be considered at all stages of the design; some good guidelines are covered in Part 1, Chaps. 3 and 4.

1.4 DIFFERENTIAL-MODE NOISE

Differential-mode noise refers to the component of high-frequency electrical noise between any two supply or output lines. For example, this would be measured between the live and neutral input lines or between the positive and negative output lines.

1.5 COMMON-MODE NOISE

For the line input, common-mode noise refers to that component of electrical noise that exists between both supply lines (in common) and the earth (ground) return.

For the outputs, the position is more complicated, as various configurations of isolated and nonisolated connections are possible. In general, output common-mode noise refers to the electrical noise between any output and some common point, usually the chassis or common return line.

Some specifications, notably those applying to medical electronics, severely limit the amount of ground return current permitted between either supply line and the earth (ground) return. A ground return current normally flows through the filter capacitors and leakage capacitance to ground, even if the insulation is perfect. The return current limitation can have a significant effect on the design of the supply and the size of input filter capacitors. In any event, capacitors in excess of 0.01 μF between the live line and ground are not permitted by many safety standards.

1.6 FARADAY SCREENS

High-frequency conducted-mode noise (noise conducted along the supply or output leads) is normally caused by capacitively coupled currents in the ground plane or between input and output circuits. For this reason, high-voltage switching devices should not be mounted on the chassis. Where this cannot be avoided, a Faraday screen should be fitted between the noise source and the ground plane, or at least the capacitance to the chassis should be minimized.

To reduce input-to-output noise coupling in isolating transformers, Faraday screens should be fitted. These should not be confused with the more familiar safety screens. (See Part 1, Chap. 4.)

1.7 INPUT FUSE SELECTION

This is an often neglected part of power supply design. Modern fuse technology makes available a wide range of fuses designed to satisfy closely defined parameters. Voltages, inrush currents, continuous currents, and let-through energy (I^2t ratings) should all be considered. (See Part 1, Chap. 5.)

Where units are dual-input-voltage-rated, it may be necessary to use a lower fuse rating for the higher input voltage condition. Standard, medium-speed glass cartridge fuses are universally available and are best used where possible. For line input applications, the current rating should take into account the 0.6 to 0.7 power factor of the capacitive input filter used in most switchmode systems.

For best protection the input fuse should have the minimum rating that will reliably sustain the inrush current and maximum operating currents of the supply at minimum line inputs. However, it should be noted that the rated fuse current given in the fuse manufacturer's data is for a limited service life, typically a thousand hours operation. For long fuse life, the normal power supply current should be well below the maximum fuse rating; the larger the margin, the longer the fuse life.

Fuse selection is therefore a compromise between long life and full protection. Users should be aware that fuses tend to age and should be replaced at routine servicing periods. For maximum safety during fuse replacement, the live input is normally fused at a point after the input switch.

To satisfy safety agency requirements and maintain maximum protection, when fuses are replaced, a fuse of the same type and rating must be used.

1.8   LINE RECTIFICATION AND CAPACITOR INPUT FILTERS

Rectifier capacitor input filters have become almost universal for direct-off-line switchmode power supplies. In such systems the line input is directly rectified into a large electrolytic reservoir capacitor.

Although this circuit is small, efficient, and low-cost, it has the disadvantage of demanding short, high-current pulses at the peak of the applied sine-wave input, causing excessive line I^2r losses, harmonic distortion, and a low power factor.

In some applications (e.g., shipboard equipment), this current distortion cannot be tolerated, and special low-distortion input circuits must be used. (See Part 1, Chap. 6.)

1.9   INRUSH LIMITING

Inrush limiting reduces the current flowing into the input terminals when the supply is first switched on. It should not be confused with "soft start," which is a separate function controlling the way the power converter starts its switching action.

In the interests of minimum size and weight, most switchmode supplies will use semiconductor rectifiers and low-impedance input electrolytics in a capacitive input filter configuration. Such systems have an inherently low input resistance; also, because the capacitors are initially discharged, very large surge currents would occur at switch-on if such filters were switched directly to the line input.

Hence, it is normal practice to provide some form of current inrush limiting on power supplies that have capacitive input filters. This inrush limiting typically takes the form of a resistive limiting device in series with the supply lines. In high-power systems, the limiting resistance would normally be removed (shorted out) by an SCR, triac, or switch when the input reservoir and/or filter capacitor has been fully charged. In low-power systems, NTC thermistors are often used as limiting devices.

The selection of the inrush-limiting resistance value is usually a compromise between acceptable inrush current amplitude and start-up delay time. Negative temperature coefficient thermistors are often used in low-power applications, but it should be noted that thermistors will not always give full inrush limiting. For example, if, after the power supply has been running long enough for the thermistor to heat up, the input is turned rapidly off and back on again, the thermistor will still be hot and hence low-resistance, and the inrush current will be large. The published specification should reflect this effect, as it is up to the user to decide whether this limitation will cause any operational problems. Since even with a hot NTC the inrush current will not normally be damaging to the supply, thermistors are usually acceptable and are often used for low-power applications. (See Part 1, Chap. 7.)

1.10   START-UP METHODS

In direct-off-line switchmode supplies, the elimination of the low-frequency (50 to 60 Hz) transformer can present problems with system start-up. The difficulty usually stems from the fact that the high-frequency power transformer cannot be used for auxiliary supplies until the converter has started. Suitable start-up circuits are discussed in Part 1, Chap. 8.

1.11 SOFT START

Soft start is the term used to describe a low-stress start-up action, normally applied to the pulse-width-modulated converter to reduce transformer and output capacitor stress and to reduce the surge on the input circuits when the converter action starts.

Ideally, the input reservoir capacitors should be fully charged before converter action commences; hence, the converter start-up should be delayed for several line cycles, then start with a narrow pulse and a progressively increasing pulse width until the output is established.

There are, in fact, a number of reasons why the pulse width should be narrow when the converter starts and progressively increase during the start-up phase. There will often be considerable capacitance on the output lines, and this should be charged slowly so that it does not reflect an excessive transient back to the supply lines. Further, where a push-pull action is applied to the main transformer, flux doubling and possible saturation of the core may occur if a wide pulse is applied to the transformer for the first half cycle of operation. (See Part 3, Chap. 7.) Finally, since an inductor will invariably appear somewhere in series with the current path, it may be impossible to prevent voltage overshoot on the output if this inductor current is allowed to rise to a high value during the start-up phase. (See Part 1, Chap. 10.)

1.12 START-UP OVERVOLTAGE PREVENTION

When the power supply is first switched on, the control and regulator circuits are not in their normal working condition (unless they were previously energized by some auxiliary supply).

As a result of the limited output range of the control and driver circuits, the large-signal slew rate may be very nonlinear and slow. Hence, during the start-up phase, a "race" condition can exist between the establishment of the output voltages and correct operation of the control circuits. This can result in excessive output voltage overshoot.

Additional fast-acting voltage clamping circuits may be required to prevent overshoot during the start-up phase, a need often overlooked in the past by designers of both discrete and integrated control circuits. (See Part 1, Chap. 10.)

1.13 OUTPUT OVERVOLTAGE PROTECTION

Loss of voltage control can result in excessive output voltages in both linear and switchmode supplies. In the linear supply (and some switching regulators), there is a direct DC link between input and output circuits, so that a short circuit of the power control device results in a large and uncontrolled output. Such circuits require a powerful overvoltage clamping technique, and typically an SCR "crowbar" will short-circuit the output to clear a series fuse.

In the direct-off-line SMPS, the output is isolated from the input by a well-insulated transformer. In such systems, most failures result in a low or zero output voltage. The need for crowbar-type protection is less marked, and indeed is often considered incompatible with size limitations. In such systems, an independent signal level voltage clamp which acts on the converter drive circuit is often considered satisfactory for overvoltage protection.

The design aim is that a single component failure within the supply will not cause an overvoltage condition. Since this aim is rarely fully satisfied by the signal level clamping techniques often used (for example, an insulation failure is not fully protected), the crowbar and fuse technique should still be considered for the most exacting switchmode designs. The crowbar also provides some protection against externally induced overvoltage conditions.

1.14 OUTPUT UNDERVOLTAGE PROTECTION

Output undervoltages can be caused by excessive transient current demands and power outages. In switchmode supplies, considerable energy is often stored in the input capacitors, and this provides "holdup" of the outputs during short power outages. However, transient current demands can still cause undervoltages as a result of limited current ratings and output line voltage drop. In systems that are subject to large transient demands, the active undervoltage prevention circuit described in Part 1, Chap. 12 should be considered.

1.15 OVERLOAD PROTECTION (INPUT POWER LIMITING)

Power limiting is usually applied to the primary circuits and is concerned with limiting the maximum throughput power of the power converter. In multiple-output converters this is often necessary because, in the interest of maximum versatility, the sum of the independent output current limits often has a total VA rating in excess of the maximum converter capability.

Primary power limiting is often provided as additional backup protection, even where normal output current limiting would prevent output overloading conditions. Fast-acting primary limiting has the advantage of preventing power device failure under unusual transient loading conditions, when the normal secondary current limiting may not be fast enough to be fully effective. Furthermore, the risk of fire or excessive power supply damage in the event of a component failure is reduced. Power supplies with primary power limiting usually have a much higher reliability record than those without this additional protection.

1.16 OUTPUT CURRENT LIMITING

In higher-power switchmode units, each output line will be independently current-limited. The current limit should protect the supply under all conditions to short-circuit. Continuous operation in a current-limited mode should not cause overdissipation or failure of the power supply. The switchmode unit (unlike the linear regulator) should have a constant current limit. By its nature, the switching supply does not dissipate excessive power under short-circuit conditions, and a constant current limit is far less likely to give the user such problems as "lockout" under nonlinear or cross-coupled load conditions. (Cross-coupled loads are loads that are connected between a positive and a negative output line without connection to the common line.)

Linear regulators traditionally have reentrant current limiting in order to prevent excessive dissipation in the series element under short-circuit conditions. Section 14.5 covers the problems associated with cross-coupled loads and reentrant current limits more fully.

1.17 BASE DRIVE REQUIREMENTS FOR HIGH-VOLTAGE BIPOLAR TRANSISTORS

In direct-off-line SMPSs the voltage stress on the main switching device can be very large, of the order of 800 to 1000 V in the case of the flyback converter.

Apart from the obvious needs for high-voltage transistors, "snubber" networks, load line shaping, and antisaturation diodes, many devices require base drive waveform shaping. In particular, the base current is often required to ramp down during the turn-off edge at a controlled rate for best performance. (See Part 1, Chap. 15.)

1.18 PROPORTIONAL DRIVE CIRCUITS

With bipolar transistors, base drive currents in excess of those required to saturate the transistor reduce the efficiency and can cause excessive turn-off storage times with reduced control at light loads.

Improved performance can be obtained by making the base drive current proportional to the collector current. Suitable circuits are shown in Part 1, Chap. 16.

1.19 ANTISATURATION TECHNIQUES

With bipolar transistors, in the switching mode, improved turn-off performance can be obtained by preventing "hard" saturation. The transistor can be maintained in a quasi-saturated state by maintaining the drive current at a minimum defined by the gain and collector current. However, since the gain of the transistor changes with device, load, and temperature, a dynamic control is required.

Antisaturation circuits are often combined with proportional drive techniques. Suitable methods are shown in Part 1, Chap. 17.

1.20 SNUBBER NETWORKS

This is a power supply engineering term used to describe networks which provide turn-on and turn-off load line shaping for the switching device.

Load line shaping is required to prevent breakdown by maintaining the switching device within its "safe operating area" throughout the switching cycle.

In many cases snubber networks also reduce RFI problems as a result of the reduced dv/dt on switching elements, although this is not their primary function.

1.21 CROSS CONDUCTION

In half-bridge, full-bridge, and push-pull applications, a DC path exists between the supply lines if the "on" states of the two switching devices overlap. This is called "cross conduction" and can cause immediate failure.

To prevent this condition, a "dead time" (a period when both devices are off) is often provided in the drive waveform. To maintain full-range pulse-width control, a dynamic dead time may be provided. (See Part 1, Chap. 19.)

1.22 OUTPUT FILTERING, COMMON-MODE NOISE, AND INPUT-TO-OUTPUT ISOLATION

These parameters have been linked together, as they tend to be mutually interdependent. In switchmode supplies, high voltages and high currents are being switched at very fast rates of change at ever-increasing frequencies. This gives rise to electrostatic and electromagnetic radiation within the power supply. The electrostatic coupling between high-voltage switching elements and the output circuit or ground can produce particularly troublesome common-mode noise problems.

The problems associated with common-mode noise are not generally recognized, and there is a tendency to leave this requirement out of the power supply specifications. Common-mode noise is a very real cause of system problems, and it should be normal practice in good power supply design to minimize the capacitance between the switching elements and chassis and to provide Faraday screens between the primary and secondary of the power transformer. Where switching elements are to be mounted on the chassis for cooling purposes, an insulated Faraday screen should be placed between the switching element and the mounting surface. This screen and any other Faraday screens in the transformer should be returned to one of the input DC supply lines so as to return capacitively coupled currents to the source. In many cases, the transformer will require an additional safety screen which will be connected to earth or chassis. This safety screen should be positioned between the RF Faraday screen and the output windings.

In rare cases (where the output voltages are high), a second Faraday screen may be required between the safety screen and the output windings to reduce output common-mode current. This screen should be returned to the common output line, as close as possible to the transformer common-line connection pin.

The screens, together with the necessary insulation, increase the spacing between the primary and secondary windings, thereby increasing the leakage inductance and degrading transformer performance. It should be noted that the Faraday screen does not need to meet the high current capacity of the safety screens and therefore can be made from lightweight material and connections. (See Part 1, Chaps. 3 and 4.)

1.23 POWER FAILURE SIGNALS

To allow time for "housekeeping" functions in computer systems, a warning of impending shutdown is often required from the power supply. Various methods are used, and typically a warning signal should be given at least 5 ms before the power supply outputs fall below their minimum specified values. This is required to allow time for a controlled shutdown of the computer.

In many cases, extremely simple power failure systems which simply recognize the presence or absence of the AC line input and give a TTL low signal within a few milliseconds of line failure are used. It should be recognized that the line input passes through zero twice in each cycle under normal conditions; since this must not be recognized as a failure, there is usually a delay of several milliseconds before a genuine failure can be recognized. When a line failure has been recognized, the normal holdup time of the power supply should provide output voltage for a further period, allowing time for the necessary housekeeping procedures.

Two undesirable limitations of these simple systems should be recognized. First, if a "brownout" condition precedes the power failure, the output voltage may fall below the minimum value without a power failure signal being generated. Second, if the line input voltage to the power supply immediately prior to failure is close to the minimum required for normal operation, the holdup time will be severely diminished, and the time between a power failure warning and supply shutdown may not be long enough for effective housekeeping.

For critical applications, more sophisticated power failure warning systems which will recognize brownout should be used. If additional holdup time is required, charge dumping techniques should be considered. (See Part 1, Chap. 12.)

1.24 POWER GOOD SIGNALS

"Power good" signals are sometimes required from the power supply. These are usually TTL-compatible outputs that go to a "power good" (high state) when all power supply voltages are within their specified operating window. "Power good" and "power failure" signals are sometimes combined. LED (light-emitting diode) status indicators are often provided with the "power good" signals, to give a visual indication of the power supply status.

1.25 DUAL INPUT VOLTAGE OPERATION

With the trend toward international trading it is becoming increasingly necessary to provide switchmode supplies for dual input, nominally 110/220-V operation. A wide variety of techniques are used to meet these dual-voltage requirements, including single or multiple, manual or automatic, transformer tap changing, and voltage doubling. If auxiliary transformers and cooling fans are to be used, these must be considered in the dual-voltage connection.

A useful method of avoiding the need for special dual-voltage fans and auxiliary transformers is shown in Part 1, Chap. 23. It should be remembered that the insulation of the auxiliary transformer and fan must meet the safety requirements for the highest-voltage input. More recently, high-efficiency "brushless" DC fans have become available; these can be driven by the supply output, overcoming insulation and tap change problems.

The voltage doubler technique with one or two link changes is probably the most cost-effective and is generally favored in switchmode supplies. However, when this method is used, the design of the filter, the input fuse, and inrush limiting should be considered. When changing the input voltage link arrangements, the low-voltage tap position gives higher current stress, whereas the higher tap position gives a greater voltage stress. The need to meet both conditions results in more expensive filter components. Therefore, dual-voltage operation should not be specified unless this is a real system requirement.

1.26 POWER SUPPLY HOLDUP TIME

One of the major advantages of switchmode supplies is their ability to maintain the output voltages constant for a short period after line failure. This "holdup time" is typically 20 ms minimum, but depends on the part of the input cycle where the power failure occurs and the loading and the supply voltage before the line failure.

A major factor controlling the holdup time is the history and amplitude of the supply voltage immediately prior to the failure condition. Most specifications define holdup time from nominal input voltage and loading. Holdup times may be considerably less if the supply voltage is close to its minimum value immediately prior to failure.

Power supplies that are specified for long holdup times at minimum input voltages are either expensive because of the increased size of input capacitors, or less efficient because the power converter must now maintain the output voltage constant for a much lower input voltage. This usually results in less efficient operation at nominal line inputs. Charge dumping techniques should be considered when long holdup times are required at low input voltages. (See Part 1, Chap. 12.)

1.27 SYNCHRONIZATION

Synchronization of the switching frequency is sometimes called for, particularly when the supply is to be used for VDU (visual display unit) applications. Although synchronization is of dubious value in most cases, as adequate screening and filtering of the supply should eliminate the need, it must be recognized that systems engineers often specify it.

The constraints placed on the power supply design by specifying synchronization are severe; for example, the low-cost variable-frequency systems cannot be used. Furthermore, the synchronization port gives access to the drive circuit of the main converter and provides a means whereby the operating integrity of the converter can be disrupted.

The possibility of badly defined or incorrect synchronization information must be considered in the design of synchronizable systems. The techniques used should be as insensitive as possible to abuse. The user should be aware that it is difficult to guarantee that a power supply will not be damaged by incorrect or badly defined synchronization signals. Because of the need to prevent saturation in wound components, most switchmode supplies use oscillator designs which can be synchronized only to frequencies higher than the natural oscillation frequency. Also, the synchronization range is often quite limited.

1.28 EXTERNAL INHIBIT

For system control, it is often necessary to turn the power supply on or off by external electronic means. Typically a TTL high signal will define the "on" condition and a TTL low the "off" condition. Activation of this electronic inhibit should invoke the normal soft-start sequence of the power supply when it is turned on. Power supplies for which this remote control function is required will often need internal auxiliary supplies which are common to the output. The auxiliary supply must be present irrespective of main converter operation. This apparently simple requirement may define the complete design strategy for the auxiliary supplies.

1.29 FORCED CURRENT SHARING

Voltage-controlled power supplies, by their very nature, are low-output-impedance devices. Since the output voltage and performance characteristics of two or more units will never be identical, the units will not naturally share the load current when they are operated in parallel.

Various methods are used to force current sharing (see Part 1, Chap. 24). However, in most cases these techniques force current sharing by degrading the output impedance (and consequently the load regulation) of the supply. Hence the load regulation performance in parallel forced current sharing applications will usually be lower than that found with a single unit.

A possible exception is the master-slave technique which tends to a voltage-controlled current source. However, the master-slave technique has fallen out of favor because of its inability to provide good parallel redundant operation. A failure of the master system usually results in a complete system failure.

More recently, interconnected systems of current-mode control topologies have shown considerable promise. The technique should in theory be quite good; however, the tendency for noise pickup on the P-terminal link between units makes it somewhat difficult to implement in practice. Further, if one unit is used to provide the control signal, failure of this unit will shut down the whole system, which is again contrary to the needs of a parallel redundant system.

The forced current sharing system described in Part 1, Chap. 24 does not suffer from these difficulties. Although the output regulation is slightly degraded, the variation in output voltage in normal circumstances is only a few millivolts, which should be acceptable for most practical applications.

Failure to provide current sharing means that one or more of the power supplies will be operating in a maximum current limited mode, while others are hardly loaded. However, so long as the current limits for the units are set at a value where continuous operation in the current limited mode gives reasonable power supply life, simple direct parallel connection can be used and should not be ignored.

1.30 REMOTE SENSING

If the load is situated some distance from the power supply, and the supply-lead voltage drops are significant, improved performance will be obtained if a remote voltage sense is used for the power supply. In principle, the reference voltage and amplifier comparator inputs are connected to the remote load by separate voltage sensing lines to remove the line-drop effects. These remote sense leads carry negligible currents, so the voltage drop is also negligible. This arrangement permits the power supply to compensate for the voltage drops in the output power leads by increasing the supply voltage as required. In low-voltage, high-current applications, this facility is particularly useful. However, the user should be aware of at least three limitations of this technique:

1. The maximum external voltage drop that can be tolerated in the supply leads is typically limited to 250 mV in both go and return leads (500 mV overall). In a 100-A 5-V application, this would represent an extra 50 W from the power supply, and it should be remembered that this power is being dissipated in the supply lines.

2. Where power supplies are to be connected in a parallel redundant mode, it is common practice to isolate each supply with a series diode. The principle here is that if one power supply should short-circuit, the diode will isolate this supply from the remaining units.

   If this connection is used, then the voltage at the terminals of the power supply must be at least 0.7 V higher than the load, neglecting any lead losses, and the required terminal voltage may exceed the power supply's design maximum unless the supply is specifically designed for this mode of operation. Furthermore, it must be borne in mind that in the event of power supply failure in this parallel redundant mode, the amplifier sense leads will still be connected to the load and will experience the load voltage. The remote sense circuit must be able to sustain this condition without further damage.

   It is common practice to link the remote sense terminals to the power supply output terminals with resistors within the supply to prevent loss of control and voltage overshoot in the event of the sense leads being disconnected. Where such resistors are used in the parallel redundant connection, they must be able to dissipate the appropriate power, V_{out}^2 /R, without failure in the event of the main terminal output voltage falling to zero.

3. Remote sense terminals are connected to a high-gain part of the power amplifier loop. Consequently, any noise picked up in the remote sense leads will be translated as output voltage noise to the power supply terminals, degrading the performance. Further, the additional phase shifts caused by lead inductance and resistance can have a destabilizing effect. Therefore, it is recommended that remote sense leads be twisted to minimize inductance and noise pickup.

   Unless they are correctly matched and terminated, coaxial leads are not recommended, as the distributed capacitance can degrade the transient performance.

1.31 P-TERMINAL LINK

In power supplies where provision is made to interlink one or more units in a parallel forced current sharing mode, current sharing communication between supplies is required. This link is normally referred to as the P-terminal link. In master-slave applications this link allows the master to control the output regulators of the slave units. In forced current sharing applications this link provides communication between the power supplies, indicating the average load current and allowing each supply to adjust its output to the correct proportion of the total load. Once again, the P-terminal link is a noise-sensitive input, so the connections should be routed so as to minimize the noise pickup. (See Part 1, Chap. 24.)

1.32 LOW-VOLTAGE CUTOUT

In most applications, the auxiliary supplies to the power unit will be derived from the same supply lines as the main converter. For the converter to start up under controlled conditions, it is necessary that the supply to both the main converter and the auxiliary circuits be correctly conditioned before the power converter action commences. It is normal practice to provide a drive inhibiting circuit which is activated when the auxiliary supply falls below a value which can guarantee proper operation. This "low-voltage inhibit" prevents the converter from starting up during the power-up phase until the supply voltage is sufficiently high to ensure proper operation. Once the converter is running, if the supply voltage falls below a second, lower value, the converter action will be inhibited; this hysteresis is provided to prevent squegging at the threshold voltage.

1.33 VOLTAGE AND CURRENT LIMIT ADJUSTMENTS

The use of potentiometers for voltage and current limit adjustments is not recommended, except for initial prototype applications. Power supply voltages and current limits, once set, are very rarely adjusted. Most potentiometers become noisy and unreliable unless they are periodically exercised, and this causes noisy and unreliable performance. Where adjustments are to be provided, high-grade potentiometers must be used.

1.34 INPUT SAFETY REQUIREMENTS

Most countries have strict regulations governing safety in electrical apparatus, including power supplies. UL (Underwriters Laboratories), VDE (Verband

Deutscher Elektrotechniker), IEC (International Electrotechnical Commission), and CSA (Canadian Standards Association) are typical examples of the bodies formulating these regulations. It should be remembered that these regulations define minimum insulation, spacing, and creepage distance requirements for printed circuit boards, transformers, and other wound components.

Meeting these specifications will have an impact on the performance and must be an integral part of the design exercise. It is very difficult to modify units to meet safety regulations after they have been designed. Consequently, drawing office and design staff should be continually alert to these requirements during the design phase. Furthermore, the technical requirements for high performance tend to be generally incompatible with the spacing requirements for the safety specifications. Consequently, a prototype unit designed without full attention to the safety spacing needs may give an excessively optimistic view of performance which cannot be maintained in the fully approvable finished product.

A requirement often neglected is that ground wires, safety screens, and screen connections must be capable of carrying the fuse fault current without rupture, to prevent loss of safety ground connections under fault conditions. Further, any removable mountings (which, for example, may have been used to provide an earth connection from the printed circuit board to earth or chassis) must have a provision for hard wiring of the ground of the host equipment main frame. Mounting screws alone do not meet the safety requirements for some authorities.

CHAPTER 2
AC POWERLINE SURGE PROTECTION

2.1 INTRODUCTION

With the advent of "direct-off-line" switchmode power supplies using sensitive electronic primary control circuits, the need for input AC powerline transient surge protection has become more universally recognized.

Measurements carried out by the IEEE over a number of years have demonstrated, on a statistical basis, the likely frequency of occurrence, typical amplitudes, and waveshapes to be expected in various locations as a result of artificial and naturally occurring electrical phenomena. These findings are published in IEEE Standard 587–1980* and are shown in Table 1.2.1. This work provides a basis for the design of AC powerline transient surge protection devices.[40]

2.2 LOCATION CATEGORIES

In general terms, the surge stress to be expected depends on the location of the equipment to be protected. When equipment is inside a building, the stress depends on the distance from the electrical service entrance to the equipment location, the size and length of connection wires, and the complexity of the branch circuits. IEEE Standard 587-1980 proposes three location categories for low-voltage AC powerlines (less than 600 V). These are shown in Fig. 1.2.1, and described as follows:

1. *Category A, Outlets and Long Branch Circuits.* This is the lowest-stress category; it applies to
   a. All outlets more than 10 m (30 ft) from Category B with #14 to #10 wires.
   b. All outlets at more than 20 m (60 ft) from the service entrance with #14 to #10 wires. In these remote locations, far away from the service entrance, the stress voltage may be of the order of 6 kV, but the stress currents are relatively low, of the order of 200 A maximum.
2. *Category B, Major Feeders and Short Branch Circuits.* This category covers the highest-stress conditions likely to be seen by a power supply. It applies to the following locations:
   a. Distribution panel devices

---

*Also issued under ANSI/IEEE Standard C64.41-1980 and IEC Publication 664-1980.

TABLE 1.2.1 Surge Voltages and Currents Deemed to Represent the Indoor Environment and Recommended for Use in Designing Protective Systems

IEEE Std. 587 location category	Comparable to IEC 664 category	Impulse			Energy (joules) deposited in a suppressor‡ with clamping voltage of:	
		Waveform	Medium exposure amplitude	Type of specimen or load circuit	500 V (120-V system)	1000 V (240-V system)
A. Long branch circuits and outlets	II	0.5 µs–100 kHz	6 kV	High impedance*	—	—
			200 A	Low impedance†	0.8	1:6
B. Major feeders, short branch circuits, and load center	III	1.2/50 µs	6 kV	High impedance*	—	—
		8/20 µs	3 kA	Low impedance†	40	80
		0.5 µs–100 kHz	6 kV	High impedance*	—	—
			500 A	Low impedance†	2	4

*For high-impedance test specimens or load circuits, the voltage shown represents the surge voltage. In making simulation tests, use that value for the open-circuit voltage of the test generator.

†For low-impedance test specimens or load circuits, the current shown represents the discharge current of the surge (not the short-circuit current of the power system). In making simulation tests, use that current for the short-circuit current of the test generator.

‡Other suppressors which have different clamping voltages would receive different energy levels.

b. Bus and feeder systems in industrial plants

c. Heavy appliance outlets with "short" connections to the service entrance

d. Lighting systems in commercial buildings

Note: Category B locations are closer to the service entrance. The stress voltages may be similar to those for category A, but currents up to 3000 A may be expected.

3. *Category C, Outside and Service Entrance.* This location is outside the building. Very high stress conditions can occur, since the line and insulator spacing is large and the flashover voltage can be greater than 6 kV. Fortunately, most power supplies will be in category B or A locations within a partially protected environment inside the building, and only protection to category A and B stress conditions is normally required.

Most indoor distribution and outlet connectors have sparkover voltages of 6 kV or less, and this, together with the inherent distribution system resistance, limits the stress conditions inside the building to much lower levels.

Where power supplies are to be provided with surge protection, the category of the protection should be clearly understood and specified in accordance with the expected location. Since the protection devices for category B locations can be large and expensive, this protection category should not be specified unless definitely required.

Where a number of supplies are to be protected within a total distributed power system, it is often more expedient to provide a single transient surge protection unit at the line input to the total system.

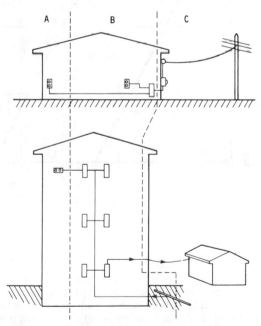

FIG. 1.2.1 Circuit location categories, as defined by IEEE Standard 587-1980. .

2.3 *LIKELY RATE OF SURGE OCCURRENCES*

Since some transient protection devices (metal oxide varistors, for example) have a limited life, dependent on the number and size of the stress surges, the likely exposure level should be considered when selecting protection devices. Figure 1.2.2 (from IEEE Standard 587) shows, statistically, the number of surges that may be expected per year, as a function of the voltage amplitude, in low-, medium-, and high-exposure locations.

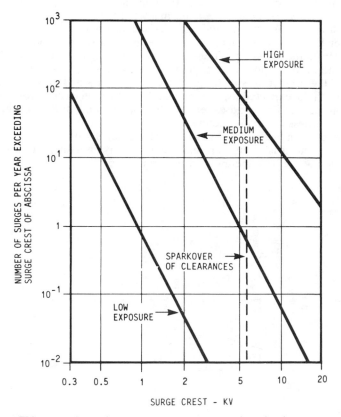

FIG. 1.2.2 Rate of surge occurrences versus voltage level at unprotected locations.

For example, in a medium-exposure location, a 5-kV spike can be expected at least once a year and, perhaps of greater concern, hundreds of transients in the range of 1 to 2 kV can occur in the same period. Since even these lower stress levels are quite sufficient to damage unprotected equipment, it is clear that some form of protection is essential in any electronic equipment to be connected to the supply lines.

IEEE Standard 587–1980 describes the exposure locations as follows:

1. *Low Exposure*. Systems in geographical areas known for low lightning activity, with little load switching activity.
2. *Medium Exposure*. Systems in geographical areas known for high lightning activity, with frequent and severe switching transients.
3. *High Exposure*. Rare but real systems supplied by long overhead lines and subject to reflections at line ends, where the characteristics of the installation produce high sparkover levels of clearances.

2.4 SURGE VOLTAGE WAVEFORMS

The IEEE investigation found that although surge voltage waveforms can take many shapes, field measurements and theoretical calculations indicate that most surge voltages in indoor low-voltage systems (AC lines less than 600 V) have a damped oscillatory shape, as shown in Fig. 1.2.3. (This is the well-known "ring wave" referred to in IEEE Standard 587.) The following quotation from this standard describes the phenomenon well:

> A surge impinging on the (distribution) system excites the natural resonant frequencies of the conductor system. As a result, not only are the surges typically oscillatory, but surges may have different amplitudes and wave shapes at different places in the system. These oscillatory frequencies of surges range from 5 kHz to more than 500 kHz. A 30 kHz–100 kHz frequency is a realistic measurement of a "typical" surge for most residential and light industrial ac line networks.

In category B locations (close to the service entrance), much larger energy levels are encountered. IEEE Standard 587 recommends two unidirectional waveforms for high- and low-impedance test specimens. These two waveforms are shown in Fig. 1.2.4a and b. For this category, the transient protection device must be able to withstand the energy specified in these two waveforms (Table 1.2.1). In addition to the unidirectional pulses, ring-wave oscillatory conditions can also occur. For these, the voltage can be of the order of 6 kV and the current 500 A. The various stress conditions are tabulated in Table 1.2.1.

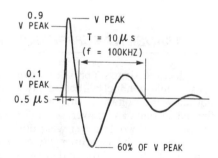

FIG. 1.2.3 Proposed 0.5-μs, 100-kHz ring wave (open-circuit voltage).

The impedance of the protection circuit is often difficult to define, since a number of devices operating in different modes and different voltages are often used in the protection unit. To satisfy both high- and low-impedance conditions, the test circuitry is usually configured to generate the voltage waveform specified on an open circuit and the current waveform specified on a short circuit before being applied to the test specimen.

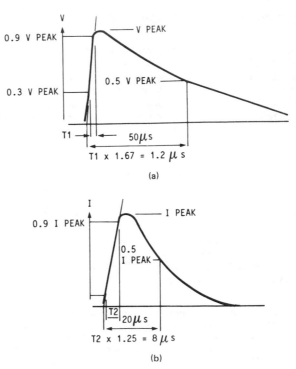

FIG. 1.2.4   Unidirectional waveshapes (ANSI/IEEE Standard 28-1974).

2.5   TRANSIENT SUPPRESSION DEVICES

The ideal transient suppression device would be an open circuit at normal voltages, would conduct without delay at some slight voltage above normal, would not allow the voltage to increase during the clamping period, would handle unlimited currents and power, would revert back to an open circuit when the stress has gone, and would never wear out.

At the time this is written, there is no single transient suppression device that approaches this ideal for all the stress conditions specified in IEEE Standard 587. Hence, at present efficient transient protection requires the use of a number of devices, carefully selected to complement each other and thus cover the full range of voltage and current stress conditions.

For the lower-stress category A locations, silicon varistors, in combination with transient suppressor diodes, filter inductors, and capacitors, are commonly used. In the higher-power category B locations, these devices are supplemented with much higher-current-rated gas-discharge tubes or spark gaps. When gas-discharge devices are used, fast-acting fuses or circuit breakers will also be fitted.

For efficient matching of the various suppressor devices, their general performance characteristics should be fully understood.

2.6 METAL OXIDE VARISTORS (MOVs, VOLTAGE-DEPENDENT RESISTORS)

As the name implies, varistors (MOVs) display a voltage-dependent resistance characteristic. At voltages below the turnover voltage, these devices have high resistance and little circuit loading. When the terminal voltage exceeds the turnover voltage, the resistance decreases rapidly and increasing current flows in the shunt-connected varistor.

The major advantages of the varistor are its low cost and its relatively high transient energy absorption capability. The major disadvantages are progressive degradation of the device with repetitive stress and a relatively large slope resistance.

The limitations of the varistor for transient suppressor applications in medium- and high-risk locations are fairly marked. Under high-exposure conditions, the device can quickly age, reducing its effective clamping action. This is a somewhat insidious process, as the degradation is not obvious and cannot be easily measured. Further, the varistor's relatively high slope resistance means that its clamping action is quite poor for high-current stress conditions (even low-voltage varistor devices have terminal voltages over 1000 V at transient currents of only a few tens of amperes). As a result, damagingly high voltages may be let through to the "protected" equipment if MOVs are used on their own. However, varistors can be of great value when used in combination with other suppressor devices.

Figure 1.2.5 shows the typical characteristics of a 275-V varistor. Note that the terminal voltage is 1250 V at a transient current of only 500 A.

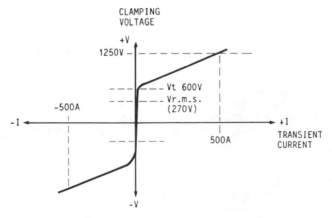

FIG. 1.2.5 Metal oxide varistor (MOV) performance characteristics.

2.7 TRANSIENT PROTECTION DIODES

Various transient suppressor diodes are available. These may be unidirectional or bidirectional as required. In general terms, silicon suppressor diodes consist of an avalanche voltage clamp device, configured for high transient capability. In a bipolar protector, two junctions are used in series "back to back." (An avalanche diode exhibits a normal diode characteristic in the forward direction.)

The transient suppressor diode has two major advantages, the first being the very high speed clamping action—the avalanche condition is established in a few nanoseconds. The second advantage is the very low slope resistance in the conduction range.

In the active region, the slope resistance can be very low, with terminal voltage increasing by only a few volts at transient currents running into hundreds of amperes. Consequently, the transient suppressor diode provides very hard and effective voltage clamping at any transient stress up to the diode's maximum current capability. The characteristics of a typical 200-V bipolar transient suppressor diode are shown in Fig. 1.2.6. Note that the terminal voltage is only 220 V at 200 A.

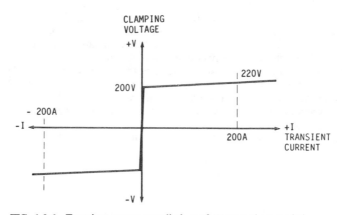

FIG. 1.2.6 Transient suppressor diode performance characteristics.

The major disadvantages of the transient suppressor diode are its relatively high cost and limited current capability. However, if the diode is overstressed, it is designed to fail to a short-circuit condition; this would normally clear the external fuse or circuit breaker, while maintaining protection of the equipment.

2.8 GAS-FILLED SURGE ARRESTERS

Much larger transient currents can be handled by the various gas-discharge suppressor devices. In such suppressors, two or more electrodes are accurately spaced within a sealed high-pressure inert gas environment. When the striking voltage of the gas tube is exceeded, an ionized glow discharge is first developed between the electrodes. As the current increases, an arc discharge is produced, providing a low-impedance path between all internal electrodes. In this mode, the device has an almost constant voltage conduction path with a typical arc drop of 25 V. The characteristics of the gas arrester are shown in Fig. 1.2.7. Note the large striking voltage and low arc voltage.

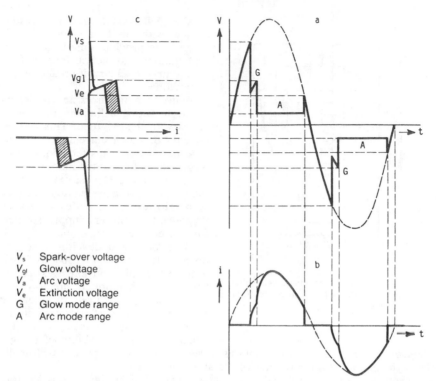

V_s Spark-over voltage
V_{gl} Glow voltage
V_a Arc voltage
V_e Extinction voltage
G   Glow mode range
A   Arc mode range

FIG. 1.2.7 Gas-filled surge arrester (SVT) performance characteristics. (*Courtesy of Siemens AG.*)

When it strikes, the gas arrester effectively short-circuits the supply, with only a small voltage being maintained across the electrodes. Because of the low internal dissipation in this mode, a relatively small device can carry currents of many thousands of amperes. With this type of suppressor, protection is provided not so much by the energy dissipated within the device itself, but by the device's short-circuit action. This forces the transient energy to be dissipated in the series resistance of the supply lines and filter.

A disadvantage of the gas arrester is its relatively slow response to an overvoltage stress. The plasma development action is relatively slow, and the striking voltage is *dv/dt*-dependent. Figure 1.2.8 shows the striking voltage as a function of *dv/dt* for a typical 270-V device. The effect is quite marked at transient edge attack rates as low as 10 V/μs. Hence, for fast transients, the gas arrester must be backed by a filter or faster-action clamp device.

A major disadvantage of the arrester is its tendency to remain in a conducting state after the transient condition has ceased. On ac lines, the recovery (blocking action) should normally occur when the supply voltage falls below the arc voltage at the end of a half cycle. However, the line source resistance can be very low, and if the current rating of the device is exceeded, the high internal temperature may prevent normal extinction of the arc, so that the device remains conducting. The follow-on current, provided by the line supply after the transient has

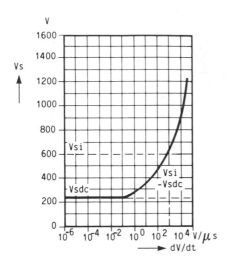

FIG. 1.2.8 Variation in sparkover voltage with applied *dv/dt* for gas-filled SVPs. (*Courtesy of Siemens AG.*)

finished, will soon destroy the arrester. Hence, with this type of device, it is essential to provide some form of current limiting, fast-acting fuse or fast-acting (magnetic) circuit breaker in the supply line.

Many manufacturers and designers advocate fitting a limiting resistor in series with the gas tube. This will reduce the follow-on current after the gas tube has struck. This technique satisfies the need to limit the follow-on current, and allows plasma extinction as the supply voltage passes through zero. However, the series resistance degrades the transient suppressor performance, since even a small (say 0.3-Ω) resistor would develop a voltage drop of 1000 V for the 3000-A IEEE Standard 587 high-current stress condition. The author prefers not to fit series resistors, but to rely on filter and external circuit resistances to limit the suppressor current; this retains the excellent clamping capability of the gas device. For extended stress conditions, a fast circuit breaker or fuse will finally clear the line input if the gas device remains conducting.

The gas arrester is still undergoing development at the time this is written, and many ingenious techniques are being developed to improve its performance.

2.9 LINE FILTER, TRANSIENT SUPPRESSOR COMBINATIONS

As mentioned above, the various transient suppressor devices have limited current capability.

Because the line impedance can be extremely low, it is often necessary to include some limiting resistance in series with the supply lines to reduce the stress on the shunt-connected suppressors. This also permits efficient voltage clamping action.

Although the series limiting may be provided by discrete resistors, in the interest of efficiency, inductors should be used. If inductors are used, it is expedient to provide additional filtering in the transient suppressor circuit at the same time. This will help to reject line-borne noise and filter out power supply–generated noise. Also, the winding resistance and inductance can provide the necessary series impedance to limit the transient current for efficient transient suppression. Consequently, transient suppression is often combined with the EMI noise filtering circuits typically required with switchmode supplies.

2.10 CATEGORY A TRANSIENT SUPPRESSION FILTERS

Figure 1.2.9 shows a typical combination of line filter and transient suppressor devices that may be found in a category A protection unit.

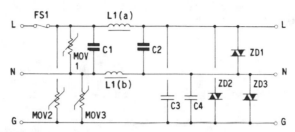

FIG. 1.2.9 Line-to-line and line-to-ground transient over-voltage protection circuit with noise filter, using MOV and SVP protection devices (low- to medium-power applications).

The inductors L1(a) and L1(b) and capacitors C1 through C4 form the normal noise filter network. At the input to this filter network, varistors MOV1 through MOV3 provide the first level of protection from line-borne transient stress. For very short lived high-voltage transients, the clamping action of the varistors, together with the voltage dropped across the series inductance, holds off the majority of the transient voltage from the output.

For more extended stress conditions, the current in L1(a) and L1(b) will increase to the point where the output capacitors C2 and C3 are charged to a voltage at which suppressor diodes ZD1, ZD2, and ZD3 are brought into conduction. These diodes prevent the output voltage from exceeding their rated clamp values for all stress currents up to the failure point of the suppression diodes. If this level is reached, the diodes fail to a short circuit, clearing the protection fuse FS1, and the unit fails to a safe condition. However, this very high level of stress should not occur in a category A location.

It should be noted that the suppressor unit also prevents voltage transients generated within the driven equipment from feeding back into the supply line. This can be an important advantage when several pieces of equipment in a system are connected to the same supply.

In this example, protection has been provided for differential- (line to neutral) and common-mode (line and neutral, to ground) stress. It will be shown later that although differential protection is often the only protection provided, common-mode stress conditions often occur in practice. Hence protection for this condition is essential for full system integrity.

The wisdom of common-mode transient suppression has been questioned as possibly being dangerous, because of the voltage "bump" on the earth return line under transient conditions. (See Sec. 2.13.) It will be shown later that this effect is almost inevitable; it should be dealt with in other ways if full protection is to be provided.

2.11 *CATEGORY B TRANSIENT SUPPRESSION FILTERS*

Although the circuit shown in Fig. 1.2.9 could be used for category B locations if suitably large devices were selected, it is more expedient to use the small low-cost gas-discharge suppressors to provide the additional protection.

Figure 1.2.10 shows a suitable circuit arrangement. This circuit combines the advantages of all three types of protection device, and also has a full common- and series-mode filter network.

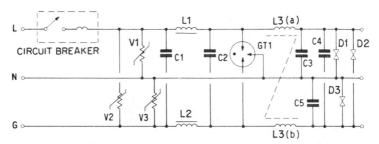

FIG. 1.2.10 Line-to-line and line-to-ground transient protection circuit with noise filter, using MOV, SVP, and transient protection diodes (for medium- to high-power applications).

The common-mode filter inductor L3(a and b) has been supplemented with additional series-mode inductors L1 and L2. These inductors, together with capacitors C1 through C5, provide a powerful filter for common- and series-mode line-conducted transient and RFI noise. This unit may be used to supplement or replace the normal line filter of the switchmode supply.

In addition to the voltage-dependent resistors (varistors) and output transient suppressor diodes, the three-terminal gas-discharge arrester tube (GT1) is shown fitted at the interface between the series-mode and common-mode inductors.

This arrangement combines the advantages of all three suppressor devices in a most effective manner. For very fast transients, once again, the input varistors V1, V2, and V3, together with L1, L2, and L3(a) and L3(b) and capacitors C1 through C5, provide efficient attenuation of the transient. For medium-stress conditions of longer duration, the current in the inductors will increase and the output voltage will also increase to the point where the output clamping diodes D1, D2, and D3 are brought into conduction protecting the load.

The major advantage of this category B suppressor is that for very large and extended stress conditions, the gas arrester GT1 will be brought to the striking voltage, effectively short circuiting all lines (and the transient) to ground.

An advantage of the three-terminal gas arrester is that, irrespective of which line the original stress appears on, all lines are shorted to ground. This tends to reduce the inevitable ground return "bump" voltage.

Extensive stress testing of this circuit has shown that in most cases, the supply line impedance, combined with the current limiting action of L1 and L2, will prevent excessive buildup of current in the gas arrester after ignition. As a result, the arrester recovers to its nonconducting state after the transient has passed, during the following zero crossover of the supply line. Hence, under the rare conditions when a gas device is called on to conduct, in most cases the power to the load is interrupted for less than a half cycle.

Because of the energy storage and holdup ability of the typical switchmode supply, a half cycle line dropout will not result in an interruption to the DC output to the loads. In the rare event of a gas tube continuing to conduct, the fast-acting magnetic circuit breaker will operate in less than a cycle, clearing the line input from the filter.

2.12 A CASE FOR FULL TRANSIENT PROTECTION

The major cause of high-voltage transients is direct or indirect lightning effects on the external power system. Irrespective of the initial cause of the transient, be it a direct strike to one or another of the supply lines or the induced effects of a near miss, the initial stress attenuation is provided by flashover between lines and from line to ground at various points throughout the distribution system.

As a result of these flashovers, the transient that arrives at a remote location will tend to be common-mode, appearing between both supply lines and ground. Even if the neutral is connected to ground near the service entrance of the building, the stress can still tend to be common-mode at the protected equipment because of flashover in the building cables, distribution boxes, and receptacles. (It is this flashover that reduces the stress between category C and category A locations.) Consequently, transient suppressors which provide protection between line and neutral only are not protecting the equipment or common-mode capacitors against the line ground, neutral ground stress conditions.

2.13 THE CAUSE OF "GROUND RETURN VOLTAGE BUMP" STRESS

A voltage stress which appears between both supply lines and the ground return is called a common-mode transient. When a common-mode transient arrives at the suppressor unit, the current is diverted to ground through one or more of the transient suppressor devices. As a result, considerable currents can flow through the ground return during a transient. Because of the resistance and inductance between the transient suppressor and the service entrance, this ground current can elevate the potential of the local system ground with respect to real ground. Hence, a possible shock hazard now exists between the case of the protected equipment and real earth. (This voltage is referred to as an earth return "bump.")

It is possible to argue, therefore, that transient suppressors which return the stress current to the ground line are a shock hazard and should not be used. This is a viable argument only if the load can be guaranteed not to break down to ground during the stress in the absence of a transient suppressor. In practice, the equipment is likely to fail in this mode, and the hazard of ground return bump will still exist, even without the suppressor. In addition, the load will not have been protected and may well be damaged.

The possibility of an earth return voltage bump under high-stress conditions should be considered an inevitable hazard with or without transient protection.

Measures should be taken to reduce the voltage by ensuring a very low resistance ground return path. If an operator has access to the equipment, all equipment within the operator's reach must be grounded to the same return. In computer rooms, the need for a good ground return may include the furniture and very fabric of the building itself.

2.14 PROBLEMS

1. Why is it important to provide AC powerline surge protection in direct-off-line switchmode power supplies?

2. Give some typical causes of AC line transients.

3. Give the number of an IEEE standard which describes the typical amplitudes and waveshapes to be expected on various line distribution systems in office and domestic locations.

4. Describe stress locations A, B, and C, as described in IEEE Standard 587–1980.

5. Explain the meaning of exposure locations, as described in IEEE Standard 587–1980.

6. How does IEEE Standard 587–1980 indicate the likely rate of surge occurrence and voltage amplitude at various locations?

7. What would be the typical waveform and transient voltage to be expected in a class A location?

8. What surge waveforms may be expected in a class B location?

9. Describe three transient protection devices commonly used in input line protection filters.

10. Describe the advantages and limitations of metal oxide varistors, transient protection diodes, and gas-filled surge suppressors.

CHAPTER 3
ELECTROMAGNETIC INTERFERENCE (EMI) IN SWITCHMODE POWER SUPPLIES

3.1 INTRODUCTION

Electromagnetic interference (EMI), otherwise referred to as radio-frequency interference (RFI), the unintentional generation of conducted or radiated energy, is indefatigable in all switchmode power supplies. The fast rectangular switching action required for good efficiency also produces a wide interference spectrum which can be a major problem.

Further, for proper operation of any electronic system, it is important that all the elements of the system be electromagnetically compatible. Also, the total system must be compatible with other adjacent systems.

As the SMPS can be such a rich source of interference, it is vital that this aspect of the design be carefully considered. Normal good design practice requires that the RF interference allowed to be conducted into the supply or output lines, or permitted to be radiated away from any power equipment, be minimized to prevent RF pollution. Further, national, federal, and international regulations limit by law the permitted interference levels.

These regulations vary according to country of origin, regulatory authority, and intended application. The power supply designer will need to study the code relevant to the proposed marketing areas. In Common Market countries, IEC BS 800, or CISPR recommendations apply. The Federal Republic of Germany requires VDE 0871 or VDE 0875, depending on the operating frequency. In the United States the Federal Communication Commission's (FCC) rules apply, and similar limits are recommended in Canada under CSA, C108.8-M1983.

In general, the range of frequencies covered by the regulations spans from 10 kHz to 30 MHz. Domestic locations have more rigorous regulations than office or industrial locations.

Figure 1.3.1 shows the FCC and VDE limits for conducted RFI emissions in force at the time of publication.

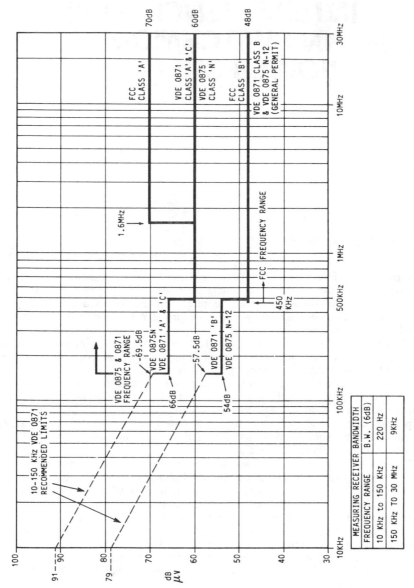

FIG. 1.3.1 Conducted-mode RFI limits as specified by FCC Part 15 (Subpart j) VDE 0871 and 0875.

3.2 EMI/RFI PROPAGATION MODES

There are two forms of propagation of interest to the power supply designer: electromagnetic radiated E and H waves and conducted interference on supply lines and interconnecting wires.

Radiated interference is normally minimized as a natural result of the layout and wiring practices required to reduce leakage inductance and improve performance. Typically the high-frequency current loops will be short, and twisted pairs will be used where possible. Transformers and chokes with air gaps will be screened to reduce radiated magnetic fields (see Part 1, Chap. 4), screened boxes or equipment enclosures will often be used.

The techniques applied to minimize conducted interference will also reduce radiated noise. The following sections concentrate on the conducted aspect of power supply interference, as once the conducted limits have been met, the radiated limits will normally be satisfied as well.

3.3 POWERLINE CONDUCTED-MODE INTERFERENCE

Two major aspects of conducted interference will be considered: differential-mode conducted noise and common-mode conducted noise.

These will be considered separately.

Differential-Mode Interference

Differential- or series-mode interference is that component of RF noise which exists between any two supply or output lines. In the case of off-line SMPS, this would normally be live and neutral ac supply lines or positive and negative output lines. The interference voltage acts in series with the supply or output voltage.

Common-Mode Interference

Common-mode interference is that component of RF noise which exists on any or all supply or output lines with respect to the common ground plane (chassis, box, or ground return wire).

3.4 SAFETY REGULATIONS (GROUND RETURN CURRENTS)

It may seem out of place to be considering safety requirements at this stage; however, this is necessary because the safety agencies specify the maximum ground return currents, so as to minimize shock hazard in the event of ground circuit faults. This requirement not only requires good attention to insulation, but also puts a severe limitation on the value of capacitors which may be fitted between the supply lines and ground. This capacitor size limitation has a profound impact on the design of the line input filters.

The permitted limits for ground return currents vary among the regulatory agencies and also depend on the intended equipment applications. For example, medical equipment, as one might expect, has a very stringent, so-called "earth leakage current" limit.

The ground return current limits, as set by some of the major regulatory agencies, that are in force at the time of printing are shown in Table 1.3.1.

TABLE 1.3.1 Maximum Ground Leakage Currents Permitted by the Safety Regulations, and the Recommended Maximum Values for Y Filter Capacitors

Country	Specification	Ground leakage current limits	Maximum value C1 and C2
U.S.A.	UL 478	5 mA 120 V 60 Hz	0.11 μF
	UL 1283	0.5–3.5 mA 120 V 60 Hz	0.011–0.077 μF
Canada	C 22.2 No 1	5 mA 120 V 60 Hz	0.11 μF
Switzerland	SEV 1054-1 IEC 335-1	0.75 mA 250 V 50 Hz	0.0095 μF
Germany	VDE 0804	3.5 mA 250 V 50 Hz	0.0446 μF
		0.5 mA 250 V 50 Hz	0.0064 μF
U.K.	BS 2135	0.25–5 mA 250 V 50 Hz	0.0032–0.064 μF
Sweden	SEN 432901	0.5 5 mA 250 V 50 Hz	0.0064 μF
		0.25–5 mA 250 V 50 Hz	0.0032–0.064 μF

Table 1.3.1 gives the maximum value of decoupling capacitance that may be fitted in positions C1 and C2 in Fig. 1.3.2, for each specification. These values assume zero contribution from insulation leakage and stray capacitance. To minimize inductor and filter size, the largest decoupling capacitor permitted by the regulations should be used. Since one side of the input is always assumed to be

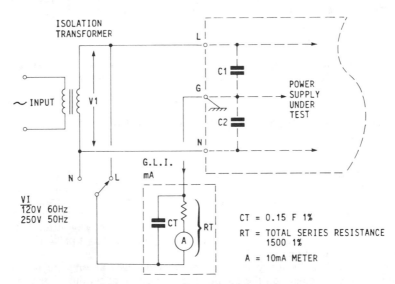

FIG. 1.3.2 Ground leakage current test circuit, as specified by CSA 22.2, part 1. (*Note:* CT and RT values depend on the equipment and agency requirements.)

neutral (connected to the ground at the service entrance), only one capacitor will be conducting at any time. However, the total leakage current should be checked to establish the total contribution from all the capacitive and insulation leakage paths.

Figure 1.3.2 shows the method of measurement for ground return currents. It is assumed that only one side of the supply could be "hot," and hence only one capacitor will be conducting to ground return at any time.

3.5 POWERLINE FILTERS

To meet the conducted-mode noise specifications, relatively powerful line filters will normally be required. However, as previously demonstrated, safety regulations severely limit the size of the capacitors fitted between the supply lines and ground plane.

Because of the limited size of the decoupling capacitors, the filter cannot easily cure the severe common-mode interference problems which can occur as a result of poor wiring, bad layout, poor screening, or bad location of the power switching elements. Hence, good EMI performance demands care and attention to all these aspects at every stage of the design and development process. There is no substitute for effective suppression of EMI at the source.

3.6 SUPPRESSING EMI AT SOURCE

Figure 1.3.3 shows several of the more common causes of EMI problems. Failure to screen the switching devices and failure to provide RF screens in the transformer are principal causes of conducted common-mode interference. This component of interference is also the most difficult to eliminate in the filter, because of the limited decoupling capacitor size.

The differential- or series-mode noise is more easily bypassed by the electrolytic storage capacitors and the relatively large decoupling capacitors C3 and C4 which are permitted across the supply lines.

Common-mode RF interference currents are introduced into the local ground plane (normally the chassis or box of the power supply) by insulation leakage and parasitic electrostatic and/or electromagnetic coupling, shown as C_{p1} through C_{p5} on Fig. 1.3.3. The return loop for these parasitic currents will be closed back to the input supply lines through the decoupling capacitors C1 and C2.

The prime mover for this loop current tends to a constant-current source, as the source voltage and source impedance are very high. Hence the voltage across the decoupling capacitors C1 and C2 tends to a voltage source proportional to the current magnitude and capacitor impedance at the interference harmonic frequency:

$$V_{hi} = I_i \times X_c$$

where V_{hi} = harmonic interference voltage
I_i = interference current at the harmonic frequency
X_c = reactance of C1 or C2 at the harmonic frequency

(It is assumed that the insulation leakage current is negligible.) This voltage source V_{hi} will now drive current into the series inductors L1, L2, and L3 and into the output lines to return via the ground line. It is this external component of RF current that will cause external interference, and hence it is this that is covered by the regulations and must be minimized.

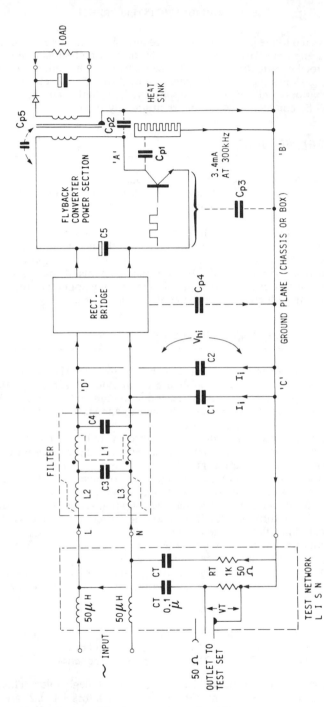

FIG. 1.3.3   Example of parasitic RFI current paths in a typical off-line switchmode power supply.

3.7 EXAMPLE

Consider the parasitic current loop A, B, C, D, and back to A, shown in Figure 1.3.3. Point A is the high-voltage switching transistor package.

For a flyback application, the voltage on this transistor may be of the order of 600 V and the switching frequency typically 30 kHz. Because of the fast switching edges, harmonics will extend up to several megahertz. Parasitic capacitive coupling (shown as C_{p1} in the diagram) will exist between the transistor case A and the ground plane B.

The tenth harmonic of the switching frequency will be 300 kHz, well inside the RF band laid down in the regulations. If square-wave operation is assumed, the amplitude of this harmonic will be approximately 20 dB down on 600 V, or 60 V. Assuming the leakage capacitance to be 30 pF, a current of 3.4 mA will flow into the ground plane at 300 kHz.

The current loop is closed back to the transistor by the filter capacitors C1 and C2.

To meet the most stringent safety regulations, the maximum capacitance allowed for C1 and C2 would be, say, 0.01 μF.

If the majority of the ground plane current returns via one of these capacitors, then the voltage V_{hi} across its terminals, nodes C to D, will be 180 mV. The inductors L1 and L2 now form a voltage divider network between point D and the simulated 50-Ω supply line resistance RT. If the voltage across RT is to be less than 250 μV (48 dB up on 1 μV, the regulation limit), then L1 and L2 must introduce an attenuation of more than 50 dB at this harmonic frequency, an almost impossible task for inductors which must also carry the supply input current.

By fitting an electrostatic screen between the transistor and the ground plane, connecting it such that the RF currents are returned to the input source, the ac voltage across the parasitic capacitance C_{p1} will be eliminated and the effective RF current from point A to ground will be considerably reduced (see Figs. 1.3.4 and 1.3.5). The demands now placed on the input filter are not so stringent.

Reducing the RF currents in the ground plane at the source is by far the best approach to EMI elimination. Once these interference currents have been introduced into the ground plane, it is very difficult to predict what path they will

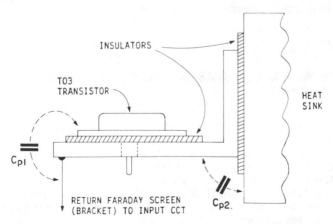

FIG. 1.3.4 TO3 mounting bracket and heat sink, with bracket configured to double as an RFI Faraday screen.

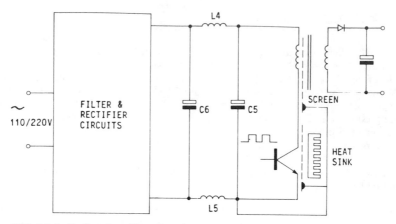

FIG. 1.3.5 Preferred positions for primary-to-ground-plane RFI screens.

take. Clearly all the high-voltage ac components should be isolated from the ground plane, or if contact cooling is required, they should be screened (see Fig. 1.3.4). Transformers should have Faraday screens, which should be returned to the input DC lines, to return capacitively coupled currents to the supply lines (see Fig. 1.3.5). These RFI screens are in addition to the normal safety screens, which must be returned to the ground plane for safety reasons.

Capacitor C4 (Fig. 1.3.3) reduces the differential- or series-mode noise applied to the terminals of L1. The major generator of noise in this part of the circuit is the input rectifier bridge (as a result of the rectifier reverse recovery current spikes). The series-mode noise generated by the power switching elements is best decoupled by a capacitor C5 close to the point where the noise is generated. In any event, the large electrolytic storage capacitors will usually effectively shunt away the majority of any series noise that appears between the high-voltage DC lines. In some cases, additional filter components L4, L5, and C6 (Fig. 1.3.5) are provided to improve the series-mode filtering.

3.8 LINE IMPEDANCE STABILIZATION NETWORK (LISN)

Figure 1.3.6 shows the standard LISN, used for the measurement of line-conducted interference, as specified by CSA C108.8-M1983 Amendment 5, 1983. (Similar networks are specified by the FCC and VDE.) In principle the wideband line chokes L1 and L2 divert any interference noise currents from the supply into the 50-Ω test receiver via the 0.1-μF capacitors C3 or C4. The line not under test is terminated in 0.1 μF and 50 Ω. It is normal to test both supply lines independently for common-mode noise, as the user can connect the input in reverse or may have isolated supplies.

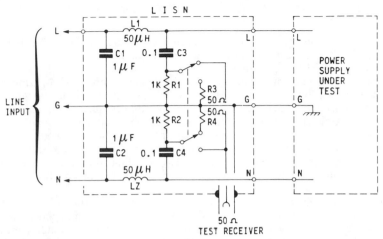

FIG. 1.3.6 Standard line impedance stabilization network (LISN) as specified for FCC, CSA, and VDE conducted-mode line interference testing.

3.9 LINE FILTER DESIGN

The design approach used in Secs. 3.4 through 3.8 was to consider the line filter as an attenuating voltage divider network for common-mode RF noise. This approach is used in preference to normal filter design techniques, as the source and load impedances are not definable in the powerline environment.

The interference noise generator, in switchmode supplies, is very often a high-voltage source in series with a high impedance; this tends to a constant-current source. To give good attenuation, one of the prime requirements is to convert the constant-current noise source into a voltage source. This is achieved by providing a low-impedance shunt path at the power supply end of the filter. Hence, powerline filters will not be symmetrical or matched networks.

"Network analysis" shows that the greater the mismatch of the filter impedance to the source or terminating impedance, the more effective the filter is in attenuating the RF interference.

Referring to Fig. 1.3.3, and assuming a constant current into nodes C and D, the attenuation into the external 50-Ω test receiver would be 12 dB/octave provided that inductors L1 and L2 and capacitors C1 and C2 have good wideband impedance characteristics. Although capacitors meeting this criterion can be easily selected, wideband inductors are not so easily found and are difficult to design, as they must also carry the supply line currents without significant power loss.

Finally, as shown in Sec. 3.4, the safety requirements set a limit on the maximum size of the decoupling capacitors C1 and C2, so that any further increase in the attenuation factor of the filter is critically dependent on the value and performance of the series inductors L1 and L2. Some design criteria for the filter inductors will now be considered.

3.10 COMMON-MODE LINE FILTER INDUCTORS

Inductor L1 in Fig. 1.3.3 should be considered a special case. For the best common-mode attenuation it must have a high common-mode inductance and also carry the 60-Hz supply current.

To provide the maximum inductance on the smallest core, a high-permeability core material will be used. It is normal practice to wind L1 with two windings. These windings carry large currents at twice the line frequency, as the rectifier diodes only conduct at the peak of the input voltage waveforms.

In more conventional choke designs, this operating condition would require a low-permeability material or air gap in the magnetic path to prevent saturation of the core. However, in this application, the two windings on L1 are phased such that they provide maximum inductance for common-mode currents but cancel for series-mode currents.

This phasing prevents the core from saturating for the normal 60-Hz differential line currents, as these flow in opposite phase in each winding, eliminating the 60-Hz induction. However, this phasing also results in negligible inductance for series-mode noise currents, and additional noncoupled inductors L2 and L3 will sometimes be required to reduce series-mode noise currents.

This is one situation in which a large leakage inductance between the two windings on L1 can be an advantage. For this reason, and to meet safety requirements, the windings will normally be physically separated and a bobbin with two isolated sections will be used. As the low-frequency induction is small, a high-permeability ferrite or iron core material may be used, without the need for an air gap.

Where this type of common-mode inductor is used for the output filter in DC applications, the series-mode DC components will also cancel, and the same conditions prevail.

The performance of L1 for common-mode noise is quite different. Common-mode noise appears on both supply lines at the same time, with respect to the ground plane. The large shunt capacitor C2 helps to ensure that the noise amplitude will be the same on both lines where they connect to the inductor. The two windings will now be in phase for this condition, and both windings behave as one, providing a large common-mode inductance.

To maintain good high-frequency rejection, the self-resonant frequency of the filter inductors should be as high as possible. To meet this need, the interwinding capacitance and capacitance to core must be as low as possible. For this reason single-layer spaced windings on insulated high-permeability ferrite toroids are often used. The effective inductance of the common-mode inductors can be quite large, typically several millihenrys.

When extra series-mode inductors are used (L2 and L3 in Fig. 1.3.3), the common-mode inductor L1 can be designed to reject the low-frequency components only, and so the interwinding capacitance is not so important. For this application ferrite E cores can be used; these have two section bobbins, giving good line-to-line insulation. Inductors L2 and L3 must provide good high-frequency attenuation and would normally be low-permeability iron powder or MPP Permalloy toroids. Single-layer wound chokes on these low-permeability cores will not saturate at the line frequency currents.

The inductance and size of the main common-mode choke L1 depends on the current in the supply lines and the attenuation required. This is best established by measuring the conducted noise with capacitors C1 and C2 in place but without inductors. The voltage and frequency of the largest harmonic are noted, and the inductance required to bring this within the limit can be calculated. It then remains to select a suitable core, wire size, and turns for the required inductance, current rating, and temperature rise.

It should be noted that the losses in L1 are nearly all resistive copper losses ($I^2 R_{Cu}$), as the core induction and skin effects are negligible. The design of L1 is an iterative process which is probably best started by selecting a core size for the current rating and required inductance using the "area-product" approach (see Part 3, Chap. 1).

3.11 DESIGN EXAMPLE, COMMON-MODE LINE FILTER INDUCTORS

Assume it has been established by calculation or measurement (Sec. 3.10) that a 100-W power supply operating from a 110-V ac supply requires a common-mode inductance of 5 mH to meet the EMI limits. Further assume the power loss in the inductor is not to exceed 1 percent (1 W) and the temperature rise is not to exceed 30 K (all typical values).

For a temperature rise of 30 K at 1 W, the thermal resistance of the finished inductor (to free air) R_0 is 30 K/W. From Table 2.19.1, at $R_0 = 30$ K/W, a core size of E25/25/7 is indicated.

For a 100-W unit with an efficiency of 70% and power factor of 0.63 (typical values for a flyback SMPS capacitor input filter), the input current will be 2 A rms at 110 V.

If the total loss (both windings) is to be 1 W, then $I^2 R = 1$ and the resistance of the total windings R_{Cu} must not exceed 0.25 Ω.

From the manufacturer's data, the copper resistance factor A_r for the E25 bobbin is 32 $\mu\Omega$. The turns to fill the bobbin and give a resistance of 0.25 Ω can now be calculated:

$$N = \sqrt{\frac{R_{Cu}}{A_r}} = \sqrt{\frac{0.25}{32 \times 10^{-6}}} = 88 \text{ turns}$$

Allowing 10% loss for the split bobbin, there will be 40 turns for each side.

The A_L factor (inductance factor) for the E25 core in the highest permeability material N30 is 3100 nH. The inductance may now be calculated:

$$L = N^2 \times A_L = 40^2 \times 3100 \times 10^{-9} = 4.96 \text{ mH}$$

The largest wire gauge that will just fill the bobbin for this number of turns (from the manufacturer's data) is AWG 20. Since the inductance is marginal, the process can be repeated with the next larger core.

3.12 SERIES-MODE INDUCTORS

The design of the series-mode iron dust or MPP cored inductors is covered in Chaps. 1, 2, and 3 of Part 3.

3.13 PROBLEMS

1. Explain and give examples of some of the typical causes of conducted and radiated RFI interference in switchmode power supplies.

2. What forms of electrical noise propagation are of most interest to the power supply designer?

3. Describe the difference between differential-mode interference and common-mode interference.

4. Why is it important to reduce interference noise to the minimum?

5. At what position in the power supply is RFI interference best eliminated?

6. Why are line filters of limited value in eliminating common-mode line-borne interference?

CHAPTER 4
FARADAY SCREENS

4.1 INTRODUCTION

One of the most difficult problems in switchmode power supply design is to reduce the common-mode conducted RFI current to acceptable limits. This conducted electrical noise problem is mainly caused by parasitic electrostatic and electromagnetic coupling between the various switching elements and the ground plane. (The ground plane can be the chassis, cabinet, or ground return wire, depending on the type of unit.)

The designer should examine the layout, identify the areas where such problems may exist, and introduce at the design stage the correct screening methods. It is very difficult to correct for poor RFI design practices at a later stage. Likely problem areas are shown in Fig. 1.4.1, a diagram of the typical problem areas for parasitic coupling in a flyback SMPS. Suitable locations for Faraday screens are shown.

In most applications, Faraday screens will be required where high-frequency, high-voltage switching waveforms can capacitively couple to the ground plane or secondary outputs. Typical positions would be where switching transistors and rectifier diodes are mounted on heat sinks which are in contact with the main chassis. Further, where components or wires carry large switching currents, noise can be coupled by both magnetic and capacitive coupling. Other likely problem areas are output rectifiers; output chassis-mounted capacitors; and capacitive coupling between the primary, secondary, and core of the main switching transformer and any other drive or control transformers.

4.2 FARADAY SCREENS AS APPLIED TO SWITCHING DEVICES

When components are mounted on heat sinks which are to be thermally linked to the chassis, the normal way of eliminating undesirable capacitive coupling is to place an electrostatic screen between the offending component and the heat sink.

This screen, normally copper, must be insulated from both the heat sink and the transistor or diode, so that it picks up the capacitively coupled ac currents and returns them to a convenient "star" point on the input circuit. For the primary components, the "star" point will usually be the common negative DC supply line, close to the switching device. For secondary components, the "star" point will normally be the common return to the transformer. Figure 1.4.1 demonstrates the principle.

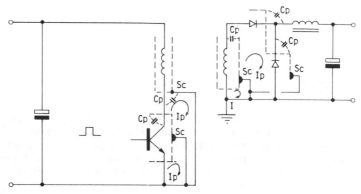

FIG. 1.4.1 Return paths for capacitively coupled Faraday screen currents in primary and secondary circuits.

One example of a TO3 transistor Faraday screen is shown in Fig. 1.4.2. The primary switching transistor, with its high voltage and high-frequency switching waveform, would couple a significant noise current through the capacitance between the transistor case and the main chassis unless a screen is fitted between them. In the mounting arrangement shown in Fig. 1.4.2, the copper screen will return this parasitic noise current to the input circuit, thus completing the current loop without introducing current into the ground plane. The screen will not inject any significant current through the capacitance to the heat sink, because it has a relatively small high-frequency ac voltage relative to the chassis or ground plane. The designer may identify other areas where problems can occur; in that event, similar screening should be used.

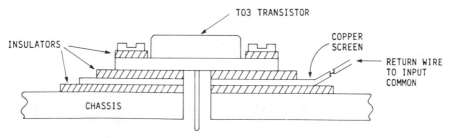

FIG. 1.4.2 Insulated Faraday screen, positioned between TO3 switching transistor and heat sink.

4.3 TRANSFORMER FARADAY SCREENS AND SAFETY SCREENS

To prevent circulation of RF currents between the primary and secondary windings or between the primary and the grounded safety screen, the main switching transformer will usually have at least one RFI Faraday screen in the primary winding. In some applications, an additional safety screen will be required between the primary and secondary windings. There are major differences in construction, location, and connection between the Faraday RFI screens and the safety screens. Safety regulations require that the safety screens be returned to the ground plane or chassis, whereas RFI screens will normally be returned to the input or output circuits. The EMI screens and connections may be made of very lightweight copper, as they carry very little current. However, for safety reasons, the safety screen must be rated for a current of at least three times the supply fuse rating.

Figure 1.4.3 shows the typical arrangement of safety and RFI screens in a switchmode transformer for "off-line" use. In the fully screened application shown, the two RFI screens will be adjacent to the primary and secondary windings, and the safety screen will be between the two RFI screens. If secondary RFI screens are not required, the safety screen will be between the primary RFI screen and any output windings. As a further insulation precaution, the primary RFI screen may be DC isolated from the input powerlines by a series capacitor. (A value of 0.01 μF at the rated isolation voltage is usually sufficient.)

The RFI screen shown on the secondary side is fitted only when maximum noise rejection is required or when output voltages are high. This screen would be returned to the common output line. Transformer screens should be fitted only

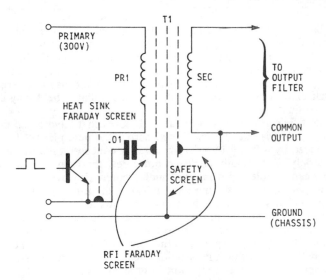

FIG. 1.4.3 Fully screened transformer, showing positions and connections of primary and secondary Faraday screens, with an additional primary-to-secondary safety screen.

when essential, as the increased buildup and winding height increase the leakage inductance and degrade the performance.

To prevent the high-frequency screen return currents (which can be considerable during the switching transient) from coupling to the secondary by normal transformer action, the screen connections should be made to the center of the screen, rather than one end. In this way, the capacitively coupled screen return currents flow in opposite directions around each half of the screen, cancelling any inductive coupling effects. Remember, the ends of the screen must be insulated to prevent a closed turn.

4.4 FARADAY SCREENS ON OUTPUT COMPONENTS

For high-voltage outputs, RFI screens may be fitted between the output rectifiers and their heat sinks. If the secondary voltages are small, say 12 V or less, the secondary transformer RFI screen and rectifier screens should not be required.

The need for Faraday screens on output rectifier diodes can sometimes be eliminated by making the diode heat sink dead to RF voltages by putting the output filter choke in the return line. Typical examples are shown in Fig. 1.4.4a and b.

If the diode and transistor heat sinks are completely isolated from the chassis (for example, mounted on the pcb), Faraday screens are unlikely to be required on these components.

4.5 REDUCING RADIATED EMI IN GAPPED TRANSFORMER CORES

Ferrite flyback transformers and high-frequency inductors will usually have a relatively large air gap in the magnetic path, to define the inductance or to prevent saturation. Considerable energy can be stored in the magnetic field associated with this air gap. Unless the transformer or choke is screened, an electromagnetic field (EMI) will be radiated from the gap, and this can cause interference to the supply itself or to local equipment. Further, this radiated field may exceed the radiated EMI limits.

The largest field radiation will occur with cores that have a gap in the outer limbs or a gap that is equally distributed across the pole pieces. This radiation may be reduced by a factor of 6 dB or more by concentrating the air gap in the center pole only. With totally enclosed pot cores, the reduction in radiation by using only a center pole gap would be much greater. However, for off-line applications, the pot core is not often used because the creepage distance requirements at the higher voltages usually cannot be satisfied.

Concentrating the air gap in the center pole alone increases the temperature rise and reduces efficiency. This increased loss is probably due to magnetic fringe effects at the edge of the pole pieces in the center of the winding. The disturbance of the magnetic field within the windings results in additional skin and eddy-current losses, and a further reduction in efficiency of up to 2%. Also, the increased losses in the region of the gap can cause a hot spot and premature failure of the insulation in this area.

In cores which are gapped in the outer legs, the addition of a copper screen

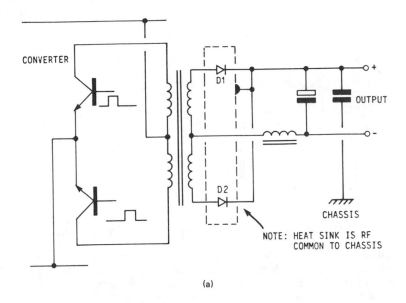

(a)

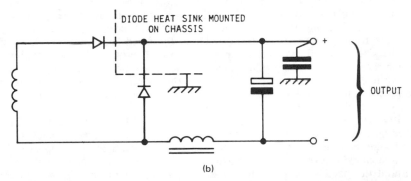

(b)

FIG. 1.4.4 A method of reducing parasitic RFI currents in chassis-mounted output diode heat sinks by fitting the output choke in the common return line. (*a*) Push-pull applications; (*b*) single-ended outputs.

around the outside of the transformer gives a considerable reduction in radiation. Figure 1.4.5 shows a typical example.

This screen should be a totally closed loop around the outside of the transformer, over the outer limbs and windings, and centered on the air gap. The width of the screen should be approximately 30% of the width of the bobbin and should be in the same plane as the windings. To be effective, it must have minimum resistance; a copper screen with a thickness of at least 0.010 in is recommended.

It would appear that this screen is effective because of both eddy-current losses and the action of the closed loop. The current induced in the closed loop will generate a back MMF to oppose radiation. In flyback transformers, the screen should not be more than 30% of the bobbin width, as problems of core

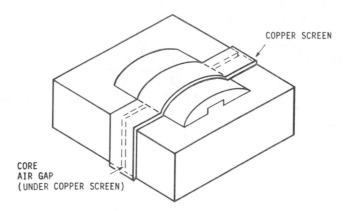

FIG. 1.4.5 Copper screen fitted to a switching transformer, to reduce RFI and EMI radiation. (*Note:* Screen goes around the outside of the core legs.)

saturation have been observed with wide screens. Although the screen is normally used for cores which are gapped in the outer legs, it will be effective for transformers with a gap in either the center pole or the outer legs. In either case, there will be a reduction in magnetic radiation of up to 12 dB.

However the application of a transformer screen results in lower transformer efficiency. This is due to the additional power losses in the screen, caused by eddy-current heating effects. If the air gap is in the outer poles, the power loss in the screen may amount to as much as 1% of the rated output power, depending on the size of the air gap and the power rating of the unit. For applications in which the air gap is in the center pole only, there will be little further increase in power loss from fitting a screen. However, the overall transformer efficiency is about the same in both cases, as the center pole gap increases the losses in the transformer windings by about the same amount.

It would seem that effective magnetic screening of the transformer can be applied only at the expense of additional power losses. Consequently, such screening should be used only where essential. In many cases, the power supply or host equipment will have a metal enclosure so that EMI requirements will be met without the need for extra transformer screening. When open-frame switching units are used in video display terminals, screening of the transformer will usually be required to prevent interference with the display by magnetic coupling to the CRT beam. The additional heat generated by the outer copper screen may be conducted away using a heat sink or a thermal shunt from the screen to the chassis. Figure 1.4.5 shows a typical example of a copper EMI screen as applied to an E core transformer with air gaps in the outer legs.

4.6 PROBLEMS

1. Why are Faraday screens so effective in reducing common-mode interference in high-voltage switching devices and transformers?
2. What is a line impedance stabilization network (LISN)?
3. What is the difference between common-mode and series-mode line filter inductors?
4. What is the difference between a Faraday screen and a safety screen in a switching transformer?

CHAPTER 5
FUSE SELECTION

5.1 INTRODUCTION

Fuses (fusible wire links) are one of the oldest and most universally used overload protection methods. However, because the function of the fuse is thought to be elementary, it tends not to get the close attention it deserves for a thorough understanding of its characteristics.

Modern fuse technology is an advanced science; new and better fuses are continually being developed to meet the more demanding requirements for protection of semiconductor circuitry. To obtain the most reliable long-term performance and best protection, a fuse must be knowledgeably chosen to suit the application.

5.2 FUSE PARAMETERS

From an electrical standpoint, fuses are categorized by three major parameters: current rating, voltage rating, and, most important, "let-through" current, or I^2t rating.

Current Rating

It is common knowledge that a fuse has a current rating and that this must exceed the maximum DC or rms current demanded by the protected circuit. However, there are two other ratings that are equally important for the selection of the correct fuse.

Voltage Rating

The voltage rating of a fuse is not necessarily linked to the supply voltage. Rather, the fuse voltage rating is an indication of the fuse's ability to extinguish the arc that is generated as the fuse element melts under fault conditions. The voltage across the fuse element under these conditions depends on the supply voltage and the type of circuit. For example, a fuse in series with an inductive circuit may see voltages several times greater than the supply voltage during the clearance transient.

Failure to select a fuse of appropriate voltage rating may result in excessive arcing during a fault, which will increase the "let-through" energy during the fuse clearance. In particularly severe circumstances, the fuse cartridge may explode, causing a fire hazard. Special methods of arc extinction are utilized in high-voltage fuses. These include sand filling and spring-loaded fuse elements.

"Let-Through" Current (I^2t Rating)

This characteristic of the fuse is defined by the amount of energy that must be dissipated in the fuse element to cause it to melt. This is sometimes referred to as the pre-arcing let-through current. To melt the fuse element, heat energy must be dissipated in the element more rapidly than it can be conducted away. This requires a defined current and time product.

For very short time periods (less than 10 ms), very little heat is conducted away from the fuse element, and the amount of energy necessary to melt the fuse is a function of the fuse element's specific heat, its mass, and type of alloy used. The heat energy dissipated in the fuse element is in the form of watt-seconds (joules), or $I^2R \times t$ for a particular fuse. As the fuse resistance is a constant, this is $\propto I^2t$, normally referred to as the I^2t rating for a particular fuse or the pre-arcing energy.

For longer periods, the energy required to melt the fuse element will vary according to the element material and the thermal conduction properties of the surrounding filler and fuse housing.

In higher-voltage circuits, an arc will be struck after the fuse element has melted and a further amount of energy will be passed to the output circuit while this arc is maintained. The magnitude of this amount of energy is dependent on the applied voltage, the characteristic of the circuit, and the design of the fuse element. Consequently, this parameter is not a function of the fuse alone and will vary with the application.

The I^2t rating categorizes fuses into the more familiar "slow-blow" normal, and "fast-blow" types. Figure 1.5.1 shows the shape of a typical pre-arcing

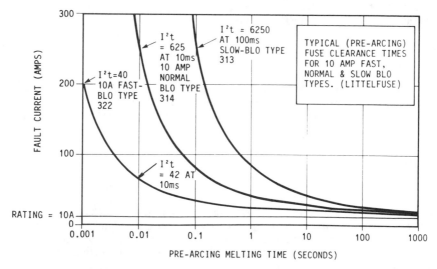

FIG. 1.5.1 Typical fuse I^2t ratings and pre-arcing fuse clearance times for fast, medium, and slow fuse links. (*Courtesy Littelfuse Inc.*)

current/time let-through characteristic for each of the three types. The curve roughly follows an I^2t law for periods of less than 10 ms. The addition of various moderators within the fuse package can greatly modify the shape of this clearance characteristic. It should be noted that the I^2t energy (and hence the energy let-through to the protected equipment) can be as much as two decades greater in a slow-blow fuse of the same DC current rating! For example, a 10-A fuse can have an I^2t rating ranging from 5A²·s for a fast fuse to 3000 A²·s for a slow fuse.

The total let-through energy of the fuse (pre-arcing plus arcing) also varies enormously. Further, it depends on the fusible link material, construction of the fusible element, applied voltage, type of fault, and other circuit-linked parameters.

5.3 TYPES OF FUSES

Time-Delay Fuse (Slow-Blow)

A time-delay fuse will have a relatively massive fuse element, usually of low-melting-point alloy. As a result, these fuses can provide large currents for relatively long periods without rupture. They are widely used for circuits with large inrush currents, such as motors, solenoids, and transformers.

Standard-Blow Fuse

These fuses are low-cost and generally of more conventional construction, using copper elements, often in clear glass enclosures. They can handle short-term high-current transients, and because of their low cost, they are widely used. Very often the size is selected for short-circuit protection only.

Very Fast Acting Fuses (HRC Semiconductor Fuses)

These fuses are intended for the protection of semiconductor devices. As such, they are required to give the minimum let-through energy during an overload condition. Fuse elements will have little mass and will often be surrounded by some form of filler. The purpose of the filler is to conduct heat away from the fuse during long-term current stress to provide good long-term reliability, and to quickly quench the arc when the fuse element melts under fault conditions. For short-term high-current transients, the thermal conductivity of the filler is relatively poor. This allows the fuse element to reach melting temperature rapidly, with the minimum energy input. Such fuses will clear very rapidly under transient current loads.

Other important fuse properties, sometimes neglected, are the long-term reliability and power loss of the fuse element. Low-cost fast-clearance fuses will often rely on a single strand of extremely thin wire. This wire is fragile and is often sensitive to mechanical stress and vibration; in any event, such fuse elements will deteriorate over the longer term, even at currents below the rated value. A typical operating life of 1000 h is often quoted for this type of fuse at its rated current.

The more expensive quartz sand-filled fuses will provide much longer life, since the heat generated by the thin element is conducted away under normal conditions. Also, the mechanical degradation of the fuse element under vibration is not so rapid, as the filling gives mechanical support.

Slow-blow fuses, on the other hand, are generally much more robust and will have longer working lives at their rated current. However, these fuses, with their high "let-through" power, will not give very effective protection to sensitive semiconductor circuits.

This brief description covers only a very few of the ingenious methods that are used in modern fuse technology to obtain special characteristics. It serves to illustrate the number of different properties that fuses can exhibit, and perhaps will draw a little more attention to the importance of correct fuse selection and replacement.

5.4 SELECTING FUSES

Off-Line Switchmode Supplies

The initial fuse selection for off-line switching supplies will be made as follows:

For the line input fuse, study the turn-on characteristics of the supply and the action of the inrush-limiting circuitry at maximum and minimum input voltages and full current-limited load. Choose a standard- or slow-blow fuse that provides sufficient current margin to give reliable operation and satisfy the inrush requirements. Its continuous current rating should be low enough to provide good protection in the event of a genuine failure. However, for long fuse life, the current rating should not be too close to the maximum rms equipment input current measured at minimum input voltage and maximum load (perhaps 150% of I_{rms} maximum). Note: Use measured or calculated rms currents, and allow for the form factor (approximately 0.6 for capacitor input filters) when calculating rms currents.

The voltage rating of the fuse must at least exceed the peak supply voltage. This rating is important, as excessive arcing will take place if the voltage rating is too low. Arcing can let through considerable amounts of energy, and may result in explosive rupture of the fuse, with a risk of fire in the equipment.

5.5 SCR CROWBAR FUSES

If SCR-type overvoltage protection is provided, it is often supplemented by a series fuse. This fuse should have an I^2t rating considerably less (perhaps 60% less) than the SCR I^2t rating, to ensure that the fuse will clear before SCR failure. Of course, a fast-blow fuse is selected in this case. The user should understand that fuses degrade with age, and there should be a periodic replacement policy. The failure of a fuse in older equipment is not necessarily an indication that the equipment has developed a fault (other than a tired fuse).

5.6 TRANSFORMER INPUT FUSES

The selection of fuses for 60-Hz transformer input supplies, such as linear regulator supplies, is not as straightforward as may have been expected.

Very often inrush limiting is not provided in linear power supply applications, and inrush currents can be large. Further, if grain-oriented C cores or similar cores are used, there is a possibility of partial core saturation during the first half cycle as a result of magnetic memory of the previous operation. These effects must be considered when selecting fuses. Slow-blow fuses may be necessary.

It can be seen from the preceding discussion that the selection of fuse rating and type for optimum protection and long life is a task to be carried out with some care. For continued optimum protection, the user must ensure that fuses are always replaced by others of the same type and rating.

5.7 PROBLEMS

1. Quote the three major selection criteria for supply or output fuses.
2. Why is the voltage rating of a fuse so important?
3. Under what conditions may the fuse voltage rating exceed the supply voltage?
4. Why is the I^2t rating of a fuse an important selection criterion?
5. Why is it important to replace a fuse with another of the same type and rating?

CHAPTER 6

LINE RECTIFICATION AND CAPACITOR INPUT FILTERS FOR "DIRECT-OFF-LINE" SWITCHMODE POWER SUPPLIES

6.1 INTRODUCTION

As previously mentioned, the "direct-off-line" switchmode supply is so called because it takes its power input directly from the ac power lines, without using the rather large low-frequency (60–50 Hz) isolation transformer normally found in linear power supplies.

In the switchmode system, the input-to-output galvanic isolation is provided by a much smaller high-frequency transformer, driven by a semiconductor inverter circuit so as to provide some form of DC-to-DC conversion. To provide a DC input to the converter, it is normal practice to rectify and smooth the 50/60-Hz ac supply, using semiconductor power rectifiers and large electrolytic capacitors. (Exceptions to this would be special low-distortion systems, where input boost regulators are used to improve the power factor. These special systems will not be considered here.)

For dual input voltage operation [nominally 120/240 V ac], it is common practice to use a full-bridge rectifier for the high-input-voltage conditions, and various link arrangements to obtain voltage doubler action for the low-input-voltage conditions. Using this approach, the high-frequency DC-to-DC converter can be designed for a nominal DC input of approximately 320 V for both input voltages.

An important aspect of the system design is the correct sizing of input inductors, rectifier current ratings, input switch ratings, filter component size, and input fuse ratings. To size these components correctly, a full knowledge of the relevant applied stress is required. For example, to size the rectifier diodes, input fuses, and filter inductors correctly, the values of peak and rms input currents will be required, while the correct sizing of reservoir and/or filter capacitors requires the effective rms capacitor current. However, these stress values are in turn a function of source resistance, loading, and actual component values.

A rigorous mathematical analysis of the input rectifier and filter is possible, but tedious.[83] Further, previous graphical methods[26] assume a resistance load with an exponential capacitor discharge. In power supply applications, the load applied to the capacitor input filter is the input loading of the regulated DC-to-DC converter section. This load is a constant-power load in the case of a switching regulator, or a constant-current load in the case of a linear regulator. Hence, this previous work is not directly applicable except where ripple voltages are relatively small.

Note: A constant-power load takes an increasing current as the input voltage falls, the reverse of a resistive load.

To meet this sizing need, a number of graphs have been empirically developed from actual system measurements. These will assist the designer in the initial component selection.

6.2 TYPICAL DUAL-VOLTAGE CAPACITOR INPUT FILTER CIRCUIT

Figure 1.6.1 shows a typical dual-voltage rectifier capacitor input filter circuit. A link option LK1 is provided which allows the rectifier capacitor circuit to be configured as a voltage doubler for 120-V operation or as a bridge rectifier for 240-V operation. The basic rectifier capacitor input filter and energy storage circuit (C5, C6, and D1 through D4) has been supplemented with an input fuse FS1, an inrush-limiting thermistor NTC1, and a high-frequency noise filter (L1, L2, L3, C1, C2, C3, and C4).

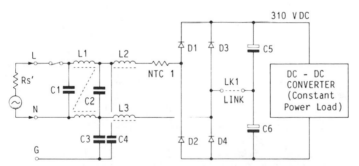

FIG. 1.6.1 Example of a direct-off-line, link-selected dual-voltage, capacitive input filter and rectifier circuit, with additional high-frequency conducted-mode input filter.

For 240-V operation, the link LK1 will not be fitted, and diodes D1 through D4 act as a full-bridge rectifier. This will provide approximately 320 V DC to the constant-power DC-to-DC converter load. Low-frequency smoothing is provided by capacitors C5 and C6, which act in series across the load.

For 120-V operation, the link LK1 is fitted, connecting diodes D3 and D4 in parallel with C5 and C6. Since these diodes now remain reverse-biased throughout the cycle, they are no longer active. However, during a positive half cycle, D1 conducts to charge C5 (top positive), and during a negative half cycle, D2 conducts to charge C6 (bottom negative). Since C5 and C6 are in series, the outputvoltage is the sum of the two capacitor voltages, giving the required voltage doubling. (In this configuration, the voltage doubler can be considered as two half-wave rectifier circuits in series, with alternate half cycle charging for the reservoir capacitors.)

6.3 EFFECTIVE SERIES RESISTANCE R_s

The effective series resistance R_s is made up of all the various series components, including the source resistance, which appear between the prime power source and the reservoir capacitors C5 and C6. To simplify the analysis, the various resistances are lumped into a single effective resistance R_s. To further reduce peak currents, additional series resistance may be added to provide a final optimum effective series resistance. It will be shown that the performance of the rectifier capacitor input filter and energy storage circuit is very much dependent on this final optimum effective series resistance.

A simplified version of the bridge circuit is shown in Fig. 1.6.2. In this simplified circuit, the series reservoir capacitors C5 and C6 are replaced by their equivalent capacitance C_e, and the effective series resistance R_s has been positioned on the output side of the bridge rectifier to further ease the analysis.

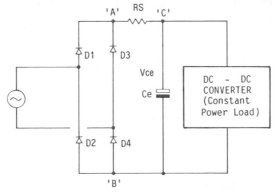

FIG. 1.6.2 Simplified capacitive input filter circuit, with full-wave bridge rectifier and lumped total effective source resistance R_s.

In the example shown in Fig. 1.6.2, the effective series resistance R_s is made up as follows:

The prime source resistance R_s' is the resistance of the power supply line itself. Its value will depend on the location of the supply, the size of utility transformer, and the distance from the service entrance. Values between 20 and 600 mΩ have been found in typical industrial and office locations. Although this may appear to be quite low, it can still have a significant effect in large power systems. In any event, the value of the source resistance is generally outside the control of the power supply designer, and at least this range must be accommodated by any practical supply design.

A second and usually larger series resistance component is usually introduced by the input fuse, filter inductors, rectifier diodes, and inrush-limiting devices. In the 100-W example shown in Fig. 1.6.1, the inrush-limiting thermistor NTC1 is the major contributor, with a "hot resistance" of typically 1 Ω. In higher-power supplies, the inrush-limiting resistor or thermistor will often be shorted out by a triac or SCR after initial start-up, to reduce the source resistance and power loss.

6.4 CONSTANT-POWER LOAD

By design, the switchmode power supply will maintain its output voltage constant for a wide range of input voltages. Since the output voltage is fixed, under steady loading conditions, the output power remains constant as the input voltage changes. Hence, since the converter efficiency also remains nearly constant, so does the converter input power.

In order to maintain constant input power as the input voltage to the converter falls, the input current must rise. Thus the voltage discharge characteristic VC_e of the storage capacitor C_e is like a reverse exponential, the voltage starting at its maximum initial value V_i after a diode conduction period.

$$VC_e = \left(V_i^2 - \frac{2Pt}{C_e} \right)^{1/2}$$

where C_e = storage capacitor value, μF
VC_e = voltage across the C_e
V_i = initial voltage on C_e at t_2
P = loading power (on converter)
t = time (μs) after t_2 but before t_3

This characteristic is shown by the solid discharge lines VC_{e1} or VC_{e2} in the period t_2–t_3 in Fig. 1.6.3.

6.5 CONSTANT-CURRENT LOAD

To complete the picture, the linear regulator must also maintain the output voltage constant as the regulator input voltage falls, between diode conduction periods. However, in the case of the linear regulator, the input current is the same as the output current, and it remains constant as the input voltage falls. Hence, for the linear regulator, the capacitor discharge characteristic is linear rather than an inverted exponential.

6.6 RECTIFIER AND CAPACITOR WAVEFORMS

Figure 1.6.3a shows the familiar full-wave rectifier waveforms that would be obtained from the circuit shown in Fig. 1.6.2. The dashed waveform is the half

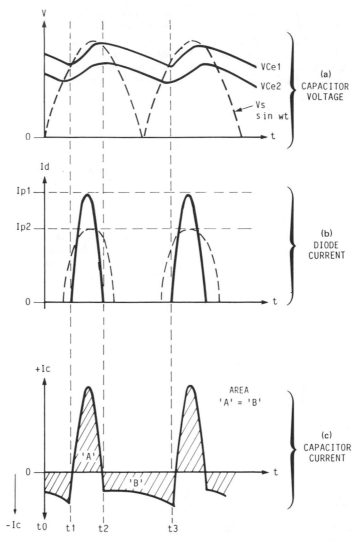

FIG. 1.6.3 Rectifier and capacitor voltage and current waveforms in a full-wave capacitor input filter. (*a*) Capacitor voltage waveform; (*b*) rectifier diode current waveform; (*c*) capacitor current waveform.

sinusoidal rectified voltage across points A–B (assuming zero diode drop). The solid line shows the capacitor voltage VC_{e1} or VC_{e2} across points C–B as applied to the load (in this case the load is the input of the regulated DC-to-DC converter section).

When the voltage applied to the bridge rectifier exceeds the previous capacitor voltage (time t_1), the rectifier diodes become forward-biased, and current flows via R_s to supply the load and charge capacitor C_e. During the conduction period (t_1–t_2), a large current flows in the rectifier diodes, input circuit, and reservoir capacitors; hence capacitors C_e will charge toward the peak voltage of the supply. However, at t_2, the applied voltage falls below the capacitor voltage, the rectifier diodes are blocked, and the input current falls to zero. Figure 1.6.3b shows the input current waveforms, and 1.6.3c, the capacitor current waveforms.

During the period t_2–t_3, the load current is supplied entirely from the storage capacitor C_e, partly discharging it. As the voltage falls, the load current increases, increasing the rate of voltage decay. At t_3, the supply voltage again exceeds the capacitor voltage, and the cycle repeats.

It should be noticed that the peak capacitor voltage is always less than the applied peak voltage as a result of the inevitable voltage drop across R_s and the rectifier diodes. This voltage drop is a function of load current and the value of R_s.

Figure 1.6.3 shows (dashed line) that increasing the effective series resistance from its minimum value to some higher value will slightly increase the voltage drop to VC_{e2}. This will reduce the peak current and increase the conduction angle of the rectifier diodes. The considerably reduced diode peak currents reduce input wiring and filter I^2r losses and improve the power factor.

The peak–peak ripple voltage is mainly a function of the capacitor size and load current. It is only slightly changed by the increased value of effective series resistance R_s.

The capacitor ripple current is shown in Fig. 1.6.3c. During the conduction period (t_1–t_2), the capacitor C_e is charging (shown as a positive current excursion); during the following diode blocking period (t_2–t_3), C_e will discharge. The peak and rms capacitor currents are a function of load, capacitor size, and the value of R_s. Under steady-state conditions, the area B (under the zero line) must equal the area A (above the line) to maintain the mean voltage across C_e constant.

6.7 INPUT CURRENT, CAPACITOR RIPPLE, AND PEAK CURRENTS

From Fig. 1.6.3, it will be clear that even if the input voltage remains sinusoidal, the input current will be very distorted, with large peak values. This distorted current waveform results in increased input I^2r power loss and low input power factors. Further, a large ripple current will flow in the filter capacitors.

Figures 1.6.4, 1.6.5, and 1.6.6 show how the rms input currents, rms capacitor currents, and peak capacitor currents are related to input power, with the value of the effective resistance factor R_{sf} as a parameter in typical applications. This information will be found useful for the correct sizing of the input components. (See Sec. 6.10.)

6.8 EFFECTIVE INPUT CURRENT I_e, AND POWER FACTOR

In Figs. 1.6.4, 1.6.5, and 1.6.6, the rms input, peak, and ripple currents are all given as a ratio to a "calculated effective input current" I_e:

$$I_e = \frac{P_{\text{in}}}{V_{\text{in}}}$$

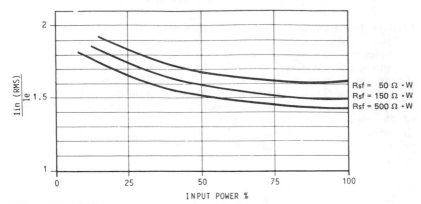

FIG. 1.6.4 RMS input current as a function of loading, with source resistance factor R_{sf} as a parameter.

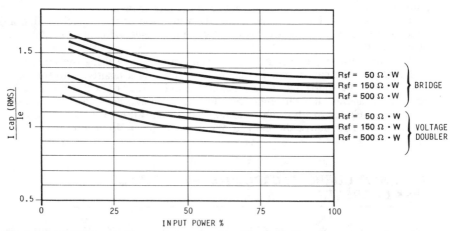

FIG. 1.6.5 RMS filter capacitor current as a function of loading, with source resistance factor R_{sf} as a parameter.

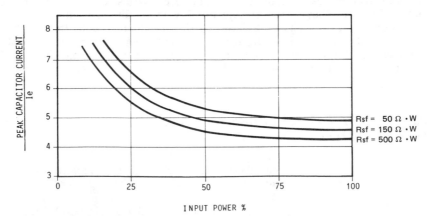

FIG. 1.6.6 Ratio of peak capacitor current to effective input current I_e as a function of loading, with source reistance factor R_{sf} as a parameter.

1.60

where I_e = calculated effective input current, A rms
  P_{in} = calculated (or measured) input power, W
  V_{in} = supply voltage, rms

Note: I_e is thus the calculated "real" component of input current (the component which produces the real power). Because of the large harmonic component in the distorted input current, the measured input rms current will be larger by an amount defined by the power factor P_f (approximately 0.63 in the case of a capacitor input filter).

Note: Although "power factor" P_f is normally defined as

$$P_f = \frac{\text{true input power}}{\text{input V·A product}}$$

in the case of the "direct-off-line" rectifier capacitor input filter, the low source resistance of the supply ensures that the input voltage remains near constant and free of distortion. Hence the power factor may be defined as the ratio of the effective input current to the rms input current, i.e.,

$$P_f = \frac{I_e}{I_{in(rms)}}$$

6.9 SELECTING INRUSH-LIMITING RESISTANCE

As previously mentioned, the effective series resistance R_s is made up of a number of factors, some of which are outside the designer's control. A large series inrush-limiting resistance has the advantage of reducing peak repetitive and inrush currents, reducing the stress on rectifier diodes, storage capacitor, and filter components. This gives a better power factor. However, it also results in a larger total power loss, reduced overall efficiency, and reduced output voltage regulation.

The inrush-limiting resistance is often a compromise selection. In low-power applications, where an inrush-limiting thermistor is used, this will usually provide sufficient "hot resistance" to limit peak currents and give the required performance. In high-power applications, where low-resistance triac or SCR inrush limiting is used, the input filter inductor often becomes the predominant series resistance and is wound to give the required resistance. The maximum value of this inductor resistance will then be limited by the permitted inductor temperature rise. However, this power loss limited approach to the inductor design has the advantage of permitting the maximum number of turns to be wound on the core, giving the maximum inductance on the selected core size. (See Chaps. 1, 2, and 3 in Part 3.)

6.10 RESISTANCE FACTOR R_{sf}

In Figs. 1.6.4, 1.6.5, and 1.6.6, the effective series resistance R_s has been converted to a resistance factor R_{sf} for more universal application, where

$$R_{sf} = R_s' \times \text{output power}$$

If specifications call for a power factor better than 0.6, it may be necessary to supplement the normal source resistance with an additional series power resistor. This has a penalty of increased power loss, with an inevitable decrease in overall efficiency. For power factors better than 0.7, a low-frequency choke input filter may be required. (Special continuous conduction boost regulator input circuits may be required in some applications.)

6.11 DESIGN EXAMPLE

The following example will serve to demonstrate the use of the graphs.

Question: For a 110-V 250-W 70% efficient "off-line" switchmode power supply using a rectifier capacitor input filter and a voltage doubler circuit, establish the fuse rating, minimum capacitor size, rms input current, and peak and rms capacitor currents.

Note: For a voltage doubler circuit, Fig. 1.6.1, the recommended minimum capacitor values are 3 µF/W (see Sec. 6.12), giving a value of 750 µF minimum for each capacitor C5 and C6.

Input Power P_{in}

Assuming an efficiency of 70%, the input power P_{in} to the converter (and filter) will be

$$P_{in} = \frac{P_{out}}{0.7} = \frac{250}{0.7} = 357 \text{ W (at 100\% load)}$$

Effective Input Current I_e

For an input voltage of 110 V, the effective input current I_e will be

$$I_e = \frac{P_{in}}{V_{in}} = \frac{357}{110} = 3.25 \text{ A}$$

Input Resistance Factor R_{sf}

Assuming a typical total effective input resistance R_s of 0.42 Ω, the resistance factor R_{sf} will be

$$R_{sf} = R_s' \times P_{in} = 0.42 \times 357 = 150 \text{ Ω} \cdot \text{W}$$

RMS Input Current $I_{in(rms)}$

Entering Fig. 1.6.4 with 100% load and a resistance factor R_{sf} of 150 yields the ratio $I_{in(rms)}/I_e = 1.48$; hence

$$I_{in} = 3.25 \times 1.48 = 4.8 \text{ A rms}$$

This rms input current will define the continuous-current rating of the input fuse at 110 V input. It is also used for the selection and losses in the input filter inductors. (Note that if the minimum input voltage is to be less than 110 V, the calculation will be done for the lowest input voltage.)

RMS Capacitor Current $I_{cap(rms)}$

Using the same 100% load and resistance factors in the voltage doubler connection, Fig. 1.6.5 gives the ratio $I_{cap(rms)} / I_e = 1$ at full load; hence

$$I_{cap} = 1 \times 3.25 = 3.25 \text{ A rms}$$

The capacitors must be chosen to meet or exceed this ripple current requirement.

Peak Input Current I_{peak}

From Fig. 1.6.6, at full load, the ratio of $I_{peak} / I_e = 4.6$, giving a peak input current of 15 A.

The rectifier diodes will be chosen to meet this peak repetitive current and the rms input current needs.

6.12 DC OUTPUT VOLTAGE AND REGULATION FOR RECTIFIER CAPACITOR INPUT FILTERS

It has been shown [26, 83] that provided that the product $\omega \times C_e \times R_L > 50$, the DC output voltage of the rectifier capacitor input filter (with a resistive load) will be defined mainly by the effective series resistance R'_s and load power. However, when the ripple voltage is low, this criterion also holds for the nonlinear converter-type load.

Figures 1.6.7 and 1.6.8 show the mean DC output voltage of the rectifier capacitive input filter as a function of load power and input rms voltage up to 1000 W, with series resistance R'_s as a parameter.

To maintain $\omega \times C_e \times R_L > 50$, the effective filter capacitor C_e must be 1.5 μF/W or greater (3 μF/W for C5 and C6 in the voltage doubler connection, remember; in this case C_e is made up of C5 and C6 in series). In general, this value of capacitance will also be found to meet ripple current and holdup time requirements.

6.13 EXAMPLE OF RECTIFIER CAPACITOR INPUT FILTER DC OUTPUT VOLTAGE CALCULATION

Consider the previous example for a 250-W unit. The input power is 357 W, and a voltage doubler circuit is to be used at 110 V input. The total series resistance

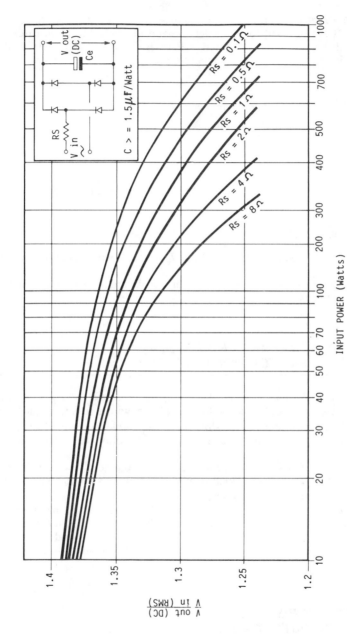

FIG. 1.6.7 Mean DC output voltage of a full-wave bridge-rectified capacitor input filter as a function of load power, with effective source resistance as a parameter. (Valid for capacitor values of 1.5 $\mu F/W$ or greater.)

1.64

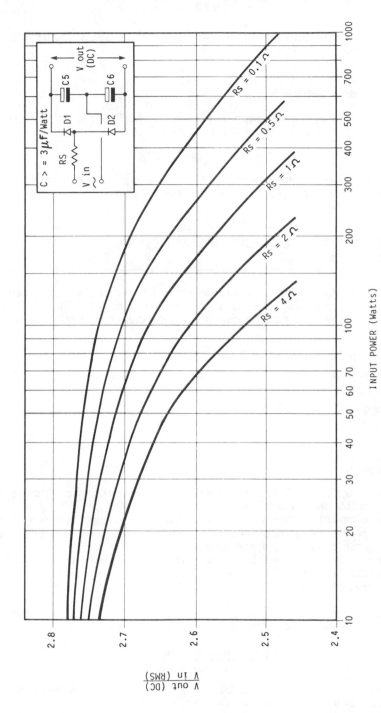

FIG. 1.6.8 Mean DC output voltage of a voltage doubler capacitor input filter as a function of load power, with effective source resistance as a parameter. (Valid for capacitor values of 3 μF/W or greater.)

INPUT POWER (Watts)

$\dfrac{V \text{ out (DC)}}{V \text{ in (RMS)}}$

1.65

R_s is 0.5 Ω, and, as previously shown, two capacitors of at least 750 μF will be used (in series).

Filter DC Output Voltage $V_{out(DC)}$

From Fig. 1.6.8, entering with a power of 357 W, the 0.5-Ω R'_s line yields the ratio $V_{out(DC)}/V_{in(rms)} = 2.6$. Hence the DC voltage is 0.5-Ω

$$2.6 \times 110 = 286 \text{ V DC}$$

The ratio improves at lower powers, and the voltage regulation may be obtained by calculating the output voltage at lower powers in a similar way.

6.14 SELECTING RESERVOIR AND/OR FILTER CAPACITOR SIZE

In the above example, the reservoir and/or filter capacitor values were chosen to meet the rather simplistic $C_e = 1.5$ μF/W criterion indicated in Sec. 6.12. In practice, one or more of the following five major factors may control the selection:

RMS ripple current rating
Ripple voltage
Voltage rating
Size and cost
Holdup time

RMS Ripple Current Rating

This rating must be satisfied to prevent excessive temperature rise in the capacitor and possible premature failure. (See Part 3, Chap. 12.)

The problem at this stage is to know what value of rms ripple current applies. As shown previously, the ripple current is already a function of capacitor value, total series resistance R_s, load, and input voltage.

However, Fig. 1.6.5 shows the measured rms ripple currents as a ratio of the "effective input current" I_e, for a range of load and source resistances, assuming that the capacitor value C_e is not less than 1.5 μF/W (3 μF/W for C5 and C6).

Note: The "effective input current" is the calculated "real" component of input current, not the measured (or calculated) rms input current; hence

$$I_e = \frac{\text{true input power}}{\text{rms input voltage}}$$

The rms input current will be greater than I_e because of the low power factor (approximately 0.63) of the rectifier capacitor input filter circuit.

Although in Fig. 1.6.5 the ratio I_c/I_e appears lower in the voltage-doubled mode, the actual ripple current will be greater, as I_e is approximately twice the value for the same output power in this mode.

(If in doubt for a particular application, check the capacitor rms current using a low-resistance true rms current meter with a high crest factor rating. See Part 3, Chaps. 12, 13, and 14.)

Ripple Voltage

This requirement will often define the minimum capacitor value when holdup time requirements are short (less than 1 cycle duration).

Large ripple voltages on C_e will reduce the range of input voltages that can be accommodated by the converter. They may also give excessive output ripple (depending on the design).

Typically, switchmode designs aim for a ripple voltage of less than 10% of V_{DC} (say 30 V_{p-p}). The ripple voltage will be maximum at minimum supply voltage, as a result of the increase of input current.

Example

Selecting C_e so as to satisfy a particular ripple voltage limit:

Consider a requirement where the primary filter ripple voltage is not to exceed 10% of V_{DC} for a 100-W supply, designed for a minimum input of 170 V rms at 60 Hz when the overall efficiency is 70%, with the effective series source resistance $R_s = 2\ \Omega$.

At 100 W output, with an efficiency of 70% the input power will be 143 W. From Fig. 1.6.7, at 143 W and $R_s = 2\ \Omega$ in the bridge-connected mode, the ratio $V_{out(DC)}/V_{in(rms)} = 1.32$, and the header voltage V_{DC} at 170 V rms input will be $1.32 \times 170 = 224$ V (DC).

The converter input power is 143 W, giving an effective DC converter input current of $P/V_{DC} = 143/224$ or 0.64 A (DC).

Extrapolating from Fig. 1.6.3, the capacitor discharge period is approximately 6 ms at 60 Hz. Since the ripple voltage is small (10% or 23 V in this example), a linear discharge will be assumed over the discharge period.

With these approximations, a simple linear equation may be used to establish the approximate value of C_e that will give the required 10% ripple voltage:

$$C_e = I \times \frac{\Delta t}{\Delta v}$$

where C_e = effective capacitor value, μF (effective value of C1 and C2 in series)

I = converter input DC current, A (0.64 A in this example)

Δt = discharge period, s (6 ms in this example)

Δv = peak–peak ripple voltage, V (in this case 10% V_{DC} = 22.4 V p-p)

Therefore

$$C_e = \frac{0.64 \times 6 \times 10^{-3}}{22.4} = 171\ \mu\text{F}$$

Since two capacitors are to be used in series, each capacitor will be 342 μF minimum. In this example, the capacitors' values exceed the minimum 3 μF/W criterion but are not clearly oversize. Hence, the ripple voltage needs may not be the dominant factor, and the ripple current rating and holdup time should also be checked.

Voltage Rating

This is perhaps an obvious parameter, but remember to consider maximum input voltages and minimum loads. Also, the voltage margin should include an allowance for temperature derating and required MTBF derating needs.

Size and Cost

High-voltage high-capacity electrolytic capacitors are expensive and large. It is not cost-effective to use oversize components.

Holdup Time

Holdup time is the minimum time period for which the supply will maintain the output voltages within their output regulation limits when the input supply is removed or falls below the input regulation limits. Although "holdup time" has been considered last, it is often the dominant factor and may even be the main reason that a switchmode supply was chosen.

In spite of its obvious importance, holdup time is often poorly specified. This parameter is a function of the size of the storage capacitor C_e, the applied load, the voltage on the capacitor at the time of line failure, and the design of the supply (dropout voltage). Note: It is difficult, inefficient, and expensive to design for a very low dropout voltage.

It is clearly very important to define the loading conditions, output voltage, and supply voltage immediately prior to failure when specifying holdup time.

It has become the industry standard to assume *nominal* input voltage and full-load operation unless otherwise stated in the specifications. In critical computer and control applications, it may be essential to provide a specified minimum holdup time from full-load and *minimum* input voltage conditions. If this is the real requirement, then it must be specified, as it has a major impact on the size and cost of the reservoir capacitors and will become the dominant selection factor. (Because of the higher cost, very few "standard off-the-shelf" supplies meet this second condition.)

In either case, if the holdup time exceeds 20 ms, it will probably be the dominant capacitor sizing factor, and C_e will be evaluated to meet this need. In this case, the minimum reservoir capacitor size $C_{e(min)}$ is calculated on the basis of energy storage requirements as follows: Let

C = minimum effective reservoir capacitor size, μF
E_o = output energy used during holdup time (output power × holdup time)
E_i = input energy used during holdup time (E_o/efficiency)
V_s = DC voltage on reservoir capacitor (at start of line failure)
E_{cs} = energy stored in reservoir capacitor (at start of line failure)
V_f = voltage on reservoir capacitor (at power supply drop-out)
E_{ef} = energy remaining in reservoir capacitor (at power supply dropout)

Now:

$$E_{cs} = \frac{1}{2} C(V_s)^2$$

$$E_{ef} = \frac{1}{2} C(V_f)^2$$

Then:

Energy used E_i = energy removed from capacitor

$$E_i = \frac{1}{2} C(V_s)^2 - \frac{1}{2} C(V_f)^2$$

$$= \frac{C(V_s^2 - V_f^2)}{2}$$

Thus:

$$C_{e(min)} = \frac{2 \times E_i}{V_s^2 - V_f^2}$$

Example

Calculate the minimum reservoir capacitor value C_e to provide 42 ms of holdup time at an output power of 90 W. The minimum input voltage prior to failure is to be 190 V.

The supply is designed for 230-V rms nominal input, with the link position selected for bridge operation. The efficiency is 70%, and the power supply drop-out input voltage is 152 V rms. The effective series resistance in the input filter (R_s) is 1 Ω.

Since the failure may occur at the end of a previous normal half cycle quiescent period, the capacitor may have already been discharging for 8 ms, so the worst-case discharge period can be (42 + 8) = 50 ms. This period must be used in the calculation.

From Fig. 1.6.7, the DC voltage across the two series storage capacitors C5 and C6 prior to line failure and at drop-out will be

$$V_s = 1.35 \times 190 = 256 \text{ V DC}$$

$$V_f = 1.35 \times 152 = 205 \text{ V}$$

During this period the energy used by the supply E_i

$$\frac{\text{Output power} \times \text{time} \times 100}{\text{Efficiency \%}} = \frac{90 \times 50 \times 10^{-3} \times 100}{70\%} = 6.43 \text{ J}$$

Therefore

$$C_{e(min)} = \frac{2 \times 6.43}{256^2 - 205^2} = 547 \text{ μF}$$

Since two capacitors in series are to be used for C_e, the value must be doubled, giving two capacitors of 1094 μF minimum. To allow for tolerance and end-of-life degrading, two standard 1500-μF capacitors would probably be used in this example.

It is clear that this is a very large capacitor for a 90-W power supply and that it is more than adequate to meet ripple current and ripple voltage requirements. This capacitor choice is clearly dominated by the holdup time needs.

6.15 SELECTING INPUT FUSE RATINGS

It has been shown in Fig. 1.6.4 that the rms input current is a function of load, source resistance R_s, and storage capacitor value. It is at a maximum at low input voltages. It is the rms input current that will cause fuse element heating and hence defines the fuse's continuous rating. Further, the fuse must withstand the inrush current on initial switch-on at maximum input voltage.

Procedure: Select the input fuse continuous rms current rating as defined by Fig. 1.6.4, allowing a 50% margin for aging effects.

Select the I^2t rating to meet the inrush needs as defined in Part 1, Chap. 7.

6.16 POWER FACTOR AND EFFICIENCY MEASUREMENTS

From Fig. 1.6.3, it can be seen that the input *voltage* is only slightly distorted by the very nonlinear load presented by the capacitor input filter. The sinusoidal input is maintained because the line input resistance is very low. The input current, however, is very distorted and discontinuous, but superficially would appear to be a part sine wave in phase with the voltage. This leads to a common error: The product $V_{in(rms)} \times I_{in(rms)}$ is assumed to give input power. *This is not so!* This product is the input volt-ampere product; it must be multiplied by the power factor (typically 0.6 for a capacitor input filter) to get true power.

The reason for the low power factor is that the nonsinusoidal current waveform contains a large odd harmonic content, and the phase and amplitude of all harmonics must be included in the measurement.

The input power is best measured with a true wattmeter with a bandwidth exceeding 1 kHz. Many moving-coil dynamometer instruments are suitable; however, beware of instruments containing iron, as these can give considerable errors at the higher harmonic frequencies. Modern digital instruments are usually suitable, provided that the bandwidth is large; they have a large crest factor, and true rms sensing is provided. Again beware of instruments which are peak or mean sensing, but only rms calibrated, as these will read correctly only for true sine-wave inputs. (Rectified moving-coil instruments fall into this category.)

When making efficiency measurements, remember that you are comparing two large numbers with only a small difference. It is the difference which defines the power loss in the system, and a small error in any reading can give a large error in the apparent loss. Figure 1.6.9 shows the possible error range as a function of real efficiency when the input and output measurements have a possible error in the range of only 2%.

In a multiple-output power supply, many instruments may be used and the po-

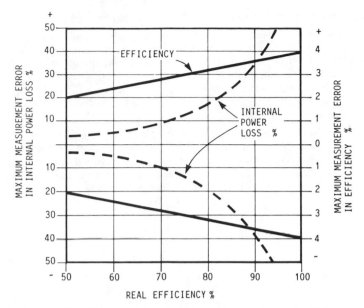

FIG. 1.6.9 Possible range of error of internal power loss and efficiency calculations as a function of real efficiency, with a measurement error of 2%.

tential for error is large. When using electrodynamic or dynamometer wattmeters, do not neglect the wattmeter burden, which is always present. This error cannot be eliminated by calibration, as it depends on the relative ratio of current to voltage, and this changes with each measurement. It also depends on the way the instrument is set up. (In general, the current shunt or coil should precede the voltage terminals for high-current, low-voltage measurements, and the reverse applies for low-current, high-voltage measurements.)

6.17 PROBLEMS

1. Why are capacitive input filters often used for direct-off-line switchmode supplies?
2. What are the major disadvantages of the capacitive input filter?
3. What is the typical power factor of a capacitive input filter, and why is it relatively poor?
4. Why must a true wattmeter be used for measuring input power?
5. Why is line inrush-current limiting required with capacitive input filter circuits?
6. Why is the input reservoir capacitor ripple current so important in the selection of input capacitor types?
7. What parameters are important in the selection of input rectifiers for capacitive input filters?

8. How can the power factor of a capacitive input filter be improved?

9. Using the nomograms shown in Sec. 1.6.9, establish the minimum input fuse rating, reservoir capacitor value, reservoir ripple current, peak current in the rectifier diodes, filter DC output voltage at full load, and voltage regulation at 10% to full load. (Assume that the output power is 150 W and the total source resistance including the inrush-limiting resistance is 0.75 Ω, the supply voltage is 100 V rms, the efficiency is 75%, and a voltage doubler circuit as shown in Fig. 1.6.8 is used.)

10. Calculate the minumum value of the reservoir capacitor needed to give a holdup time of one half cycle at 60 Hz if the SMPS is 70% efficient and the output power is to be 200 W. (Assume that the supply voltage just before line failure is 90 V rms and the dropout voltage is 80 V rms. The supply has a voltage doubler input as shown in Fig. 1.6.8, and the source resistance R_{sf} is 0.5 Ω.)

CHAPTER 7
INRUSH CONTROL

7.1 INTRODUCTION

In "direct-off-line" switchmode supplies, where minimum size and cost are a major consideration, it is common practice to use direct-off-line semiconductor bridge rectification with capacitive input filters to produce the high-voltage DC supply for the converter section.

If the line input is switched directly to this type of rectifier capacitor arrangement, very large inrush currents will flow in the supply lines, input components, switches, rectifiers, and capacitors. This is not only very stressful on these components, it may also cause interference with other equipment sharing a common supply line impedance.

Various methods of "inrush current control" are used to reduce this stress. Normally these methods include some form of series limiting resistive device in one or more of the supply lines between the input point and the reservoir capacitors.

These limiting devices usually take one of the following three forms: series resistors, thermistor inrush limiting, and active limiting circuits.

7.2 SERIES RESISTORS

For low-power applications, simple series resistors may be used, as shown in Fig. 1.7.1. However, a compromise must be made, as a high value of resistance, which will give a low inrush current, will also be very dissipative under normal operating conditions. Consequently, a compromise selection must be made between acceptable inrush current and acceptable operating losses.

The series resistors must be selected to withstand the initial high voltage and high current stress (which occurs when the supply is first switched on). Special high-current surge-rated resistors are best suited for this application. Adequately rated wirewound types are often used, however. If high humidity is to be expected, the wirewound types should be avoided. With such resistors, the transient thermal stress and wire expansion tend to degrade the integrity of the protective coating, allowing the ingress of moisture and leading to early failure.

Figure 1.7.1 shows the normal positions for the limiting resistors. Where dual input voltage operation is required, two resistors should be used in positions R1 and R2. This has the advantage of effective parallel operation for low-voltage link

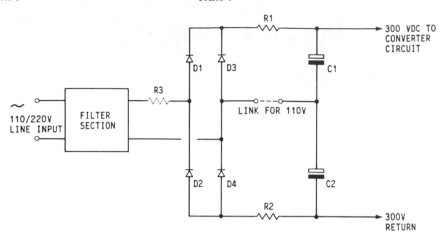

FIG. 1.7.1 Resistive inrush limiting circuit. (Suitable for bridge and voltage doubler operation, maintaining the inrush current at the same value.)

positions and series operation for high-voltage link positions. This limits the inrush current at similar values for the two conditions.

Where single-range input voltages are used, then a single inrush-limiting device may be fitted at position R3 at the input of the rectifiers.

7.3 THERMISTOR INRUSH LIMITING

Negative temperature coefficient thermistors (NTC) are often used in the position of R1, R2, or R3 in low-power applications. The resistance of the NTCs is high when the supply is first switched on, giving them an advantage over normal resistors. They may be selected to give a low inrush current on initial switch-on, and yet, since the resistance will fall when the thermistor self-heats under normal operating conditions, excessive dissipation is avoided.

However, a disadvantage also exists with thermistor limiting. When first switched on, the thermistor resistance takes some time to fall to its working value. If the line input is near its minimum at this time, full regulation may not be established for the warmup period. Further, when the supply is switched off, then rapidly turned back on again, the thermistor will not have cooled completely and some proportion of the inrush protection will be lost.

Nevertheless, this type of inrush limiting is often used for small units, and this is why it is bad practice to switch SMPSs off and back on rapidly unless the supply has been designed for this mode of operation.

7.4 ACTIVE LIMITING CIRCUITS (TRIAC START CIRCUIT)

For high-power converters, the limiting device is better shorted out to reduce losses when the unit is fully operating.

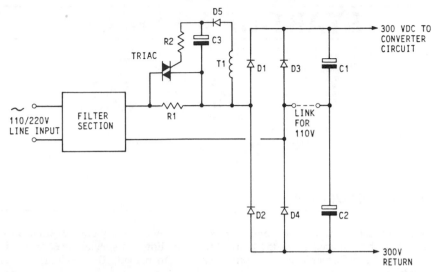

FIG. 1.7.2 Resistive inrush-limiting circuit with triac bypass for improved efficiency. (Note: Higher inrush current for bridge operation.)

Position R1 will normally be selected for the start resistor so that a single triac or relay may be used. R1 can be shunted by a triac or relay after start-up, as shown in Fig. 1.7.2. Since the start resistance can have a much higher value in this type of start-up circuit, it is not normally necessary to change the start resistor for dual input voltage operation.

Although Fig. 1.7.2 shows an active limiting arrangement in which a resistor is shunted by a triac, other combinations using thyristors or relays are possible.

On initial switch-on, the inrush current is limited by the resistor. When the input capacitors are fully charged, the active shunt device is operated to short out the resistor, and hence the losses under normal running conditions will be low.

In the case of the triac start circuit, the triac may be conveniently energized by a winding on the main converter transformer. The normal converter turn-on delay and soft start will provide a delay to the turn-on of the triac. This will allow the input capacitors to fully charge through the start resistor before converter action starts. This delay is important, because if the converter starts before the capacitors are fully charged, the load current will prevent full charging of the input capacitors, and when the triac is energized there will be a further inrush current.

For high-power or low-voltage DC-to-DC converter applications (where the power loss in the triac is unacceptable), a relay may be used. However, under these conditions, it is very important that the input capacitors be fully charged before the relay is operated. Consequently, converter action must not commence until after relay contact closure, and suitable timing circuits must be used.

7.5 PROBLEMS

1. What are three typical methods of inrush control used in switchmode supplies?
2. Describe the major advantages and limitations of each method.

CHAPTER 8
START-UP METHODS

8.1 INTRODUCTION

If the auxiliary supply is used only to power the power supply converter circuits, it will not be required when the converter is off. For this special case, the main converter transformer can have extra windings to provide the auxiliary power needs.

However, for this arrangement, some form of start-up circuit is required. Since this start circuit only needs to supply power for a short start-up period, very efficient start systems are possible.

8.2 DISSIPATIVE (PASSIVE) START CIRCUIT

Figure 1.8.1 shows a typical dissipative start system. The high-voltage DC supply will be dropped through series resistors R1 and R2 to charge the auxiliary storage capacitor C3. A regulating zener diode ZD1 prevents excessive voltage being developed on C3. The charge on C3 provides the initial auxiliary power to the control and drive circuits when converter action is first established. This normally occurs after the soft-start procedure is completed.

The auxiliary supply is supplemented from a winding on the main transformer T1 when the converter is operating, preventing any further discharge of C3 and maintaining the auxiliary supply voltage constant.

A major requirement for this approach is that sufficient start-up delay must be provided in the main converter to permit C3 to fully charge. Further, C3 must be large enough to store sufficient energy to provide all the drive needs for correct start-up of the converter.

In this circuit, R1 and R2 remain in the circuit at all times. To avoid excessive dissipation the resistance must be high, and hence the standby current requirements of the drive circuit must be low, prior to converter start-up. Since C3 may be quite large, a delay of two or three hundred milliseconds can occur before C3 is fully charged. To ensure a good switching action for the first cycle of operation, C3 must be fully charged before start-up, and this requires a low-voltage inhibit and delay on the start-up control and drive circuits.

To its advantage, the technique is very low cost, and resistors R1 and R2 can replace the normal safety discharge resistors which are inevitably required across the large storage capacitors C1 and C2.

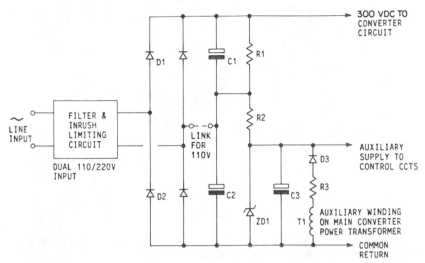

FIG. 1.8.1 Resistive, dissipative start circuit, providing initial low-voltage auxiliary power needs from the 300-V DC supply.

8.3 TRANSISTOR (ACTIVE) START CIRCUIT

Figure 1.8.2 shows the basic circuit of a more powerful and fast-acting start system, incorporating a high-voltage transistor Q1. In this arrangement, the resistance of R1 and R2 and the gain of Q1 are chosen such that transistor Q1 will be biased into a fully saturated "on" state soon after initial switch-on of the supply.

As C1 and C2 charge, current flows in R1 and R2 to the base of Q1, turning Q1 fully on. Zener diode ZD1 will not be conducting initially, as the voltage on C3 and the base of Q1 will be low. With Q1 turned on, a much larger current can flow in the low-resistance R3 to charge C3.

In this circuit, resistor R3 can have a much lower value than R1 and R2 in the circuit shown in Fig. 1.8.1. This will not result in excessive dissipation or degrade the efficiency, as current will flow in R3 only during the start-up period. Transistor Q1 will turn off after C3 has charged and will be operating in a saturated "on" state throughout the start-up period; hence its dissipation will also be very low. R3 should be chosen to have a high surge rating (i.e., it should be wirewound or carbon composition).

After switch-on, capacitor C3 will charge up relatively quickly and the voltage on Q1 emitter and base will track this rising voltage $+V_{be}$ until the voltage on the base of Q1 approaches the zener voltage ZD1. At this point ZD1 starts to conduct, tending to pinch off Q1 and reducing the charge current into C3. The voltage and dissipation will now build up across Q1. However, once converter action is established, regenerative feedback from the auxiliary winding on the main transformer will provide current via D6 and resistor R4 to capacitor C3. Hence the voltage on C3 will continue to increase until the base-emitter of Q1 is reversed-biased and it is fully turned off.

At this point, diode D5 is brought into conduction, and the voltage across C3 will now be clamped by the zener diode ZD1 and diode D5. The dissipation in

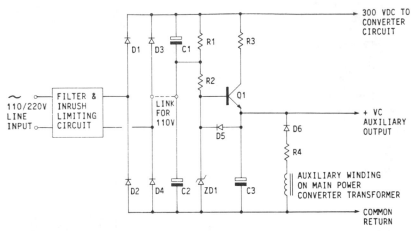

FIG. 1.8.2 Lower-dissipation, active transistor start circuit, providing initial low-voltage auxiliary supply needs from the 300-V DC supply.

ZD1 depends on the values of R4 and the maximum auxiliary current. With Q1 off, the current in R3 ceases, and its dissipation and that of Q1 will fall to zero.

As the start-up action is fast, much smaller components can be used for R3 and Q1 than would otherwise be necessary, and heat sinks will not be required. To prevent hazardous dissipation conditions in Q1 and R3 in the event of failure of the converter, R3 should be able to support continuous conduction, or "fail safe." Fusible resistors or PTC thermistors, with their inherent self-protection qualities, are ideal for this application.

This circuit is able to supply considerably more start-up current and gives greater freedom in the design of the drive circuit.

8.4 IMPULSE START CIRCUITS

Figure 1.8.3 shows a typical impulse start circuit which operates as follows.

Resistors R1 and R2 (normally the discharge resistors for the reservoir capacitors C1 and C2) feed current into capacitor C3 after switch-on. The auxiliary supply capacitor C4 will be discharged at this time.

The voltage on C3 will increase as it charges until the firing voltage of the diac is reached. The diac will now fire and transfer part of the charge from C3 into C4, the transfer current being limited by resistor R3.

The values of capacitors C3 and C4 and the diac voltage are chosen such that the required auxiliary voltage will be developed across C4 and the converter will start via its normal soft-start action.

Once again, by regenerative feedback (via D5 and the auxiliary winding), the auxiliary power is now provided from the main transformer. As C4 is further charged and its voltage increases, the diac will turn off since the voltage across it can no longer reach the firing value (because of the clamping action of ZD1 on C3).

This arrangement has the advantage of supplying a high current during the

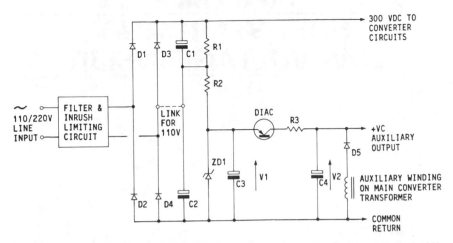

FIG. 1.8.3 Diac impulse start circuit, providing initial low-voltage auxiliary needs from the 300-V DC supply.

turn-on transient, without excessive dissipation in the feed resistors R1 and R2. In the rare event of the converter failing to start on the first impulse, the start-up action will repeat as soon as capacitor C4 has discharged and C3 recharged to the appropriate firing value for the diac.

The choice of diac is important. It must be able to deliver the required turn-on current, and its firing voltage must be less than $V_1 - V_{start}$ and greater than $V_1 - V_2$;* otherwise lockout can occur after the first impulse. It is possible to replace the diac with a small SCR and the appropriate gate drive circuit.

*Where V_1 is the ZD1 clamp voltage, V_{start} the control circuit start voltage, and V_2 the voltage on C4 when the converter is running.

CHAPTER 9
SOFT START AND LOW-VOLTAGE INHIBIT

9.1 INTRODUCTION

Soft-start action is quite different from the inrush limiting discussed in Sec. 7.1, although the two functions are complementary. Both actions reduce the inrush current to the supply during the initial switch-on period. However, whereas inrush limiting directly limits the current into the input capacitors, soft start acts upon the converter control circuit to give a progressively increasing pulse width. This progressive start not only reduces the inrush current stress on the output capacitors and converter components, it also reduces the problems of transformer "flux doubling" in push-pull and bridge topologies. (See Part 3, Chap. 7.)

It is normal practice with switchmode supplies to take the line input directly to the rectifier and a large storage and/or filter capacitor via a low-impedance noise filter. To prevent large inrush currents on initial switch-on, inrush-control circuitry is normally provided. In large power systems, the inrush limiting often consists of a series resistor which is shorted out by a triac, SCR, or relay when the input capacitors are fully charged. (Part 1, Chap. 7 shows typical inrush-control circuits.)

To allow the input capacitors to fully charge during start-up, it is necessary to delay the start-up of the power converter so that it does not draw current from the input capacitors until these are fully charged. If the capacitors have not been fully charged, there will be a current surge when the inrush-control SCR or triac operates to bypass the inrush-limiting series resistor. Furthermore, if the converter was allowed to start up with maximum pulse width, there would be a large current surge into the output capacitors and inductors, resulting in overshoot of the output voltage because of the large current in the output inductor, and possibly saturation effects in the main transformer.

To deal with these start-up problems, a start-up delay and soft-start procedure is usually provided by the control circuit. This will delay the initial switch-on of the converter and allow the input capacitors to fully charge. After the delay, the soft-start control circuit must start the converter from a narrow pulse condition and slowly increase. This will allow the transformer and output inductor working conditions to be correctly established. This will prevent "flux doubling" in push-pull circuits (see Part 3, Chap. 7). At the same time, the output voltages will be more slowly established, reducing the secondary inductor current surge and the tendency for output voltage overshoot. (See Part 1, Chap 10.)

9.2 SOFT-START CIRCUIT

A typical soft-start circuit is shown in Fig. 1.9.1. This operates as follows:

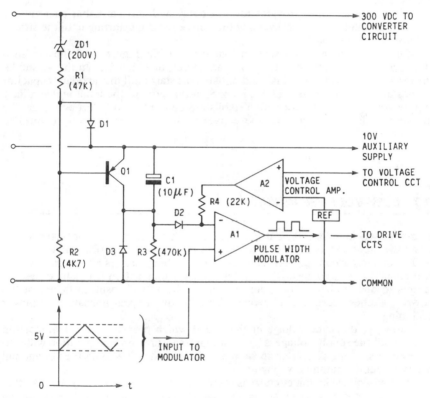

FIG. 1.9.1 Soft-start circuit for duty-cycle-controlled SMPS.

When the supply is first switched on, C1 will be discharged. The increasing voltage on the 10-V supply line will take the inverting input of amplifier A1 positive, inhibiting the output of the pulse-width modulator. Transistor Q1 will be turned on via R2, keeping C1 discharged until the 300-V DC line to the converter circuit has been established to a voltage exceeding 200 V.

At this point ZD1 will start to conduct and Q1 will be turned off. C1 will now charge via R3, taking the voltage on the inverting input of A1 toward zero and allowing the output of the pulse-width modulator to provide progressively increasing pulses to the drive circuit until the required output voltage has been developed.

When the correct output voltage has been established, amplifier A2 takes over control of the voltage at the inverting input of amplifier A1. C1 will continue charging via R3, reverse-biasing diode D2 and removing the influence of C1 from the modulator action. When the supply is turned off, C1 will quickly discharge through D3, resetting C1 for the next start action. D1 prevents Q1 being reversed-biased by more than a forward diode drop when the input voltage is high.

This circuit not only provides turn-on delay and soft start, but also gives a low-voltage inhibit action, preventing the converter from starting until the supply voltage is fully established.

Many variations of this basic principle are possible. Figure 1.9.2 shows a soft-start system applied to the transistor start circuit of Fig. 1.8.2. In this example, the input to ZD2 will not go high and initiate soft start until the auxiliary capacitor C3 has charged and Q1 turned off. Hence, in this circuit, the input and auxiliary supply voltages must be correctly established before the soft-start action can be initiated. This will ensure that the converter starts under correctly controlled conditions.

9.3 LOW-VOLTAGE INHIBIT

In many switchmode designs it is necessary to prevent power converter action when the input supply voltage is too low to ensure proper performance.

The converter control, drive, and power switching circuits all require the correct supply voltage to ensure a well-defined switching action. In many cases, attempts to operate below the minimum input voltage will result in failure of the power switches because of ill-defined drive conditions and nonsaturated power switching.

Normally, the same voltage inhibit signal which prevents the initial start-up action until the supply voltage is high enough to ensure correct operation will also be used to shut the converter down in a well-defined way should the voltage fall below a second minimum voltage.

The low-voltage inhibit circuitry is often linked to the soft-start system, so that the unit will not turn on by normal soft-start action until the correct operating voltage has been established. This also provides the delay required on the soft-start action and prevents start-up race conditions.

A typical soft-start circuit with a low-voltage inhibit is shown in Fig. 1.9.2. In this circuit, Schmitt trigger action is provided by the auxiliary winding with sufficient hysteresis to prevent squegging at the turn-on threshold. (In this context, "squegging" refers to the rapid "on-off" switching action that would occur at the low-voltage threshold as a result of load-induced input voltage changes.)

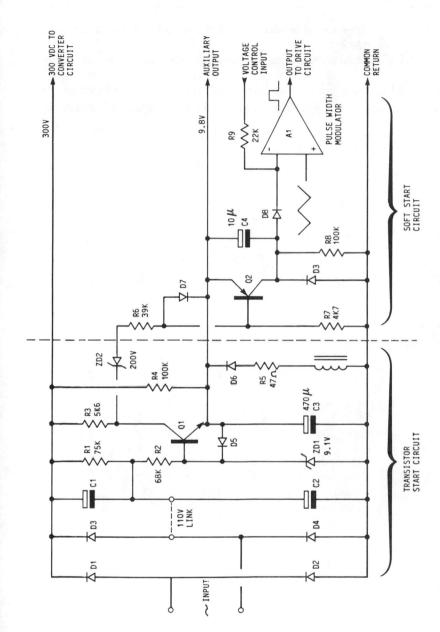

FIG. 1.9.2 Combined low-dissipation transistor auxiliary start circuit, with duty ratio control (pulse-width modulator) and soft-start characteristic.

9.4 PROBLEMS

1. Under what conditions may an impulse-type start circuit, be considered a suit-able start technique?
2. Under what conditions would impulse start circuits not be considered suit-able?
3. What is the function of a soft-start circuit as opposed to inrush limiting?
4. What is the function of input low-voltage inhibit in switchmode applications?

CHAPTER 10
TURN-ON VOLTAGE OVERSHOOT PREVENTION

10.1 INTRODUCTION

When a power supply is first switched on, either from the line input switch or by electronic means (say from a TTL logic "high" signal), there will be a delay while the power and control circuits establish to their correct working conditions. During this period, it is possible for the output voltage to exceed its correct working value before full regulation is established, giving a "turn-on voltage overshoot."

10.2 TYPICAL CAUSES OF TURN-ON VOLTAGE OVERSHOOT IN SWITCHMODE SUPPLIES

In most switchmode power supplies, a controlled start-up sequence is initiated at switch-on. Should the turn-on be from a line input switch, the first action will be "inrush limiting," where a resistive element in series with the line input reduces the peak inrush currents for a few cycles while the input capacitors are charged up.

Following this inrush limiting, there will be a soft-start action. For soft start, the pulse width to the power switching devices is progressively increased to establish the correct working conditions for transformers, inductors, and capacitors. The voltage on the output capacitors is progressively increased with the intention of smoothly establishing the required output voltage. However, even under this controlled turn-on condition, it is possible for the output voltage to overshoot, as a result of race conditions in the control circuit as follows.

Figure 1.10.1 shows the output filter and control amplifier of a typical duty-cycle-controlled switchmode power supply. The control amplifier has a simple pole-zero compensation network to stabilize the loop.

When the input is first applied to this supply, and throughout the start-up phase, the control amplifier A1 will recognize the output voltage as being low, and will demand maximum output and hence maximum pulse width from the ramp comparator A2. The high-gain-control amplifier A1 will be operating in a saturated "high" state, with its output near +5 V. Hence, during this start-up phase, the compensation capacitor C1 will be charged to +5 V.

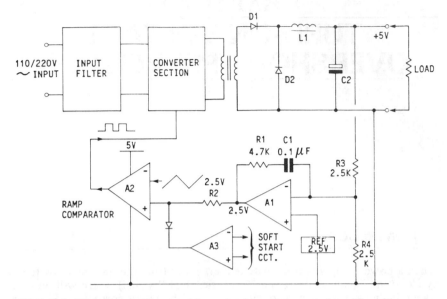

FIG. 1.10.1 Typical duty ratio control loop, showing voltage control amplifier with compensation components R1 and C1.

During this start-up phase, the pulse width and hence the output voltage will be under the control of the soft-start circuit and amplifier A3. Therefore the control amplifier will remain in its saturated "high" state until the output voltage is within 1 or 2 mV of the required value. At this point, the output capacitors have been charged and a considerable current has been established in the output inductor L1.

As the output voltage passes through the required value, the control amplifier A1 will start to respond. However, a considerable delay will now ensue while the compensation network R1, C1 establishes its correct DC bias. Since the output voltage of amplifier A1 starts near +5 V (far away from the correct mean working point of 2.5 V), and the slew rate of the amplifier is defined by the time constant of R1, C1, the correct amplifier working conditions are not established for a considerable period. (In this example, the delay will be approximately 500 μs.) During the delay period, the pulse width will not be significantly reduced, as the output of amplifier A1 must be close to 2.5 V before it comes within the control range of the pulse-width modulator A2. This delay, together with the excess current now flowing in the output inductor L1, will cause a considerable overshoot. (The output voltage will go to 7.5 V in this example, as shown in Fig. 1.10.2.)

10.3 OVERSHOOT PREVENTION

The overshoot can be considerably reduced by making the soft-start action very slow, allowing the amplifier to take over before the overshoot is too large. This has the disadvantage that the turn-on delay can be unacceptably long.

A much better arrangement is the linear power control circuit shown in Fig.

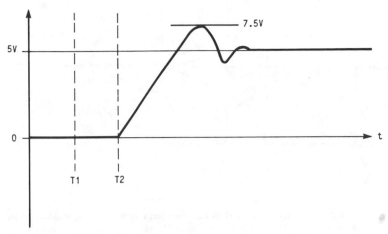

FIG. 1.10.2 Output voltage characteristic of the circuit in Fig. 1.10.1 during the "turn-on" transient, showing output voltage overshoot.

1.10.3. In this circuit the 2.5-V reference voltage for the control amplifier will be near zero at the noninverting input to the amplifier when first switched on, as C1 will be discharged prior to initial switch-on. The voltage on C1 will progressively increase as C1 charges via R1 and R4. Thus the reference voltage is arranged to increase at a rate somewhat slower than the soft-start action. As a result, the control amplifier will establish its normal working conditions at a much lower output voltage so that the latter part of the turn-on action is under full control of the voltage control amplifier A1.

The output voltage now increases progressively, as shown in Fig. 1.10.4, in response to the increasing reference voltage, under the full command of the control amplifier. Since the correct bias conditions for C2 and amplifier A1 were es-

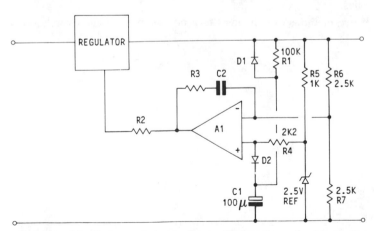

FIG. 1.10.3 Modified control circuit, showing "turn-on" overshoot prevention components R1, D1, D2, and C1.

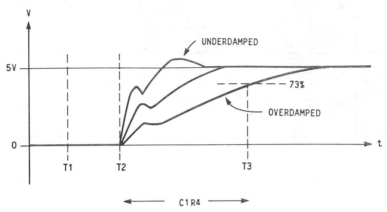

FIG. 1.10.4 "Turn-on" characteristics of modified circuit, showing underdamped, overdamped, and optimum response.

tablished at a much lower voltage, there will not be an overshoot when the correct voltage has been established. For optimum selection of R1, R4, and C1, the change in the reference and hence the output voltage is nearly asymptotic to the required 5-V value. The turn-on characteristic can be changed by adjusting R1, R4, and C1. Typical turn-on characteristics of this type of circuit are shown in Fig. 1.10.4. Small values of C1 will give underdamped and large values of C1 overdamped performance. The same principle can be applied to any switchmode or linear control circuit.

10.4 PROBLEMS

1. Give a typical cause of "turn-on" output voltage overshoot in switchmode supplies.
2. Give two methods of reducing "turn-on" output voltage overshoot.

CHAPTER 11
OVERVOLTAGE PROTECTION

11.1 INTRODUCTION

During fault conditions, most power supplies have the potential to deliver higher output voltages than those normally specified or required. In unprotected equipment, it is possible for output voltages to be high enough to cause internal or external equipment damage. To protect the equipment under these abnormal conditions, it is common practice to provide some means of overvoltage protection within the power supply.

Because TTL circuits are very vulnerable to overvoltages, it is becoming industry standard practice to provide overvoltage protection on all 5-V outputs. Protection for other output voltages is usually provided as an optional extra, to be specified if required by the systems engineer (user).

11.2 TYPES OF OVERVOLTAGE PROTECTION

Overvoltage protection techniques fall broadly into three categories:

Type 1, simple SCR "crowbar" overvoltage protection

Type 2, overvoltage protection by voltage clamping techniques

Type 3, overvoltage protection by voltage limiting techniques

The technique chosen will depend on the power supply topology, required performance, and cost.

11.3 TYPE 1, SCR "CROWBAR" OVERVOLTAGE PROTECTION

As the name implies, "crowbar" overvoltage protection requires the short-circuiting of the offending power supply output in response to an overvoltage condition on that output. The short-circuiting device, usually an SCR, is activated when the overvoltage stress exceeds a preset limit for a defined time period. When the SCR is activated, it short-circuits the output of the power supply to the common return line, thus collapsing the output voltage. A typical simple SCR "crowbar" overvoltage protection circuit connected to the output of a linear regulator is shown in Fig. 1.11.1a. It is important to appreciate that under fault conditions, the SCR "crowbar" shunt action does not necessarily provide good long-term protection of the load. Either the shunt device must be sufficiently powerful to sustain the short-circuit current condition for extended periods, or some external current limit, fuse, or circuit breaker must be actuated to remove the stress from the SCR.

With linear regulator-type DC power supplies, SCR "crowbar" overvoltage protection is the normal protection method, and the simple circuit shown in Fig. 1.11.1a is often used. The linear regulator and "crowbar" operate as follows:

The unregulated DC header voltage V_H is reduced by a series transistor Q1 to provide a lower but regulated output voltage V_{out}. Amplifier A1 and resistors R1 and R2 provide the regulator voltage control, and transistor Q2 and current limiting resistor R1 provide the current limit protection.

The most catastrophic failure condition would be a short circuit of the series

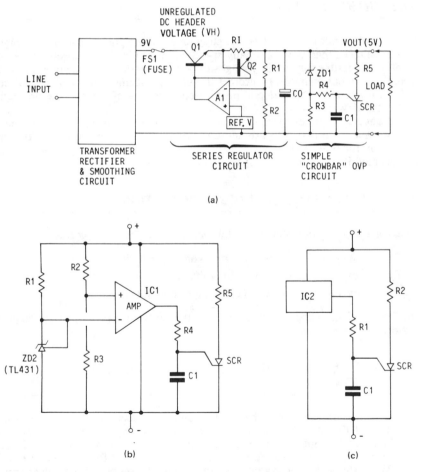

(a)

(b)   (c)

FIG. 1.11.1 (a) SCR "crowbar" overvoltage protection circuit, applied to a simple linear regulator. (b) A more precise SCR "crowbar" protection circuit using a voltage comparator IC. (c) A specialized control IC driving an SCR "crowbar."

regulating device Q1, so that the higher unregulated header voltage V_H is now presented to the output terminals. Under such fault conditions, both voltage control and current limit actions are lost, and the "crowbar" SCR must be activated to short-circuit the output terminals.

In response to an overvoltage fault, the "crowbar" circuit responds as follows: As the voltage across the output terminals rises above the "crowbar" actuation voltage, zener diode ZD1 conducts driving current via R4 into the SCR gate delay capacitor C1. After a short delay period defined by the values of C1, R4 and the applied voltage C1 will have charged to the gate firing voltage (0.6 V), and the SCR will conduct to short-circuit the output terminals via the low-value limiting resistor R5. However, a large current now flows from the unregulated DC input through the shunt-connected "crowbar" SCR. To prevent over-dissipation in the SCR, it is normal, in linear regulators, to fit a fuse FS1 or circuit breaker in the unregulated DC supply. If the series regulator device Q1 has failed, the fuse or circuit breaker now clears, to disconnect the prime source from the output before the "crowbar" SCR is destroyed.

The design conditions for such a system are well defined. It is simply necessary to select an SCR "crowbar" or other shunt device which is guaranteed to be more powerful than the fuse or circuit breaker's "let-through" energy. With SCRs and fuses, this "let-through" energy is normally defined in terms of the I^2t product, where I is the fault current and t the fuse or breaker clearance time. (See Part 1, Chap. 5.)

Crowbar protection is often preferred and hence specified by the systems engineer because it is assumed to provide full protection (even for externally caused overvoltage conditions). However, full protection may not always be provided, and the systems engineer should be aware of possible anomalous conditions.

In standard, "off-the-shelf" power supply designs, the crowbar SCR is chosen to protect the load from internal power supply faults. In most such cases, the maximum let-through power under fault conditions has been defined by a suitably selected internal fuse. The power supply and load are thus 100% protected for internal fault conditions. However, in a complete power supply system, there may be external sources of power, which may become connected to the terminals of the SCR-protected power supply as a result of some system fault. Clearly, the fault current under these conditions can exceed the rating of the "crowbar" protection device, and the device may fail (open circuit), allowing the overvoltage condition to be presented to the load.

Clearly, such external fault loading conditions cannot be anticipated by the power supply designer, and it is the responsibility of the systems engineer (user) to specify the worst-case fault condition so that suitable "crowbar" protection devices can be provided.

11.4 "CROWBAR" PERFORMANCE

More precise "crowbar" protection circuits are shown in Fig. 1.11.1b and c. The type of circuit selected depends on the performance required. In the simple "crowbar," there is always a compromise choice to be made between ideal fast protection (with its tendency toward nuisance operation) and delayed operation (with its potential for voltage overshoot during the delay period).

For optimum protection, a fast-acting, nondelayed overvoltage "crowbar" is required. This should have an actuation voltage level that just exceeds the normal power supply output voltage. However, a simple fast-acting "crowbar" of this type will often give many "nuisance" operations, since it will respond to the slightest transient on the output lines. For example, a sudden reduction in the load on a normal linear regulator will result in some output voltage overshoot. (The magnitude of the overshoot depends on the transient response of the power supply and the size of the transient load.) With a very fast acting "crowbar," this common transient overvoltage condition can result in unnecessary "crowbar" operation and shutdown of the power supply. (The current limiting circuit would normally limit the fault current in this type of nuisance operation, so it usually would only require a power on–off recycling to restore the output.) To minimize such nuisance shutdowns, it is normal practice to provide a higher trip voltage and some delay time. Hence, in the simple "crowbar" circuit, a compromise choice must be made between operating voltage, delay time, and required protection.

Figure 1.11.1d^1 shows the response of a typical delayed "crowbar" to an overvoltage fault condition in a linear regulator. In this example, the regulator transistor Q1 has failed to a short circuit at instant t_1. In this failure mode, the output voltage is rapidly increasing from the normal regulated terminal voltage V_0 toward the unregulated header voltage V_H at a rate defined by the loop inductance, the source resistance, and the size of the output capacitors C0. The crowbar has been set to operate at 5.5 V, which occurs at instant t_2; however, because of the crowbar delay (t_2 to t_3) of 30 μs (typical values), there is a voltage overshoot. In the example shown, the rate of change of voltage on the output terminals is such that the crowbar operates before the output voltage has reached 6 V. At this time the output voltage is clamped to a low value V_c during the clearance time of the fuse (t_3 to t_4), at which time the voltage falls to zero. Hence, full protection of an external IC load would be provided.

In this example the SCR delay time was selected to be compatible with the 20-μs transient response typical of a linear regulator. Although this delay will prevent nuisance shutdowns, it is clear that if the maximum output voltage during the delay period is not to exceed the load rating (normally 6.25 for 5-V ICs), then the maximum dv/dt (rate of change of output voltage under fault conditions) must be specified. The power supply designer should examine the failure mode, because with small output capacitors and low fault source resistance, the dv/dt requirements may not be satisfied. Fortunately the source resistance need is often met by the inevitable resistance of the transformer, rectifier diodes, and current sense resistors and the intrinsic resistance of the series fuse element.

11.5 LIMITATIONS OF "SIMPLE" CROWBAR CIRCUITS

The well-known simple crowbar circuit shown in Fig. 1.11.1a is popular for many noncritical applications. Although this circuit has the advantages of low cost and circuit simplicity, it has an ill-defined operating voltage, which can cause large operating spreads. It is sensitive to component parameters, such as temperature coefficient and tolerance spreads in the zener diode, and variations in the gate-cathode operating voltage of the SCR. Furthermore, the delay time provided by C1 is also variable, depending upon the overvoltage stress value, the parameters of the series zener diode ZD1, and the SCR gate voltage spreads.

When an overvoltage condition occurs, the zener diode conducts via R4, to charge C1 toward the SCR gate firing voltage. The time constant of this charge action is a function of the slope resistance of ZD1. This is defined by the device

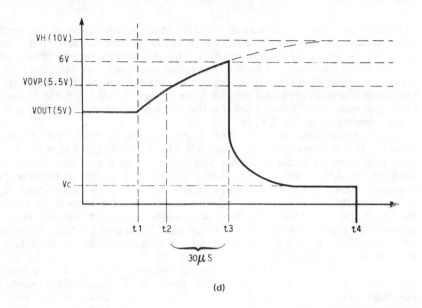

(d)

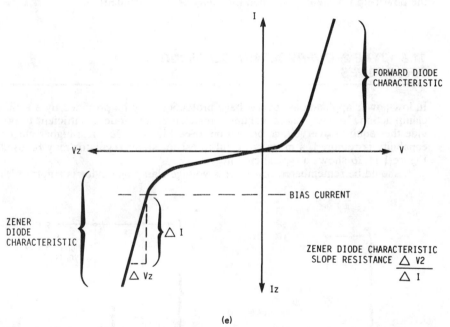

(e)

Figure 1.11.1 (d) Typical performance characteristic of a delayed "crowbar" circuit. (e) Typical zener diode characteristic.

parameters and the current flowing in ZD1, which is a function of the applied stress voltage. Hence, the slope resistance of ZD1 is very variable, giving large spreads in the operating delay of the SCR. The only saving grace in this circuit is that the delay time tends to be reduced as the overvoltage stress condition increases. Resistor R1 is fitted to ensure that the zener diode will be biased into its linear region at voltages below the gate firing voltage to assist in the definition of the output actuating voltage. A suitable bias point is shown on the characteristics of the zener diode in Fig. 1.11.1e.

A much better arrangement is shown in Fig. 1.11.1b. In this circuit a precision reference is developed by integrated circuit reference ZD2 (TL 431 in this example). This, together with comparator amplifier IC1 and the voltage divider network R2, R3, defines the operating voltage for the SCR. In this arrangement, the operating voltage is well defined and independent of the SCR gate voltage variations. Also, R4 can have a much larger resistance, and the delay (time constant R4, C1) is also well defined. Because the maximum amplifier output voltage increases with applied voltage, the advantage of reduced delay at high overvoltage stress conditions is retained. This second technique is therefore recommended for more critical applications.

Several dedicated overvoltage control ICs are also available; a typical example is shown in Fig. 1.11.1c. Take care to choose an IC specifically designed for this requirement, as some voltage control ICs will not operate correctly during the power-up transient (just when they may be most needed).

11.6 TYPE 2, OVERVOLTAGE CLAMPING TECHNIQUES

In low-power applications, overvoltage protection may be provided by a simple clamp action. In many cases a shunt-connected zener diode is sufficient to provide the required overvoltage protection. (See Fig. 1.11.2a.) If a higher current capability is required, a more powerful transistor shunt regulator may be used. Figure 1.11.2b shows a typical circuit.

It should be remembered that when a voltage clamping device is employed, it

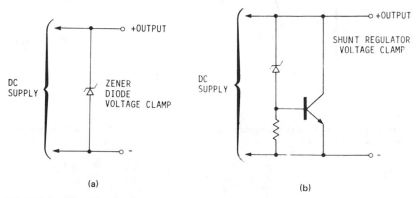

(a)

(b)

FIG. 1.11.2 Shunt regulator-type voltage clamp circuits.

is highly dissipative, and the source resistance must limit the current to acceptable levels. Hence, shunt clamping action can be used only where the source resistance (under failure conditions) is well specified and large. In many cases shunt protection of this type relies on the action of a separate current or power limiting circuit for its protective performance.

An advantage of the clamp technique is that there is no delay in the voltage clamp action, and the circuit does not require resetting upon removal of the stress condition. Very often, overvoltage protection by clamp action is better fitted at the load end of the supply lines. In this position it becomes part of the load system design.

11.7 OVERVOLTAGE CLAMPING WITH SCR "CROWBAR" BACKUP

It is possible to combine the advantages of the fast-acting voltage clamp with the more powerful SCR crowbar. With this combination, the delay required to prevent spurious operation of the SCR will not compromise the protection of the load, as the clamp circuit will provide protection during this delay period.

For lower-power applications, the simple expedient of combining a delayed crowbar as shown in Fig. 1.11.1a with a parallel zener clamp diode (Fig. 1.11.2a) will suffice.

In more critical high-current applications, simple zener clamp techniques would be excessively dissipative, but without voltage clamping the inevitable voltage overshoot caused by the delay in the simple crowbar overvoltage protection circuit would be unacceptable. Furthermore, nuisance shutdowns caused by fast-acting crowbars would also be undesirable.

For such critical applications, a more complex protection system can be justified. The combination of an active voltage clamp circuit and an SCR crowbar circuit with self-adjustable delay can provide optimum performance, by eliminating nuisance shutdowns and preventing voltage overshoot during the SCR delay period. The delay time is arranged to reduce when the stress is large to prevent excessive dissipation during the clamping period. (Figure 1.11.3a shows a suitable circuit, and Fig. 1.11.3b the operating parameters.)

In the circuit shown in Fig. 1.11.3a, the input voltage is constantly monitored by comparator amplifier A1, which compares the internal reference voltage ZD1 with the input voltage (V_{out} power supply), using the divider chain R1, R2. (Voltage adjustment is provided by resistor R1.) In the event of an overvoltage stress, A1+ goes high and the output of A1 goes high; current then flows in the network R4, ZD2, Q1 base-emitter, and R6. This current turns on the clamp transistor Q1.

Q1 now acts as a shunt regulator and will try to maintain the terminal voltage at the clamp value by shunting away sufficient current to achieve this requirement. During this clamping action, zener diode ZD2 is polarized, and point A goes high by an amount defined by the zener diode voltage, the base-emitter voltage of Q1, and a further voltage defined by the clamp current flowing in R6. This total voltage is applied to the SCR via the series network R7, C1, R8 such that C1 will be charging toward the gate firing voltage of the SCR. If the overvoltage stress condition continues for a sufficient period, C1 will charge to 0.6 V, and SCR1 will fire to short-circuit the supply to the common line. (Resistor R9 limits the peak current in SCR1.)

The performance parameters of this circuit are shown in Fig. 1.11.3b. For a limited stress condition, trace A will be produced as follows: At time t_1 an

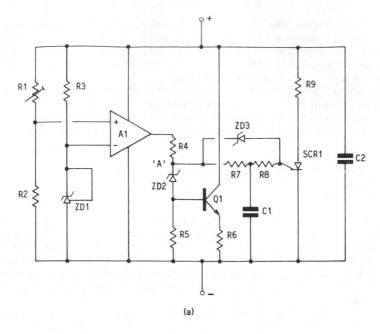

(a)

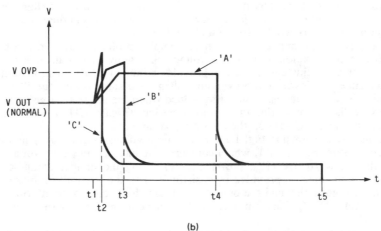

(b)

FIG. 1.11.3 (*a*) OVP combination circuit, showing an active voltage clamp combined with an SCR crowbar. (*b*) Operating characteristics for the OVP combination circuit shown in (*a*).

overvoltage fault condition occurs and the voltage rises to the voltage clamp point V_{ovp}. At this point, Q1 conducts to shunt away sufficient current to maintain the voltage constant at V_{ovp} until time t_4. At this instant, SCR1 is fired, to reduce the output voltage to a low value defined by the SCR saturation voltage. At time t_5 the external fuse or circuit breaker operates to disconnect the supply. It is clear from this diagram that if the clamping action were not provided, the voltage could have risen to an unacceptably high value during the delay period as a result of the long delay and the rapidly rising edge on the stress voltage condition.

If the current flowing in Q1 during a clamping period is large, the voltage across emitter resistor R6 will rapidly increase, increasing the voltage at point A. As a result, the delay time for SCR1 will be reduced to t_3, and the shorter delay reduces the stress and overvoltage excursion on Q1. This is depicted by trace B in Fig. 1.11.3b.

Finally, for highly stressful conditions where the current during the clamping period is very large, the voltage across R6 will be high enough to bring zener diode ZD3 into conduction, bypassing the normal delay network. SCR1 will operate almost immediately at t_2, shutting down the supply. This is shown by trace C in the diagram.

This circuit provides the ultimate in overvoltage protection, minimizing nuisance shutdowns by providing maximum delay for small, low-stress overvoltage transient conditions. The delay time is progressively reduced as the overvoltage stress becomes larger, and for a genuine failure, very little delay and overshoot is allowed. This technique should be considered as part of an overall system strategy, and the components selected to satisfy the maximum stress conditions.

11.8 SELECTING FUSES FOR SCR "CROWBAR" OVERVOLTAGE PROTECTION CIRCUITS

In the event of an overvoltage stress condition caused by the failure of the series regulator in a linear power supply, the "crowbar" SCR will be required to conduct and clear the stress condition by blowing the series protection fuse. Hence, the designer must be confident that the fuse will open and clear the faulty circuit before the SCR is destroyed by the fault current.

If a large amount of energy is dissipated in the junction of the SCR within a short period, the resultant heat cannot be conducted away fast enough. As a result, an excessive temperature rise occurs, and thermal failure soon follows. Hence, the failure mechanism is not simply one of total dissipation, but is linked to the time period during which the energy is dissipated.

For periods below 10 ms, very little of the energy generated at the junction interface will be conducted away to the surrounding package or heat sink. Consequently, for a very short transient stress, the maximum energy limit depends on the mass of the junction; this is nearly constant for a particular device. For SCRs, this energy limit is normally specified as a 10-ms I^2t rating. For longer-duration lower-stress conditions, some of the heat energy will be conducted away from the junction, increasing the I^2t rating.

In the SCR, the energy dissipated in the junction is more correctly $(I^2 \times R_j + V_d \times I) \times t$ joules, where R_j is the junction slope resistance and V_d is the diode voltage drop. However, at high currents, I^2R_j losses predominate, and since the slope resistance R_j tends to be a constant for a particular device, the failure energy tends to $K \times I^2t$.

The same general rules as were considered for the SCR failure mechanism apply to the fuse clearance mechanism. For very short time periods (less than 10 ms), very little of the energy dissipated within the fuse element will be conducted away to the case, the fuse clips, or the surrounding medium (air, sand, etc.). Once again, the fusing energy tends to be constant for short periods, and this is defined in terms of the 10-ms I^2t rating for the fuse. For longer-duration lower-stress conditions, some of the heat energy will be conducted away, increasing the I^2t rating. Figure 1.5.1 shows how the I^2t rating of a typical fast fuse changes with stress duration.

Modern fuse technology is very sophisticated. The performance of the fuse can be modified considerably by its design. Fuses with the same long-term fusing current can behave entirely differently for short transient conditions. For motor starting and other high-inrush loading requirements, "slow-blow" fuses are chosen. These fuses are designed with relatively large thermal mass fuse elements which can absorb considerable energy in the short term without fuse rupture. Hence they have very high I^2t ratings compared with their longer-term current ratings.

At the other end of the scale, fast semiconductor fuses have very low fuse element mass. These fuses are often filled with sand or alumina so that the heat generated by normal loading currents can be conducted away from the low-mass fuse element, giving higher long-term current ratings. As previously explained, in the short term, the heat conduction effects are negligible, and very small amounts of total energy, if dissipated rapidly within the fuse element, are sufficient to cause fuse rupture. Such fuses have very low I^2t ratings compared with their longer-term current ratings, and will more effectively protect the SCR and the external load.

Figure 1.5.1 shows examples of the clearance current–time characteristics for typical "slow-blow," "normal-blow," and "fast-blow" fuses. It should be noted that although the long-term fusing current is 10 A in all cases, the short-term I^2t ratings range from 42 at 10 ms for the fast fuse to over 6000 at 100 ms for the slow fuse. Since the "crowbar" SCR I^2t rating must exceed the fuse I^2t rating, it is clearly important to select both with care. It is also important to remember that in the linear regulator, the output capacitor must be discharged by the crowbar SCR and is not within the fused part of the loop. Since the maximum current and di/dt of the SCR must also be satisfied, it is often necessary to fit a series limiting inductor or resistor in the anode of the SCR. (See R9 in Fig. 1.11.3a.)

The I^2t rating of the SCR must include sufficient margin to dissipate the energy $\frac{1}{2}CV^2$ stored in the output capacitor, in addition to the fuse let-through energy. Finally, the possibility of a short circuit to other sources of power external to the supply must be considered when selecting SCR ratings.

It has been assumed in this example that the fuse is in a noninductive low-voltage loop. Hence the example has considered only the pre-arcing or melting energy.

In high-voltage circuits or loops with high inductance, an arc will be drawn during clearance of the fuse element, increasing the I^2t let-through energy. This effect must be considered when selecting the fuse and SCR.

11.9   TYPE 3, OVERVOLTAGE PROTECTION BY VOLTAGE LIMITING TECHNIQUES

In switchmode power supplies, the crowbar or clamp voltage protection techniques tend to be somewhat less favored because of their relatively large size and dissipation.

By its nature, the off-line switchmode power supply tends to "fail safe"—that is, to a zero or low-voltage condition. Most failure modes tend to result in zero output voltage. Since the high-frequency transformer provides galvanic isolation between the input supply and the output lines, the need for crowbar-type overvoltage protection is considerably less than would be the case with the linear regulator. Hence, in switchmode supplies, overvoltage protection by converter voltage limiting or shutdown is more usually provided. Normally an independent voltage control circuit is energized if the main voltage control loop fails. (A possible exception to this would be the DC-to-DC switchmode regulator, where galvanic isolation may not be provided.)

Many types of converter voltage limiting circuits are used; Fig. 1.11.4 shows a typical example. In this circuit, a separate optocoupler is energized in the event of an overvoltage condition. This triggers a small-signal SCR on the primary circuit to switch off the primary converter. The main criterion for such protection is that the protection loop should be entirely independent of the main voltage control loop. Unfortunately, this requirement is often violated; for example, a separate amplifier within the same voltage control IC package would not be acceptable as a control amplifier in the overvoltage control loop. The normal criterion is that the system should not produce an overvoltage for any single component failure. In the previous example, this criterion is violated because if the IC package were to fail, both control and protection amplifiers would be lost and overvoltage protection would not be provided.

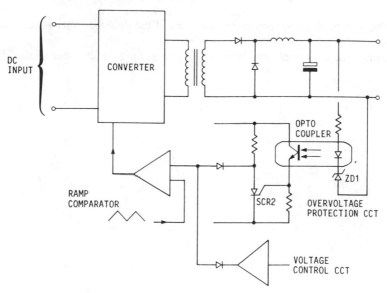

FIG. 1.11.4 Typical overvoltage shutdown protection circuit for SMPS. This circuit operates on the control circuit of the switchmode supply to shut down the converter during an overvoltage stress.

Voltage limiting circuitry may either latch, requiring a cycling of the supply input to reset, or be self-recovering, depending on application requirements. For example, the circuit shown in Fig. 1.11.4 may be made self-recovering by replacing SCR2 with a clamp transistor. Voltage limiting circuits come in many forms and must be configured to suit the overall circuit topology. In multiple-output applications, where independent secondary current limits or regulators are provided, the voltage limit circuit may act upon the current limit circuit to provide the overvoltage protection. Once again, the criterion is that a single component failure should not result in an overvoltage condition. Many techniques are used, and it is beyond the scope of this book to cover more than the bare essentials.

11.10 PROBLEMS

1. Why is output overvoltage protection often considered necessary?
2. Name three types of overvoltage protection in common use.
3. Explain where the three types of overvoltage protection may be used.
4. What is the industry standard criterion for the reliability of overvoltage-protected circuits?
5. Describe what is meant by crowbar overvoltage protection.
6. Describe the problems normally encountered with a fast-acting crowbar protection circuit.
7. List the disadvantages and advantages of a delayed overvoltage protection circuit.
8. What can be done to reduce the problems of the delayed overvoltage protection circuit while retaining the advantages?
9. Explain the important criteria in fuse selection for SCR crowbar applications.

CHAPTER 12
UNDERVOLTAGE PROTECTION

12.1 INTRODUCTION

The need for undervoltage protection is often overlooked in system design. In most power systems, a sudden and rapid increase in load current (for example, inrush currents to disk drives) results in a power supply line voltage dip. This is due to the rapid increase in current during the transient demand and the limited response time of the power supply and its connections.

Even when the transient performance of the power supply itself is beyond reproach, the voltage at the load can still dip when the load is remote from the supply, as a result of line resistance and inductance.

When the load variations are relatively small and short-lived, it is often sufficient to provide a low-impedance capacitor at the load end of the supply lines to "hold up" the voltage during transient loading. However, for large load variations lasting several milliseconds, extremely large shunt capacitors would be required if the voltage is to be maintained close to its nominal value.

It is possible, by fitting an active "undervoltage suppression circuit," to prevent the undervoltage dip at the load without needing excessively large storage capacitors. The following describes a suitable system.

12.2 UNDERVOLTAGE SUPPRESSOR PERFORMANCE PARAMETERS

Figure 1.12.1a, b, and c shows the typical current and voltage waveforms that may be expected at the load end of the DC output lines from a power supply when a large transient load is applied by the load.

Figure 1.12.1a shows a large transient load current demand during the period from t_1 to t_2. Figure 1.12.1b shows the undervoltage transient that might typically be expected at the load during this transient. (Assume that the voltage dip is caused by the resistance and inductance of the supply lines in this example.)

Figure 1.12.1c shows the much-reduced undervoltage transient that would be

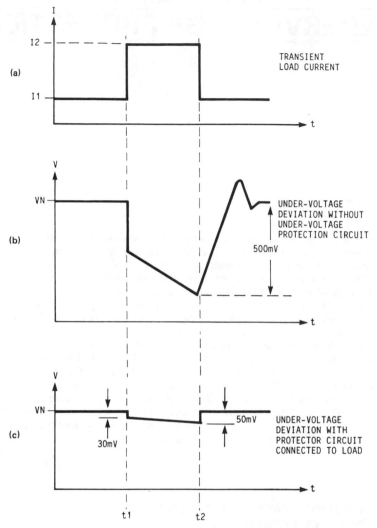

FIG. 1.12.1 Characteristics of a typical "undervoltage transient protection" circuit. (*a*) Load current transient. (*b*) Typical undervoltage transient excursion without protection circuit fitted. (*c*) Undervoltage transient excursion with protection circuit fitted.

seen at the load when the undervoltage suppressor circuit shown in Fig. 1.12.4 is fitted at the load end of the supply lines as shown in Fig. 1.12.2.

Figure 1.12.2 shows how an undervoltage suppressor should be connected to the load at the end of the power supply distribution lines. This circuit stores the energy required to eliminate the undervoltage transient in two small capacitors C1 and C2. An active circuit supplies the required current during the transient demand, preventing any large deviations in the supply voltage at the load. C1 and

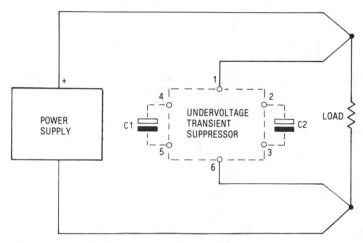

FIG. 1.12.2 Position and method of connection for undervoltage protection circuit.

C2 can be quite small because in this circuit 75% of the stored energy is available for use. How this is achieved is shown in principle in Figure 1.12.3*a, b,* and *c*.

12.3 BASIC PRINCIPLES

Figure 1.12.3*a* shows the method of energy storage and delivery. When SW1 is open, capacitors C1 and C2 are charged in parallel from the supply lines via resistors R1 and R2. They will eventually charge to the supply voltage V_s.

If this circuit is now removed from the supply lines and SW1 is closed, C1 and C2 will be connected in series, and a voltage of $2 \times V_s$ will be provided at the terminals.

In Figure 1.12.3*b*, this circuit (in its charged state) is shown connected to a linear regulator circuit at the input to the regulator transistor Q1. Now if, during an undervoltage condition, SW1 is closed, capacitors C1 and C2 will be connected in series and provide a voltage of $2 \times V_s$ at point A in the circuit.

Since the header voltage at point A (the input of the linear regulator) now exceeds the required output voltage V_s, Q1 can operate as a linear regulator, supplying the required transient current and maintaining the output voltage across the load nearly constant. This can continue until C1 and C2 have discharged to half their initial voltage.

In the active state, C1, C2, SW1, and Q1 form a series circuit. The position of individual components in a series chain has no effect on the overall function of the network; further, SW1 and Q1 can both act as switches, and one of them is redundant. In this example SW1 is to be made redundant.

Figure 1.12.3*c* shows a practical development of the circuit; SW1 has been removed, and Q1 has been moved to the original position of SW1. Q1 now carries out both the original switch functions of SW1 and the linear regulator functions of Q1. Although it is perhaps not obvious, examination will show that this circuit has the same properties as the circuit shown in Fig. 1.12.3*b*.

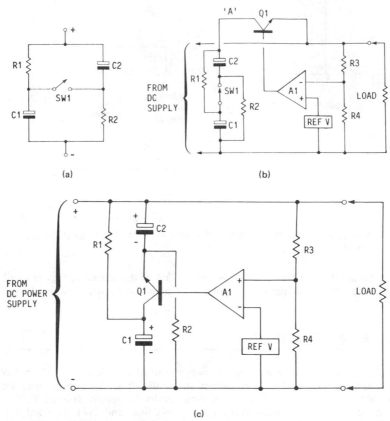

FIG. 1.12.3 Undervoltage circuit development steps.

As previously stated, voltage regulation can continue as long as the charge in C1 and C2 can maintain the required header voltage. Clearly this depends on the load current and the size of C1 and C2. The voltage at point A will not fall below a value where Q1 goes out of regulation until the voltage across each capacitor is approximately half its original value. Since the energy stored in the capacitors is proportional to V^2, three-quarters of the stored energy is available for use.

Because of the efficient use of the stored energy, very much smaller capacitors can be used (compared with what would be required if normal shunt capacitors were used on their own). Further, the load voltage can be maintained within a few millivolts of normal operation throughout the undervoltage stress period, even though the capacitor voltages are falling. Hence much better performance can be provided with active transient suppression.

It should be noticed that resistors R1 and R2 form undesirable loads on C1 and C2 when the circuit is active (SW1 or Q1 closed); hence their resistance is a compromise selection. A high value of resistance presents minimum loading but requires a longer charging time.

12.4 PRACTICAL CIRCUIT DESCRIPTION

Figure 1.12.4 shows a practical implementation of this technique. In this circuit, switch SW1 or Q1 is replaced by Darlington-connected transistors Q3 and Q4. These transistors operate as a switch and linear regulator.

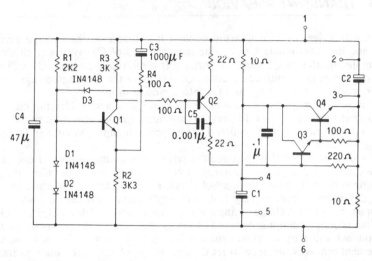

FIG. 1.12.4 Example of an undervoltage protection circuit.

Although Q3 and Q4 are now shown positioned between the two capacitors C1 and C2, it has been demonstrated above that since they still form a series network, their position in the series chain does not change the function of the circuit.

Q1 and Q2 are part of the drive and linear regulator control circuit. The control circuit is not easily identified as a linear regulator, because it appears to lack the normal reference voltage. However, a relative reference voltage which is proportional to the mean (normal) supply voltage V_s is set up on C3. An absolute reference voltage is not what is required here; setting a relative reference voltage on C3 makes the unit self-voltage-tracking. Hence the circuit responds to any transient deviation which is below normal; it does not require presetting to a particular voltage.

12.5 OPERATING PRINCIPLES (PRACTICAL CIRCUIT)

Initial Conditions

A bias voltage is set up on the base of Q1 by the current in R1, D1, and D2. Q1 conducts to develop a second bias voltage across R2 of approximately one diode drop (0.6 V). The current flow in R3 is similar to that in R2, and a third bias voltage is set up across R3 which is slightly less than that across R2, since the resistance of R3 is lower than that of R2.

Hence, under quiescent conditions, transistor Q2 is close to conducting. At the same time, capacitor C3 will charge through R4, R2, D3, D1, and D2 so that the voltage on its negative terminal will end up at the same value as the emitter voltage of Q1. Also, C1 and C2 will charge up to the input voltage V_s via the 10-Ω resistors.

12.6 TRANSIENT BEHAVIOR

When a transient current demand occurs, it will reduce the voltage across the load and input terminals 1 to 6. The negative end of C3 will track this change, taking the emitter of Q1 negative. After a few millivolts change, Q1 will start to turn on, bringing Q2 into conduction. Q2 will drive the Darlington-connected linear regulator transistors Q3 and Q4 into conduction.

This action progressively connects C1 and C2 in series, driving current into the output terminals 1 to 6 to prevent any further reduction in terminal voltage. Hence the circuit can be considered to be "propping up" the voltage by using the charge on C1 and C2.

It should be noted that the circuit is self-tracking—during normal operation the voltage across C3 adjusts to respond to any transient deviation below the normal working voltage. Because the control circuit is always active and close to conduction, the response time is very fast. The small shunt-connected capacitor C4 can maintain the output voltage during the very short turn-on delay of Q3 and Q4.

Undervoltage clamping occurs as soon as the output voltage has dropped from its nominal value by a defined margin (typically 30 mV). This self-tracking arrangement removes the need to set the operating voltage of the undervoltage protection circuit to suit the power supply output voltage.

This protection circuit can be extremely useful where load transients are a problem. It is best positioned close to the transient demand, to eliminate the effects of voltage drop in the supply lines. In some applications, extra capacitors may be required to extend the holdup time; these can be connected to terminals 2, 3, 4, and 5 across C1 and C2.

A further possible advantage to be gained from this technique is that the peak current demand on the power supply can be reduced. This may permit a lower current rating (more cost-effective) power supply to be used.

The decision to use such systems becomes part of the total power system design philosophy. Since this is not part of the power supply, it is the system designer who should consider such needs.

Figure 1.12.1b and c shows the performance that may be expected at the load with and without the undervoltage protector. It is clear that even if the power supply has a very fast transient response, the improvement in performance at the load with the UVP unit can be very significant.

12.7 PROBLEMS

1. Even if the power supply has an ideal transient response, it is still possible for undervoltage transients to occur at the load. Under what conditions would this occur?

2. What advantages does an active undervoltage protection circuit have over a decoupling capacitor?

CHAPTER 13
OVERLOAD PROTECTION

13.1 INTRODUCTION

In computer and professional-grade power supplies, it is normal practice to provide full overload protection. This includes short-circuit protection and current limits on all outputs. The protection methods take many forms, but in all cases the prime function is to protect the power supply, irrespective of the value or duration of the overload, even for continuous short-circuit conditions.

Ideally the load will also be protected. To this end the current limit values should not exceed the specified current rating of the load supply by more than 20%, and the user should choose a supply rating to suit the application. This will usually ensure that the power supply, connectors, cabling, printed circuit tracks, and loads are fully protected for fault conditions.

Full protection is relatively expensive, and for small, low-power units (particularly flyback supplies) full protection is not always essential. Such units may use simple primary power limiting, and have some areas of vulnerability for unusual partial overloading conditions.

13.2 TYPES OF OVERLOAD PROTECTION

Four types of overload protection are in general use:

1. Overpower limiting
2. Output constant-current limiting
3. Fuses or trip devices
4. Output foldback (reentrant) current limiting

13.3 TYPE 1, OVERPOWER LIMITING

The first type is a power-limiting protection method, often used in flyback units or supplies with a single output. It is primarily a power supply short-circuit protection technique. This and the methods used in types 2 and 4 are electronic, and depend on the power supply remaining in a serviceable condition. The supply may be designed to shut down or self-reset if the overload is removed.

In this type of protection, the power (usually in the primary side of the converter transformer) is constantly monitored. If this power exceeds a predetermined limit, then the power supply shuts down or goes into a power-limited mode of operation. In a multiple-output unit, the power would be the sum of the individual outputs.

The power limiting action would normally take one of five forms:

A. Primary overpower limiting

B. Delayed overpower shutdown

C. Pulse-by-pulse overpower/overcurrent limiting

D. Constant-power limiting

E. Foldback (reentrant) overpower limiting

13.4 TYPE 1, FORM A, PRIMARY OVERPOWER LIMITING

In this form of power limiting, the primary power is constantly monitored. If the load tries to exceed a defined maximum, the input power is limited to prevent any further increase.

Usually, the shape of the output current shutdown characteristic is poorly defined when primary power limiting is used on its own. However, because of its low cost, primary power limiting has become generally accepted in lower-power, low-cost units (particularly in multi-output flyback power supplies).

It should be noted that when a load fault develops in a multiple-output system, a line which has been designed to provide only a small proportion of the total power may be expected to support the full output power if it is the only line which is overloaded.

Often these simple primary power limiting systems give full protection only for short-circuit conditions. An area of vulnerability can exist when partial overloads are applied, particularly when these are applied to a single output of a multiple-output system. Under these conditions, partial overloads may result in eventual failure of the power supply if they persist for long periods; hence it is better to remove this stress as soon as possible by turning the supply off. For this reason the delayed overpower trip technique (form b) is recommended.

13.5 TYPE 1, FORM B, DELAYED OVERPOWER SHUTDOWN PROTECTION

One of the most effective overload protection methods for low-power, low-cost supplies is the delayed overpower shutdown technique. This operates in such a way that if the load power exceeds a predetermined maximum for a duration beyond a short defined safe period, the power supply will turn off, and an input power off-on cycle will be required to reset it to normal operation.

Not only does this technique give the maximum protection to both power supply and load, but it is also the most cost-effective for small units. Although this method seems generally unpopular with most users, it should not be neglected, as it makes good sense to turn the power supply off when overloads occur. A persistent power overload usually indicates a fault within the equipment, and the shutdown method will provide full protection to both load and supply.

Unfortunately, many specifications eliminate the possibility of using a simple trip type of protection by demanding automatic recovery from an overload condition. It is possible that the user has specified automatic recovery because of previous bad experience (e.g., "lockout" or nuisance shutdowns) with reentrant or trip-type systems which did not have a sufficient current margin or a delayed shutdown. The power supply designer should question such specifications. Modern switchmode supplies are capable of delivering currents well in excess of their continuous rated value for short periods of time, and with delayed shutdown they will not "lock out" even if a shutdown system has been used.

In the delayed trip system, short transient current requirements are accommodated, and the supply will shut down only if the stress exceeds safe amplitudes for long periods. Short-lived transient currents can be provided without jeopardizing the reliability of the power supply or having a very significant impact on the cost of the unit. It is the long-term continuous current requirements that affect cost and size. There will usually be some degradation in the performance of the unit during the high-current transient. Specified voltage tolerances and ripple values may be exceeded. Typical examples of loads subject to large but short transients would be floppy disks and solenoid drivers.

13.6   TYPE 1, FORM C, PULSE-BY-PULSE OVERPOWER/CURRENT LIMITING

This is a particularly useful protection technique that will often be used in addition to any secondary current limit protection.

The input current in the primary switching devices is monitored on a real-time basis. If the current exceeds a defined limit, the "on" pulse is terminated. With discontinuous flyback units, the peak primary current defines the power, and hence this type of protection becomes a true power limit for such units.

With the forward converter, the input power is a function of input current and voltage; hence this type of protection provides a primary current limit in this type of circuit. However, this technique still provides a useful measure of power limit protection so long as the input voltage is constant.

A major advantage of the fast pulse-by-pulse current limit is that it provides protection to the primary switching devices under unusual transient stress, for example, transformer staircase-saturation effects.

Current-mode control provides this primary pulse-by-pulse current limiting as a normal function of the control technique, one of its major advantages. (See Part 3, Chap. 10.)

13.7   TYPE 1, FORM D, CONSTANT POWER LIMITING

Constant input power limiting will protect the primary circuit by limiting the maximum transmitted power. However, in the case of the flyback converter, this technique does little to protect the secondary output components. For example, consider a discontinuous flyback converter for which the peak primary current has been limited, giving limited transmitted power.

When the load exceeds this limit (load resistance reducing), the output voltage begins to fall. However, since it is the input (and hence output) volt-ampere product that has been defined, as the output voltage starts to fall, the output current will increase. (On short circuit the secondary current will be large and the total power must be dissipated within the power supply.) Hence this form of power limiting is normally used to supplement some other form of limiting, such as secondary current limits.

13.8 TYPE 1, FORM E, FOLDBACK (REENTRANT) OVERPOWER LIMITING

This technique is an extension of form d in which a circuit monitors primary current and secondary voltage, and reduces the power as the output voltage falls. By this means, the output current can be reduced as the load resistance falls, preventing excessive stress on secondary components. It has the possible disadvantage of "lockout" with nonlinear loads. (See Part 1, Chap. 14.)

13.9 TYPE 2, OUTPUT CONSTANT CURRENT LIMITING

Power supplies and loads can be very effectively protected by limiting the maximum current allowed to flow under fault conditions. Two types of current limiting are in common use, constant current and foldback current limiting. The first type, constant current limiting, as the name implies, limits the output current to a constant value if the load current tries to exceed a defined maximum. A typical characteristic is shown in Fig. 1.13.1.

From this diagram, it can be seen that as the load current increases from a low value (R1, high resistance) to its maximum normal current value (R3, median resistance), the current will increase at constant voltage along the characteristic P1–P2–P3, which are all currents and voltages within the normal working range of the supply.

When the limiting current is reached at P3, the current is not allowed to increase any further. Hence, as the load resistance continues to fall toward zero, the current remains nearly constant and the voltage must fall toward zero, characteristic P3–P4. The current-limited area is often not well specified, and the working point will be somewhere in the range P4 to P4 at a load resistance of R4.

Since the current limit is normally provided as a protection mechanism for the power supply, the characteristic in the current-limited range may not be well defined. The limit current range, P4–P4, may change by as much as 20% as the load resistance is taken toward zero (a short circuit). If a well-defined constant current range is required, a "constant current power supply" should be specified (see Part 2, Chap. 22).

Current limiting will normally be applied to the secondary of the power converter. In a multiple-output system, each output will have its own individual current limiting. The current limits will normally be set at some independent maxi-

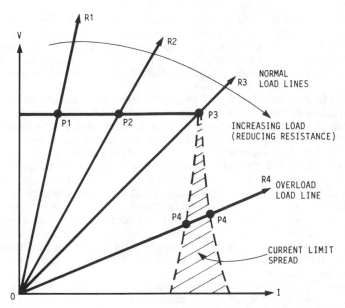

FIG. 1.13.1 Typical *V/I* characteristics of a "constant-current-limited" power supply, showing linear (resistive) load lines.

mum value for each output line, irrespective of the power rating of the supply. If all outputs are fully loaded simultaneously, the total loading may exceed the maximum power rating of the supply. Hence, a primary power limit will often be provided to supplement the secondary current limits. Under fault conditions, both primary and secondary components are fully protected, and the loads will all have limited currents within their design maximums at all times.

This method of current limiting undoubtedly gives the user and the supply the best protection. Not only are currents limited to values consistent with the design ratings for each line, but minimum problems occur with nonlinear or cross-connected loads. The lockout difficulties often associated with foldback limit systems are completely eliminated. Also, automatic recovery is provided when an overload is removed. Moreover, such units may be operated in parallel, the only proviso being that the current limit should be set to some value within the continuous working range. This method of protection is recommended for professional-grade supplies, although it is more expensive.

13.10   TYPE 3, OVERLOAD PROTECTION BY FUSES, CURRENT LIMITING, OR TRIP DEVICES

Type 3 employs mechanical or electromechanical current protection devices, and these will normally require operator intervention to be reset. In modern electronic switchmode power supplies, this type of protection is normally used only as a backup to the self-recovery electronic protection methods. Hence it is a "last ditch" protection method. It is required to operate only if the normal electronic protection fails. In some cases a combination of methods may be used.

Included in type 3 protection methods are fuses, fusible links, fusible resistors, resistors, thermal switches, circuit breakers, PTC thermistors, and so on. These devices all have their place, and should be considered for specific applications.

Where fuses are used, it should be remembered that currents well in excess of the fuse rating can be taken through the fuse for considerable periods before fuse clearance. Also, fuses running at or near their rated value have a limited life and should be periodically replaced. Remember also that fuses dissipate power and have considerable resistance; when used in output circuits, they will often have resistance values well above the normal output resistance of the supply.

However, fuses do have good applications. For example, when a small amount of logic current (say a few hundred milliamperes) is required from a high-current output, this may be a good application for a fuse. Clearly, it would not be sensible to design a printed circuit board and connections to withstand the high current that would flow on this low-power logic board in the event of a short circuit, and a fuse could be used in this application, providing protection without excessive voltage drop. More sophisticated protection techniques may not be justified in this situation.

Fuses or circuit breakers will also be used to back up the electronic overload protection, such as, SCR "crowbar" protection in linear power supplies, in many applications. In such applications the performance of the fuse is critical, and the fuse type and rating must be carefully considered. (See Part 1, Chap. 5.)

13.11 PROBLEMS

1. What is the normal overload protection criterion for professional-grade power supplies?
2. Give four types of overload protection in common use.
3. Give the main advantages and limitations of each of the four types of protection.

CHAPTER 14
FOLDBACK (REENTRANT) OUTPUT CURRENT LIMITING

14.1 INTRODUCTION

Foldback current limiting, sometimes referred to as reentrant current limiting, is similar to constant current limiting, except that as the voltage is reduced as a result of the load resistance moving toward zero, the current is also induced to fall. However, this aparently minor change in the characteristic has such a major impact on the performance that it justifies special attention. To introduce the principle, a linear power supply will be considered.

In linear power supplies, the purpose of foldback current limiting is to prevent damage to the power supply under fault conditions. With foldback limiting, the current is reduced under overload conditions, reducing the power stress on the linear regulator transistors. Because of the high dissipation that would otherwise occur, some form of foldback current limiting is almost universal in linear power supplies.

14.2 FOLDBACK PRINCIPLE

Figure 1.14.1 shows a typical reentrant characteristic, as would be developed *measured* at the output terminals of a foldback-limited power supply.

A purely resistive load will develop a straight load line (for example, the 5-Ω load line shown in Fig. 1.14.1). A resistive load line has its point of origin at zero, and the current is proportional to voltage.

As a resistive load changes, the straight line (which will start vertically at zero load—i.e., infinite resistance) will swing clockwise around the origin to become horizontal for a short circuit (zero resistance). It should be noted that a straight resistive load line can cross the reentrant characteristic of the power supply at only one point, for example, point P1 in Fig. 1.14.1 or 1.14.3. Consequently, "lockout" cannot occur with linear resistive loads, even if the shutdown characteristic is reentrant.

In the example shown in Fig. 1.14.1, as the load current increases from zero, the voltage initially remains constant at the stabilized 5-V output. However, when the maximum limiting current I_{max} has been reached at P2, any further attempt to increase the load (reduction of load resistance) results in a reduction in

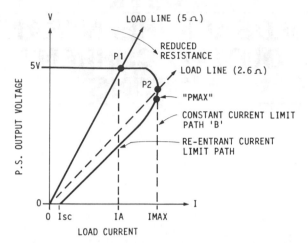

FIG. 1.14.1 Current overload reentrant shutdown characteristic of foldback current-limited supply.

both output voltage and current. Hence, under short-circuit conditions, only a small current I_{sc} flows in the output terminals.

14.3 FOLDBACK CIRCUIT PRINCIPLES AS APPLIED TO A LINEAR SUPPLY

In the simple linear regulator shown in Fig. 1.14.2*a,* a typical foldback current limit circuit is shown (in dashed outline). The output parameters are shown in Fig. 1.14.1, and the regulation dissipation in Fig. 1.14.2*b.*

This circuit operates as follows: When the main series regulator transistor Q1 is conducting, a voltage proportional to the output current I_{load} is developed across the current limit resistor R1. This voltage, together with the base-emitter voltage of Q1, is applied to the base of the current limit transistor Q2 through the divider network R2, R3.

Since the base-emitter voltage drop V_{be} of Q1 will be approximately the same as that of Q2 at the point of current limit, the voltage across R2 is the same as that across R1 but is level-shifted by $+V_{be}$. At the point of transition into current limit, the current flow in R2 is the same as that in R3 (neglecting small base currents), and Q2 is on the threshold of conduction.

Any further increase in the load current at this point will increase the voltage across R1 and hence across R2, and Q2 will be progressively turned on. As Q2 conducts, it diverts the drive current from Q3 away from Q1 into the output load, Q1 starts to turn off, and the output voltage falls. Note: Q3 is a constant-current source.

As the output voltage falls, the voltage across R3 decreases and the current in R3 also decreases, and more current is diverted into the base of Q2. Hence, the current required in R1 to maintain the conduction state of Q2 is also decreased.

Consequently, as the load resistance is reduced, the output voltage and current fall, and the current limit point decreases toward a minimum when the out-

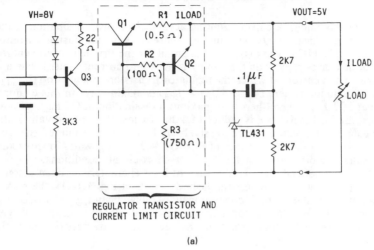

REGULATOR TRANSISTOR AND
CURRENT LIMIT CIRCUIT

(a)

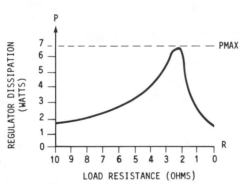

(b)

FIG. 1.14.2 (*a*) Foldback current limit circuit. (*b*) Regulator dissipation with re-entrant proteciton.

put voltage is zero (output short circuit). At short circuit, the current in R1 is very small, and the voltage across R1 and R2 will also be small.

The short-circuit current is not well defined, as the base current of Q2 depends on its current gain, which will vary between devices; also, the V_{be} of Q1 and Q2 are temperature-dependent. These variations can be minimized by mounting Q2 on the same heat sink as Q1 and by using relatively low values for R1 and R2 (typically R1 would be of the order of 100 Ω in the example shown).

In Fig. 1.14.1, it should be noted that a current "foldback" occurs when the current tries to exceed I_{max}. This characteristic is developed as follows:

If the 5-Ω load line is allowed to swing clockwise (resistance being reduced toward zero), the current path shown in Fig. 1.14.1 will be traced out. From its initial working current (say 1 A at point P1), the current first rises to its limiting value I_{max}, then falls toward zero as the load resistance is further reduced. For short-circuit conditions, the current falls to a low value I_{sc}.

Since the header voltage of the linear regulator V_H remains relatively constant throughout this foldback current limiting action, the power dissipation in the series regulator transistor Q1 will initially increase with increasing currents, as shown in Fig. 1.14.2b. The dissipation is small for the first part of the characteristic, but increases rapidly as the supply moves into current limit. It has a maximum value at a current at which the regulator transistor voltage drop (I_{load} product) is a maximum [where ($V_H - V_{\text{out}}$) × I_{load} is maximum]—in this example, at a current of 2.2 A, where there is a maximum dissipation in Q1 P_{max} of 6.8 W.

When the load resistance is reduced further (below this critical value), the series regulator dissipation is progressively reduced as a result of the current foldback. It has a minimum value of $P \times Q1 = I_{sc} \times V_H$ watts in this example. This results in a dissipation of 1.8 W under short-circuit conditions.

It should be noticed that had the current limit characteristic been a constant-current type (shown by the vertical dashed path B in Fig. 1.14.1), the maximum dissipation under short-circuit conditions would have been $I_{\text{max}} \times V_H$, or 12.8 W in this example. Hence the constant current limit places considerably greater stress on the regulator transistor than the reentrant characteristic, in the linear regulator example.

14.4 "LOCKOUT" IN FOLDBACK CURRENT-LIMITED SUPPLIES

With the resistive load (the straight-line loads depicted in Figs. 1.14.1 and 1.14.3), there can only be one stable point of operation, defined by the intersection of the

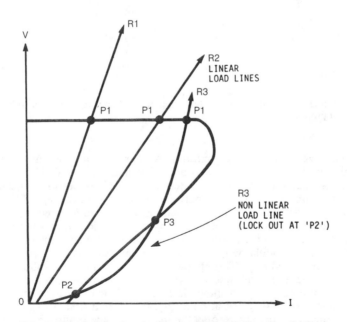

FIG. 1.14.3 Overload and start-up characteristics of a foldback, current-limited supply, showing performance for linear and nonlinear load lines.

load line for a range of given loads with the power supply characteristic (for example, all points P1). Hence, the reentrant characteristic shown would be swept out as the load resistance is varied from maximum to zero. This characteristic is swept out without instability or "lockout"; however, this smooth shutdown may not occur with nonlinear loads.

Figure 1.14.3 shows a very nonlinear load line R3 (such as may be encountered with tungsten filament lamps) impressed on the power supply reentrant current limit characteristic.

It should be understood that a tungsten filament lamp has a very low resistance when it is first switched on (because of the low temperature of the filament wire). Consequently, a relatively large current flows at low applied voltages. As the voltage and current increase, the temperature and resistance of the filament increase, and the working point changes to a higher resistance. A nonlinear characteristic is often found in active semiconductor circuits.

It should be noticed that this nonlinear load line crosses the power supply reentrant current characteristic at three points. Points P1 and P2 are both stable operating points for the power supply. When such a supply-load combination is first switched on, the output voltage is only partially established to point P2, and lockout occurs. (It is interesting to note that if the supply is switched on before the load is applied, it may be expected that the correct working point P1 will be established.) However, point P1 is a stable operating point only for a lamp that was previously working. When the lamp is first switched on, lockout will still occur at point P2, during the lamp power-up phase. This is caused because the slope resistance of the lamp load line at point P2 is less than the slope of the power supply reentrant characteristic at the same point. Since P2 is a stable point, lockout is maintained, and in this example the lamp would never be fully turned on.

Reentrant lockout may be cured in several ways. The reentrant characteristic of the power supply may be modified to bring it outside the nonlinear load line of the lamp, as shown in plots B and C in Fig. 1.14.4. This characteristic now provides only one stable mode of operation at point P1. However, modifying the cur-

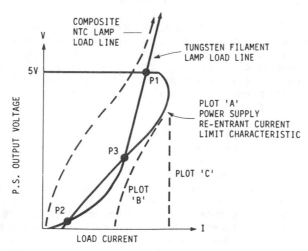

FIG. 1.14.4 Nonlinear load line, showing "lockout" and modified characteristics to prevent lockout.

rent limit characteristic means that under short-circuit conditions the current is increased, with a corresponding increase in regulator transistor dissipation. This increase may not be within the design parameters of the power supply. For this reason, one of the more complex current limit circuits may be preferred. These change the shape of the limit characteristic during the turn-on phase, then revert to the normal reentrant shape.

Other methods of curing lockout include modifying the shape of the nonlinear load line of the lamp itself—for example, by introducing a nonlinear resistor in series with the lamp circuit. NTCs (negative coefficient resistors) are particularly suitable, as the resistance of the load will now be high when the lamp is first switched on, and low in the normal operating mode. The NTC characteristic is the inverse of the lamp characteristic, so that the composite characteristic tends to be linear or even overcompensated, as shown in Fig. 1.14.4. However, a slightly higher voltage is now required from the power supply to offset the voltage drop across the NTC.

NTCs are the preferred cure, since they not only cure the "lockout" but also prevent the large inrush current to the lamp which would normally occur when the lamp is switched on. This limiting action can considerably increase the lamp life.

Nonlinear loads come in many forms. In general, any circuit that demands a large inrush current when it is first switched on may be subject to lockout when reentrant current protection is used.

14.5 REENTRANT LOCKOUT WITH CROSS-CONNECTED LOADS

Lockout problems can occur even with linear resistive loads when two or more foldback-limited power supplies are connected in series. (This series connection is often used to provide a positive and negative output voltage with respect to a common line.) In some cases series power supplies are used to provide higher output voltages.

Figure 1.14.5a shows a series arrangement of foldback-limited supplies. Here, positive and negative 12-V outputs are provided. The normal resistive loads R1 and R2 would not present a problem on their own, provided that the current is within the reentrant characteristic, as shown by load lines R1 and R2 in Fig. 1.14.5b. However, the cross-connected load R3 (which is connected across from the positive to the negative output terminals) can cause lockout depending on the load current magnitude.

Figure 1.14.5b shows the composite characteristic of the two foldback-protected supplies. The load lines for R1 and R2 start at the origin for each supply and can cross the reentrant characteristics at only one point. However, the cross-connected load R3 can be assumed to have its origin at $V+$ or $V-$. Hence, it can provide a composite loading characteristic which is inside or outside of the reentrant area, depending on its value. In the example shown, although the sum of the loads is within the characteristic at point P1, a possible lockout condition occurs at point P2, when the supplies are first switched on. Once again, one cure is to increase the short-circuit current for the two power supplies to a point beyond the composite load line characteristic.

In Fig. 1.14.5a, shunt-connected clamp diodes D1 and D2 must be fitted to prevent one power supply reverse-biasing its complement during the power-up

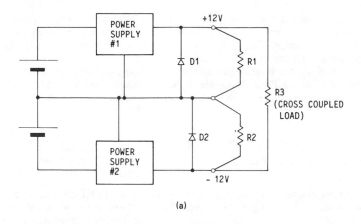

(a)

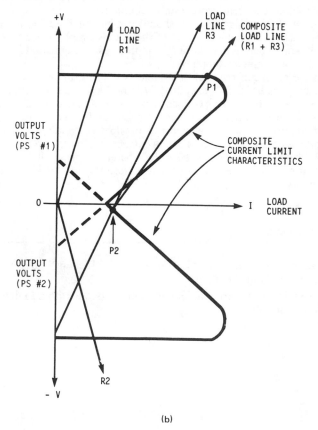

(b)

FIG. 1.14.5 (*a*) Bipolar connection with cross-coupled load. (*b*) Composite characteristic with bipolar load connections.

phase. With foldback protection, if a reverse voltage bias is applied to the output terminals of the power supply, the reentrant characteristic is deepened and the current is even lower. This effect is shown in the dashed extension to the reentrant characteristics in Fig. 1.14.5b.

In conclusion, it can be seen that there are many possible problems in the application of foldback-limited supplies. Clearly, these problems are best avoided by not using the foldback method if it is not essential.

14.6 FOLDBACK CURRENT LIMITS IN SWITCHMODE SUPPLIES

The previous limitations would also apply to the application of foldback protection in switchmode supplies. However, in switchmode units, the dissipation in the control element is no longer a function of the output voltage and current, and the need for foldback current protection is eliminated.

Consequently, foldback protection should not be specified for switching supplies. It is not necessary for protection of the supply and is prone to serious application problems, such as "lockout." For this reason, constant current limits are preferred in switchmode supplies.

Although the nonlinear reentrant characteristic has little to recommend it for switchmode supplies, it is often specified. It is probable that its introduction and continued use stems from the experience with the linear dissipative regulator, where excessive internal dissipation would occur under short-circuit conditions with a constant current limit. However, this dissipative condition does not occur in switchmode supplies, and since a reentrant characteristic can cause problems for the user, there would seem to be little reason to specify it for switchmode applications. It makes little sense to put extra circuitry into the power unit which only degrades its utility.

14.7 PROBLEMS

1. Explain in simple terms the phenomenon of "lockout" and its cause in foldback current-limited supplies.

2. How is it possible to ensure that lockout will not occur with a foldback current limited power supply?

CHAPTER 15
BASE DRIVE REQUIREMENTS FOR HIGH-VOLTAGE BIPOLAR TRANSISTORS

15.1 INTRODUCTION

Where high-voltage bipolar transistors are used in off-line flyback converters, stress voltages of the order of 800 V may be encountered. Higher-voltage transistors with V_{ceo} ratings in the range 400 to 1000 V generally behave somewhat differently from their lower-voltage counterparts. This is due to a fundamental difference in the construction of high-voltage devices.

To obtain the most efficient, fast, and reliable switching action, it is essential to use correctly profiled base drive current waveforms. To explain this, a simplified review of the physical behavior of high-voltage bipolar transistors would be useful. (A full examination of the physics of high-voltage transistors is beyond the scope of this book, but excellent explanations are provided by W. Hetterscheid[49] and D. Roark.[81])

High-voltage devices will generally have a relatively thick region of high-resistivity material in the collector region, and low-resistance material in the base-emitter region. As a result of this resistance profile, it is possible (with an incorrectly profiled base drive) to reverse-bias the base-emitter region during the turn-off edge. This reverse-bias voltage effectively cuts off the base-emitter diode, so that transistor action stops. The collector current is now diverted into the base connection during the turn-off edge, giving diodelike turn-off switching action. That is, the collector-base region now behaves in the same way as a reversed-biased diode. It displays a slow recovery characteristic and has a large recovered charge.

15.2 SECONDARY BREAKDOWN

The slow recovery characteristic described above is particularly troublesome during the turn-off edge with inductive collector loads (such as would be presented by the normal leakage inductance of a power transformer).

As a result of the current forcing action of the collector inductance, any part of the chip which remains conducting during a turn-off edge must continue to carry the previously established collector current. Hence, the slow blocking action of the reverse-biased diodelike collector-base recovery not only results in slow and dissipative turn-off, but also gives rise to "hot spots" on the chip as the current is forced into a progressively small conduction area during the turn-off edge.

It is these "hot spots" which overstress the chip and may cause premature failure. The effect is often referred to as "reversed-biased secondary breakdown."[80]

15.3 INCORRECT TURN-OFF DRIVE WAVEFORMS

Surprisingly, it is the application of energetic and rapid reverse base drive during the turn-off edge which is the major cause of secondary breakdown failure of high-voltage transistors with inductive loads.

Under aggressive negative turn-off drive conditions, carriers are rapidly removed from the area immediately adjacent to the base connections, reverse-biasing the base-emitter junction in this area. This effectively disconnects the emitter from the remainder of the chip. The relatively small high-resistance area in the collector junction will now grow relatively slowly (1 or 2 μs), crowding the collector current into an ever-diminishing portion of the chip.

As a result, not only will the turn-off action be relatively slow, but progressively increasing stress is put on the conducting region of the chip. This leads to the formation of the hot spots and possible device failure, as previously explained.

15.4 CORRECT TURN-OFF WAVEFORM

If the base current is reduced more slowly during the turn-off edge, the base-emitter diode will not be reverse-biased, and transistor action will be maintained throughout turn-off. The emitter will continue to conduct, and carriers will continue to be removed from the complete surface of the chip. As a result, all parts of the chip discontinue conducting at the same instant.

This gives a much faster turn-off collector-current edge, gives lower dissipation, and eliminates hot spots. However, the storage time (the delay between the start of base turn-off and the collector-current edge) with this type of drive will be longer.

15.5 CORRECT TURN-ON WAVEFORM

During the turn-on edge, the reverse of the above turn-off action occurs. It is necessary to get as much of the high-resistance region of the collector conducting as quickly as possible. To achieve this, the base current should be large, with a fast-rising edge; thus carriers are injected into the high-resistance region of the collector as quickly as possible.

The turn-on current at the beginning of the "on" period should be considerably higher than that necessary to maintain saturation during the majority of the remaining "on" period.

15.6   ANTISATURATION DRIVE TECHNIQUES

To reduce the storage time, it is a good practice to inject only sufficient base current toward the end of the "on" period to just ensure that the transistor remains near but not into saturation. Self-limiting antisaturation networks ("Baker clamps") are recommended for this. (See Part 1, Chap. 17.)

With inductive loads, in addition to the base-current shaping, it is usually necessary to provide "snubber networks" between collector and emitter. This snubbing also helps to prevent secondary breakdown.[80, 82] (See Part 1, Chap. 18.)

It should be remembered that low-voltage power transistors will not necessarily display the same behavior. These transistors often have a much more heavily doped collector region, and the resistance is much lower. Applying a rapid reverse-bias voltage to these devices during turn-off is unlikely to generate a high-resistance area. Hence, with low-voltage transistors, fast switching action and short storage times are best achieved by using fast reverse-biased base drive during the turn-off edge.

15.7   OPTIMUM DRIVE CIRCUIT FOR HIGH-VOLTAGE TRANSISTORS

A fully profiled base drive circuit is shown in Fig. 1.15.1a, and the associated drive waveforms are shown in Fig. 1.15.1b. This drive circuit operates as follows.

When the drive input to point A goes positive, current will initially flow via C1 and D1 into the base-emitter junction of the switching transistor Q1. The initial current is large, limited only by the source resistance and input resistance to Q1, and Q1 will turn on rapidly.

As C1 charges, the voltage across R1, R2, C2, and Lb will increase, and current will build up in Lb during the remainder of the "on" period.

Note: While current is flowing in Lb, C2 will continue to charge until the voltage across it equals the zener voltage (D2). D2 now conducts, and the drive current will be finally limited by R1. (R2 has a relatively large resistance, and the current flow in R2 is small.)

When the drive goes low, D1 blocks, and C1 discharges into R2. The forward current in Lb decays to zero and then reverses under the forcing action of the reverse voltage at point B. (C2 is large and maintains its charge during the "off" period.)

Hence during the turn-off edge current builds up progressively in the reverse direction in the base-emitter of Q1 until the excess carriers are removed and the base-emitter diode blocks. At this instant the voltage on Q1 base flies negative under the forcing action of Lb, forcing the transistor into reverse base-emitter breakdown. This reverse breakdown of the base-emitter diode is a nondamaging action and clamps off the base emitter voltage at the breakdown value until the energy in Lb has been dissipated.

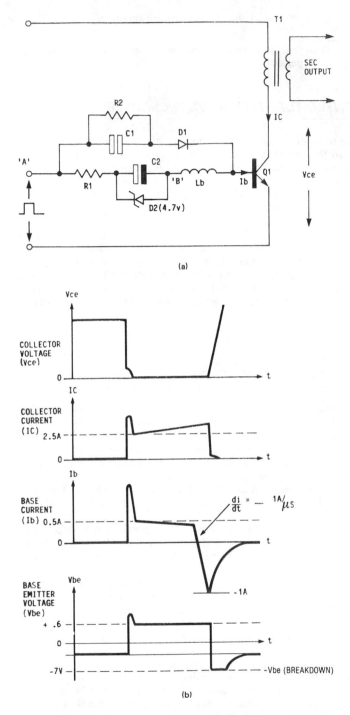

FIG. 1.15.1 (*a*) Base drive current shaping for high-voltage bipolar transistors. (*b*) Collector voltage, collector current, base drive current, and base emitter voltage waveforms.

1.124

The base drive current waveforms are shown in Fig. 1.15.1b.

Although it is not essential to profile the drive current waveform for all types of high-voltage transistor, most types will respond well to this type of drive. If the selected transistor is not rated for reverse base-emitter breakdown, then the values of Lb and R3 should be selected to prevent this action, or clamp zeners should be fitted across the base-emitter junctions.

Since switching device secondary breakdown is probably the most common cause of failure in switchmode power supplies, the designer is urged to study appropriate references.[49, 50, 79, 80, 81, 82]

15.8 PROBLEMS

1. Why do some high-voltage bipolar transistors require specially profiled base drive current waveforms?

2. Explain one cause of secondary breakdown in a high-voltage bipolar transistor.

3. Draw the typical ideal base drive current waveform for high-voltage bipolar transistors with inductive loads.

CHAPTER 16
PROPORTIONAL DRIVE CIRCUITS FOR BIPOLAR TRANSISTORS

16.1 INTRODUCTION

It has been shown (see Part 1, Chap. 15) that to obtain the most efficient performance from bipolar power switching transistors, the base drive current must be correctly profiled to suit the characteristics of the transistor and the collector-current loading conditions. If the base drive current remains constant, problems can arise in applications where the collector current (load) is not constant.

When the drive current has been chosen for optimum performance at full load, if it then remains the same for light loading conditions, the excessive drive will give long storage times, which can lead to a loss of control in the following way. Under light loading (when narrow pulses are most required), the long storage time will give an excessively wide pulse. The control circuit now reverts to a "squegging" control mode. (This is the cause of the well-known "frying-pan noise," a nondamaging instability common to many switchmode supplies at light loads.)

Hence, to prevent overdrive and squegging when the load (collector current) is variable, it is better to make the amplitude of the base drive current proportional to the collector current. Many proportional drive circuits have been developed to meet this requirement. A typical example follows.

16.2 EXAMPLE OF A PROPORTIONAL DRIVE CIRCUIT

Figure 1.16.1 shows a typical proportional drive circuit applied to a single-ended forward converter. In this arrangement, a proportion of the collector current is current-transformer-coupled by T1 into the base-emitter junction of the main switching transistor Q1, providing positive proportional feedback. The drive ratio I_b/I_c is defined by the turns ratio of the drive transformer P1/S1 to suit the gain characteristics of the transistor (typically a ratio of between $\frac{1}{10}$ and $\frac{1}{5}$ will be used).

Because the drive power during most of the "on" period is provided from the

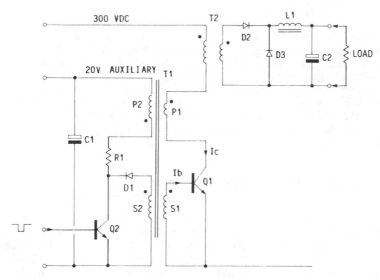

FIG. 1.16.1   Single-ended forward converter with single-ended proportional base drive circuit.

collector circuit, by the coupling from P1 to S1, the drive requirements from Q2 and the auxiliary drive circuit are quite small.

16.3 TURN-ON ACTION

During the previous "off" period of Q1, energy has been stored in T1, since Q2, R1, and P2 have been conducting during this period. When Q2 turns off, the drive transformer T1 provides the initial turn-on of Q1 by transformer flyback action. Once Q1 is conducting, regenerative feedforward from P1 provides and maintains the drive to Q1. Hence, Q2 is turned off for the conducting ("on") period of Q1, and on for the "off" period of Q1.

16.4 TURN-OFF ACTION

When Q2 is turned on again, at the end of a conducting period of Q1, the voltage on all windings is taken to near zero by the clamping action of Q2 and D1 across the clamp winding S2. The previous proportional drive current from P1 is now transformed into the loop S2, D1, and Q2, together with any reverse recovery current from the base-emitter junction of Q1 via S1 (less the current transformed from P2 as a result of conduction in R1). Hence the base drive is removed, and Q1 turns off.

As positive feedback from P1 to S1 is provided in this drive circuit, some care must be taken to prevent high-frequency parasitic oscillation of Q1 during its intended "off" state. This is achieved by making the "off" state of Q1 the low-impedance "on" state of Q2, and by making the leakage inductance between S1 and S2 small. Consequently, any tendency for feedback from P1 to S1 will be clamped by the drive transistor Q2, D1, and S2, which will not allow the start of any winding to go positive.

To prevent Q2 from turning off when it should be on during the power-down phase (leading to loss of control during input power-down), the auxiliary supply to the drive circuit must be maintained during the system power-down phase. (Large capacitors may be required on the auxiliary supply lines.)

16.5   DRIVE TRANSFORMER RESTORATION

For the first part of the "on" period of the driver transistor Q2, D1 and S2 will be conducting. However, when Q1 has turned off and the recovery current in the base-emitter junction of Q1 has fallen to zero, S2 and hence D1 will become reversed-biased as a result of the voltage applied to winding P2 via R1. The start of all windings will now go negative, and current will build up in winding P2, resetting the core back toward negative saturation.

At saturation, the current in P2 and Q2 is limited only by resistor R1, the voltage on all windings is zero, and the circuit has been reset ready for the next "on" cycle.

The need for minimum leakage inductance between S1 and S2 tends to be incompatible with the need for primary-to-secondary isolation and creepage distance. Hence if T1 is used to provide such primary-to-secondary circuit isolation in direct-off-line applications, the transformer may need to be considerably larger than the power needs alone would dictate.

16.6   WIDE-RANGE PROPORTIONAL DRIVE CIRCUITS

Where the range of input voltage and load are very wide, the circuit shown in Fig. 1.16.1 will have some limitations, as follows.

When the input voltage is low, the duty cycle will be large, and Q1 may be "on" for periods considerably exceeding 50% of the total period. Further, if the minimum load is small, L1 will be large to maintain continuous conduction in the output filter. Under these conditions, the collector current is small, but the "on" period is long.

During the long "on" period, a magnetizing current builds up in the drive transformer T1 as a result of the constant base drive voltage V_{be} of Q1 which appears across winding S1. Since the drive transformer is a current transformer during this period, the magnetizing current is subtracted from the output current. Hence, the intended proportional drive ratio is not maintained throughout the long "on" period (the drive falls toward the end of the period). To minimize this effect, a large inductance is required in the drive transformer T1.

However, at the end of the "on" period, Q2 must reset the drive transformer core during the short "off" period which now remains. To allow a quick reset,

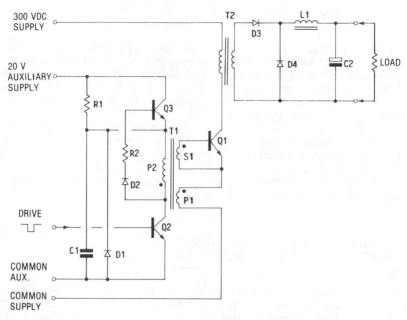

FIG. 1.16.2   Single-ended forward converter with push-pull proportional base drive circuits.

the volts per turn on P2 must be large. This requires either a small number of turns on P2 (with a large reset current) or a large auxiliary voltage. In either case, the power loss on R1 will be relatively large.

Hence, a compromise must be made in inductance turns and auxiliary voltage which is difficult to optimize for wide-range control at high frequencies. This conflict can be solved by the circuit shown in Fig. 1.16.2.

In the circuit shown in Fig. 1.16.2, capacitor C1 charges rapidly when Q2 is off via R1 and Q3. Q3 will be turned on hard by the base drive loop P2, D2, R2 (the starts of all windings being positive when Q2 is "off" and Q1 "on").

16.7   TURN-OFF ACTION

When Q2 is turned on, the voltage across P2 is reversed, and the transferred current from S1 and P1 flows in the low-impedance loop provided by C1, P2, and Q2. The voltage on all windings is reversed rapidly, turning off Q1. At the same time, Q3 is turned off, so that as the core is reset and C1 discharges, only a small current is taken from the supply via R1, which is now much higher resistance than the similar resistor shown in Fig. 1.16.1.

If Q2 is "on" for a long period and C1 is fully discharged, a flywheel action will be provided by D1, preventing reversal of voltage on P2 by more than a diode drop. The turns ratio is such that Q1 will not be turned on under these conditions. Finally the core will return to a reset point defined by the current in R1.

16.8   TURN-ON ACTION

When Q2 is turned off, the starts of all windings will go positive by flyback action, and Q1 will be turned on. Regenerative drive from P1 and S1 maintains the drive, holding Q1 and Q3 on and rapidly recharging C1. This action is maintained until Q2 is turned on again to complete the cycle. The advantage of this arrangement is that the core can be reset rapidly by using a high auxiliary supply voltage without excessive dissipation in R1 and Q2.

Hence, in this circuit the conflict between transformer inductance and reset requirements is much reduced; however, the inductance will be made only just large enough to limit the magnetizing current to acceptable limits. Sufficient drive must be available to ensure correct switching action under all conditions. If the magnetizing current component in the drive transformer is allowed to exceed the collector current, then positive feedback action will be lost.

16.9   PROPORTIONAL DRIVE WITH HIGH-VOLTAGE TRANSISTORS

If Q1 is a high-voltage transistor, it is probable that some shaping of the base drive current will be required for reliable and efficient operation, as shown in Sec. 15.1 of Part 1.

Figure 1.16.3 shows a suitable modification to the drive circuit in Figure 1.16.2 for high-voltage transistors; base drive shaping has been provided by R4, D3, C2, R3, and Lb.

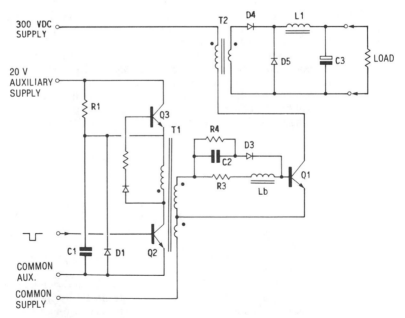

FIG. 1.16.3 Push-pull-type proportional drive circuit with special drive current shaping for high-voltage transistors.

16.10 PROBLEMS

1. What are the major advantages of proportional drive?
2. Why does the drive transformer in a proportional drive circuit tend to be larger than the power requirements alone would indicate?
3. The maximum duty ratio for a transformer-coupled proportional drive circuit tends to be limited to less than 80%. Why is this?
4. What controls the minimum and maximum inductance of the proportional drive transformer?

CHAPTER 17

ANTISATURATION TECHNIQUES FOR HIGH-VOLTAGE TRANSISTORS

17.1 INTRODUCTION

In high-voltage bipolar switching transistors, whereas the "fall time" (speed or dv/dt of the turn-off edge) is mainly determined by the shape of the base drive current turn-off characteristic (see Chap. 15), the storage time (delay between the application of the base turn-off drive and the start of the turn-off edge) is dependent on the minority carrier concentration in the base region immediately prior to turn-off action.

The storage time will be minimized by minimizing the minority carrier concentration, that is, by ensuring that the base current is only just sufficient to maintain the transistor in a quasi-saturated state prior to turn-off.

One method often used to achieve this is the "Baker (diode) clamp." This circuit has the advantage that, because it is an active drive clamp (with negative feedback), it compensates for the inevitable variations in gain and saturation voltage of the various devices. Also, it responds to changes of parameters within the switching transistor that occur as a result of temperature and load variations.

17.2 BAKER CLAMP

Figure 1.17.1 shows a typical Baker clamp circuit. It operates as follows.

Diodes D1 and D2, in series with the base drive to Q1, provide a voltage drop in addition to the transistor V_{be}, so that the drive voltage at node A will rise to approximately 2 V when Q1 is driven on.

As Q1 turns on, the voltage on its collector will fall toward zero. When the voltage reaches approximately 1.3 V, diode D3 will conduct and divert drive current away from the base and into the collector of Q1. As this clamping action is subject to negative feedback, it will self-adjust until the collector voltage is effectively clamped at 1.3 V.

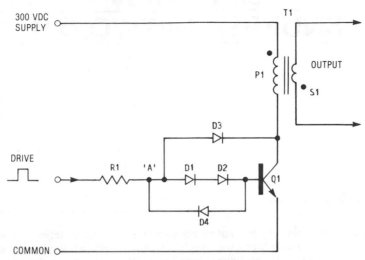

FIG. 1.17.1 "Baker clamp" antisaturation drive clamp circuit.

As a result, the transistor is maintained in a quasi-saturated "on" state with just sufficient base drive current to maintain this condition. This quasi-saturated state maintains minimum minority carriers in the base region during the "on" period, giving minimum storage time during the turn-off action. During turn-off, D4 provides a path to Q1 base for the reverse turn-off current.

The number of series diodes in the base circuit, D1, D2, ... D*n,* will be selected to suit the transistor saturation voltage. The clamp voltage must be above the normal saturation voltage of the transistor at the working current, to ensure that true transistor action is maintained in the quasi-saturated "on" state.

A disadvantage of the technique is that the collector voltage during the "on" period is somewhat larger than it would be for a fully saturated state, which increases the power loss in the transistor.

The Baker clamp arrangement combines ideally with the low-loss "Weaving snubber diode" shown in Fig. 1.18.3. (See Part 1, Chap. 18.)

17.3 PROBLEMS

1. What would be the main advantage of using an antisaturation drive technique in high-voltage switching transistor applications?

2. Describe the action of a typical antisaturation clamp circuit used for bipolar transistors.

CHAPTER 18
SNUBBER NETWORKS

18.1 INTRODUCTION

Snubber networks (usually dissipative resistor-capacitor diode networks) are often fitted across high-voltage switching devices and rectifier diodes to reduce switching stress and EMI problems during turn-off or turn-on of the switching device.

When bipolar transistors are used, the snubber circuit is also required to give "load line shaping" and ensure that secondary breakdown, reverse bias, "safe operating area" limits are not exceeded. In off-line flyback converters, this is particularly important, as the flyback voltage can easily exceed 800 V when 137-V (ac) voltage-doubled input rectifier circuits or 250-V (ac) bridge rectifier circuits (dual input voltage circuits) are used.

18.2 SNUBBER CIRCUIT (WITH LOAD LINE SHAPING)

Figure 1.18.1*a* shows the primary of a conventional single-ended flyback converter circuit P1, Q1 with a leakage inductance energy recovery winding and diode P2, D3. Snubber components D1, C1, and R1 are fitted from the collector to the emitter of Q1. Figure 1.18.1*b* shows the voltage and current waveforms to be expected in this circuit. If load line shaping is required, then the main function of the snubber components is to provide an alternative path for the inductively maintained primary current I_P as Q1 turns off. With these components fitted, it is now possible to turn off Q1 without a significant rise in its collector voltage during the turn-off edge. (The actual voltage increase on the collector of Q1 during the turn-off edge depends on the magnitude of the diverted current I_s, the value of the snubber capacitor C1, and the turn-off time t_1 to t_2 of Q1.) Without these components, the voltage on Q1 would be very large, defined by the effective primary leakage inductance and the turn-off di/dt. Because the snubber network also reduces the rate of change of collector voltage during the turn-off edge, it reduces RFI problems.

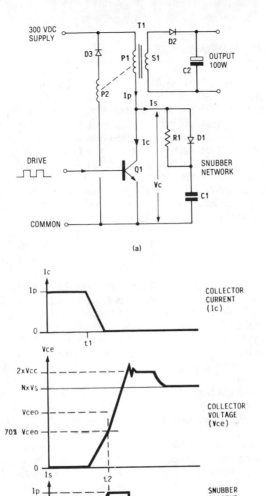

FIG. 1.18.1 (*a*) Conventional dissipative *RC* snubber circuit applied to a flyback. (*b*) Current and voltage waveforms of *RC* snubber circuit.

1.135

18.3 OPERATING PRINCIPLES

During the turn-off edge of Q1, under steady-state conditions, the action of the circuit is as follows.

As Q1 starts to turn off at t_i (Fig. 1.18.1b), the primary and leakage inductance of T1 will maintain a constant primary current I_P in the transformer primary winding. This will cause the voltage on the collector of Q1 to rise (t_1 to t_2), and the primary current will be partly diverted into D1 and C1 (I_s) (C1 being discharged at this time). Hence, as the current in Q1 falls, the inductance forces the difference current I_s to flow via diode D1 into capacitor C1.

If transistor Q1 turns off very quickly (the most favorable condition), then the rate of change of the collector voltage dV_C/dt will be almost entirely defined by the original collector current I_P and the value of C1.

Hence

$$\frac{dV_c}{dt} = \frac{I_P}{C1}$$

With Q1 off, the collector voltage will ramp up linearly (constant-current charge) until the flyback clamp voltage ($2 \times V_{DC}$), is reached at t_3, when D3 will conduct. Shortly after this (the delay depends on the primary-to-secondary leakage inductance), the voltage in the output secondary winding will have risen to a value equal to that on the output capacitor C2. At this point, the flyback current will be commutated from the primary to the secondary circuit to build up at a rate controlled by the secondary leakage inductance and the external loop inductance through D2, C2 (t_3 to t_4).

In practice Q1 will not turn off immediately; hence, *if secondary breakdown is to be avoided, the choice of snubber components must be such that the voltage on the collector of Q1 does not exceed V_{ceo} before the collector current has dropped to zero.*

Figure 1.18.2a and b shows the relatively high edge dissipation and secondary breakdown load line stress, when snubber components are not fitted. Figure 1.18.2c and d shows the more benign turn-off waveforms obtained from the same circuit when optimum snubber values are fitted.

18.4 ESTABLISHING SNUBBER COMPONENT VALUES BY EMPIRICAL METHODS

Referring again to Fig. 1.18.1a,, unless the turn-off time of Q1 is known (for the maximum collector current conditions and selected drive circuit configuration), the optimum choice for C1 will be an empirical one, based upon actual measurements of collector turn-off voltages, currents, and time.

The minimum value of C1 should be such as to provide a safe voltage margin between the V_{ceo} rating of the transistor and the actual measured collector voltage at the instant the collector current reaches zero at t_2. A margin of at least 30% should be provided to allow for component variations and temperature effects.

The design of the drive circuit, collector current loading, and operating temperatures have a considerable influence on the switching speed of Q1. A very large value of C1 should be avoided, since the energy stored in this capacitor at

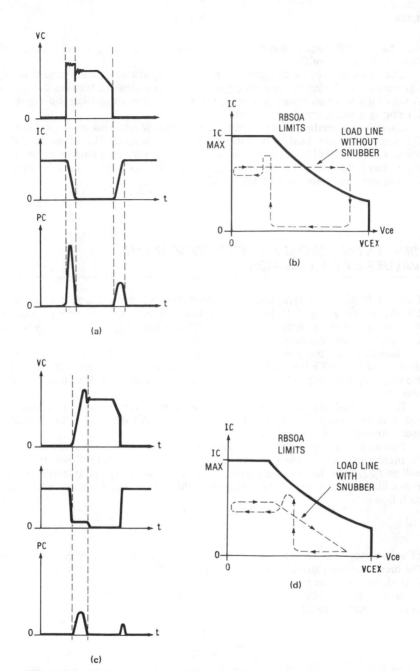

FIG. 1.18.2 "Safe operating area" characterisitics, with and without snubber circuits. (*a*) Turn-on and turn-off voltage, current, and dissipation stress without load line shaping. (*b*) Active load line imposed on "reverse base safe operating area" (RBSOA) limits without load line shaping. (Note secondary breakdown stress.) (*c*) Turn-on and turn-off voltage, current, and dissipation stress with load line shaping. (*d*) Load line and (RBSOA) limits with load line shaping.

1.137

the end of the flyback action must be dissipated in R1 during the first part of the next "on" period of Q1.

The value of R1 is a compromise selection. A low resistance results in high currents in Q1 during the turn-on edge. This gives excessive turn-on dissipation. A very high resistance, on the other hand, will not provide sufficient discharge of C1 during a minimum "on" period.

Careful examination of the voltage and current waveforms on the collector of Q1, under dynamic loading conditions, is recommended. These should include initial turn-on at full-load maximum input voltage, wide and narrow pulse conditions, and output short circuit. The selection of R1 and C1 for this type of snubber network must always be a compromise.

18.5 ESTABLISHING SNUBBER COMPONENT VALUES BY CALCULATION

Figure 1.18.1*b* shows typical turn-off waveforms when the snubber network D1, C1, R1 shown in Fig. 1.18.1 is fitted. In this example, C1 was chosen such that the voltage on the collector V_{ce} will be 70% of the V_{ceo} rating of Q1 when the collector current has dropped to zero at time t_2.

Assuming that the primary inductance maintains the primary current constant during the turn-off edge, and assuming a linear decay of collector current in Q1 from t_1 to t_2, the snubber current I_s will increase linearly over the same period, as shown.

It is assumed that the fall time of the collector current (t_1 to t_2) is known from the manufacturer's data or is measured under active drive conditions at maximum collector voltage and current.

During the collector-current fall time of Q1 (t_1 to t_2), the current in C1 (I_s) will be increasing linearly from zero to I_P. Hence the mean current over this period will be $I_P/2$. Provided that the maximum primary current I_P and turn-off time t_1 to t_2 are known, the value of the optimum snubber capacitor C1 may be calculated as follows:

$$\frac{dV_C}{dt} = \frac{1}{2}\frac{I_P}{C1}$$

(The ½ factor assumes a linear turn-off ramp on the collector current I_C such that the mean current flowing into C1 is ½ the turn-off peak value during the turn-off period, as shown in Fig. 1.18.1*b*.)

Hence, if the collector voltage is to be no more than 70% of V_{ceo} when the collector current reaches zero at time t_2, then

$$C1 = \frac{I_P \times t_f}{2 \times (70\% \ V_{ceo})}$$

where I_P = maximum primary current, A
$\quad\ t_f$ = Q1 collector current fall time (t_1 to t_2), μs
$\ V_{ceo}$ = V_{ceo} rating of selected transistor, V
$\quad\ C1$ = snubber capacitance, μF

18.6 TURN-OFF DISSIPATION IN TRANSISTOR Q1

By the same logic as used above (although the waveform is inverted), C1 and transistor Q1 both see the same mean current and voltage during the turn-off period. Hence, the dissipation in the transistor during the turn-off period t_1 to t_2 will be the same as the energy stored in C1 at the end of the turn-off period (t_2).
Hence

$$P_{Q1(off)} = \frac{1}{2} C1 \times (70\% \ V_{ceo})^2 \times f$$

where $P_{Q1(off)}$ = power dissipated in Q1 during the off period, mW
$\quad$ C1 = snubber capacitance, μF
$\quad$ V_{ceo} = V_{ceo} rating of transistor (70% V_{ceo} is the chosen maximum voltage at $I_C = 0$)
$\quad$ f = frequency, kHz

18.7 SNUBBER RESISTOR VALUES

The snubber discharge resistor R1 is chosen to discharge the snubber capacitor C1 in the minimum selected "on" period. The minimum "on" period is given by the designed minimum load at maximum input voltage and operating frequency.
The *CR* time constant should be less than 50% of the minimum "on" period to ensure that C1 is effectively discharged before the next "off" period. Hence

$$R1 = \frac{1}{2} \frac{t_{off, (min)}}{C1}$$

18.8 DISSIPATION IN SNUBBER RESISTOR

The energy dissipated in the snubber resistor during each cycle is the same as the energy stored in C1 at the end of the "off" period. However, the voltage across C1 depends on the type of converter circuit. With complete energy transfer, the voltage on C1 will be the supply voltage V_{cc}, as all flyback voltages will have fallen to zero before the next "on" period. With continuous-mode operation, the voltage will be the supply voltage plus the reflected secondary voltage.
Having established the voltage across C1 immediately before turn-on (V_C), the dissipation in R1 (P_{R1}) may be calculated as follows:

$$P_{R1} = \frac{1}{2} C1 \ V_C^2 f$$

18.9 MILLER CURRENT EFFECTS

When measuring the turn-off current, the designer should consider the inevitable Miller current that will flow into the collector capacitance during the turn-off edge. This effect is often neglected in discussions of high-voltage transistor action. It results in an apparent collector-current conduction, even when Q1 is fully turned off. Its magnitude depends on the rate of change of collector voltage (dV_C/dt) and collector-to-base depletion capacitance. Further, if the switching transistor Q1 is mounted on a heat sink, there may be considerable capacitance between the collector of Q1 and the common line, providing an additional path for apparent collector current. This should not be confused with Miller current proper, as its magnitude can often be several times greater than the Miller current.

These capacitive coupling effects result in an apparent collector current throughout the turn-off edge, giving a plateau on the measured collector current. Hence, the measured current can never be zero as the collector voltage passes through V_{ceo}. Figure 1.18.2c shows the plateau current. This effect, although inevitable, is generally neglected in the published secondary breakdown characteristics for switching transistors. Maximum collector dV_C/dt values are sometimes quoted, and this can be satisfied by a suitable selection of C1. When power FET switches are used, the maximum dV/dt values must be satisfied to prevent parasitic transistor action; hence, snubber networks must still be used in most high-voltage power FET applications.

18.10 THE WEAVING LOW-LOSS SNUBBER DIODE*

As shown above, to reduce secondary breakdown stress during the turn-off of high-voltage bipolar transistors, it is normal practice to use a snubber network.

Unfortunately, in normal snubber circuits, a compromise choice must be made between a high-resistance snubber (to ensure a low turn-on current) and a low-resistance snubber (to prevent a race condition at light loads where narrow pulse widths require a low CR time constant). This paradox often results in a barely satisfactory compromise. The "Weaving snubber diode" provides an ideal solution.

The circuit for this snubber arrangement is shown in Fig. 1.18.3. It operates as follows.

Assume that transistor Q1 is on so that the collector voltage is low. Current will be flowing from the supply line through the transformer primary P1, and also from the auxiliary supply through resistor R2 and snubber diode D5 into the transistor collector.

At the end of the "on" period, Q1 will start to turn off. As the collector current falls, the transformer primary leakage inductance will cause the collector voltage to rise. However, when the collector voltage is equal to the auxiliary supply voltage, the primary current will be diverted into the snubber diode D5 (flowing in the reverse recovery direction in D5) and back into the auxiliary supply through D6. This reverse current flow in D5 will continue for its reverse recovery time.

During this reverse recovery period, Q1 will continue to turn off, its collector current falling to zero, while the collector stress voltage remains clamped by D5 at a value only slightly above the auxiliary supply voltage. Consequently, Q1 turns off under negligible stress conditions.

The reverse recovery time of the snubber diode must be longer than the turn-

---

*The "snubber diode" was patented by Rodney J. Weaving in 1979.

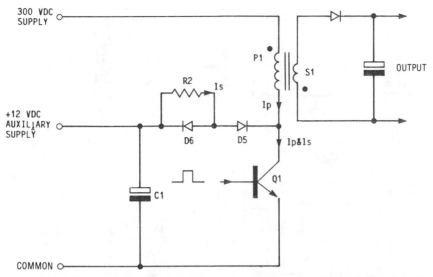

FIG. 1.18.3 The "Weaving snubber diode" low-loss switching stress reduction (snubber) circuit.

off time of transistor Q1. Special medium-speed soft recovery diodes are manu-factured for this purpose (for example, Philips® Type #BYX 30 SN).

During the turn-off action, the recovered charge from the snubber diode D5 is stored in the auxiliary capacitor C1, to be used to polarize D5 during the next "on" period; consequently, very little turn-off energy is lost to the system.

When Q1 turns on again, very little charge will be extracted from the cathode of D5 during the turn-on edge, because the diode depletion layer is wide and the capacitance low (the normal variable-capacitance behavior of the diode). Hence the turn-on stress of Q1 is not significantly increased.

When Q1 is in its saturated "on" state, a current will flow from the auxiliary supply and capacitor C1 to reestablish the forward-bias condition of the snubber diode D5, part of this energy being the previous recovered charge. As soon as the snubber diode is conducting, it is conditioned for a further turn-off cycle.

18.11 "IDEAL" DRIVE CIRCUITS FOR HIGH-VOLTAGE BIPOLAR TRANSISTORS

Figure 1.18.4 shows a combination of the "snubber diode" and "Baker clamp" circuits, with a push-pull base drive to Q1.

This arrangement is particularly suitable for high-voltage flyback converters where the collector voltage may be of the order of 800 V or more during the flyback period. It operates as follows.

When the drive voltage goes high, Q2 is turned on and Q3 off. Current flows via R3, Q2, C2, and D7 to the base of the power transistor Q1. The overdrive provided by the low-impedance R3, C2 network turns Q1 on rapidly.

As Q1 turns on, the collector voltage falls. When this reaches 12 V (the aux-

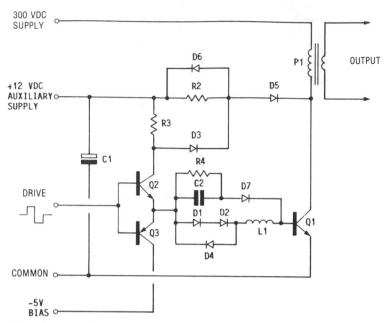

FIG. 1.18.4 Snubber diode and Baker antisaturation clamp combination.

iliary supply voltage), the snubber diode D5 will be forward-biased and current will flow via R2, D5 into the collector of Q1.

Q1 continues to turn on, taking the collector voltage toward zero, until the Baker clamp voltage is reached. At this value D3 becomes forward-biased, diverting part of the base drive current into D3, D5, and the collector of Q1.

At this point C2 will have charged to a voltage such that the drive current will be diverted via D1, D2, and L1 into the base of Q1. The voltage on the collector of Q2 will now be defined by the sum of the voltage drops across Q1 (V_{be}), D1, D2, and Q2 (V_{sat})—say, 2.5 V. The collector clamp voltage will be this value less the voltage drop across D3, D5—say, 1 V. This voltage can be increased by introducing more diodes in series with D1 and D2. (The voltage across L1 is negligible.)

Hence, during the remainder of the "on" period, the main drive current path is via R3, Q2, D1, D2, L1, into the base emitter of Q1. Baker clamp action is provided by D3, D5.

At the end of the "on" period, the drive voltage goes low, turning Q2 off and Q3 on and clamping the cathode of D4 to the −5-V bias line. Diodes D7, D1, and D2 will be reverse-biased, and the turn-off current path is via D4 and L1.

However, L1 was conducting current in the forward direction before turn-off and will continue to maintain this forward (but now decaying) current for the first part of the turn-off action. Hence the turn-off current in L1 will decay to zero and then reverse via D4, providing the ideal turn-off current ramp specified for high-voltage transistors in Chap. 15. Resistor R4 discharges C2 during the "off" period.

When all carriers have been removed from the base-emitter junction of Q1, the junction will block, and the flyback action of L1 will force the base-emitter into reverse breakdown. The breakdown voltage (approximately -7.5 V) is less than the -5-V bias, and this breakdown action stops when the energy in L1 is dissipated. Note: Many high-voltage transistors are designed for this breakdown mode of operation during turn-off.

At the same time, as Q1 turns off, the collector voltage will be rising toward the flyback voltage (800 V). However, when the collector voltage reaches 12 V (the auxiliary voltage), the snubber diode D5 will be reversed-biased, and the collector current will be diverted into D5, D6, and the auxiliary line. The reverse recovery time of D5 is longer than the turn-off time of Q1, and Q1 turns off under low-stress conditions with only 12 V on the collector. When Q1 has turned off and D5 blocks, the collector voltage will rise to the flyback value. The recovered charge of D5 is stored on C1 for the next forward drive pulse.

Although this circuit does not provide proportional drive current in the conventional way, the Baker clamp adjusts the current into the base of the power device to suit the gain and collector current. Hence the action is similar to that of the proportional drive circuit except that the drive power needs are greater.

In conclusion, this drive circuit combines most of the advantages of the proportional drive circuit, the snubber diode, and the Baker clamp. It also provides a correctly profiled drive current to give low stress and fast and efficient switching action in high-voltage, high-power bipolar switching applications.

18.12 PROBLEMS

1. Explain what is meant by the term "snubber network."
2. Explain the two major functions of a typical snubber network.
3. Discuss the criteria for selecting snubber components, for a bipolar transistor with an inductive load, if secondary breakdown is to be avoided.
4. Why is a large snubber capacitor undesirable?
5. Describe a low-loss snubber technique that may be used in place of the conventional RC snubber network.
6. Using the snubber network shown in Fig. 1.18.1, calculate the minimum snubber capacitance required to prevent the collector voltage on Q1 exceeding 70% of V_{ceo} during Q1 turn-off. (Assume that the fall time of Q1 is 0.5 μs, the collector current I_P is 2 A, and the V_{ceo} rating is 475 V.)

CHAPTER 19
CROSS CONDUCTION

19.1 INTRODUCTION

The term "cross conduction" is used to describe a potentially damaging condition that can arise in half-bridge and full-bridge push-pull converters.

The problem is best explained with reference to the circuit shown in Fig. 1.19.1. It can be clearly seen that in this half-bridge configuration, if Q1 and Q2 are both turned on at the same time, they will provide a direct short circuit across the supply lines (transformers T1 and T2 are current transformers and have little resistance). This will often result in immediate failure, as a result of the damagingly high currents that will flow in the switching devices.

Clearly the transistors would not normally be driven such that they would both be on at the same time. The cause of cross conduction can normally be traced to excessive storage time in the switching transistors. Figure 1.19.2 shows typical base drive and collector-current waveforms for the two half-bridge transistors Q1 and Q2 under square-wave (100% duty cycle), full-conduction conditions. As may be expected, because of the storage time t_1–t_3, cross conduction occurs.

In the top waveform, the base drive to Q1 is shown being removed at time t_1 (the beginning of the "off" period for Q1 and the "on" period for Q2). However, because of the inevitable storage time of transistor Q1, its collector current is not blocked until a somewhat later time t_3. At the same time, the lower transistor Q2 is turning on, as shown in the lower waveform. In bipolar transistors, the turn-on delay is typically less than the storage time; hence, with a full 100% duty cycle (push-pull base drive), there will be a short period (t_2 to t_3) when both devices will be conducting. Since these are directly across the supply lines, the low source impedance allows very large collector currents to flow. This effect is shown as current spikes on the waveforms for Q1 and Q2 in Fig. 1.19.2.

If the source impedance of the supply lines is very low, and no series current limiting is provided, damagingly large cross-conduction currents will flow between Q1 and Q2 under the above conditions, and the excessive stress may cause failure of the transistors.

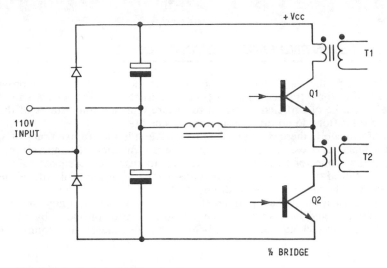

FIG. 1.19.1 Basic half-bridge circuit.

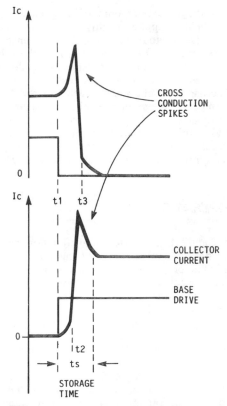

FIG. 1.19.2 Typical cross-conduction current waveforms.

19.2 PREVENTING CROSS CONDUCTION

Traditionally, the method used to prevent cross conduction is to provide a "dead" time (both transistors off), between alternate on drive pulses. This "dead time" must be of sufficient duration to ensure that the "on" states of the two power transistors do not overlap under any conditions.

Unfortunately, there is a considerable variation in the storage times of apparently similar devices. Also, the storage time is a function of temperature, drive circuit, and collector-current loading. Hence, to ensure an adequate safety margin, the "dead time" will need to be considerable, and this will reduce the efficiency and the range of pulse-width control.

Clearly, a system which permits 100% pulse width without any risk of cross conduction would be preferred. The dynamic control provided by the cross-coupled inhibit technique described below admirably meets this requirement.

19.3 CROSS-COUPLED INHIBIT

Figure 1.19.3 shows the basic elements of a dynamic cross-coupled, cross-conduction inhibit technique, applied in this example to a push-pull converter.

In a similar way to the previous example, if, in the push-pull converter, transistors Q1 and Q2 are turned on at the same instant, the primary winding of the transformer T2 will be short-circuited and very large collector currents will flow in the transistors, probably with catastrophic results.

In Figure 1.19.3, cross conduction is prevented by the AND gates U2 and U3. (These gates are often part of the main control IC.) The circuit is shown operating with full duty cycle square-wave base drive. Previously, this would result in severe cross-conduction problems. However, in this circuit, cross conduction is prevented by the cross-coupled inhibit input to the gates, provided by resistors R3 or R4 (depending on the state of conduction Q1 and Q2).

19.4 CIRCUIT OPERATION

Consider Figs. 1.19.2 and 1.19.3 for the initial condition when Q2 is just about to turn on (point t_1 in the drive waveform). At this instant, input 1 of gate U3 is enabled for an "on" state of Q2. However, as a result of its storage time, Q1 will still be conducting and its collector voltage will be low. Hence, input 2 of U3 will be low. As a result of the gating action of U3, the turn-on of Q2 is delayed until the voltage on the collector of Q1 goes high. This does not occur until the end of its storage period, when Q1 turns fully off. As a result, cross conduction is actively prevented; Q2 turns on only after Q1 has fully turned off. The same action occurs when the drive is applied to U2, except that in this case the turn-on of Q1 is delayed until Q2 turns off.

It should be noted that this gating action is self-adjusting and will accommodate variations in the storage times of the two switching devices. Being dynamic, it always permits full conduction angle, while entirely eliminating the possibility of cross conduction.

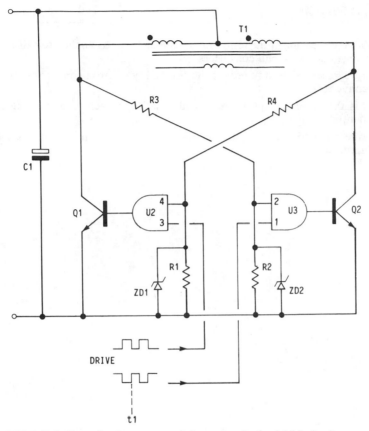

FIG. 1.19.3 Example of a cross-coupled cross-conduction inhibit circuit.

In principle the same technique can be applied to the half-bridge and full-bridge converters, although the drive circuits are somewhat more complex, as the switching devices do not share a common line.

Because the collector voltage swing of Q1 and Q2 would normally exceed the voltage rating of the control circuit, some form of voltage clamping is normally required. In this example, zener diodes ZD1 and ZD2 provide the required clamping action.

Not all control ICs provide the necessary inhibit inputs. In this case, the function may be provided externally to the IC; otherwise a dead band must be provided.

The major advantage of the cross-coupled inhibit technique is that it extends the pulse-width control range from zero to 100%, while maintaining complete integrity as far as cross-conduction problems are concerned. Its advantages should not be overlooked.

19.5 PROBLEMS

1. Explain the meaning of the term "cross conduction" as applied to half-bridge, full-bridge, and push-pull converters.
2. Describe a method used to reduce the possibility of cross conduction in push-pull converters.
3. What is the disadvantage of the "dead time" approach to preventing cross conduction?
4. Describe a method of preventing cross conduction which does not rely on a built-in dead time.

CHAPTER 20
OUTPUT FILTERS

20.1 INTRODUCTION

Undoubtedly, one of the most objectionable properties of switchmode supplies is their predilection for high-frequency radiated and conducted ripple and noise (RF interference).

To keep this interference within reasonable bounds, there must be strict attention to noise reduction techniques throughout the electrical and mechanical design. Faraday screens would be used in transformers and between high-frequency high-voltage components and the ground plane. (These screening methods are more fully covered in Part 1, Chap. 4.) In addition, to reduce conducted-mode noise, low-pass input and output filters will be required.

20.2 BASIC REQUIREMENTS

Output Low-Pass Filters

The following section on output-filter design assumes that normal good design practice has already been applied to minimize conducted-mode noise and that RFI filters have been fitted to the input supply lines, as specified in Sec. 3.1.

To provide a steady DC output, and reduce ripple and noise, LC low-pass filters (as shown in Fig. 1.20.1a) will normally be provided on switching supply outputs. In forward converters, these filters carry out two main functions. The prime requirement is one of energy storage, so as to maintain a nearly steady DC output voltage throughout the power switching cycle. A second, and perhaps less obvious, function is to reduce high-frequency conducted series and common-mode output interference to acceptable limits.

Unfortunately, these two requirements are not compatible. To maintain a nearly constant DC output voltage, the current in the output capacitor must also be nearly constant; hence a considerable inductance will be required in the output inductor. Since the inductor must also carry the DC output current, it is often large and may have many turns. This results in a large interwinding capacitance, giving a relatively low self-resonant frequency. Such inductors will have a low impedance at frequencies above self-resonance and will not provide very effective attenuation of the high-frequency components of the conducted interference currents.

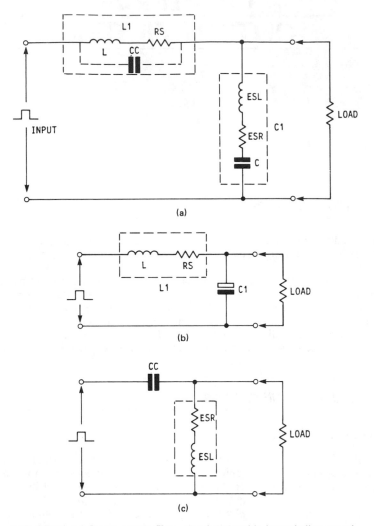

FIG. 1.20.1 (*a*) Power output filter, showing parasitic interwinding capacitance C_C and series resistance R_s for L1, and series inductance ESL and resistance ESR for C1. (*b*), (*c*) Output filter equivalent circuits at low frequency (*b*) and high frequency (*c*).

Further, the filter capacitors will have been chosen primarily for energy storage capability and ripple rating, rather than for high-frequency impedance. Often the effective series resistance (ESR) and effective series inductance (ESL) of large electrolytic capacitors can be significant. Hence, unless the more expensive low-ESR capacitor types are used, the output-capacitor high-frequency noise attenuation can be very poor.

These unwanted, and hence "parasitic," ESR, ESL, and interwinding capacitance effects deserve further examination.

20.3 PARASITIC EFFECTS IN SWITCHMODE OUTPUT FILTERS

Figure 1.20.1a shows a single-stage LC output filter (such as might be found in a typical forward converter. It includes the parasitic elements C_C, R_s, ESL, and ESR.

The series inductor arm L1 shows an ideal inductor L in series with the inevitable winding resistance R_s. The parasitic distributed interwinding capacitance is included as lumped equivalent capacitor C_C.

The shunt capacitor C1 includes the effective series inductance ESL and the effective series resistance ESR.

The equivalent circuit of this network at low and medium frequencies is shown in Fig. 1.20.1b. The effect of C_C, ESL, and ESR is small at low frequencies and may be neglected. From this equivalent circuit, it is clear that the filter will be effective as a low-pass filter for the low and medium end of the frequency range.

A second equivalent circuit for high frequencies is shown in Fig. 1.20.1c. At high frequencies, the ideal inductance tends to high impedance, taking out the L–R_s arm, and the ideal capacitor C tends to zero, taking out C. Thus, the parasitic components become predominant, changing the single-stage low-pass LC filter to an effective high-pass filter. This occurs at some high frequency, where the interwinding capacitance C_C and effective series inductance ESL become predominant. Hence, this type of power output filter is not very effective in attenuating high-frequency conducted-mode noise.

20.4 TWO-STAGE FILTERS

As shown above, attempts to satisfy all the voltage averaging and noise rejection requirements in a single LC filter would require the selection of expensive components, particularly in flyback converters. Even then, only mediocre high-frequency performance would be obtained.

Figure 1.20.2 shows how a far more cost-effective wideband filter can be produced, using a second-stage, much smaller, LC filter to reject the high-frequency noise. The second stage L2, C2, may be quite small and inexpensive because only small inductance and capacitance values are required in this second stage. At the same time, much lower cost standard electrolytic capacitors and inductors may be used in the first stage (L1, C1), thus reducing the overall cost and improving the performance.

In Fig. 1.20.2, the first capacitor C1 is selected for the required ripple current

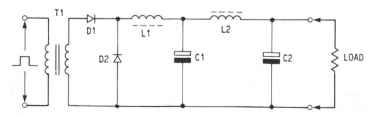

FIG. 1.20.2 Two-stage output filter.

rating and energy storage needs. (This depends on the load current and the operating frequency.) C1 will often be quite large, but does not need to be a low-ESR type when a two-stage filter is used.

The first inductor L1 is designed to carry the maximum load current with minimum loss and without saturation. To obtain the maximum inductance and minimum resistance in the smallest size, L1 will have a multiple-turn multilayer winding. Although this gives the maximum inductance, it results in a relatively large interwinding capacitance and low self-resonant frequency. Suitable core materials for L1 include gapped ferrites, Permalloy, iron-dust toroids, or gapped silicon iron in "E-I" shapes. L1 will have the majority of the inductance required for energy storage considerations.

The second inductor L2 is designed to have the maximum impedance at high frequency, and requires a low interwinding capacitance. This will provide a high self-resonant frequency. L2 may take the form of a small ferrite rod, a ferrite bobbin, small iron-dust toroids, or even an air-cored coil. Since the AC voltage across L2 is small (of the order of 500 mV), the magnetic radiation from an incomplete magnetic path will be quite small and should not present an EMI problem. Normal ferrite materials may be used for a ferrite rod inductor, as the large air gap will prevent DC saturation of the core.

The second capacitor C2 is much smaller than C1. It is selected for low impedance at the switching and noise frequencies (rather than for its energy storage ability). In many cases C2 will consist of a small electrolytic shunted by a low-inductance foil or ceramic capacitor. Since L1 and L2 conduct a large DC current component, the term "choke" is more correctly applied to these items. A design example follows.

20.5   HIGH-FREQUENCY CHOKE EXAMPLE

To get the best performance from the high-frequency choke L2, the interwinding capacitance should be minimized.

Figure 1.20.3a shows a 1-in-long ferrite rod choke with a 5/16-in diameter, wound with 15 turns of closely packed #17 AWG wire. Figure 1.20.3b shows a plot of phase shift and impedance as a function of frequency for this choke. The phase shift is zero at the self-resonant frequency, which in this case is 4.5 MHz.

The impedance plot in Fig. 1.20.3c shows the improvement obtained by reducing the interwinding capacitance. This plot was obtained from the same choke after spacing the windings and insulating them from the rod with 10-mil Mylar tape.

In this second example, 15 turns of 20-gauge wire are used, with a space between each turn. The plot shows that the reduction in interwinding capacitance has increased the impedance and shifted the self-resonant frequency to 6.5 MHz. This will result in a reduction in high-frequency noise in the final filter.

A small proportion of the high-frequency interference will bypass the filter by inductive and capacitive coupling in the pcb or supply leads. The effect of this will be reduced by fitting the smaller capacitor C2 as close as possible to the output terminals of the supply.

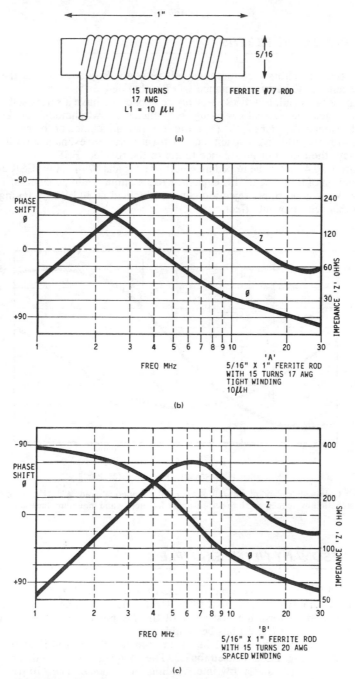

FIG. 1.20.3 (*a*) Ferrite rod choke. (*b*) Impedance and phase shift of ferrite rod choke with tight winding, as a function of frequency. Note self-resonant frequency at 4 MHz. (*c*) Impedance and phase shift of spaced winding (low interwinding capacitance) ferrite rod choke. Note self-resonant frequency at 6 MHz.

1.153

20.6   RESONANT FILTERS

By selecting capacitors such that their self-resonant frequency is near the switching frequency, the best performance will be obtained.

Many of the small, low-ESR electrolytic capacitors have a series self-resonant frequency near the typical operating frequencies of switchmode converters. At the self-resonant frequency, the parasitic internal inductance of the capacitor resonates with the effective capacitance to form a series resonant circuit. At this frequency, the capacitor impedance tends to the residual ESR.

Figure 1.20.4 shows the impedance plot of a typical 470-µF low-ESR capacitor as a function of frequency. This capacitor has a minimum impedance of 19 mΩ at 30 kHz. Very good ripple rejection can be obtained at 30 kHz by taking advantage of this self-resonant effect.

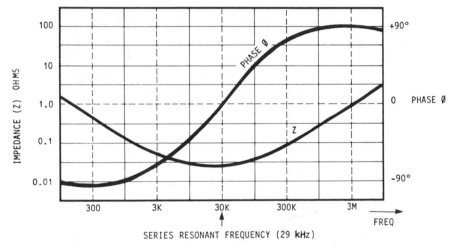

FIG. 1.20.4 Impedance and phase shift of a typical commerical-grade 470-µF electrolytic capacitor as a function of frequency. Note self-resonant frequency and minimum impedance at 29 kHz.

20.7   RESONANT FILTER EXAMPLE

Figure 1.20.5 shows a typical output stage of a small 30-kHz, 5-V, 10-A flyback converter with a two-stage output filter. (In flyback converters, the transformer inductance and C1 form the first stage of the *LC* power filter.) A second stage high-frequency filter L2, C2 has been added.

For this example, the same 1 in, 5⁄16-in-diameter ferrite rod inductor used to obtain plot *c* in Fig. 1.20.3 is used for L2. The 15 spaced turns on this rod give an inductance of 10 µH and a low interwinding capacitance. The 470-µF low-ESR capacitor used for the impedance plot in Fig. 1.20.4 is fitted in position C2.

Note: The minimum impedance of this capacitor occurs at 30 kHz, where the phase shift is zero. This is the series self-resonant frequency for this capacitor. Its im-

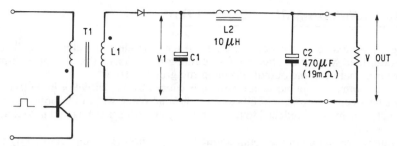

FIG. 1.20.5 Example of resonant output filter applied to a flyback converter secondary.

pedance will be predominately resistive with a value of 19 mΩ, as shown in Fig. 1.20.4.

The attenuation provided by this *LC* network at 30 kHz (the switching frequency) may now be very easily calculated, since the capacitor C2 looks predominately resistive and forms a simple divider network with the series impedance of inductor L2. (The small phase shift can be neglected, as $X_{L2} \geq$ ESR, C2.)

The ratio of the output voltage ripple (V_1) to the ripple voltage across the first capacitor C1 is

$$\frac{V_{C1}}{V_1} = \frac{\text{ESR}}{X_{L2} + \text{ESR}}$$

As $X_{L2} \geq$ ESR, the attenuation ratio A_r tends to

$$A_r = \frac{X_{L2}}{\text{ESR}}$$

where X_L = inductive reactance, $2\pi f L$
    ESR = effective series resistance of capacitor at resonance

At 30 kHz, X_L will be

$$X_L = 2\pi f l = 2\pi \times 30 \times 10^3 \times 10 \times 10^{-6} = 1.9 \ \Omega$$

From Fig. 1.10.4, the ESR of C2 at 30 kHz is 0.019 Ω. Hence, the attenuation ratio A_r will be

$$A_r = \frac{1.9}{0.019} = 1:100 \text{ ratio}$$

This gives a ripple rejection ratio of 100:1 at the switching frequency.

The switching frequency ripple is normally the predominant ripple component in flyback converters. By making use of the self-resonant properties of the electrolytic capacitor, an extremely good ripple rejection of 40 dB is obtained with very small, low-cost components. Further, the improved high-frequency noise rejection is obtained without compromise to the medium-frequency transient response, because the series inductance has not been increased significantly.

20.8   COMMON-MODE NOISE FILTERS

The discussion so far has been confined to series-mode conducted noise. The filter described will not be effective for common-mode noise, that is, noise voltages appearing between the output lines and the ground plane.

The common-mode noise component is caused by capacitive or inductive coupling between the power circuits and the ground plane within the power supply. Initially this must be reduced to a minimum by correct screening and layout at the design stage.

Further reduction of the common-mode output noise may be obtained by splitting inductor L1 or L2 into two parts to form a balanced filter, as shown in Fig. 1.20.6. Additional capacitors C3 and C4 are then required between each output line and the ground plane to provide a return path for the residual common-mode noise current. In effect, L1(a) and C3 form a low-pass filter from the positive output, and L1(b) and C4 form the filter for the negative output, with the ground plane as the return path.

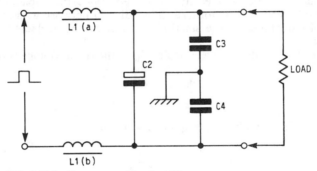

FIG. 1.20.6   Common-mode output filter.

Because of the decoupling provided by the much larger capacitor C2, acceptable results will often be obtained by fitting a single common-mode decoupling capacitor in position C3 or C4.

20.9   SELECTING COMPONENT VALUES FOR OUTPUT FILTERS

The size and value of the main output inductor L1 and storage capacitor C1 (Fig. 1.20.1a) depends on a number of factors:

Type of converter

Operating frequency

Maximum load current

Minimum load current

Mark space ratio (duty cycle)

Ripple current

Ripple voltage

Transient response

Output voltage

The requirements of L1 will now be considered in terms of the requirement for this type of converter.

20.10 MAIN OUTPUT INDUCTOR VALUES (BUCK REGULATORS)

In general, the main inductance L1 in the output of a buck regulator filter circuit should be as small as possible to give the best transient response and minimum cost. If a large inductance is used, then the power supply cannot respond rapidly to changes in load current. At the other extreme, too low an inductance will result in very large ripple currents in the output components and converter circuits which will degrade the efficiency. Further, discontinuous operation will occur at light loads.

One approach is to select L1 such that the inductor will remain in continuous conduction for the minimum load current (often specified as 10% of I_{max}).

Keeping the inductance in continuous conduction has two advantages. First, the control circuit is only required to make small changes in pulse width to control the output voltage as the load changes (provided the inductor remains in conduction throughout the operating cycle). Second, the output ripple voltage will remain small over this range of load changes.

The main disadvantage of this approach is that the inductance can be quite large; moreover, the rule cannot be used if the load current must be controlled right down to zero.

A second, more universal rule is to choose the inductance value such that the ripple current has an acceptable peak-to-peak limit, say 10% to 30% of the maximum load current at nominal input voltages.

Note: In flyback converters, the main inductance L1 is integral to the transformer, and its value is defined by the power transfer requirements. In this type of converter, high ripple currents must be accommodated in the filter components, particularly for complete energy transfer systems.

20.11 DESIGN EXAMPLE

Assume that a design is required for the main output inductor L1 for a single-ended forward converter and filter, as shown in Fig. 1.20.1a. The specification for the converter is as follows:

Output power = 100 W

Output voltage = 5 V

Output current = 20 A

Operating frequency = 30 kHz

Minimum load = 20%

The design approach will assume that the output ripple current must not exceed 30% of I_{load} (6 A p-p in this example).

Also, to allow for a range of control, the pulse width at nominal input will be 30% of the total period (that is 10 μs).

To provide an output of 5 V at a pulse width of 30%, the transformer secondary voltage will be

$$V_s = \frac{V_{out} \times t_p}{t_{on}} = \frac{5 \times 33.33}{10} = 16.66 \text{ V}$$

where t_p = total period (at 30 kHz), μs
$\quad t_{on}$ = "on" time, μs
$\quad V_s$ = secondary voltage

The voltage V_L across the inductor L1 during the forward "on" period is the secondary voltage less the output voltage, assuming that the output capacitor C1 is large and the voltage change during the "on" period is negligible.

Then

$$V_L = V_s - V_{out} = 16.66 - 5 = 11.66 \text{ V}$$

For steady-state conditions, the current change for the "on" period must equal the current change during the "off" period (in this example, 6 A). Neglecting second-order effects, the inductance may be calculated as follows:

$$L = \frac{V_L \times \Delta t}{\Delta i}$$

where L = required inductance, μH
$\quad \Delta t$ = "on" time, μs
$\quad \Delta i$ = current change during "on" time
$\quad V_L$ = voltage across inductor

Therefore

$$L = 11.66 \times \frac{10}{6} = 19.4 \text{ μH}$$

Note: A simple linear equation can be used, as the voltage across the inductance is assumed not to change during the "on" time and di/dt is constant.

In this example, the inductance is large because sufficient energy must be stored during the "on" period to maintain the current during the "off" period. In push-pull forward converters the "off" period is much smaller, so that the secondary voltage and hence the inductance value would also be smaller.

20.12 OUTPUT CAPACITOR VALUE

It is normally assumed that the output capacitor size will be determined by the ripple current and ripple voltage specifications only. However, if a second-stage output filter L2, C2 is used, a much higher ripple voltage could be tolerated at the terminals of C1 without compromising the output ripple specification. Hence if ripple voltage were the only criterion , a much smaller capacitor could be used.

For example, assume that the ripple voltage at the terminals of C1 can be 500 mV. The current change in L1 during the "on" period will mainly flow into C1, and hence the capacitance value required to give a voltage change of 500 mV can be calculated as follows (the following equation assumes a perfect capacitor with zero ESR):

$$C = \frac{\Delta I \times t_{on}}{\Delta V_o}$$

where C = output capacitance value, μF
    ΔI = current change in L1 during "on" period, A
    t_{on} = "on" time, μs
    ΔV_o = ripple voltage, V p-p

Therefore

$$C = \frac{6 \times 10}{0.5} = 120 \ \mu F$$

Hence, just to meet the ripple voltage requirements, a very small capacitor of only 120 μF would be required. However, in applications in which the load current can change rapidly over a large range (transient load variations), a second transient load variation criterion may define the minimum output capacitor size.

Consider the condition when the load suddenly falls to zero after a period of maximum load. Even if the control circuit responds immediately, the energy stored in the series inductor ($\frac{1}{2} LI^2$) must be transferred to the output capacitor, increasing its terminal voltage. In the above example, with an output capacitor of only 120 μF, a series inductance of 19.4 μH, and a full-load current of 20 A, the voltage overshoot on load removal would be nearly 100%. This would probably be unacceptable, and hence the maximum acceptable voltage overshoot on load removal may become the controlling factor.

The minimum output capacitor value to meet the voltage overshoot requirements using the transferred energy criteria can be calculated as follows:

Energy in output inductor when full load is suddenly removed:

$$\frac{1}{2} LI^2$$

The energy change in the output capacitor after the event will be

$$\frac{1}{2} C(V_P)^2 - \frac{1}{2} C(V_o)^2$$

where V_P = maximum output voltage = 6 V
    V_o = normal output voltage = 5 V

Hence

$$\frac{1}{2} LI^2 = \frac{1}{2} C(V_P^2 - V_o^2)$$

Rearranging for C,

$$C = \frac{LI^2}{V_P^2 - V_o^2}$$

If the maximum voltage in this example is not to exceed 6 V, then the minimum value of output capacitance will be

$$C = \frac{19.5 \times (20)^2}{36 - 25} = 709 \ \mu F$$

Further, the ripple current requirements may demand that a larger capacitor be used. Some allowance should also be made for the effects of the capacitor ESR, which will increase the ripple voltage by about 20% typically, depending on the ESR and ESL of the capacitor and the size, shape, and frequency of the ripple current (Part 3, Chap. 12).

In conclusion, it has been shown that very effective series- and common-mode conducted ripple rejection can be obtained by the addition of a relatively small additional *LC* output filter network. This relatively simple change allows good ripple and noise rejection to be obtained using lower-cost medium-grade electrolytic capacitors and conventional inductor designs.

20.13 PROBLEMS

1. Discuss the major disadvantage of switchmode power supplies compared with the older linear regulator types.
2. Is the design of the output filter the only *most* important factor in reducing output ripple noise?
3. Explain the meaning of the term "choke" as applied to output filters.
4. Why are power output filters often relatively ineffective in dealing with high-frequency noise?
5. Why are two-stage filters sometimes used in output filter applications?
6. What is the difference between common-mode and differential-mode noise filters?
7. In what way does the design of a common-mode choke differ from that of a series-mode choke?

CHAPTER 21
POWER FAILURE WARNING CIRCUITS

21.1 INTRODUCTION

Many instrument and computer systems require early warning of imminent power failure, to provide sufficient time for an organized system shutdown. To maintain the output voltages above the minimum specified values during this "housekeeping" process, sufficient energy must be stored in the power supply. A minimum holdup time (after power failure warning) of between 2 and 10 ms is usually specified.

21.2 POWER FAILURE AND BROWNOUT

Line failure can, of course, take many forms, but it will normally fall into one of the following three categories.

1. *Total Line Failure:* Instantaneous and catastrophic failure to zero or near zero voltage.
2. *Partial Brownout:* A fall in line voltage to a value below the normal minimum (but not zero), followed by a recovery to normal.
3. *Brownout failure:* A brownout condition followed by eventual failure.

21.3 SIMPLE POWER FAILURE WARNING CIRCUITS

Figure 1.21.1 shows a simple optically coupled circuit typical of those often used for power failure warning. However, it will be shown that this type of circuit is suitable only for type 1 failures, that is, total line failure conditions. It operates as follows.

The ac line input is applied to the network R1 and bridge rectifier D1 such that unidirectional current pulses flow in the optical coupler diode. This maintains a pulsating conduction of the optical coupler transistor Q1. While this pulsating

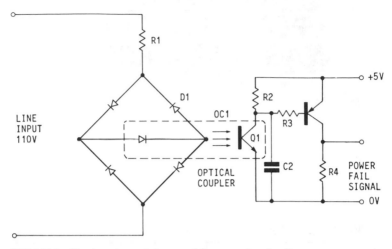

FIG 1.21.1 Simple optocoupled power failure warning circuit.

condition continues, C2 will be "pumped" low and maintain Q2 "on." Hence the output power failure signal will remain high all the time the ac supply voltage is high enough to drive current into D1.

When the ac line input fails, D1 no longer provides current to OC1, and Q1 turns off. C2 will charge via R2, and Q2 turns off. The power failure signal then goes low.

Because this circuit does not have a defined threshold voltage, it will give the required advance warning correctly only for condition 1 (a complete or nearly complete line failure). It will not necessarily give the advance warning correctly for condition 2 or 3 because during brownout the voltage may still be high enough to maintain D1 conducting. Further, there is a delay between line failure and a warning signal as C2 charges.

During a line failure, the energy stored in the power supply will maintain the output voltage for a time period that depends on the input voltage prior to failure, the part of the cycle in which the failure occurs, the loading conditions, and the design of the supply. This holdup time can be more, but must not be less, than the power failure warning period required plus the delay period of the warning circuit.

With the simple circuit of Fig. 1.21.1, during a brownout condition as specified for condition 2, the voltage may fall low enough for power supply output regulation to be lost, but not low enough for a power failure signal to be given. For brownout (condition 3), even if the power supply maintains the required output voltages, at the end of the brownout period, when the line voltage eventually fails, the circuit will respond with a failure signal, and there will not be sufficient energy remaining in the power supply to maintain the output voltage for the prescribed warning period. Hence this type of circuit is not always fully satisfactory for brownout conditions.

Since brownout conditions occur most often, the simple type of power failure warning shown in Fig. 1.21.1, although often used, may be of little value.

21.4 DYNAMIC POWER FAILURE WARNING CIRCUITS

The more complex dynamic power failure warning circuits are able to respond to brownout conditions. Many types of circuit are in use, and it may be useful to examine some of the advantages and disadvantages of some of the more common techniques.

Figures 1.21.2 and 1.21.3 show two circuits that will ensure that sufficient warning of failure is given for all conditions.

In the first example, a fraction of the DC voltage on the power converter reservoir capacitors C1 and C2 is compared with a reference voltage by comparator amplifier A1. If this voltage falls to a value at which the power supply (if it were operating at full load) would only just provide the prescribed hold time, then the

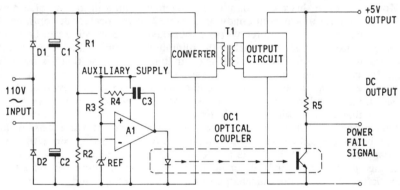

FIG. 1.21.2 More precise "brownout" power failure warning circuits.

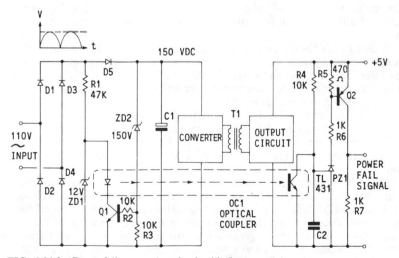

FIG. 1.21.3 Power failure warning circuit with "brownout" detection.

output of amplifier A1 goes high, energizing the optical coupler, and a failure warning will be generated; the output signal goes low ("true low" logic).

This is a well-defined and reliable warning system, but it requires that the power supply be designed to provide sufficient holdup time from a minimum defined input voltage (below the normal minimum working voltage), to ensure that the specified warning period is satisfied before the output voltage falls. To meet this need, a larger and more expensive supply is required for the following reasons.

Since a warning must not be given at or above the specified minimum working line voltage for the power supply, the selected warning voltage value must be lower than the minimum DC voltage normally found on C1 and C2 under fully loaded minimum line input voltage conditions.

To provide the required holdup time under fully loaded conditions, from this lower capacitor voltage, the converter must continue to give full output for a supply voltage which is even lower than normal; hence larger reservoir capacitors and larger-current-rated input components will be required. This makes the power supply larger and more expensive.

Moreover, even this more complex arrangement can still give a false power failure warning for a brownout condition of type 2. If the brownout continues for a period and then the supply recovers, a spurious failure warning can be caused by the capacitor voltage falling below the minimum warning value before the line recovers, initiating a failure signal. It is clear that in this case there is no option but to indicate a failure signal when the stored energy on the storage capacitors reaches the critical value. Although the line may recover before eventual failure of the outputs, a failure signal must be given at this time because the system cannot know that the line will recover in time.

The arrangement has the advantage that short transient variations in input voltage below the critical limit will not cause a failure warning, since the reservoir capacitors will not discharge to the critical voltage very rapidly. A further advantage is that at lower loads or higher input voltages, there will be a longer delay before the capacitors discharge to the critical voltage and a power failure warning is generated.

This system provides the maximum rejection of input transient conditions, eliminating spurious and unnecessary failure warnings. The delay time adjusts "dynamically" in response to the loading and input voltage conditions; hence the name.

Figure 1.21.3 shows a circuit that has advantages similar to those of the previous dynamic system, but does not require an auxiliary supply or comparator amplifier. This circuit can be used in the rectified supply to the main converter, as shown, or, with appropriate component adjustments, in the supply to the auxiliary converter. It operates as follows.

The bridge rectifier D1–D4 will provide a unidirectional half-sine-wave input to the divider chain R1, ZD1. At the same time, this input is applied to diode D5.

The peak ac input voltage is rectified by D5 and stored on capacitor C1. The DC voltage on C1 is monitored via ZD2 and Q1 and would normally bias Q1 "on".

The rectifier diode D5 blocks the DC voltage on C1 and allows the voltage across R1, OC1, and Q1 to fall to zero each half cycle; that is, the voltage across R1, OC1, and Q1 follows the input voltage. Hence, OC1 must turn off for a short period each cycle, even if the DC voltage on C1 is high and Q1 is on.

A failure warning will be given if OC1 turns off for more than 3 ms. This occurs if the voltage on C1 falls to a value at which Q1 and hence OC1 turns off; this critical voltage is defined by ZD1. Also, if the input supply fails completely for more than 3 ms, a power failure warning will be given irrespective of the state of charge on C1 (the main reservoir capacitor in the supply). In this case OC1 is off because the supply to R1 and OC1 is missing if the input supply is missing.

As long as the voltage on C1 is above the minimum value required to give the required minimum holdup time, the zener diode ZD2 will be conducting and Q1 will be on. During each half cycle, when the supply voltage to R1 exceeds a few volts, OC1 will turn on, providing a discharge pulse to C2 and preventing C2 from charging to the 2.5-V reference voltage PZ1. (At 2.5 V, PZ1 and Q2 would turn on, giving a fail signal.) This discharge "pumping" action will continue as long as the supply voltage on C1 is above the minimum value and the supply does not fail.

If the line input fails or the voltage on C1 falls below the minimum value required to maintain ZD2 conducting (brownout), Q1 and OC1 will remain off and the pulse discharge of C2 will stop. C2 will now charge, turning PZ1 and Q1 on and giving a power failure warning. This warning will be given if OC1 is off for more than 3 ms. The delay period is well defined. When OC1 is off, C2 charges via R4 until the threshold voltage of PZ1 is reached (2.5 V). At this voltage PZ1 conducts, turning on Q2 and generating a power failure "high" signal.

If the line input fails, even if C1 remains charged and Q1 remains on, there is no supply to R1 and OC1 and a failure indication is given. This fast response provides and earlier warning of line failure so that the power supply holdup time need not be so long.

21.5 INDEPENDENT POWER FAILURE WARNING MODULE

The previous two power failure circuits must be part of the power supply, as they depend on the internal DC header voltage for their operation. Figure 1.21.4 shows a circuit that will operate directly from the line input and is independent of any power supply.

This circuit has its own bridge rectifier D1–D4, which again provides a unidirectional half-sine-wave input to the feed resistor R1, ZD1, and the optical cou-

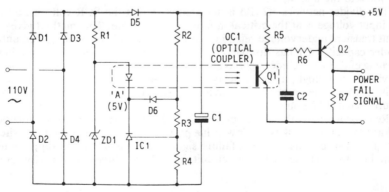

FIG. 1.21.4 Independent power failure module for direct operation from ac line inputs.

pler diode OC1 and IC1. Provided that the DC voltage on C1 is above the critical minimum value, IC1 (a TL431 shunt regulator IC) will be turned on, regulating the voltage at point A at 5 V. (Note: Diode D6 conducts, clamping the voltage across R3, R4 and maintaining point A at 5 V.)

When the rectified input to R1 rises above 5 V, during each half cycle, the OC1 diode will conduct, turning the OC1 transistor Q1 on and providing a discharge pulse to C2. This "pumping" action prevents C2 from charging; R5 and R6 will be conducting, and Q2 will be on. The output warning signal will remain "high," in this case the normal power good indication state.

As before, a failure (low) signal will be given if OC1 is off for more than 3 ms, allowing C2 to charge. This occurs if the voltage on C1 falls below the critical value required to maintain IC1 "on" or if the line input fails.

This circuit is more precise than the previous systems, with a better temperature coefficient. The shunt regulator IC1 has a more precise internal voltage reference. Otherwise the function is similar to that shown for Fig. 1.21.3.

The time constant for the divider network R2, R3, R4, and C1 should be much less than the discharge time constant of power supply primary capacitors, to ensure that a warning is given for brownout conditions before the power supply drops out of regulation.

21.6   POWER FAILURE WARNING IN FLYBACK CONVERTERS

Very simple power failure warning circuits can be fitted to flyback converters, because in the forward direction the flyback transformer is a true transformer, providing an isolated and transformed output voltage which is proportional to the applied DC.

Figure 1.21.5 shows the power section of a simple single-output flyback supply providing a 5-V output. Diode D1 conducts in the flyback mode of T1 to charge C2 and deliver the required 5-V output. The control circuit adjusts the duty cycle in the normal way to maintain the output voltage constant.

An extra diode and capacitor D2, C3 have been added such that D2 conducts in the forward mode of T1, developing a voltage V_f on C3 of V_p/n, which is proportional to the line input.

The divider network R2, R3 is selected such that the SCR will turn on when the input voltage is at the critical minimum value. Note: This method gives good input transient undervoltage rejection, as a warning will not be generated until the header capacitor C1 has discharged to the critical value required for minimum warning of dropout. Under light loading conditions, or when the input voltage has previously been high, a longer delay is provided.

In this example an option is provided for a "true high" or "true low" power failure signal (PFS) output.

Resistor R1 limits the charge current into C3 and prevents peak rectification of leakage inductance spikes. Its low value prevents any race condition at switch-on and gives fast response. After a failure signal has been given, the supply must be turned off to reset the SCR. The circuit is simple but gives good performance.

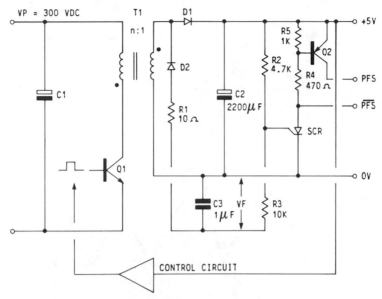

FIG. 1.21.5 A simple power failure warning circuit for flyback converters.

21.7 FAST POWER FAILURE WARNING CIRCUITS

The previous systems shown in this section respond quite slowly to brownout conditions, because they are sensing peak or mean voltages. The filter capacitor in the warning circuit introduces a delay. Its value is a compromise, being low enough to prevent a race between the holdup time of the power supply and the time constant of the filter capacitor, but large enough to give acceptable ripple voltage reduction.

It is possible to detect the imminent failure of the line before this has fully developed by looking directly at the rectified line input. The circuit can respond to the reduction in the dv/dt (rate of change of input voltage), which occurs at the beginning of a half cycle of operation if the peak voltage is going to be low. Hence the system is able to give more advanced warning of impending low-voltage conditions.

The circuit recognizes very early that the rate of change of input voltage is below the value necessary to generate the correct peak ac voltage. If the dv/dt as the supply passes through zero is low, failure is assumed, and a warning signal is generated before the half cycle is complete. This provides a useful extra few milliseconds of warning.

Figure 1.21.6a shows a simulated line brownout characteristic, in which the applied sine-wave input suffers a sudden reduction in voltage on the second cycle.

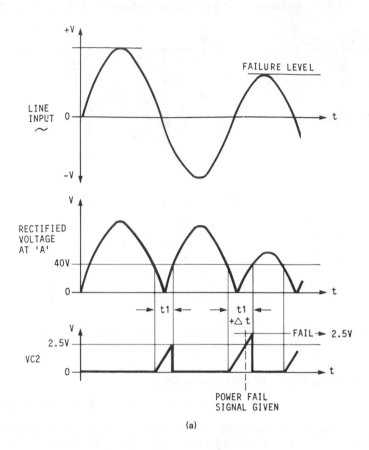

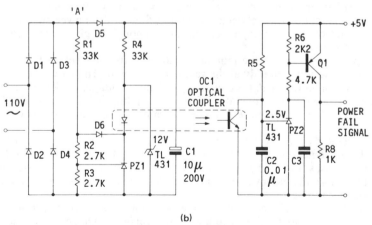

FIG. 1.21.6 (*a*) "Brownout" ac line voltage waveforms, showing "optimum speed" circuit action. (*b*) "Optimum speed" power failure warning circuit for direct operation from ac line inputs.

When the rectified waveform is compared to a reference voltage, the change in supply voltage shows up as an increase in the time Δt taken for the rectified voltage to exceed the reference value. This change can be used to indicate a probable failure before the full half cycle has been established. This method gives the earliest possible warning of power brownout or failure. Figure 1.21.6b shows a suitable circuit.

This circuit operates as follows. The line input is bridge-rectified by diodes D1 through D4. A divider network of resistors R1, R2, and R3 is placed across the bridge output, and this load ensures a clean rectified half-cycle waveform at point A, as shown in Fig. 1.21.6a. This waveform is applied to the input of the comparator amplifier of PZ1 by the network R1, R2, and R3. As the supply voltage at A passes through 50 V during the falling second half of a half cycle, the voltage applied to PZ1 passes through 2.5 V and the shunt regulator PZ1 and optical coupler OC1 will turn off.

This starts a timing sequence on C2 such that unless the supply voltage rises through 50 V once again during the next positive-going edge of a half cycle within a prescribed time, then PZ2 and Q1 are turned on, giving a power failure signal.

The timing is defined by C2, R5, and the secondary voltage (5 V in this example). During each half cycle, OC1 turns off and C2 will be charged from the time the input supply falls below 50 V to the time it returns above 50 V. If the "off" time of OC1 gets longer (as would be the case for a low input voltage, as shown in Fig. 1.21.6a), the voltage ramp across C2 will exceed 2.5 V, and PZ2 will be turned on. Q1 then indicates a power failure. An optocoupler is incorporated to isolate the sensing circuit from the output signal.

In this circuit the operating voltage is well defined. It may be adjusted so that a failure will be indicated only for line voltage variations which fall below the critical value required to provide the power supply holdup time.

The circuit is very fast and will give a brownout power failure warning within 1 to 8 ms, depending on where in a cycle a failure occurs.

21.8 PROBLEMS

1. Explain the purpose of a power failure warning circuit.
2. How is a power failure warning signal developed in a flyback switchmode supply?
3. What is meant by brownout power failure warning?
4. Describe the principle employed in a fast power failure warning circuit.

CHAPTER 22
CENTERING (ADJUSTMENT TO CENTER) OF AUXILIARY OUTPUT VOLTAGES ON MULTIPLE-OUTPUT CONVERTERS

22.1 INTRODUCTION

When more than one winding is used on a converter transformer to provide auxiliary outputs, a problem can sometimes arise in obtaining the correct output voltages. Because the transformer turns can only be adjusted in increments of one turn (or in some cases a half turn; see Part 3, Chap. 4) it may not be possible to get exact voltages on all outputs.

When output auxiliary regulators (often three-terminal series regulators) are to be used, the secondary output voltage error is generally not a problem. However, in many cases additional regulation is not provided, and it is desirable to "center" the output voltage (set it to an absolute value).

The following method describes a way of achieving this voltage adjustment in a loss-free manner, using small saturable reactors.

22.2 EXAMPLE

Consider the triple-output forward-converter secondary circuit shown in Fig. 1.22.1. Assume that the 5-V output is a closed-loop regulated output, fully stabilized and adjusted.

There are two auxiliary 12-V outputs, positive and negative, which are now semiregulated as a result of the closed-loop control on the 5-V line. Assume that the regulation performance required from the 12-V outputs is such that additional series regulators would not normally be required (say, ±6%).

Further assume that to obtain 12 V out, the transformer in this example re-

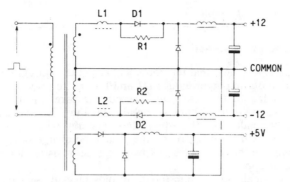

FIG. 1.22.1 Saturating-core "centering inductors" applied to a Multiple-output push-pull converter.

quires 11.5 turns and the half turn is not possible for flux balancing reasons. If 12 turns are used on the transformer, the output voltage on the 12-V lines will be high by approximately 0.7 V. (Remember, this output is obtained with a predefined pulse width which was set by the main control loop for the 5-V output.) Assume also that under these conditions, the pulse width is 15 μs on and 18 μs off, giving a total period of 33 μs.

It is not possible to reduce the overall pulse width to obtain the correct output on the 12-V lines, as this will also reduce the 5-V output. If, on the other hand, the pulse width to the 12-V outputs could be reduced without changing the pulse width to the 5-V line, then it would be possible to produce the required output voltage on all lines. It is possible to achieve this with a saturable reactor.

22.3 SATURABLE REACTOR VOLTAGE ADJUSTMENT

Consider the effect of placing a saturable reactor toroid (as described in Part 2, Chap. 21) on the output lines from the transformer to the 12-V rectifiers D1 and D2.

These reactors L1 and L2 are selected and designed so that they take a time-delay period t_d to saturate, specified by

$$t_d = t_{on} - \frac{\text{required } V_{out} \times t_{on}}{\text{actual } V_{out}} \ \mu s$$

In this case,

$$t_d = 15 - \frac{12 \times 15}{12.7} = 0.827 \ \mu s$$

The extra time delay t_d is introduced on the leading edge of the output power pulse by the saturable reactor, and the 12.7-V output would be adjusted back to 12 V. This action is more fully explained in Part 2, Chap. 21.

It remains only to design the reactors to obtain the above conditions.

22.4 REACTOR DESIGN

Step 1, Selection of Material

From Fig. 1.22.1, it is clear that the cores will be set to saturation during the forward conduction of the output diodes D1 and D2 and to provide the same delay on the leading edge for the next "on" period, the cores must reset during the "off" period. When D1 and D2 are not conducting the "flywheel diodes" D3 and D4 are normally conducting. If a square loop material with a low remanence is chosen, the cores will often self-reset, the recovered charge of D1 and D2 being sufficient to provide the reset action. However, reset resistors R1 and R2 may be required in some applications.

A number of small square-loop ferrite toroids meet these requirements, and the TDK H5B2 material in a toroidal form is chosen for this example.

Step 2, Obtaining the Correct Delay Time

Prior to saturation, the wound toroid will conduct only magnetization current and, therefore, will be considered in its "off" state.

The time taken for the core to saturate when the "on" period starts (diodes forward-biased) will depend on the applied voltage, the number of turns, the required flux density excursion, and the area of the core, as defined by the following equation:

$$t_d = \frac{N_P \times \Delta B \times A_e}{V_s}$$

where t_d = required time delay, μs
$\quad N_P$ = turns
$\quad \Delta B$ = change in flux density from B_r to B_{sat}, T
$\quad B_r$ = flux remanence at $H = 0$
$\quad B_s$ = flux density at saturation, T
$\quad A_e$ = effective area of core, mm^2
$\quad V_s$ = secondary voltage, V

In this example, the secondary voltage V_s applied to the core at the start of the "on" period may be calculated from the duty ratio and the output voltage as follows:

$$V_s = \frac{V_{out}(t_{on} + t_{off})}{t_{on}}$$

where V_{out} = required output voltage, V
$\quad t_{on}$ = "on" period, μs
$\quad t_{off}$ = "off" period, μs

In this example,

$$V_s = \frac{12.7(15 + 18)}{15} = 27.9 \text{ V}$$

There are now two variables available for final voltage adjustments: turns and core area. Assume, for convenience, that a single primary turn is to be used; that is, the output wire from the transformer is simply passed through the toroid. There is now only one variable, the core area, and the required core cross-sectional area may be calculated as follows:

$$A_e = \frac{V_s \times t_d}{N \times \Delta B}$$

In this example,

$$A_e = \frac{27.9 \times 0.827}{1 \times 0.4} = 57.7 \text{ mm}^2$$

This is a relatively large core, and for economy in low-current applications more primary turns may be used. For example, 5 turns on the primary and a core of ⅕ of the previous area will give the same delay. The area would now be $A_e = 11.4$ mm^2, and a TDK T7-14-3.5 or similar toroid would be suitable.

It may be necessary to fit a resistor (R1, R2) across the rectifier diodes D1 and D2 to allow full restoration of the core during the "off" period, as the leakage current and recovered charge from D1 and D2 may not be sufficient to guarantee full recovery of the core during the nonconducting (reverse-voltage) period.

Note: This method of voltage adjustment will hold only for loads exceeding the magnetizing current of the saturable reactor; hence the voltage tends to rise at light loads. Where control is required to a very low current, it is better to use a small, high-permeability core with more turns, as the inductance increases as N^2 while the delay is proportional to N (giving lower magnetization current and control to lower currents).

A further advantage of the saturable reactor used in this way is that it reduces the rectifier diode reverse recovery current, an important advantage in high-frequency forward and continuous-mode flyback converters.

22.5 PROBLEMS

1. What is meant by the term "centering" as applied to multiple-output converters?

2. Why is centering sometimes required in multiple-output applications?

3. Describe a method of nondissipative voltage centering commonly used in ratio-controlled converters.

4. Explain how saturable reactors L1 and L2 in Fig. 1.22.1 reduce the output voltages of the 12-V outputs.

5. Assume that the single-ended forward converter shown in Fig. 1.22.1 gives the required 5-V output when the duty ratio is 40% at a frequency of 25 kHz. The 5-V secondary has 3 turns, the 12-V secondaries have 9 turns each, and the rectifier drop is 0.7 V.

   If L1 and L2 have 3 turns on a T8-16-4 H5B2 toroid core (see Fig. 2.15.4 and Table 2.15.1), calculate the output voltage with and without L1 and L2). Is there a better turns selection for 12 V?

CHAPTER 23
AUXILIARY SUPPLY SYSTEMS

23.1 INTRODUCTION

Very often, an auxiliary power supply will be required, to provide power for control and drive circuits within the main switchmode unit.

Depending on the chosen design approach, the auxiliary supply will be common to either input or output lines, or in some cases will be completely isolated. A number of ways of meeting these auxiliary requirements are outlined in the following sections.

The method chosen to provide the auxiliary needs should be considered very carefully, as this choice will often define the overall design strategy. For example, in "off-line" supplies, if the internal auxiliary supply to the control and drive circuits is common to the input line, then some method is required to isolate the control signal developed at the output from the high-voltage input. Often optical couplers or transformers will be used for this purpose.

Alternatively, if the internal auxiliary supply is common to the output circuit, then the drive transformer to the power transistors may be required to provide the isolation. For this application, it must meet the creepage distance and isolation requirements for the various safety specifications. This makes the design of the drive transformers more difficult.

When the specification requires power good and power failure signals or remote control functions, it may be necessary to have auxiliary power even when the main converter is not operating. For these applications, impulse start techniques and auxiliary supply methods that require the power converter to be operating would not be suitable. Hence, all the ancillary requirements must be considered before choosing the auxiliary supply method.

23.2 60-Hz LINE TRANSFORMERS

Very often small 60-Hz transformers will be used to develop the required auxiliary power. Although this may be convenient, as it allows the auxiliary circuits to be energized before the main converter, the 60-Hz transformer tends to be rather large, as it must be designed to meet the insulation and creepage requirements of the various safety specifications. Hence, the size, cost, and weight of a 60-Hz auxiliary supply transformer tends to make it less attractive for the smaller switchmode applications.

In larger power systems, where the auxiliary transformer size would not have a very dramatic effect on the overall size and cost of the supply, the 60-Hz transformer can be an expedient choice.

Some advantages of the transformer approach are that fully isolated auxiliaries can easily be provided. Hence, the control circuitry may be connected to input or output lines, and the need for further isolation may be eliminated. Further, the auxiliary supply is available even when the main switching converter is not operating.

23.3 AUXILIARY CONVERTERS

Very small, lightweight auxiliary power supplies can be made using self-oscillating high-frequency flyback converters. The output windings on the converter can be completely isolated and provide both input and output auxiliary needs, in the same way as the previous 60-Hz transformers.

Because auxiliary power requirements are usually very small (5 W or less), extremely small and simple converters can be used. A typical example of a nonregulated auxiliary converter is shown in Fig. 1.23.1. In this circuit, a self-oscillating flyback converter operates from the 150-V center tap of the voltage doubler in the high-voltage DC supply to the main converter.

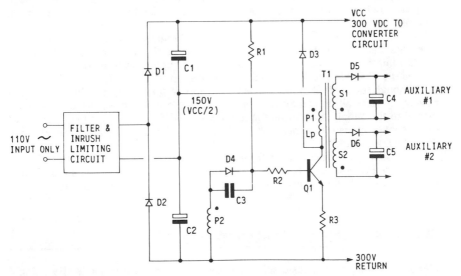

FIG. 1.23.1 Auxiliary power supply converter of the single-transformer, self-oscillating flyback type, with energy recovery catch diode D3.

23.4 OPERATING PRINCIPLES

Initially, Q1 starts to turn on as a result of the base drive current in resistors R1 and R2. As soon as Q1 starts to turn on, regenerative feedback via winding P2 will assist the turn-on action of the transistor, which will now latch to an "on" state.

With Q1 on, current will build up linearly in the primary winding at a rate defined by the primary inductance and applied voltage ($dI/dt \propto V_{cc}/L_P$). As the current builds up in the collector and emitter of Q1, the voltage across R3 will increase. The voltage on the base of Q1 will track the emitter voltage (plus V_{be}), and when the base voltage approaches the voltage developed across the feedback winding P2, the current in R2 will fall toward zero, and Q1 will start to turn off.

Regenerative feedback from P2 will now reverse the base drive voltage, turning Q1 off more rapidly. By flyback action, the collector of Q1 will fly positive until the clamp diode D3 is brought into conduction. Flyback action will continue until most of the energy stored in the transformer is returned to the 300-V supply line.

However, at the same time, a small amount of energy will be transferred to the output via D5 and D6. Because D3 conducts throughout the flyback period, and the same primary winding is used for both forward and flyback actions, the flyback voltage will be equal to the forward voltage, and the output voltage will be defined by the supply voltage. Also, the flyback period will be the same as the forward period, giving a 50% mark space ratio, that is, a square-wave output.

The inductance of the transformer primary should be chosen by gapping the core such that the stored energy at the end of an on period $\frac{1}{2} LI_P^2$ is at least three or four times greater than that required for the auxiliary outputs, so that clamping diode D3 will always be brought into conduction during the flyback period. This way, the secondary flyback voltage will be defined by the turns ratio and the primary voltage. This can be an advantage when the auxiliary voltage is to be used for power failure/power good indications and to provide low-input-voltage inhibit actions.

As the primary turns are very large (300 to 500 turns typically), there is a considerable distributed interwinding capacitance in the primary. The relatively small primary inductance, given by the large air gap in the transformer core, improves the switching action in these small converters.

Although the current in the primary may appear quite large for the transmitted power, the overall efficiency remains high, as the majority of the energy is returned to the supply line during the flyback period. The mean off-load current of these converters is often only 2 or 3 mA, although the peak primary current may be as high as 50 mA.

Because a DC current is taken from the center tap of the input capacitors C1 and C2, this simple converter is suitable only for voltage doubler applications. If full-wave input rectification is used, a DC restoration resistor is required across C1.

23.5 STABILIZED AUXILIARY CONVERTERS

Many variations of this basic self-oscillating converter are possible. By using a high-voltage zener on the input side, it is possible to provide stabilized auxiliary outputs and also maintain a constant operating frequency.

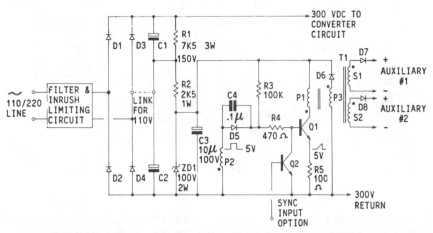

FIG. 1.23.2 Stabilized auxiliary power converter of the self-oscillating flyback type, with energy recovery winding P3 and synchronization input Q2.

Figure 1.23.2 shows one such modification to the basic circuit. This circuit is also more suitable for dual-voltage operation, as the flyback energy is returned to the same input line as the primary load.

The input voltage is stabilized by ZD1 and gives constant-frequency operation. This auxiliary converter may be used as the basic clock for the control circuit, providing drive directly to the power switching transistors. Very simple and effective switchmode supplies may be designed using this principle.

An energy recovery winding P3 and diode D6 have been added to the transformer so that the spare flyback energy is returned to the same supply capacitor as the primary winding P1 (C3). This makes the mean loading current on ZD1 very low, allowing simple and efficient zener diode preregulation. Both forward and flyback voltages are regulated by ZD1, providing a regulated flyback voltage and hence regulated outputs. It is important to use bifilar windings for P1 and P3 and to fit the energy recovery diode D6 in the top end of the flyback winding P3. In this position it isolates the collector of the switching transistor from the interwinding capacitance in T1 during the turn-on edge of Q1.

The DC supply to the auxiliary converter is taken from the main 300-V line via R1 and R2, which have been selected so that input link changes for dual input voltage operation will not affect the operating conditions of ZD1. Further, an extra transistor Q2 has been added to the base of the converter transistor Q1 to permit external synchronization of the converter frequency.

It should be noted that the frequency may only be synchronized to a higher value, as Q2 can terminate an "on" period early, but cannot extend an "on" period. Turning on Q2 results in immediate flyback action for each sync pulse, giving higher-frequency operation.

23.6 HIGH-EFFICIENCY AUXILIARY SUPPLIES

Figure 1.23.3 shows a more efficient version of the previous circuit, in which the loss incurred in the feed resistors R1 and R2 has been eliminated by using a sep-

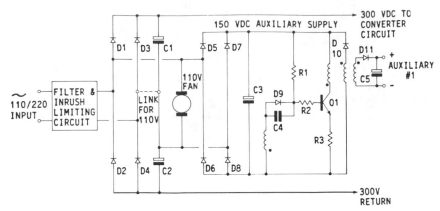

FIG. 1.23.3 Auxiliary power converter of the self-oscillating flyback type with a 110-V ac cooling fan supply, suitable for use with 110-V ac fans in dual input voltage applications.

arate bridge rectifier D5–D8 to supply the converter. This arrangement is particularly useful for dual 110–220-V applications, as the rectifiers D5–D8 are effectively fed with 110 V for both positions of the voltage selector link, that is, for both 110- and 220-V operation. This effective 110-V ac line is also used in this example to supply the 110-V cooling fan. Hence, the same fan may be used for both input voltages. (To meet the safety requirements, the insulation rating for the fan must be suitable for the higher-voltage conditions.)

Note: When operating from 220-V line inputs, the link to the center of C1–C2 is removed. Under this condition, the load on the 300-V DC line must exceed the fan and auxiliary loading, to ensure DC restoration of the center point of C1–C2. Hence this circuit is suitable only for applications in which a minimum load is maintained on the output. Capacitors C1 and C2 must be selected to accommodate the additional ripple current loading provided by the fan and the auxiliary converter, although this will probably be a small percentage of the total loading in most applications.

23.7 AUXILIARY SUPPLIES DERIVED FROM MAIN CONVERTER TRANSFORMER

When the main converter is operating, it is clear that a winding on the main converter transformer can provide the auxiliary supply needs. However, some means is required to provide the auxiliary power to the control circuits during the start-up phase. The following chapter describes a number of starting methods.

23.8 PROBLEMS

1. Explain why the characterictics of the auxiliary power supply systems are sometimes fundamental to the operation of the main power section.

2. What is the major disadvantage of using small 60-Hz transformers for auxiliary power systems?

CHAPTER 24
PARALLEL OPERATION OF VOLTAGE-STABILIZED POWER SUPPLIES

24.1 INTRODUCTION

Stabilized-voltage power supplies, both switching and linear, have extremely low output resistances, often less than 1 mΩ. Consequently, when such supplies are connected in parallel, the supply with the highest output voltage will supply the majority of the output current. This will continue until this supply goes into current limit, at which point its voltage will fall, allowing the next highest voltage supply to start delivering current, and so on.

Because the output resistance is so low, only a very small difference in output voltage (a few millivolts) is required to give large current differences. Hence, it is impossible to ensure current sharing in parallel operation by output voltage adjustment alone. Generally any current imbalance is undesirable, as it means that one unit may be overloaded (operating all the time in a current-limited mode), while a second parallel unit may be delivering only part of its full rating.

Several methods are used to make parallel units share the load current almost equally.

24.2 MASTER-SLAVE OPERATION

In this method of parallel operation, a designated master is selected, and this is arranged to provide the voltage control and drive to the power sections of the remainder of the parallel units.

Figure 1.24.1 shows the general arrangement of the master-slave connection. Two power supplies are connected in parallel. (They could be switching or linear supplies.) Both supplies deliver current to a common load. An interconnection is made between the two units via a link (this is normally referred to as a P-terminal link). This terminal links the power stages of the two supplies together.

The master unit defines the output voltage, which may be adjusted by VR2. The slave unit will be set to a much lower voltage. (Alternatively, the reference will be linked out, LK1.) The output of amplifier A1′ will be low, and diode D1′ is reversed-biased. Q3′ will not be conducting, and the drive to Q2′ will be pro-

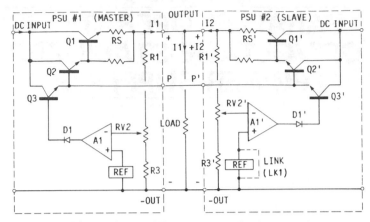

FIG. 1.24.1   Linear voltage-stabilized power supplies in master-slave connection.

vided by Q3 in PSU1 via the P-terminal link. The drive transistor Q3 must have sufficient spare drive current to provide the needs of all the parallel units; hence there is a limit to the number of units that can be connected in parallel. Drive accommodation is normally provided for a minimum of five parallel supplies.

In this arrangement, the slave supplies are operating as voltage-controlled current sources. Current sharing is provided by the voltage drop across the emitter sharing resistors R_s and R_s'. The current-sharing accuracy is not good because of the rather variable base-emitter voltages of the power transistors. A sharing accuracy of 20% would be typical for this type of connection.

The major disadvantage of master-slave operation is that if the master unit fails, then all outputs will fail. Further, if a power section fails, the direct connection between the two units via the P terminal tends to cause a failure in all units.

24.3   VOLTAGE-CONTROLLED CURRENT SOURCES

This method of parallel operation relies on a principle similar to that of the master-slave, except that the current-sharing P-terminal connection is made at a much earlier signal level in the control circuit. The control circuit is configured as a voltage-controlled current source. The voltage applied to the P terminal will define the current from each unit, the total current being the sum of all the parallel units. The voltage on the P terminal, and hence the total current, is adjusted to give the required output voltage from the complete system. Figure 1.24.2 shows the general principle.

In this arrangement the main drive to the power transistors Q1 and Q1' is from the voltage-controlled current amplifiers A1 and A1'. This operates as follows.

Assume that a reference voltage REF has been set up by one of the amplifiers. (REF2 and REF2' must be equal, as they are connected by the P terminals.) The conduction of transistors Q1 and Q1' will be adjusted by the amplifiers so that the

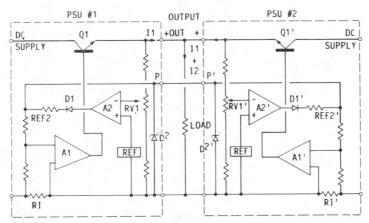

FIG. 1.24.2 Parallel operation of current-mode-controlled linear power supplies, showing natural current-sharing ability.

currents in the two current-sensing resistors R1 and R1′ will be well defined and equal. The magnitude of the currents depends on the reference voltage on P and the resistor values.

The dominant control amplifier, A2 or A2′ (the one set to the highest voltage), will now adjust the current to obtain the required output voltage. The other amplifier will have its output diode reverse-biased.

The major advantage of this arrangement is that a failure in the power section is less likely to cause a fault in the P-terminal interconnection, and the current sharing is well defined.

This circuit lends itself well to parallel redundant operation. See Sec. 24.5.

24.4 FORCED CURRENT SHARING

This method of parallel operation uses a method of automatic output voltage adjustments on each power supply to maintain current sharing in any number of parallel units. This automatic adjustment is obtained in the following way.

Because the output resistance in a constant-voltage supply is so low (a few milliohms or less), only a very small output voltage change is required to make large changes in the output current of any unit.

With forced current sharing, in principle any number of units can be connected in parallel. Each unit compares the current it is delivering with the average current for the total setup and adjusts its output voltage so as to make its own output current equal to the average current.

Figure 1.24.3 shows the principle used for this type of system. Amplifier A1 is the voltage control amplifier of the supply. It operates in the normal way, comparing the output voltage from the divider network R3, R4 with an internal reference voltage V'_{ref} and controlling the power stage so as to maintain the output voltage constant. However, V'_{ref} is made up of the normal reference voltage V_{ref} in series with a small adjustable reference V2 developed by the divider network R1, R2 from the current sense amplifier A2. V2, and hence V'_{ref}, may be in-

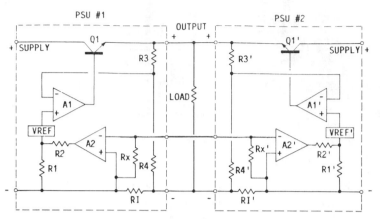

FIG. 1.24.3 Parallel operation of voltage-stabilized linear power supplies, showing forced current-sharing circuit.

creased or decreased in response to the output of amplifier A2. The maximum range of adjustment is limited, being typically 1% or less.

Amplifier A2 compares the output current of its own power supply with the average output current of all the power supplies by comparing the voltage analogue across the internal current shunt R1 with the average voltage analogue generated by all the shunts and averaged by the interconnection resistors Rx. A2 will increase or decrease the second reference voltage V2, and hence the output voltage of its supply, so as to maintain its current on a par with the average.

An interconnection between the power supplies must be provided to carry the information on the average current. (This is sometimes known as a P-terminal link.)

Any number of such supplies can be directly connected in parallel. All that will be required from the supplies is that their output voltages must be adjusted to be within the voltage capture range (better than 1% of the required output voltage in this example).

The major advantage of this technique for parallel redundant operation is that in the event of one power supply failing, the remaining working units will redistribute the load current equally among them without interruption to the output.

The output voltage of the combination will adjust itself to the average value of the independent units.

A more practical arrangement of this circuit principle is demonstrated in Fig. 1.24.4. This circuit has the advantage that the reference voltage can be increased or decreased as required.

The output voltage of amplifier A2 (node A) will normally be equal to the reference voltage V_{ref}. Hence there will be no corrective action so long as the output current is equal to the average current of the combination. Under these conditions, the voltages at node B and node C are the same. If the current is not balanced, then the voltage at node B will not be the same as that at node C, and the output of amplifier A2 will change to adjust the reference voltage. This will result in a change in output voltage and a correction in the output current, to recover a balanced condition.

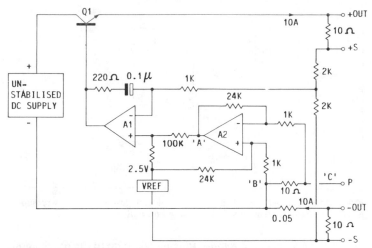

FIG. 1.24.4   Example of a forced current-sharing circuit.

24.5 PARALLEL REDUNDANT OPERATION

The purpose of parallel redundant operation is to ensure maintenance of power even in the event of one power supply failure. In principle, n supplies (where n is two or more) are connected in parallel to supply a load that has a maximum demand that is $n - 1$ of the total combination rating. Hence, if a supply fails, the remainder of the units will take up the load without an interruption in the service.

In practice, the failed supply may short-circuit (for example, the SCR overvoltage crowbar may fire). To prevent this supply from overloading the remainder of the network, the power supplies will usually be rectifier-diode OR-gated into the output line. Figure 1.24.5 shows a typical arrangement.

Remote voltage sensing is not recommended for parallel redundant operation, as the remote connections provide alternative current paths in the event of a power supply failure. If line voltage drops are a problem, then the diodes should

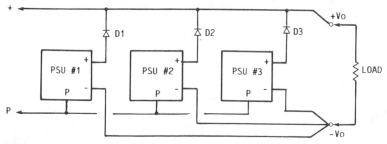

FIG. 1.24.5   Parallel redundant connection of stabilized voltage power supplies.

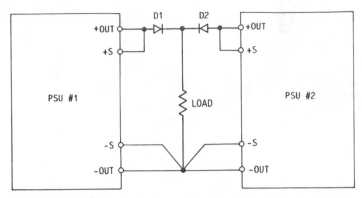

FIG. 1.24.6 Parallel redundant operation of voltage-stabilized power supplies, showing quasi-remote voltage sensing connections.

be mounted at the load end and remote sensing taken up to the diode anode only, as shown in Fig. 1.24.6.

Power supplies of the forced-current-sharing type are most suitable for this type of parallel redundant mode operation, as the P-terminal link provides current sharing and does not compromise the operation if a supply fails. In fact, the technique ensures that the remainder of the supplies share the load equally, increasing their output currents as required to maintain a constant output voltage.

24.6 PROBLEMS

1. Why does operating constant-voltage power supplies in parallel present a problem?
2. What is meant by parallel master-slave operation?
3. Explain the major disadvantage of master-slave operation.
4. What is meant by forced current sharing for parallel operation?
5. What is the major disadvantage of forced current sharing?
6. What is meant by parallel redundant operation?

P · A · R · T · 2

DESIGN: THEORY AND PRACTICE

CHAPTER 1

MULTIPLE-OUTPUT FLYBACK SWITCHMODE POWER SUPPLIES

1.1 INTRODUCTION

Figure 2.1.1 shows the basic circuit of a triple-output flyback power supply.

The flyback unit combines the actions of an isolating transformer, an output inductor, and a flywheel diode in a single transformer. As a result of this magnetic integration, the circuit provides extremely cost-effective and efficient stabilized DC outputs.

The technique is particularly useful for multiple-output applications, where several semistabilized outputs are required from a single supply. The major disadvantage is that high ripple currents flow in transformer and output components, reducing their efficiency. As a result of this limitation, the flyback converter is usually restricted to power levels below 150 W.

1.2 EXPECTED PERFORMANCE

In the example shown in Fig. 2.1.1, the main output is closed-loop-controlled and is thus fully regulated. The auxiliary outputs are only semiregulated and may be expected to provide line and load regulation of the order of ±6%. Where better regulation is required, additional secondary regulators will be needed.

In flyback supplies, secondary regulators are often linear dissipative types, although switching regulators may be used for higher efficiency. For low-current outputs, the standard three-terminal IC regulators are particularly useful. The dissipation in the linear regulators is minimized as a result of the preregulation provided by the closed-loop control of the main output. In some applications, the closed-loop control regulation may be shared between two or more outputs.

Since the most cost-effective flyback converters will not have additional secondary regulators, overspecifying the requirements is a mistake. The essential attractions of this type of converter—simplicity and low cost—will be lost if additional circuitry is required to meet very critical specifications. For such

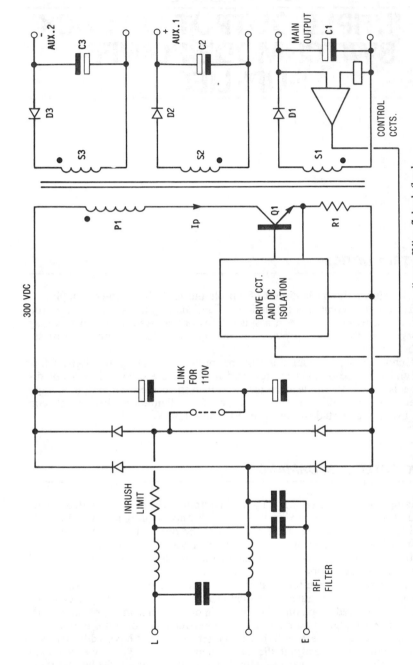

FIG. 2.1.1 Power rectifier and converter section of a typical triple-output, direct-off-line flyback (buck-boost) switchmode power supply.

applications, the designer might do well to consider one of the more sophisticated multiple-output topologies with their inherently higher performance.

1.2.1 Output Ripple and Noise

Where very low levels of output ripple are required, the addition of a small *LC* noise filter near the output terminals will often eliminate the need for expensive low-ESR capacitors in the main secondary reservoir positions.

For example, a typical 5-V 10-A supply may use the highest-quality low-ESR capacitors in positions C1, C2, and C3 of the single-stage filter shown in Fig. 2.1.1, but this will rarely give a ripple figure of less than 100 mV. However, it is relatively easy to keep ripple figures below 30 mV when low-cost standard electrolytic capacitors are used in positions C1, C2, and C3 by adding a high-frequency *LC* output filter. This approach can be very efficient and cost-effective. (See Part 1, Chap. 20.) It should be understood that in a flyback converter the inductor can be quite small, since it is not required for energy storage (as it would be in a forward converter).

1.2.2 Synchronization

In fixed-frequency flyback units, some means of synchronizing the switching frequency to an external clock is often provided. This synchronization can lead to fewer interference problems in some applications.

1.3 OPERATING MODES

Two modes of operation are clearly identifiable in the flyback converter:

1. "Complete energy transfer" (discontinuous mode), in which all the energy that was stored in the transformer during an energy storage period ("on" period) is transferred to the output during the flyback period ("off" period).

2. "Incomplete energy transfer" (continuous mode), in which a part of the energy stored in the transformer at the end of an "on" period remains in the transformer at the beginning of the next "on" period.

1.3.1 Transfer Function

The small-signal transfer functions for these two operating modes are quite different, and they are dealt with separately in this section. In practice, when a wide range of input voltages, output voltages, and load currents is required, the flyback converter will be required to operate (and be stable) in both complete and incomplete energy transfer modes, since both modes will be encountered at some point in the operating range.

As a result of the change in transfer function at the point where there is a move from one mode to the other, together with the merging into one component of the transformer, output inductor, and flywheel diode actions, flyback converters can be among the most difficult to design.

1.3.2 Current-Mode Control

The introduction of current-mode control to the pulse-width modulation action has very much reduced the control-loop problems, particularly for the complete energy transfer mode. Hence current-mode control is recommended for flyback systems. However, current-mode control does not eliminate the stability problems inherent in the incomplete energy transfer mode, because of the "right-half-plane zero" in the transfer function. This will require the gain of the control loop to roll off at a low frequency, degrading the transient response. (See Part 3, Chap. 9.)

1.4 OPERATING PRINCIPLES

Consider Fig. 2.1.1. In this circuit, the high-voltage-rectified 300-V DC line is switched across the primary winding of a transformer P1, using a single switching device Q1. The control circuit has a fixed frequency, and the duty ratio of Q1 is adjusted to maintain the output voltage constant on the main output line. It will be shown that the unit may operate in a complete or incomplete energy transfer mode, depending on the duty ratio and load.

1.5 ENERGY STORAGE PHASE

The energy storage phase is best understood by considering the action of the basic single-output flyback converter shown in Fig. 2.1.2.

When transistor Q1 is turned on, the start of all windings on the transformer

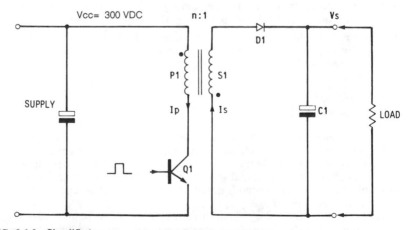

FIG. 2.1.2 Simplified power section of a flyback (buck-boost) converter.

will go positive. The output rectifier diode D1 will be reverse-biased and will not conduct; therefore current will not flow in the secondary while Q1 is conducting.

During this energy storage phase only the primary winding is active, and the transformer may be treated as a simple series inductor; hence the circuit can be further simplified to that shown in Fig. 2.1.3a.

From Fig. 2.1.3a it is clear that when Q1 turns on, the primary current I_p will increase at a rate specified by

$$\frac{di_p}{dt} = \frac{V_{cc}}{L_p}$$

where V_{cc} = supply voltage
L_p = primary inductance

This equation shows that there will be a linear increase of primary current during the time Q1 is conducting, (t_{on}). During this period the flux density in the core will increase from the residual value B_r to its peak working value B_w. The corresponding current waveforms and flux density changes are shown in Fig. 2.1.3b.

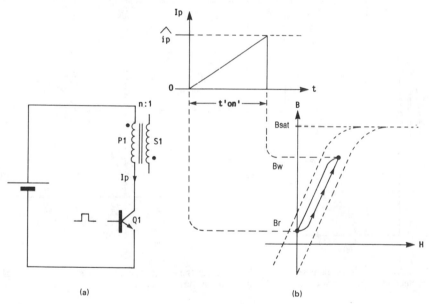

(a)   (b)

FIG. 2.1.3 (a) Equivalent primary circuit during the energy storage phase. (b) Primary current waveform and magnetization during the energy storage phase.

1.6 ENERGY TRANSFER MODES (FLYBACK PHASE)

When Q1 turns off, the primary current must drop to zero. The transformer ampere-turns cannot change without a corresponding change in the flux density $-\Delta B$. As the change in the flux density is now negative-going, the voltages will reverse on all windings (flyback action). The secondary rectifier diode D1 will conduct, and the magnetizing current will now transfer to the secondary. It will continue to flow from start to finish in the secondary winding. Hence, the secondary (flyback) current flows in the same direction in the windings as the original primary current, but has a magnitude defined by the turns ratio. (The ampere-turns product remains constant.)

Under steady-state conditions, the secondary induced emf (flyback voltage) must have a value in excess of the voltage on C1 (the output voltage) before diode D1 can conduct. At this time the flyback current will flow in the secondary winding starting at a maximum value I_s, where $I_s = n \times I_p$. (n is the transformer turns ratio and I_p is the primary current at the instant of turn-off of Q1.) The flyback current will fall toward zero during the flyback period. Since during the flyback period Q1 is "off" and the primary is no longer conducting, the primary winding can now be neglected, and the circuit simplifies to that shown in Fig. 2.1.4a. The flyback secondary current waveform is shown in Fig. 2.1.4b.

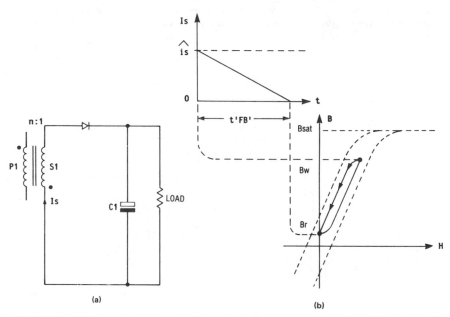

FIG. 2.1.4 (a) Equivalent secondary circuit during the energy transfer phase (flyback period). (b) Secondary current waveform and magnetization during the flyback period.

For complete energy transfer conditions, the flyback period is always less than the "off" period, and the flux density in the core will fall from its peak value B_w to its residual value B_r during the flyback period. The secondary current will also decay at a rate specified by the secondary voltage and secondary inductance; hence

$$\frac{di_s}{dt} = \frac{V_s}{L_s}$$

where V_s = secondary voltage
L_s = inductance of transformer referred to secondary

1.7 FACTORS DEFINING OPERATING MODES

1.7.1 Complete Energy Transfer

If the flyback current reaches zero before the next "on" period of Q1, as shown in Fig. 2.1.5a, the system is operating in a complete energy transfer mode. That

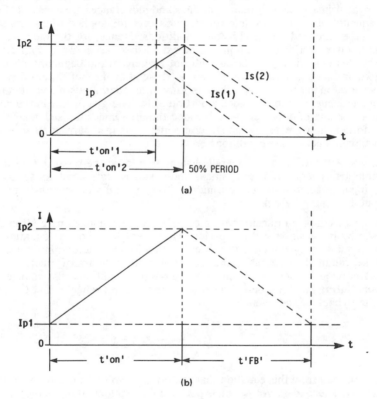

FIG. 2.1.5 (a) Primary current waveform I_p and secondary current waveforms I_s (discontinuous-mode) operation. (b) Primary and secondary waveforms for incomplete energy transfer (continuous-mode) operation.

is, all the energy that was stored in the transformer primary inductance during the "on" period will have been transferred to the output circuit during the flyback period, before the next storage period starts. If the flyback current does not reach zero before the next "on" period (Fig. 2.1.5b), then the system is operating in the incomplete energy transfer mode.

1.7.2 Incomplete Energy Transfer

If, in the circuit example shown in Fig. 2.1.2, the "on" period is increased and the "off" period correspondingly decreased, more energy is stored in the transformer during the "on" period. For steady-state operation, this extra energy must be extracted during the "off" period. If the input and output voltages are to be maintained constant, it will be shown that the load current must be increased to remove the extra energy.

The slope of the input and output current characteristics cannot change, because the primary and secondary voltages and inductances are constant. Further, the equality of the forward and reverse volt-seconds applied to the transformer must be maintained for steady-state conditions. Hence, for the increased "on" period, a new working condition will be established, as shown in Fig. 2.1.5b.

For this condition, the current will not be zero at the beginning of an "on" period, and an equal value will remain at the end of the "off" period (with due allowance for the turns ratio). This is known as continuous-mode operation or incomplete energy transfer, since a portion of the energy remains in the magnetic field at the end of a flyback period. Since the area under the secondary-current waveform is now greater by the DC component, the load current must be greater to maintain steady-state conditions.

Note: The behavior of the overall system should not be confused by the term "incomplete energy transfer," since, under steady-state conditions, all the energy input to the transformer during the "on" period will be transferred to the output during the flyback period.

In this example, a transition from complete to incomplete energy transfer was caused by increasing the "on" period. However, the following equation shows that the mode of operation is in fact controlled by four factors: input and output voltage, the mark space ratio, and the turns ratio of the transformer.

As previously mentioned, under steady-state conditions, the change in flux density during the "on" period must equal the return change in flux density during the flyback period. Hence

$$\Delta \Phi = \frac{V_{cc} \cdot t_{on}}{N_p} = \frac{V_s \cdot t_{off}}{N_s}$$

It will be seen from this equation that the primary volt-seconds per turn must be equal to the secondary volt-seconds per turn if a stable working point for the flux density is to be established.

In the forward direction, the "on" period can be adjusted by the control circuit to define the peak primary current. However, during the flyback period, the output voltage and secondary turns are constant, and the active flyback period must self-adjust until a new stable working point for the transformer flux density is established. It can continue to do this until the flyback period extends to meet the beginning of the next "on" period (Fig. 2.1.5b).

At the critical point where the flyback current has just reached zero before the next "on" period, any further increase in duty ratio or load will result in the unit moving from the complete to the incomplete energy transfer mode. At this point, no further increase in pulse width is required to transfer more current, and the output impedance becomes very low. Hence, the transfer function of the converter changes to a low-impedance two-pole system.

1.8 TRANSFER FUNCTION ANOMALY

The flyback converter operating in open loop and in the complete energy transfer mode (discontinuous mode) has a simple single-pole transfer function and a high output impedance at the transformer secondary. (To transfer more power requires an increase in pulse width.)

When this system reverts to the incomplete energy transfer mode (continuous-mode operation), the transfer function is changed to a two-pole system with a low output impedance (the pulse width is only slightly increased when more power is demanded). Further, there is a "right-half-plane zero" in the transfer function, which will introduce an extra 180° of phase shift at high frequency; this can cause instability. The loop stability must be checked for both modes of operation if it is possible for both modes to occur in normal use. To determine the need for this, consider light loading, normal loading, and short-circuit conditions. In many cases, although complete energy transfer may have been the design intention, incomplete transfer may occur under overload or short-circuit conditions at low input voltages, leading to instability. (See Part 3, Chap. 9 and Sec. 10.6.)

1.9 TRANSFORMER THROUGHPUT CAPABILITY

It is sometimes assumed that a transformer operating in the complete energy transfer mode has greater transmissible power than the same transformer operating in an incomplete transfer mode. (It sounds as if it should.) However, this is true only if the core gap remains unchanged.

Figure 2.1.6a and b shows how, by using a larger air gap, the same transformer may be made to transfer more power in the incomplete transfer mode than it did previously in the complete transfer mode (even with a smaller flux excursion). In applications in which the transformer is "core loss limited" (usually above 60 kHz for typical ferrite transformers), considerably more power may be transmitted in the incomplete energy transfer mode, because the reduced flux excursion results in lower core losses and reduced ripple currents in both primary and secondary.

Figure 2.1.6a shows the B/H curve for a core with a small air gap and a large flux density change. Figure 2.1.6b shows the B/H curve for the same core with a larger air gap and a smaller flux density change.

In general, the power available for transfer is given by

$$P = f \cdot V_e \int_{B_r}^{B_w} H \, dB$$

where f = frequency
V_e = effective volume of core and air gap

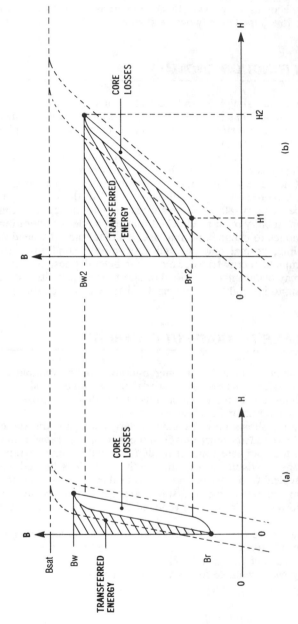

FIG. 2.1.6 (*a*) Magnetization loop and energy transferred in a flyback converter transformer when the core air gap is small (high-permeability magnetic path). (*b*) Magnetization loop and energy transferred in a flyback converter transformer when a large air gap is used in the core.

This power is proportional to the shaded area to the left of the B/H curve in Fig. 2.1.6; it is clearly larger for the example in Fig. 2.1.6b (the incomplete energy transfer case). Much of the extra energy is stored in the air gap; consequently, the size of the air gap will have a considerable effect upon the transmissible power. Because of the very high reluctance of the air gap, it is quite usual to have more energy stored in the gap than in the transformer core itself.

At the end of the "on" period, energy of $\frac{1}{2}LP \cdot I_{p2}^2$ will have been stored in the transformer magnetic field. This energy, less the energy remaining in the core ($\frac{1}{2}LP \cdot I_{p1}^2$) is transferred to the output circuit each cycle.

In conclusion, the designer must choose the mode of operation depending on the performance required and the power to be transferred, be aware of the need to check the mode of operation under all possible loading conditions, and be prepared to design the control loop to deal with all realistic conditions.

1.10 SPECIFICATION NOTES

The designer should be alert to the tendency for specifications to escalate. When a flyback converter is to be considered and potential requirements are large, costs are often particularly sensitive. The designer should establish with the customer the real limitations of the application. It may well be that a typical performance of 6% regulation on the auxiliary outputs of a multiple-output unit would be acceptable. This allows a semiregulated flyback system to be used. To guarantee a result of 5% (hardly better), a secondary regulator would be required, with consequent loss of efficiency and increased cost.

Very often specifications call for fixed frequency, or even a synchronized operating condition. This synchronization is often specified when a power supply is to be used for video display terminals or computer applications. Very often, in specifying such requirements, the user is making an assumption that the switching noise or magnetic field generated by the power supply will in some way interfere with the system performance. However, in a well-designed, well-filtered, and well-screened modern switching supply, the noise level is unlikely to be sufficiently high to cause interference. Moreover, in many cases, synchronization makes the noise even more noticeable. In any event, synchronization is a poor substitute for eliminating the noise problem altogether.

If the specification calls for a fixed frequency or synchronization, the designer would do well to check this requirement with the user. Have available for demonstration a well-screened variable-frequency unit. This should have a copper screen on the transformer and a second-stage output LC filter. If possible, try the sample in the actual application. The author has found that the user is often well satisfied with the result, and of course the cost of the supply would be much lower.

In some applications in which a number of switching supplies are to be operated from the same input supply (more usual with DC-to-DC converters), the input filter requirements can be reduced by using synchronized and phase-shifted clock systems. This approach also eliminates low-frequency intermodulation components, and in this application the extra cost of a synchronized unit may well be justified.

Having fully researched the application, the designer is in a position to confidently select the most effective approach to meet the final specification requirements.

1.11 SPECIFICATION EXAMPLE FOR A 110-W DIRECT-OFF-LINE FLYBACK POWER SUPPLY

For the following example, a fixed-frequency single-ended bipolar flyback unit with three outputs and a power of 110 W is to be considered. It will be shown later that the same design approach is applicable to variable-frequency self-oscillating units.

Although most classical design approaches assume that the mode of operation will be either entirely complete energy transfer (discontinuous mode) or entirely incomplete energy transfer (continuous mode), in practice a system is unlikely to remain in either of these two modes for the complete range of operation. Consequently, in the simplified design approach used here, it will be assumed that both modes of operation will exist at some point within the working range. This approach also tends to yield higher efficiency, as the peak primary and secondary currents are reduced.

1.11.1 Specification

Output power:	110 W
Input voltage range:	90–137/180–250 (user selectable)
Operating frequency:	30 kHz
Output voltages:	5 V, 10 A
	12 V, 3 A
	−12 V, 2 A
Line and load regulation:	1% for main 5-V output
	6% typical for a 40% load change (from 60% nominal)
Output current range:	20% to full load
Output ripple and noise:	1% maximum
Output voltage centering:	± 1% on 5-V lines
	± 3% on 12-V lines
Overload protection:	By primary power limit and shutdown requiring power on/off reset cycle
Overvoltage protection:	5-V line only by converter shutdown, i.e., crowbar not required

1.11.2 Power Circuit

The above specification requirements can be met using a single-ended flyback system without secondary regulators (see Fig. 2.1.1). To meet the need for dual input voltage by a link change, voltage doubling techniques can be employed for the input line rectifiers when they are set for 110-V operation. Consequently, the rectified DC line will be approximately 300 V for either 110-V or 220-V nominal inputs.

A voltage analogue of the primary current for primary power limiting is available across the emitter resistor R1. This waveform may also be used for control purposes when current-mode control is to be used. (See Part 3, Chap. 10.) A separate overvoltage protection circuit monitors the 5-V output and shuts the converter down in the event of a failure in the main control loop.

To meet the requirements for low output ripple, a two-stage *LC* filter will be fitted in this example. This type of filter will allow standard medium-grade electrolytic output capacitors to be used, giving a lower component cost. (A suitable filter is shown in Part 1, Chap. 20.) The control circuit is assumed to be closed to the 5-V output to give the best regulation on this line. The details of the drive circuitry have been omitted; suitable systems will be found in Part 1, Chaps. 15 and 16.

1.11.3 Transformer Design

The design of the transformer for this power supply is shown in Part 2, Chap. 2.

1.12 PROBLEMS

1. From what family of converters is the flyback converter derived?
2. During which phase of operation is the energy transferred to the secondary in a flyback converter?
3. Describe the major advantages of the flyback technique.
4. Describe the major disadvantages of the flyback technique.
5. Why is the transformer utilization factor of a flyback converter often much lower than that of a push-pull system?
6. Under what operating conditions will the flyback converter give a core utilization factor similar to that of a forward converter?
7. Why is an output inductor not required in the flyback system?
8. Describe the two major modes of operation in the flyback converter.
9. What are the major differences in the transfer functions between continuous- and discontinuous-mode operation?
10. Why would an air gap normally be required in the core of a flyback transformer when ferrite core material is used?
11. Why is primary power limiting alone usually inadequate for full short-circuit protection of a flyback converter?

CHAPTER 2

FLYBACK TRANSFORMER DESIGN

For a Direct-Off-Line
Flyback Switchmode Power Supply

2.1 INTRODUCTION

Because the flyback converter transformer combines so many functions (energy storage, galvanic isolation, current-limiting inductance), and also because it is often required to support a considerable DC current component, it can be rather more difficult to design than the more straightforward push-pull transformer. For this reason, the following section is entirely devoted to the design of such transformers.

To satisfy the design requirement, many engineers prefer to use an entirely mathematical technique. This is fine for the experienced engineer. However, because it is difficult to get a good working feel for the design by using this approach, it will not be used here.

In the following transformer design example, the chosen process will use an iterative technique. No matter where the design is started, a number of approximations must be made initially. The problem for the inexperienced designer is to get a good feel for the controlling factors. In particular, the selection of core size, the primary inductance, the function of the air gap, the selection of primary turns, and the interaction of the ac and DC current components within the core are often areas of much confusion in flyback transformer design.

To give the designer a better feel for the controlling factors, the following design approach starts with an examination of the properties of the core material and the effect of an air gap. This is followed by an examination of the ac and DC core polarization conditions. Finally, a full design example for a 100-W transformer is given.

2.2 CORE PARAMETERS AND THE EFFECT OF AN AIR GAP

Figure 2.2.1a shows a typical B/H (hysteresis) loop for a transformer-grade ferrite core, with and without an air gap. It should be noted that although the permeability (slope) of the B/H loop changes with the length of the air gap, *the saturation flux density of the combined core and gap remains the same.* Further, the magnetic field intensity H is much larger, and the residual flux density B_r much lower, in the gapped case. These changes are very useful for flyback transformers, which use only the first quadrant of the B/H loop.

Figure 2.2.1b shows only the first quadrant of the hysteresis loop, the quadrant used for flyback converter transformers. It also shows the effect of introduc-

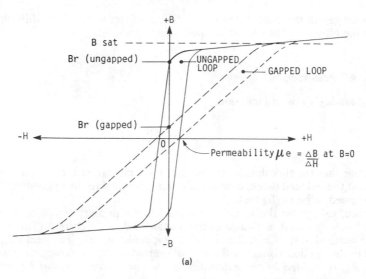

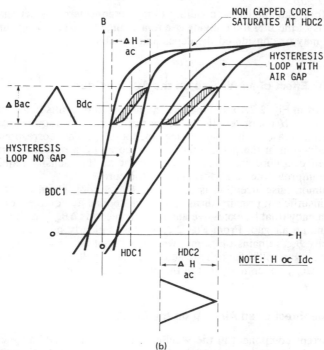

FIG. 2.2.1 (*a*) Total magnetization loops for a ferrite transformer, with and without an air gap. (*b*) First-quadrant magnetization loops for a typical ferrite core in a single-ended flyback converter when large and small air gaps are used. Notice the increased transferred energy ΔH when a large air gap is used.

ing an air gap in the core. Finally, this diagram demonstrates the difference between the effects of the ac and DC polarizing conditions.

2.2.1 AC Polarization

From Faraday's law of induction,

$$\text{emf} = \frac{N\,d\Phi}{dt}$$

It is clear that the flux density in the core must change at a rate and amplitude such that the induced (back) emf in the winding is equal to the applied emf (losses are assumed to be negligible).

Hence, to support the ac voltage applied to the primary (more correctly, the applied volt-seconds), a change in flux density ΔB_{ac} is required. (This is shown on the vertical axis in Fig. 2.2.1b.) The amplitude of ΔB_{ac} is therefore proportional to the applied voltage and the "on" period of the switching transistor Q1; hence B_{ac} is *defined by the externally applied ac conditions, not by the transformer air gap.*

Therefore, the applied ac conditions may be considered as acting on the vertical B axis of the B/H loop, giving rise to a change in magnetizing current ΔH_{ac}. Hence H may be considered the dependent variable.

2.2.2 The Effect of an Air Gap on the AC Conditions

It is clear from Fig. 2.2.1b that increasing the core gap results in a decrease in the slope of the B/H characteristic but does not change the required ΔB_{ac}. Hence there is an increase in the magnetizing current ΔH_{ac}. This corresponds to an effective reduction in the permeability of the core and a reduced primary inductance. Hence, a core gap does not change the ac flux density requirements or otherwise improve the ac performance of the core.

A common misconception is to assume that a core which is saturating as a result of insufficient primary turns, excessive applied ac voltages, or a low operating frequency (that is, excessive applied volt-seconds ΔB_{ac}) can be corrected by introducing an air gap. From Fig. 2.2.1b, this is clearly not true; the saturated flux density B_{sat} remains the same, with or without an air gap. However, introducing an air gap will reduce the residual flux density B_r and increase the working range for ΔB_{ac}, which may help in the discontinuous mode.

2.2.3 The Effect of an Air Gap on the DC Conditions

A DC current component in the windings gives rise to a DC magnetizing force H_{DC} on the horizontal H axis of the B/H loop. (H_{DC} is proportional to the mean DC ampere-turns.) For a defined secondary current loading, the value of H_{DC} is defined. Hence, for the DC conditions, B may be considered the dependent variable.

It should be noted that the gapped core can support a much larger value of H (DC current) without saturation. Clearly, the higher value of H, H_{DC2}, would be sufficient to saturate the ungapped core in this example (even without any ac component). Hence an air gap is very effective in preventing core saturation that would be caused by any DC current component in the windings. When the flyback converter operates in the continuous mode, a considerable DC current component is present, and an air gap must be used.

Figure 2.2.1b shows the flux density excursion ΔB_{ac} (which is required to support the applied ac voltage) applied to the mean flux density B_{dc} developed by the DC component H_{DC} for the nongapped and gapped example. For the nongapped core, a small DC polarization of H_{DC1} will develop the flux density B_{dc}. For the gapped core, a much larger DC current (H_{DC2}) is required to produce the same flux density B_{dc}. Further, it is clear that in the gapped example the core will not be saturated even when the maximum DC and ac components are added.

In conclusion, Fig. 2.2.1b shows that the change in flux density ΔB_{ac} required to support the applied ac conditions does not change when an air gap is introduced into the core. However, the mean flux density B_{dc} (which is generated by the DC current component in the windings) will be very much less if a gap is used.

The improved tolerance to DC magnetization current becomes particularly important when dealing with incomplete energy transfer (continuous-mode) operation. In this mode the current in the core never falls to zero, and clearly the ungapped core would saturate.

Remember, the applied volt-seconds, turns, and core *area* define the required ac change in flux density ΔB_{ac} applied to the vertical B axis, while the mean DC current, turns, and magnetic path *length* set the value of H_{DC} on the horizontal axis. Sufficient turns and core area must be provided to support the applied ac conditions, and sufficient air gap must be provided in the core to prevent saturation and support the DC current component.

2.3 GENERAL DESIGN CONSIDERATIONS

In the following design, the ac and DC conditions applied to the primary are dealt with separately. Using this approach, it will be clear that the applied ac voltage, frequency, area of core, and maximum flux density of the core material control the *minimum* primary turns, irrespective of core permeability, gap size, DC current, or required inductance.

It should be noted that the primary inductance will not be considered as a transformer design parameter in the initial stages. The reason for this is that the inductance controls the mode of operation of the supply; it is not a fundamental requirement of the transformer design. Therefore, inductance will be considered at a later stage of the design process. Further, when ferrite materials are used at frequencies below 60 kHz, the following design approach will give the maximum inductance consistent with minimum transformer loss for the selected core size. Hence, the resulting transformer would normally operate in an incomplete energy transfer mode as a result of its high inductance. If the complete energy transfer mode is required, this may be obtained by the simple expedient of increasing the core gap beyond the minimum required to support the DC polarization, thereby reducing the inductance. This may be done without compromising the original transformer design.

When ferrite cores are used below 30 kHz, the minimum obtainable copper loss will normally be found to exceed the core loss. Hence maximum (but not optimum) efficiency will be obtained if maximum flux density is used. Making B large results in minimum turns and minimum copper loss. Under these conditions, the design is said to be "saturation limited." At higher frequencies, or when less efficient core materials are used, the core loss may become the predominant factor, in which case lower values of flux density and increased turns would be used and the design is said to be "core loss limited." In the first case the design efficiency is limited; optimum efficiency cannot be realized, since this requires core and copper losses to be nearly equal. Methods of calculating these losses are shown in Part 3, Chap. 4.

2.4 DESIGN EXAMPLE FOR A 110-W FLYBACK TRANSFORMER

Assume that a transformer is required for the 110-W flyback converter specified in Part 2, Sec. 1.11.

2.4.1 Step 1, Select Core Size

The required output power is 110 W. If a typical secondary efficiency of 85% is assumed (output diode and transformer losses only), then the power transmitted by the transformer would be 130 W.

We do not have a simple fundamental equation linking transformer size and power rating. A large number of factors must be considered when making this selection. Of major importance will be the properties of the core material, the shape of the transformer (that is, its ratio of surface area to volume), the emissive properties of the surface, the permitted temperature rise, and the environment under which the transformer will operate.

Many manufacturers provide nomograms giving size recommendations for particular core designs. These recommendations are usually for convection cooling and are based upon typical operating frequencies and a defined temperature rise. Be sure to select a ferrite that is designed for transformer applications. This will have high saturation, low residual flux density, low losses at the operating frequency, and high curie temperatures. High permeability is not an important factor for flyback converters, as an air gap will always be used with ferrite materials.

Figure 2.2.2 shows the recommendations for Siemens N27 Siferrit material at an operating frequency of 20 kHz and a temperature rise of 30 K. However, most real environments will not be free air, and the actual temperature rise may be greater where space is restricted or less when forced-air cooling is used. Hence some allowance should be made for these effects. Manufacturers usually provide nomograms for their own core designs and materials. For a more general solution, use the "area-product" design approach described in Part 3, Sec. 4.5.

In this example an initial selection of core size will be made using the nomogram shown in Fig. 2.2.2. For a flyback converter with a throughput power of 130 W, an "E 42/20" is indicated. (The nomogram is drawn for 20-kHz operation; at 30 kHz the power rating of the core will be higher.)

The static magnetization curves for the N27 ferrite (a typical transformer material) are shown in Fig. 2.2.3.

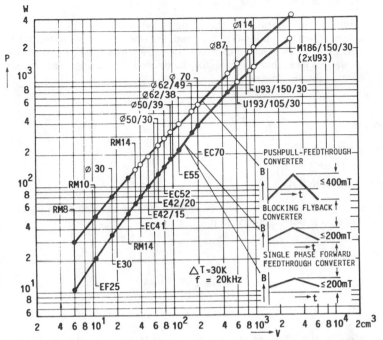

FIG. 2.2.2 Nomogram of transmissible power P as a function of core size (volume), with converter type as a parameter. (*Courtesy of Siemens AG.*)

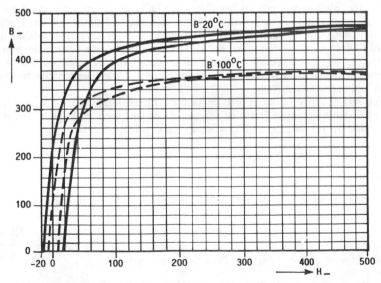

FIG. 2.2.3 Static magnetization curves for Siemens N27 ferrite material. (*Courtesy of Siemens AG.*)

2.4.2   Step 2, Selecting "on" Period

The maximum "on" period for the primary power transistor Q1 will occur at minimum input voltage and maximum load. For this example, it will be assumed that the maximum "on" period cannot exceed 50% of a total period of operation. (It will be shown later that it is possible to exceed this, using special control circuits and transformer designs.)

Example

   Frequency 30 kHz
   Period 33 μs
   Half period 16.5 μs

Allow a margin so that control will be well maintained at minimum input voltage; hence, the usable period is say 16 μs.

Hence

$$t_{on\ (max)} = 16\ \mu s$$

2.4.3   Step 3, Calculate Minimum DC Input Voltage to Converter Section

Calculate the DC voltage V_{cc} at the input of the converter when it is operating at full load and minimum line input voltage.

For the input capacitor rectifier filter, the DC voltage cannot exceed 1.4 times the rms input voltage, and is unlikely to be less than 1.2 times the rms input voltage. The absolute calculation of this voltage is difficult, as it depends on a number of factors which are not well defined—for example, the source impedance of the supply lines, the rectifier voltage drop, the characteristics and value of the reservoir capacitors, and the load current. Part 1, Chap. 6 provides methods of establishing the DC voltage.

For this example, a fair approximation of the working value of V_{cc} at full load will be given by using a factor of 1.3 times the rms input voltage. (This is again multiplied by 1.9 when the voltage doubling connection is used.)

Example

At a line input of 90 V rms, the DC voltage V_{cc} will be approximately $90 \times 1.3 \times 1.9 = 222$ V.

2.4.4   Step 4, Select Working Flux Density Swing

From the manufacturer's data for the E42/20 core, the effective area of the center leg is 240 mm². The saturation flux density is 360 mT at 100°C.

The selection of a working flux density is a compromise. It should be as high as possible in medium-frequency flyback units to get the best utility from the core and give minimum copper loss.

With typical ferrite core materials and shapes, up to operating frequencies of 30 kHz, the copper losses will normally exceed the core losses for flyback transformers, even when the maximum flux density is chosen; such designs are "saturation limited." Hence, in this example maximum flux density will be chosen; however, to ensure that the core will not saturate under any conditions, the lowest operating frequency with maximum pulse width will be used.

With the following design approach, it is likely that a condition of incomplete energy transfer will exist at minimum line input and maximum load. If this occurs, there will be some induction contribution from the effective DC component in the transformer core. However, the following example shows that as a large air gap is required, the contribution from the DC component is usually small; therefore the working flux density is chosen at 220 mT to provide a good working margin. (See Fig. 2.2.3.)

Hence, for this example the maximum peak-to-peak ac flux density B_{ac} will be chosen at 220 mT.

The total ac plus DC flux density must be checked in the final design to ensure that core saturation will not occur at high temperatures. A second iteration at a different flux level may be necessary.

2.4.5 Step 5, Calculate Minimum Primary Turns

The minimum primary turns may now be calculated using the volt-seconds approach for a single "on" period, because the applied voltage is a square wave:

$$N_{min} = \frac{V \cdot t}{\Delta B_{ac} \cdot A_e}$$

where N_{min} = minimum primary turns
$\quad\quad V = V_{cc}$ (the applied DC voltage)
$\quad\quad\ t$ = "on" time, μs
$\quad\ \Delta B_{ac}$ = maximum ac flux density, T
$\quad\quad A_e$ = minimum cross-sectional area of core, mm^2

Example

For minimum line voltage (90 V rms) and maximum pulse width of 16 μs

$$N_{min} = \frac{V \cdot t}{b \cdot A_e} = \frac{222 \times 16}{0.220 \times 181} = 89 \text{ turns}$$

Hence $\quad\quad\quad\quad\quad\quad N_{p(min)} = 89 \text{ turns}$

2.4.6 Step 6, Calculate Secondary Turns

During the flyback phase, the energy stored in the magnetic field will be transferred to the output capacitor and load. The time taken for this transfer is, once again, determined by the volt-seconds equation. If the flyback voltage referred to the primary is equal to the applied voltage, then the time taken to extract the energy will be equal to the time to input this energy, in this case 16 μs, and this is the criterion used for this example. Hence the voltage seen at the collector of the switching transistor will be twice the supply voltage, neglecting leakage inductance overshoot effects.

Example

At this point, it is more convenient to convert to volts per turn.

$$\text{Primary V/turn} = \frac{V_{cc}}{N_p} = \frac{222}{89} = 2.5 \text{ V/}N$$

The required output voltage for the main controlled line is 5 V. Allowing for a voltage drop of 0.7 V in the rectifier diode and 0.5 V in interconnecting tracks and the transformer secondary, the voltage at the secondary of the transformer should be, say, 6.2 V. Hence, the secondary turns would be

$$N_s = \frac{V_s}{V/N} = \frac{6.2}{2.5} = 2.48 \text{ turns}$$

where V_s = secondary voltage
N_s = secondary turns
V/N = volts per turn

For the low-voltage, high-current secondaries, half turns are to be avoided unless special techniques are used because saturation of one leg of the E core might occur, giving poor transformer regulation. Hence, the turns should be rounded up to the nearest integer. (See Part 3, Chap. 4.)

In this example the turns will be rounded up to 3 turns. Hence the volts per turn during the flyback period will now be less than during the forward period (if the output voltage is maintained constant). Since the volt-seconds/turn are less on the secondary, a longer time will be required to transfer the energy to the output. Hence, to maintain equality in the forward and reverse volt-seconds, the "on" period must now be reduced, and the control circuitry is able to do this. Also, because the "on" period is now less than the "off" period, the choice of complete or incomplete energy transfer is left open. Thus the decision on operating mode can be made later by adjusting primary inductance, that is, by adjusting the air gap.

It is interesting to note that in this example, if the secondary turns had been adjusted downward, the volts per turn during the flyback period would always exceed the volts per turn during the forward period. Hence, the energy stored in the core would always be completely transferred to the output capacitor during the flyback period, and the flyback current would fall to zero before the end of a period. Therefore, if the "on" time is not permitted to exceed 50% of the total period, the unit will operate entirely in the complete energy transfer mode, *irrespective of the primary inductance value*. Further, it should be noted that if the turns are rounded downward, thus forcing operation in the complete energy transfer mode, the primary inductance in this example will be too large, and this results in the inability to transfer the required power. In the complete energy transfer mode, the primary current must always start at zero at the beginning of the energy storage period, and with a large inductance and fixed frequency, the current at the end of the "on" period will not be large enough to store the required energy ($\frac{1}{2}LI^2$). Hence, the system becomes self power limiting, a sometimes puzzling phenomenon. The problem can be cured by increasing the core air gap, thus reducing the inductance. This limiting action cannot occur in the incomplete energy transfer mode.

Hence $$N_s = 3 \text{ turns}$$

2.4.7 Step 7, Calculating Auxiliary Turns

In this example, with three turns on the secondary, the flyback voltage will be less than the forward voltage, and the new flyback volts per turn V_{fb}/N is

$$\frac{V_{fb}}{N} = \frac{V_s}{3} = \frac{6.2}{3} = 2.06 \quad \text{V/turn}$$

The mark space ratio must change in the same proportion to maintain volt-seconds equality:

$$t_{on} = \frac{P \times V_{fb}/N}{V_{fb}/N + V/N} = \frac{33 \times 2.06}{2.06 + 2.5} = 14.9 \ \mu s$$

where t_{on} = "on" time of Q1
  P = total period, μs
  V_{fb}/N = new secondary flyback voltage per turn
  V/N = primary forward voltage per turn

The remaining secondary turns may then be calculated to the nearest half turn.

Example

For 12-V outputs, $$N_s = \frac{V_s}{V_{fb}/N} = \frac{13}{2.06} = 6.3 \text{ turns}$$

where V_s = 13 V for the 12-V output (allowing 1 V for the wiring and rectifier drop)
  V_{fb}/N = adjusted secondary volts per turn

Half turns may be used for these additional auxiliary outputs provided that the current is small and the mmf is low compared with the main output. Also, the gap in the outer limbs will ensure that the side supporting the additional mmf will not saturate. If only the center leg is gapped, half turns should be avoided unless special techniques are used. (See Part 3, Sec. 4.14.)

In this example, 6 turns are used for the 12-V outputs, and the output will be high by 0.4 V. (This can be corrected if required. See Part 1, Chap. 22.)

2.4.8 Step 8, Establishing Core Gap Size

General Considerations. Figure 2.2.1*a* shows the full hysteresis loop for a typical ferrite material with and without an air gap. It should be noted that the gapped core requires a much larger value of magnetizing force H to cause core saturation; hence, it will withstand a much larger DC current component. Furthermore, the residual flux density B_r is much lower, giving a larger usable working range for the core flux density, ΔB. However, the permeability is lower, resulting in a smaller inductance per turn (smaller A_L value) and lower inductance.

With existing ferrite core topologies and materials, it will be found that an air gap is invariably required on flyback units operating above 20 kHz.

In this design, the choice between complete and incomplete energy transfer has yet to be made. This choice may now be made by selecting the appropriate primary inductance, which may be done by adjusting the air gap size. Figure 2.2.1*b* indicates that increasing the air gap will lower the permeability and reduce the inductance. A second useful feature of the air gap is that at $H = 0$, flux retention B_r is much lower in the gapped case, giving

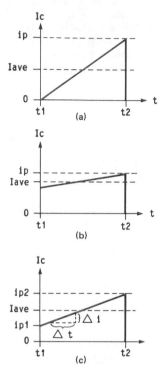

FIG. 2.2.4 Primary current waveforms in flyback converters. (*a*) Complete energy transfer mode; (*b*) incomplete energy transfer mode (maximum primary inductance); (*c*) incomplete energy transfer mode (optimum primary inductance).

a larger working range ΔB for the flux density. Finally, the reduced permeability reduces the flux generated by any DC component in the core; consequently, it reduces the tendency to saturate the core when the incomplete energy transfer mode is entered.

The designer now chooses the mode of operation. Figure 2.2.4 shows three possible modes. Figure 2.2.4*a* is complete energy transfer. This may be used; however, note that peak currents are very high for the same transferred energy. This mode of operation would result in maximum losses on the switching transistors, output diodes, and capacitors and maximum I^2R (copper) losses within the transformer itself. Figure 2.2.4*b* shows the result of having a high inductance with a low current slope in the incomplete transfer mode. Although this would undoubtedly give the lowest losses, the large DC magnetization component and high core permeability would result in core saturation for most ferrite materials. Figure 2.2.4c shows a good working compromise, with acceptable peak currents and an effective DC component of one-third of the peak value. This has been found in practice to be a good compromise choice, giving good noise margin at the start of the current pulse (important for current-mode control), good utilization of the core with reasonable gap sizes, and reasonable overall efficiency.

2.4.9 Step 9, Core Gap Size (The Practical Way)

The following simple, practical method may be used to establish the air gap.

Insert a nominal air gap into the core, say, 0.020 in. Run up the power supply with manual control of pulse width and a current probe in the transformer primary. Nominal input voltage and load should be used. Progressively increase the pulse width, being careful that the core does not saturate by watching the shape of the current characteristic, until the required output voltage and currents are obtained. Note the slope on the current characteristic, and adjust the air gap to get the required slope.

This gives a very quick method of obtaining a suitable gap that does not require Hanna curves. Even when gaps are calculated by other methods, some adjustment similar to the preceding will probably be required. This check is recommended as a standard procedure, as many supplies have failed at high temperature or under transient conditions because the transformer did not perform as intended.

2.4.10 Calculating the Air Gap

Using Fig. 2.2.4, the primary inductance may be established from the slope of the current waveform ($\Delta i / \Delta t$) as follows:

$$V_{cc} = L_p \cdot \frac{\Delta i_c}{\Delta t}$$

Example

From Fig. 2.2.4,

$$i_{p2} = 3i_{p1} \quad \text{(by choice)}$$

Therefore

$$I_m \text{ (mean current during the ''on'' period)} = 2 \cdot i_{p1}$$

The input power is 130 W, and therefore the average input current for the total period I_a may be calculated:

$$I_a = \frac{\text{input power}}{V_{cc}} = \frac{130}{222} = 0.586 \text{ A}$$

Therefore the mean current I_m for the ''on'' period will be

$$I_m = \frac{I_a \times \text{total period}}{\text{``on'' time}} = \frac{0.586 \times 33}{14.9} = 1.3 \text{ A}$$

The change of current Δi during the "on" period is $2 \cdot i_{p1} = I_m = 1.3$ A and the primary inductance may be calculated as follows:

$$L_p = \frac{V_{cc} \times \Delta t}{\Delta i} = \frac{222 \times 14.9 \times 10^{-6}}{1.3} = 2.54 \text{ mH}$$

Once the primary inductance L_p and number of turns N_p are known, the gap may be obtained using the Hanna curves (or A_L/DC bias curves) for the chosen core, if these are available. Remember,

$$A_L = \frac{L_p}{N_p^2}$$

If no data is available and the air gap is large (more than 1% of magnetic path length), assume that all reluctance is in the air gap, and calculate a conservative gap size using the following formula:

$$\alpha = \frac{\mu_r \times N_p^2 \times A_e}{L_p}$$

where α = total length of air gap, mm
　　　$\mu_r = 4\pi \times 10^{-7}$
　　　N_p = primary turns
　　　A_e = area of core, mm^2
　　　L_p = primary inductance, mH

Example

$$\alpha = \frac{4\pi \times 10^{-7} \times 89^2 \times 181}{2.54} = 0.7 \text{ mm or } 0.027 \text{ in}$$

Note: Use $\alpha/2$ if the gap goes right across the core. (In some cases, the area of the outer limbs is not equal to the area of the center core, in which case an adjustment must be made for this.)

2.4.11 Step 10, Check Core Flux Density and Saturation Margin

It is now necessary to check the maximum flux density in the core, to ensure that an adequate margin between the maximum working value and saturation is provided. It is essential to prevent core saturation under any conditions, including transient load and high temperature. This may be checked in two ways: by measurement in the converter, or by calculation.

Core Saturation Margin by Measurement

Note: It is recommended that this check be carried out no matter what design approach was used, as it finally proves all is as intended.

1. Set the input voltage to the minimum value at which control is still maintained—in this example, 85 V.
2. Set output loads to the maximum power limited value.
3. With a current probe in series with the primary winding P1, reduce the operating frequency until the beginning of saturation is observed (indicated by an upturn of current at the end of the current pulse). The percentage increase in the "on" time under these conditions compared with the normal "on" time gives the percentage flux density margin in normal operation. This margin should allow for the reduction in flux level at high temperatures (see Fig. 2.2.3), and an extra 10% should be allowed for variations among cores, gap sizes, and transient requirements. If the margin is insufficient, increase the air gap.

Core Saturation Margin by Calculation

1. Calculate the peak AC flux contribution B_{ac}, using the volt-seconds equation, and calculate or measure the values of "on" time and applied voltage, with the power supply at maximum load and minimum input voltage, as follows:

$$B_{ac} = \frac{V \cdot t}{N_p \cdot A_e}$$

where $V = V_{cc}$, V
$\quad\quad t =$ "on" time, μs
$\quad\quad N_p =$ primary turns
$\quad\quad A_e =$ area of core, mm^2
$\quad\quad B_{ac} =$ peak flux density change, T

Note: B_{ac} is the change in flux density required to sustain the applied voltage pulse and does not include any DC component. It is therefore independent of gap size.

Example

$$B_{ac} = \frac{222 \times 14.9}{89 \times 181} = 205 \text{ mT}$$

2. Calculate the contribution from the DC component B_{DC}, using the solenoid equation and the effective DC component I_{DC} indicated by the amplitude of the current at the beginning of an "on" period.
   By assuming that the total reluctance of the core will be concentrated in the air gap, a conservative result showing an apparently higher DC flux density will be obtained. This approximation allows a simple solenoid equation to be used.

$$B_{DC} = \mu_0 \cdot H = \frac{\mu_0 \cdot N_p \cdot I_{DC}}{\alpha \times 10^{-3}}$$

where $\mu_0 = 4\pi \times 10^{-7}$ H/m
N_p = primary turns
I_{DC} = effective DC current, A
α = air gap, mm
B_{DC} = DC flux density, T

Example

$$B_{DC} = \frac{4\pi \times 10^{-7} \times 89 \times 0.65}{0.7 \times 10^{-3}} = 103 \text{ mT}$$

The sum of the ac and DC flux density gives the peak value for the core. Check the margin against the core material characteristics at 100°C.

Example

$$B_{max} = B_{ac} + B_{DC} = 205 + 103 = 308 \text{ mT maximum}$$

2.5 FLYBACK TRANSFORMER SATURATION AND TRANSIENT EFFECTS

Note: The core flux level has been chosen for minimum input voltage and maximum pulse width conditions. It can be seen that this leaves a vulnerability to core saturation at high input voltages. However, under high-voltage conditions, the pulse width required for the transmitted power will be proportionately narrower, and the transformer will not be saturated.

Under transient load conditions, when the power supply has been operating at light loads with a high input voltage, if a sudden increase in load is demanded, the control amplifier will immediately widen the drive pulses to supply this extra power. A short period will now ensue during which both input voltage and pulse width will be maximum and the transformer could saturate, causing failure.

The following options should be considered to prevent this condition.

1. The transformer may be designed for the higher-voltage maximum-pulse-width condition. This will require a lower flux density and more primary turns. This has the disadvantage of reducing the efficiency of the transformer.

2. The control circuit can be made to recognize the high-stress condition and maintain the pulse width at a safe value during the transient condition. This is also somewhat undesirable, since the response time to the applied current demand will be relatively slow.

3. A third option is to provide a pulse-by-pulse current limit on the drive transistor Q1. This current-limiting circuit will recognize the onset of core saturation resulting from the sudden increase in primary current and will prevent any further increase in pulse width. This approach will give the fastest response

time and is the recommended technique. Current-mode control automatically provides this limiting action.

2.6 CONCLUSIONS

The preceding sections gave a fast and practical method for flyback transformer design. Many examples have shown that the results obtained by this simple approach are often close to the optimum design. The approach quickly provides a working prototype transformer for further development and evaluation of the supply.

In this design example, no attempt has been made to specify wire sizes, wire shapes, or winding topology. It is absolutely essential that these be considered, and the designer should refer to Part 3, Chap. 4, where these factors are discussed in more detail. It is important to realize that just filling the available bobbin area with the largest gauge of wire which will fit simply will not do for these high-frequency transformers. Because of proximity and skin effects (see App. 4.B on p. 3.99), the copper losses obtained in this way can quite easily exceed the optimum design values by a factor of 10 or more.

2.7 PROBLEMS

1. Calculate the minimum number of primary turns required on a complete energy transfer (discontinuous-mode) flyback transformer if the optimum flux density swing is to be 200 mT. (The core area is 150 mm, the primary DC voltage is 300 V, and the maximum "on" period is 20 μs.)

2. For the conditions in Prob. 2.7.1, calculate the secondary turns required to give an output voltage of 12 V if the flyback voltage is not to exceed 500 V (neglecting any overshoot). Assume the rectifier diode drop is 0.8 V.

3. Calculate the maximum operating frequency if complete energy transfer (discontinuous-mode) operation is to be maintained.

4. Calculate the required primary inductance, and hence the air gap length, if the transferred power is to be 60 W. (Assume maximum operating frequency, complete energy transfer, and no transformer loss. A transformer-grade ferrite core is used, and all reluctance is concentrated in the air gap.)

CHAPTER 3
REDUCING TRANSISTOR SWITCHING STRESS

3.1 INTRODUCTION

There are two major causes of high switching stress in the flyback converter. Both are associated with the turn-off behavior of the transistor with an inductive load. The most obvious effect is the tendency for the collector voltage to overshoot during the turn-off edge, caused mainly by the transformer leakage inductance. The second, less obvious effect is the high secondary breakdown stress that will occur during the turn-off edge if load line shaping is not used.

The voltage overshoot problem is best dealt with by ensuring that the leakage inductance is as small as possible, then clamping the tendency to overshoot by dissipative or energy recovery methods. The following section describes a dissipative clamp system. A more efficient energy recovery method using an extra winding is described in Part 2, Sec. 8.5.

If the energy recovery winding method is to be used with the flyback converter, the clamp voltage should be at least 30% higher than the reflected secondary voltage, to ensure efficient transfer of energy to the secondary. (The extra flyback voltage is required to drive current more rapidly through the secondary leakage inductance).

3.2 SELF-TRACKING VOLTAGE CLAMP

When a transistor in a circuit with an inductive or transformer load is turned off, the collector will tend to fly to a high voltage as a result of the energy stored in the magnetic field of the inductor or leakage inductance of the transformer.

In the flyback converter, the majority of the energy stored in the transformer will be transferred to the secondary during the flyback period. However, because of the leakage inductance, there will still be a tendency for the collector voltage to overshoot at the beginning of the flyback period unless some form of voltage clamp is provided.

In Fig. 2.3.1, the cumulative effects of transformer leakage inductance, the inductance of output capacitor, and loop inductance of the secondary circuit have been lumped together as L_{LT} and referred to the primary side of the transformer in series with the main primary inductance L_p.

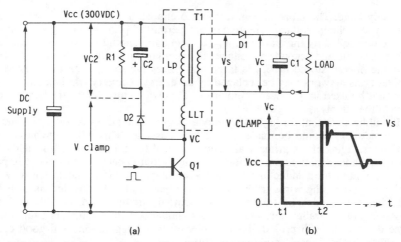

FIG. 2.3.1 (*a*) Stress-reducing self-tracking collector voltage clamp applied to the primary of a flyback converter. (*b*) Collector voltage waveform, showing voltage clamp action.

Consider the action during turn-off following an "on" period during which a current has been established in the primary winding of T1. When transistor Q1 turns off, all transformer winding voltages will reverse by flyback action. The secondary voltage V_s will not exceed the output V_c, except by the output rectifier diode drop D1. The collector of Q1 is partly isolated from this clamp action by the leakage inductance L_{LT}, and the energy stored in L_{LT} will take the collector voltage more positive.

If the clamp circuit, D2, C2, were not provided, then this flyback voltage could be damagingly high, as the energy stored in L_{LC} would be redistributed into the leakage capacitance seen at the collector of Q1.

However, in Fig. 2.3.1, under steady-state conditions, the required clamping action is provided by components D2, C2, and R1, as follows.

C2 will have been charged to a voltage slightly more positive than the reflected secondary flyback voltage. When Q1 turns off, the collector voltage will fly back to this value, at which point diode D2 will conduct and hold the voltage constant (C2 being large compared with the captured energy). At the end of the clamping action, the voltage on C2 will be somewhat higher than its starting value.

During the remainder of the cycle, the voltage on C1 will return to its original value as a result of the discharge current flowing in R1. The spare flyback energy is thus dissipated in R1. This clamp voltage is self-tracking, as the voltage on C2 will automatically adjust its value, under steady-state conditions, until all the spare flyback energy is dissipated in R1. If all other conditions remain constant, the clamp voltage may be reduced by reducing the value of R1 or the leakage inductance L_{LT}.

It is undesirable to make the clamp voltage too low, as the flyback overshoot has a useful function. It provides additional forcing volts to drive current into the secondary leakage inductance during the flyback action. This results in a more rapid increase in flyback current in the transformer secondary, improving the transfer efficiency and reducing the losses incurred in R1. This is particularly important for low-voltage, high-current outputs, where the leakage inductance is relatively large. Therefore, it is a mistake to choose too low a value for R1, and hence a low clamp voltage. The maximum permitted primary-voltage overshoot will be controlled by the transistor V_{CEX} rating and should not be less than 30% above the reflected secondary voltage. If necessary, use fewer secondary turns.

If the energy stored in L_{LT} is large and excessive dissipation in R1 is to be avoided, this network may be replaced by an energy recovery winding and diode, as would be used in a forward converter. This will return the spare flyback energy to the supply.

It will be clear that for high efficiency and minimum stress on Q1, the leakage inductance L_{LT} should be made as small as possible. This will be achieved by good interleaving of the primary and secondary of the transformer. It is also necessary to choose minimum inductance in the output capacitor, and, most important, minimum loop inductance in the secondary circuits. The latter may be achieved by keeping wires from the transformer as closely coupled as possible and ideally twisted. The tracks on the printed circuit board should run as a closely coupled parallel pair, and distances should be kept small. It is attention to these details that will provide high efficiency, good regulation, and good cross regulation in the flyback-mode power supply.

3.3 FLYBACK CONVERTER "SNUBBER" NETWORKS

The turn-off secondary breakdown stress problem is usually dealt with by "snubber networks"; a typical circuit is shown in Fig. 2.3.2. The design of the snubber network is more fully covered in Part 1, Chap. 18.

Snubber networks will be required across the switching transistor in off-line

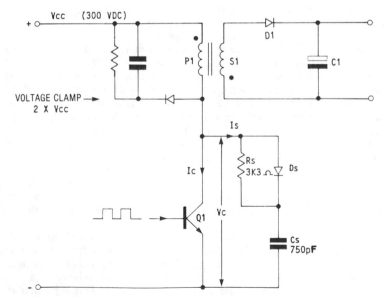

FIG. 2.3.2 Dissipative "snubber" circuit applied to the collector of an off-line flyback converter.

flyback converters to reduce secondary breakdown stress. Also, it is often necessary to snub rectifier diodes to reduce the switching stress and RF radiation problems.

In Fig. 2.3.2, snubber components D_s, C_s, and R_s are shown fitted across the collector and emitter of Q1 in a typical flyback converter. Their function is to provide an alternative path for the inductively driven primary current and reduce the rate of change of voltage (dv/dt) on the collector of Q1 during the turn-off action of Q1.

The action is as follows: As Q1 starts to turn off, the voltage on its collector will rise. The primary current will now be diverted via diode D_s into capacitor C_s. Transistor Q1 turns off very quickly, and the dv/dt on the collector will be defined by the original collector current at turn-off and the value of C_s.

The collector voltage will now ramp up until the clamp value ($2 \times V_{cc}$) is reached. Shortly after this, because of leakage inductance, the voltage in the output secondary winding will have risen to V_{sec} (equal to the output voltage plus a diode drop), and the flyback current will be commutated from the primary to the secondary via D1 to build up at a rate controlled by the secondary leakage inductance.

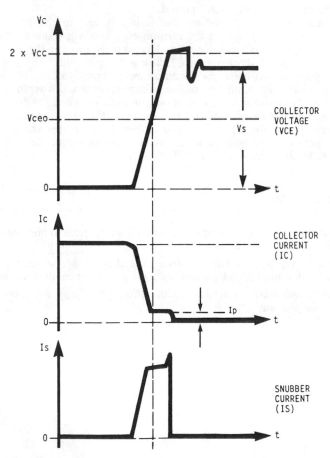

FIG. 2.3.3 Collector voltage and collector current waveforms, showing phase shift when dissipative snubber components are fitted. Also snubber current waveform during Q1 turn-off edge.

In practice, Q1 will not turn off immediately, and if secondary breakdown is to be avoided, the choice of snubber components must be such that the voltage on the collector of Q1 will not exceed V_{ceo} before the collector current has dropped to zero, as shown in Fig. 2.3.3.

Unless the turn-off time of Q1 is known, the optimum choice for these components is an empirical one, based upon measurements of collector turn-off voltage and current. Part 1, Chap. 18 and Fig. 1.18.2a, b, c, and d show typical turn-off waveforms and switching stress with and without snubber networks.

A safe voltage margin should be provided on the collector voltage when the current is zero, say at least 30% below V_{ceo}, as there is a considerable influence on these parameters from operating temperatures, loads, the spread of transistor parameters and the drive design. Figure 2.3.3 shows the limiting condition; in this example, the collector current has just dropped to zero as the collector voltage "hits" V_{ceo}.

On the other hand, too large a value of C3 should be avoided, since the energy stored in this capacitor at the end of the flyback period must be dissipated in R2 during the first part of the "on" period.

The value of R2 is a compromise selection. A very low resistance results in excessive current in Q1 during the turn-on edge and will result in excessive dissipation during the "on" transient. Too large a resistance, on the other hand, will not provide sufficient discharge of C3 during the minimum "on" period.

The values shown are a good compromise choice for the 100-W example. However, a careful examination of the voltage and current waveforms on the Q1 collector, under narrow pulse conditions, is recommended. The selection must always be a compromise for this type of snubber. The optimum selection of snubber components is more fully covered in Part 1, Chap. 18, and more effective snubber methods may be used which avoid a compromise. (See "The Weaving Low-Loss Snubber Diode," Part 1, Sec. 18.10.)

3.4 PROBLEMS

1. Why is the switching transistor particularly susceptible to high-voltage switching stress in the flyback converter?

2. Why does the flyback voltage often exceed the value which would be indicated by the turns ratio between the primary and secondary circuits?

3. Describe two methods used to reduce the high-voltage stress on the flyback switching element.

CHAPTER 4
SELECTING POWER COMPONENTS FOR FLYBACK CONVERTERS

4.1 INTRODUCTION

In general, a flyback converter is much more demanding on component ratings than would be a forward converter of the same power. In particular, the ripple current requirements for output rectifiers, output capacitors, transformers, and switching transistors are much larger. However, the increased cost incurred for the larger components will be offset by a reduction in circuit complexity, since output inductors will not be required and there is only one rectifier diode for each output line.

In flyback applications, components will be selected to meet the particular voltage and current needs of each unit. The designer must remember, however, that even for the same output power rating, different modes of operation impose different stress conditions on the components. The recommendations for the selection of power components in the following section, although particularly suitable for the flyback converter shown in Part 2, Chaps. 1 and 2, generally apply to all flyback converters.

The graphs and components shown are included for illustration only; it is not intended to suggest that they are necessarily the most suitable. Similar suitable components are available from many manufacturers.

4.2 PRIMARY COMPONENTS

4.2.1 Input Rectifiers and Capacitors

There are no special requirements imposed on the input rectifiers and storage capacitors in the flyback converter. Hence these will be similar to those used in other converter types and will be selected to meet the power rating and hold-up requirements. (See Part 1, Chap. 6.)

4.2.2 Primary Switching Transistors

The switching transistor in a flyback supply is very highly stressed. The current rating depends on the maximum load, efficiency, input voltage, operating mode, and converter design. It will be established by calculating the peak collector current at minimum input voltage and maximum load. In the examples shown in Fig. 2.2.4, the peak collector current ranges from three to six times the mean current, depending on the operating mode.

The maximum collector voltage is also quite high. It is defined by the maximum input voltage (off load), the flyback factor, the transformer design, the inductive overshoot, and the snubbing method.

For example, when working from a nominal 110-V ac supply line, the maximum input is typically specified as 137 V rms. For this input voltage the maximum off-load DC header voltage V_{cc} (using the voltage doubler input circuit) will be

$$V_{cc} = V_{in} \times \sqrt{2} \times 2$$

where V_{cc} = DC header voltage
$\quad\quad V_{in}$ = maximum ac input voltage, rms

In this example,

$$V_{cc} = 137 \times 1.42 \times 2 = 389 \text{ V}$$

Typically the flyback voltage is at least twice V_{cc}, or 778 in this example. Hence, allowing a 25% margin for inductive overshoot, the peak collector voltage will be 972 V, and a transistor with V_{cex} rating of 1000 V would be selected.

In addition to meeting these stressful conditions, the flyback transistor must provide good switching performance, low saturation voltage and have a useful gain margin at the peak working current. Since the selection of the transistor will also qualify the gain, it defines the requirements of the drive circuit. Hence, the selection of a suitable power transistor is probably the most important parameter for efficiency and long-term reliability in flyback converters.

Note: To avoid secondary breakdown, current tailing, and overdissipation in high-voltage bipolar transistors, it is essential that the correct drive and load line shaping be used.

Suitable drive circuits, waveforms, secondary breakdown, and tailing problems are discussed in Part 1, Chaps. 15, 16, 17, and 18.

4.3   SECONDARY POWER COMPONENTS

4.3.1   Rectifiers

The output rectifier diodes in flyback converters are subject to a large peak and rms current stress. The actual values depend on the load, conduction angle, leakage inductance, operating mode, and output capacitor ESR. Typically the rms current will be 1.6 to 2 × I_{DC}, while peak currents may be as high as 6 × I_{DC}. Because the precise conditions are often unknown, the calculation of diode currents is difficult, and empirical methods are recommended.

For the initial prototype breadboarding, select diodes of adequate mean and peak rating. Fast diodes with a reverse recovery time not exceeding 75 ns should be used. The final selection of the optimum rectifier diodes should be made after measurement of the prototype secondary rectifier currents. Attempts to calculate the rms and peak diode currents are generally not very accurate, as it is difficult to predict the various effects of leakage inductance, output loop inductance, pcb track and wiring resistance, and the ESR and ESL of the output capacitors. These parameters have a very significant effect on the rectifier rms and peak current requirements, particularly at low output voltages, high frequency, and high currents. These measurements can be made in the following way.

4.3.2 Rectifier Ripple Current Measurement Procedure

1. Connect a current probe of adequate rating in series with the output rectifier to be measured. (See Part 3, Chaps. 13 and 14 for suitable current probe design.)
2. Using the oscilloscope, observe the current waveform and note the peak current value.
3. Transfer the current probe to a true rms ammeter (e.g., a thermocouple instrument or true rms-reading instrument with a crest factor of at least 10/1) and measure the rms current, making due allowance for the current probe and meter multiplying factors. These measurements should be carried out at maximum input voltage and maximum load.

Select diodes with appropriate peak and rms ratings.

4.3.3 Rectifier Losses

The actual power loss in the output rectifier diode of a flyback supply depends on a number of factors, including forward dissipation, reverse leakage, and recovery losses. The forward dissipation depends on the effective forward resistance of the diode throughout its forward conduction and the shape of the current pulse, both of which are nonlinear. (In practice the secondary current waveform is often very different from the ideal triangular shape normally assumed in calculations.) Consequently, it is often more expedient to measure the temperature rise of the diode in the prototype, and from this calculate the junction temperatures and heat sink requirements for worst-case conditions.

From temperature rise measurements carried out on rectifiers in several flyback supplies (comparing the temperature rise caused by DC stress with ac stress conditions), it has been found that an approximate rectifier dissipation may be calculated, using the measured rectifier rms current (approximately $1.6\ I_{DC}$) and an assumed forward voltage drop of 800 mV for silicon diodes or 600 mV for Schottky diodes. Adequate heat sinks based on these calculations should be provided for initial prototypes. (See Part 3, Chap. 16.)

4.4 OUTPUT CAPACITORS

Output capacitors are also highly stressed in flyback converters. Normally the output capacitors will be selected for three major parameters: absolute capacitance value, capacitor ESR and ESL, and capacitor ripple current ratings.

4.4.1 Absolute Capacitance Value

When ESR and ESL are low, the capacitance value will control the peak-to-peak ripple voltage at the switching frequency. Because the ripple voltage is normally small compared with the mean output voltage, a linear decay of voltage across the output capacitors may be assumed during the "off" period. During this period, the capacitor must deliver all the output current, and the voltage across the capacitor terminals will decay at approximately 1 V/µs/A (for a 1-µF capacitor). Hence, when the maximum "off" time, the load current, and the required peak-to-peak ripple voltage are known, the minimum output capacitance may be calculated as follows:

$$C = \frac{t_{\text{off}} \times I_{\text{DC}}}{V_{p-p}}$$

where C = output capacitance, µF

t_{off} = off time, µs

I_{DC} = load current, A

V_{p-p} = ripple voltage p-p

For example, for a 5-V, 10-A output line and an output ripple of 100 mV,

$$C = \frac{18 \times 10^{-6} \times 10}{0.1} = 1800 \ \mu\text{F}$$

Note: Attempts to keep the peak-to-peak ripple voltage below 100 mV in a single-stage output filter will not be cost-effective. To obtain a lower output ripple, an extra LC stage should be fitted.

4.4.2 Capacitor ESR and ESL

Figure 2.4.1*a* and *b* shows the effect of the ESR and ESL (effective series resistance and inductance) of the output capacitor on the output ripple. In practice, the ripple voltage will be much greater than would be expected from the selection of the output capacitance alone, and an allowance for this effect should be made when selecting capacitor size. If a single-stage output filter is used (no extra series choke), then the ESR and ESL of the output capacitors will have a significant effect on the high-frequency ripple voltage, and the best low-ESR capacitors should be used.

The beneficial effects of an additional output LC filter should not be neglected in flyback systems. Such filtering reduces output noise and allows the use of much lower cost ordinary-grade electrolytics of adequate ripple rating as the major energy storage elements. (See Part 1, Chap. 20.)

4.4.3 Capacitor Ripple Current Ratings

In a flyback converter, the typical rms ripple current in the output capacitors will be 1.2 to 1.4 times the DC output current. (See Part 1, Chap. 20.) The output

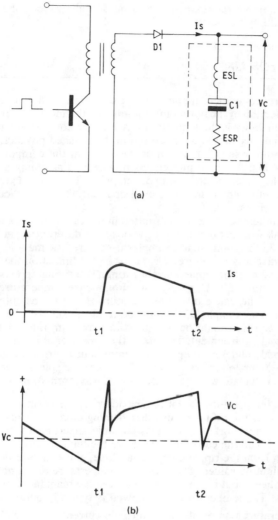

FIG. 2.4.1 (*a*) Secondary circuit of a flyback converter, showing parasitic series components ESL and ESR. (*b*) Output voltage waveforms, showing effect of parasitic components.

capacitors must be capable of conducting the output ripple current without excessive temperature rise.

For a more accurate assessment of the ripple current, the following measurement procedure is recommended. Using a current probe of adequate rating, measure the rms current in the output capacitor leads under full load at maximum line input. (A true rms meter should be used with the current probe.) Select a capacitor to meet the ripple requirements, making due allowance for frequency and temperature multiplying factors. (See Part 3, Chap. 12.)

4.5 CAPACITOR LIFE

Although the preceding measurements and calculations will give a good starting point for the selection of the optimum components, the most important parameter for long-term reliability is the temperature rise of the components in the working environment, and this should be measured in the finished product.

The temperature rise is a function of the stress in the component, heat sink design, air flow, and the proximity effects of surrounding components. Radiated and convected heat from nearby components will often cause a greater temperature rise in a component than internal dissipation. This is particularly true for electrolytic capacitors.

The maximum temperature rise permitted in the capacitor, as a result of ripple current and peak working temperature, varies with different capacitor types and manufacturers. In the component examples used here, the maximum rise permitted for ripple current is 8°C in free air, and it is this limitation that the manufacturer uses to establish the ripple current rating. The rating applies up to an ambient air temperature of 85°C, giving a maximum case temperature of 93°C.

Irrespective of the cause of the temperature rise, the absolute limit of case temperature (in this example, 93°C) must be used to establish the limits of operation of the unit. This should be measured at maximum rated temperature and load, in its normal environment. The life of the capacitor at its maximum temperature is not good, and lower operating temperatures are recommended. If in doubt as to the temperature rise caused by internal ripple current (this can be very difficult to calculate with complex flyback waveforms), proceed as follows:

1. Measure the temperature rise of the capacitor under normal operating conditions away from the influence of other heating effects. (If necessary, mount the capacitor on a short length of twisted cable, inserting a thermal barrier between the unit and the capacitor.) Measure the temperature rise of the capacitor resulting from the ripple component alone, and compare this with a manufacturer's limiting values. The permitted temperature rise is not always given on the data sheets, but it can be obtained from the manufacturer's test and QA departments. The temperature rise allowed is typically between 5 and 10°C.

2. If the temperature rise resulting from ripple current is acceptable, mount the capacitor in its normal position and subject the power supply to its highest-temperature stress and load conditions. Measure the surface temperature of the capacitor, and ensure that it is within the manufacturer's rating. This way you can be sure to avoid thermal runaway, a possibility with electrolytic capacitors.

4.6 GENERAL CONCLUSIONS CONCERNING FLYBACK CONVERTER COMPONENTS

The power elements of a flyback system have been discussed in considerable detail. Careful attention to the ratings and operating conditions for every component is essential for good performance and reliable operation. To the power supply engineer, this will become second nature. The selection is a laborious process that cannot be avoided if the most cost-effective and reliable selections are to be made. Calculations can take the designer only part of the way, as much of the information critical to these selections is just not available without making the appropriate measurements.

The leakage inductance of the transformer, the track layout and size, the values of ESR and ESL of the output components, component layout, and cooling arrangements have a considerable influence on the stress and ratings of the components. These effects cannot be reliably predicted. When actual measurements are not made, a wide safety margin must be applied to the calculated values in the selection of component ratings.

Much of the optimization and proving measurements can be more easily carried out at the design approval stage and will be limited to those prototypes which are destined for final production.

If long-term reliable operation and cost-effective design are to be achieved, it is incumbent upon the cognizant engineer to ensure that the design is optimized before the product is finalized and that all necessary approval testing is carried out.

4.7 PROBLEMS

1. Explain the major parameters that control the selection of the switching transistor in a flyback converter.
2. What controls the selection of secondary rectifier diodes in a flyback converter?
3. Which parameters of the flyback converter control the selection of the output capacitors?

CHAPTER 5
THE DIAGONAL HALF-BRIDGE FLYBACK CONVERTER

5.1 INTRODUCTION

This converter, also known as the two-transistor converter, is particularly suitable for power field-effect transistor (FET) operation. Hence, FET devices are shown in the example used here, but the same design procedure would apply for transistor operation.

The topology also lends itself to all the previous modes of flyback operation—that is, fixed frequency, variable frequency, and complete or incomplete energy transfer operation. However, there are cost penalties incurred for the additional power device and its isolated drive.

5.2 OPERATING PRINCIPLE

In the circuit shown in Fig. 2.5.1, the high-voltage DC line is switched to the primary of a transformer by two power FET transistors, FT1 and FT2. These switches are driven by the control circuitry such that they will both be either "on" or "off" together. Flyback action takes place during the "off" state, as in the previous flyback examples.

The control, isolation, and drive circuitry will be very similar to that previously used for single-ended flyback converters. A small drive transformer is used to provide the simultaneous but isolated drives to the two FET switches.

It should be noted that the cross-connected diodes D1 and D2 return excess flyback energy to the supply lines and provide hard voltage clamping of FT1 and FT2 at a value of only one diode drop above and below the supply-line voltages. Hence switching devices with a 400-V rating may be confidently used, and this topology lends itself very well to power FETs. Moreover, the energy recovery action of diodes D1 and D2 eliminates the need for an energy recovery winding or excessively large snubbing components. The voltage and current waveforms are shown in Fig. 2.5.2.

Because the transformer leakage inductance plays an important role in the action of the circuit, the distributed primary and secondary leakage inductive components have been lumped into effective total inductances L_{Lp} and L_{Ls} and shown external to the "ideal" transformer for the purpose of this explanation.

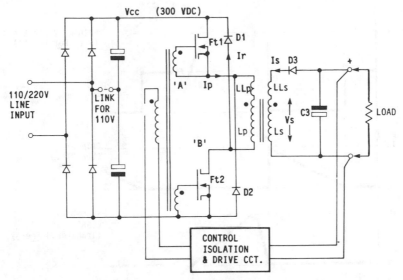

FIG. 2.5.1 Diagonal half-bridge (two-transistor) single-ended flyback converter using power FET primary switches.

The power section operates as follows: When FT1 and FT2 are on, the supply voltage will be applied across the transformer primary L_p and leakage inductance L_{Lp}. The starts of all windings will go positive, and the output rectifier diode D3 will be reverse-biased and cut off; therefore secondary current will not flow during the "on" period and the secondary leakage inductance L_{Ls} can be neglected.

During the "on" period current will increase linearly in the transformer primary (see Fig. 2.5.2) as defined by the equation:

$$\frac{dI_p}{dt} = \frac{V_{cc}}{L_p}$$

Energy of $\frac{1}{2} L_p \cdot I_p^2$ will be stored in the coupled magnetic field of the transformer, and energy of $\frac{1}{2} L_{L_p} \cdot I_p^2$ in the effective leakage inductances.

At the end of the "on" period, FT1 and FT2 will turn off simultaneously, and the primary supply current in the FET will fall to zero. However, the magnetic field strength cannot change without a corresponding change in the flux density, and by flyback action all voltages on the transformer will reverse. Initially diodes D1 and D2 are brought into conduction clamping the primary flyback voltage (developed by the primary leakage inductance) to the supply-line voltage. Since the polarity is reversed on all windings, the secondary emf V_s will also bring the output rectifier diode D3 into conduction, and current I_s builds up in the secondary winding, as defined by the secondary leakage inductance L_{L_s}.

When the secondary current has built up to a value of $n \times I_p$, where n is the turns ratio and the energy stored in the primary leakage inductance L_{Lp} has been transferred back to the supply line, the energy recovery clamp diodes D1 and D2 will cease conduction, and the primary voltage V_p will fall back to the reflected

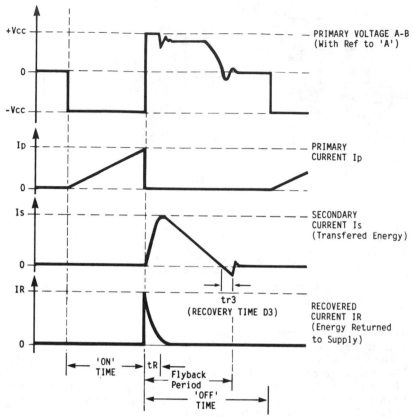

FIG. 2.5.2 Primary and secondary waveforms for diagonal half-bridge flyback converter, showing "recovered" energy (energy returned to the supply).

secondary voltage. At this time the voltage across the primary will be the voltage across C3 (as referred to the primary by normal transformer action). This clamped flyback voltage must by design be less than the supply voltage V_{cc}; otherwise the flyback energy will all be returned to the supply. However, under normal conditions, in a complete energy transfer system, the remaining energy stored in the transformer magnetic field will be transferred to the output capacitor and load during the remaining "off" period of FT1 and FT2. At the end of the "off" period, a new power cycle will start, and the process continues.

5.3 USEFUL PROPERTIES

This type of converter has a number of useful properties that should not be overlooked.

First (and particularly important for power FET operation), the voltages on the two power devices cannot exceed the supply voltage by more than two diode drops for any operating condition, provided that fast-action clamping diodes are used for D1 and D2. This very hard voltage clamping action is ideal for power FET operation, as these devices are particularly vulnerable to overvoltage stress.

Second, any energy stored in primary leakage inductance will be returned to the supply line by D1 and D2 at the beginning of the flyback period and is not lost to the system.

Third, under transient loading conditions, if excessive energy has been stored in the transformer primary during the previous "on" period, this will also be returned to the supply line during the flyback period.

Fourth, compared with the single-ended flyback converter, the power devices may be selected for a much lower operating voltage, since the doubling effect which occurs with a single-ended system is absent in this topology.

Finally, a major advantage of this technique is that a bifilar-wound energy recovery winding will not be required; hence the cost and a possible source of unreliability are eliminated.

5.4 TRANSFORMER DESIGN

The hard voltage clamping action of the cross-connected primary energy recovery diodes (D1 and D2), and the preference to operate at higher frequency with FET devices, means that the primary and secondary leakage inductances of the transformer will play an important role in the operation of the supply.

The energy stored in the primary leakage inductance L_{Lp} cannot be transferred to the output circuit; it gets returned to the supply. Hence the leakage inductance results in a useless (loss-generating) interchange of energy in the primary circuit. Also, the secondary leakage inductance results in a slow buildup of current in the secondary rectifiers during the flyback period. This delay means that an additional proportion of the stored energy is returned to the primary circuit and will not be transferred to the output. This proportion increases if the frequency is increased, and clearly the leakage inductance must be minimized for best performance.

A further basic difference between the performance of this arrangement and that of the normal single-ended flyback converter must be considered in the transformer design. In the single-ended flyback converter, it is common practice to allow the flyback voltage to be as large as possible so as to drive the secondary current more rapidly through the output leakage inductance. In the diagonal half-bridge flyback converter, the flyback voltage cannot exceed the forward voltage, since the same primary winding carries out the forward polarization and reverse flyback energy return functions. Hence, because of the hard clamping provided by the primary diodes D1 and D2, it is not possible to increase the primary flyback voltage above the supply lines, and for this application, it is particularly important to design the transformer for minimum leakage inductance.

When selecting secondary turns, the transferred secondary flyback voltage as applied to the primary should be at least 30% lower than the minimum applied primary voltage; otherwise an excessive proportion of the stored energy will be returned to the input line via D1 and D2 at the beginning of the flyback period.

In all other respects, the transformer design procedure is identical to that in the single-ended flyback case, Part 2, Chap. 2, and the same procedure should be adopted.

5.5 DRIVE CIRCUITRY

To ensure rapid and efficient switching of the power FETs, the drive circuit must be capable of charging and discharging the relatively large gate input capacitance of the FETs quickly. Special low-resistance drive circuitry should be used for this application.

5.6 OPERATING FREQUENCY

The use of power FETs permits efficient high-frequency operation of the primary power switches. The size of the transformer and output capacitors may be reduced at high frequencies, but the leakage inductance of the transformer, the ESR of the output capacitors, and fast recovery of the rectifiers now becomes particularly important. Therefore, for high-frequency operation, not only must the transformer be correctly designed, but the external components must also be selected correctly.

5.7 SNUBBER COMPONENTS

Because power FET devices are not subject to the same secondary breakdown mechanisms that occur with bipolar devices, from a reliability standpoint, it is often considered that snubber components are not essential. However, in most FET applications, a small *RC* snubber network will still be fitted across the FETs to reduce RF radiation and meet the *dv/dt* limitations of the FET. (With very high *dv/dt,* some power FETs display a failure mode resulting from conduction of the internal parasitic transistor.) However, it is true that the larger snubber components normally associated reducing secondary breakdown stress for bipolar transistors are not required with power FETs.

To reduce the length of the primary HF current path, a low-inductance capacitor should be fitted across the supply lines as close to the power switches and energy recovery diodes D1 and D2. This is particularly important in high-frequency converters.

5.8 PROBLEMS

1. How does the primary topology of the diagonal half-bridge flyback converter differ from that of a single-ended flyback converter?
2. What is the major advantage of the diagonal half-bridge topology?
3. Why is the diagonal half-bridge flyback converter topology particularly suitable for power FET operation?
4. Why is the leakage inductance in the diagonal half-bridge topology particularly important to its performance?

CHAPTER 6

SELF-OSCILLATING DIRECT-OFF-LINE FLYBACK CONVERTERS

6.1 INTRODUCTION

The class of converters considered in this section relies on positive feedback from the power transformer to provide the oscillatory behavior.

Because of their simplicity and low cost, these converters can provide some of the most cost-effective solutions for multiple-output low-power requirements. With good design, extremely efficient switching action and reliable performance are obtained. A number of problems which beset the driven converter—in particular, cross conduction and transformer saturation—are overcome as a natural consequence of the self-oscillating topology. As the mode of operation is always complete energy transfer, current-mode control can very easily be applied, giving fast, stable single-pole loop response.

Self-oscillating converters are extremely low cost because few drive and control components are required. Attention to good filter design and screening of the transformer make this converter suitable for computers, video display terminals, and similar demanding applications.

There is a tendency to assume that these simple units are not real contenders in professional power supply applications. This misconception is probably due to the rather poor performance of some of the early self-oscillating designs. Also, the variation of operating frequency with load and input voltage has been considered undesirable in some applications. However, since the output is DC, there should not be a problem with the operating frequency in most applications, so long as efficient input and output filtering and magnetic screening have been provided.

6.2 CLASSES OF OPERATION

There are three classes of operation:

Type A, fixed "on" time, variable "off" time
Type B, fixed "off" time, variable "on" time
Type C, variable "on" time, "off" time, and repetition rate (frequency)

The major differences in the performances of these classes are as follows:

Type A will operate at an extremely low frequency when the load is light.

Type B will have a low frequency when the load is maximum.

Type C has a more desirable characteristic, as the frequency remains reasonably constant from full load down to approximately 20% load. Below 20% load the frequency will usually become progressively higher. (See Fig. 2.6.1.)

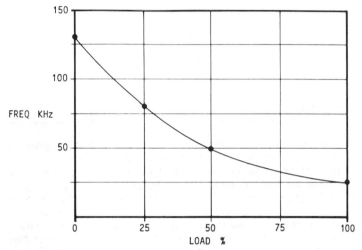

FIG. 2.6.1 Typical frequency variation of self-oscillating converter as a function of load.

6.3 GENERAL OPERATING PRINCIPLES

In the self-oscillating converters considered here, the switching action is maintained by positive feedback from a winding on the main transformer. The frequency is controlled by a drive clamping action which responds to the increase in magnetization current during the "on" period. The amplitude at which the primary current is cut off, and hence the input energy, is controlled to maintain the output voltage constant. The frequency is subject to variations caused by changes in the magnetic properties of the core, the loading, or the applied voltage.

Figure 2.6.2 shows the major power components of a unit of type C. The converter is self-oscillating as a result of the regenerative feedback to the power transistor base from the feedback winding P2. The circuit functions as follows.

After switch-on, a voltage is developed across C1, and a current will flow in R1 to initiate turn-on of transistor Q1. As Q1 starts to turn on, regenerative feedback is applied via the feedback winding P2 to increase the positive drive to the base of Q1. The base current flowing initially in C2 and then in D1 as the drive voltage is established. Consequently, Q1 will turn on very rapidly, and its maxi-

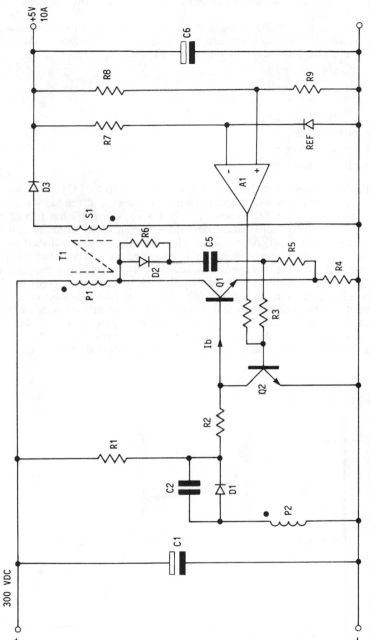

FIG. 2.6.2 Nonisolated, single-transformer, self-oscillating flyback converter, with primary current-mode control.

2.51

mum drive current will be defined by resistors R2 and R1 and the voltage across the feedback winding P2.

As these units are operating in the complete energy transfer mode, when Q1 turns on, current will build up from zero in P1 (the primary winding of the main transformer) at a rate defined by the primary inductance L_p. Hence

$$\frac{dI_p}{dt} = \frac{V_{cc}}{L_p}$$

where I_p = primary current
V_{cc} = primary voltage
L_p = primary inductance

As the collector current, and therefore the emitter current, of Q1 increases, a voltage will be developed across R4 which will also increase at the same rate toward the turn-on voltage of Q2 (approximately 0.6 V). When Q2 has turned on sufficiently to divert the majority of the base drive current away from the base of Q1, Q1 will begin to turn off. At this point, the collector voltage will start to go positive, and regenerative turn-off action is provided by the snubber current flowing in D2, C5, R5. The voltage developed across R5 will assist the turn-on action of Q2 and turn-off of Q1. Further, by flyback action, all voltages on the transformer T1 will reverse, providing additional regenerative turn-off of Q1 by P2 going negative; the reverse current flow in C2 assists the turn-off action of Q1.

Although this drive system is extremely simple, it operates in a very well defined way. Examination of the base current of Q1 will show almost ideal drive waveforms (see Fig. 2.6.3). The reason for the turn-off slope shown in Fig. 2.6.3

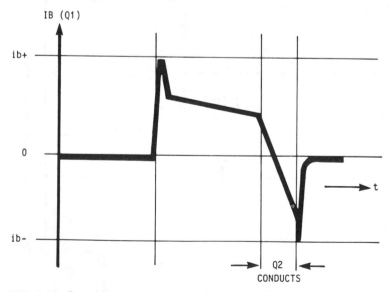

FIG. 2.6.3 Base drive current waveform of self-oscillating converter.

when Q1 conducts is that toward the end of the "on" period Q2 responds to a base drive voltage which is ramping upward; therefore Q2 turns on progressively, giving the very desirable downward ramp to the base drive current of Q1. This is the ideal drive waveform for most high-voltage transistors, as regenerative turn-off action does not occur until all carriers have been removed from the base of Q1 and the collector current has started to fall. This turn-off waveform prevents hot-spot generation in the transistor Q1 and secondary breakdown problems. (See Part 1, Chap. 15.)

The system also has automatic primary power limiting qualities. The maximum current that can flow in R4 before transistor Q2 turns on is limited to $V_{be}/R4$, even without drive from the control circuit. Consequently, automatic overpower limitation is provided without the need for further current-limiting circuitry.

In normal operation, the control circuit will respond to the output voltage and apply a drive signal to the base of Q2, taking the Q2 base more positive. This will reduce the current required in R4 to initiate turn-off action. Consequently, the output power can be continuously controlled so as to maintain the output voltage constant in response to load and input variations.

In foldback current-limiting applications, additional information on output current and voltage is processed by the control circuit to reduce the power limit under short-circuit conditions. Note that a constant primary power limit (on its own) would provide very little protection for the output circuitry, as there would be a large current flowing in the output when the output voltage is low or a short circuit is applied.

6.4 ISOLATED SELF-OSCILLATING FLYBACK CONVERTERS

A more practical implementation of the self-oscillating technique is shown in Fig. 2.6.4. In this example, the input and output circuits are isolated, and feedback is provided by an optical coupler OC1.

Components D3, C4, and R8 form a self-tracking voltage clamp (see Sec. 3.2). This clamp circuit prevents excessive collector voltage overshoot (which would have been generated by the primary leakage inductance) during the turn-off action of Q1.

Components D1 and C3 are the rectifier and storage capacitor for the auxiliary supply line which provides the supply to the control optocoupler OC1.

6.5 CONTROL CIRCUIT (BRIEF DESCRIPTION)

A very simple control circuit is used. The diode of the optical coupler OC1 is in series with a limiting resistor R9 and a shunt regulator U1 (Texas Instruments TL430).

When the reference terminal of the shunt regulator V1 is taken to 2.5 V, current will start to flow into the cathode of V1 via the optocoupler diode, and control action is initiated. The ratio of R12 and R11 is selected for the required output, in this case 12 V.

The optocoupler transistor responds to the output control circuit so as to apply a bias current to R3. A voltage divider network is formed by OC1 and R3 and

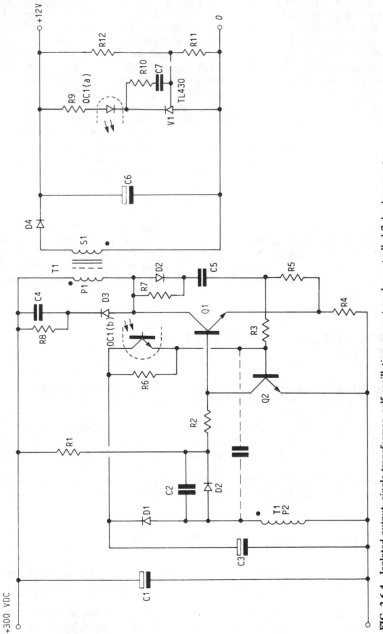

FIG. 2.6.4 Isolated-output, single-transformer, self-oscillating, current-mode-controlled flyback converter. The control loop is closed to the output using an optical coupler.

the base of Q2 as the optocoupler current increases, and so the ramp voltage required across R4, and hence the collector current required to turn Q2 on and Q1 off, will be reduced. (A more complete description of this control circuit is given in Sec. 7.4.)

As Q1 starts to turn off, its collector voltage will go positive, and the collector current will be diverted into the snubber components D2, C5, and R5. The voltage across R5 results in an increase in base drive voltage to R3 and Q2, as R5 has a higher resistance than R4, more than compensating for the drop in voltage on R4. This gives a further regenerative turn-off action to Q1. (The action of the snubber components is more fully described in Part 1, Chap. 18.)

This simple circuit has a number of major advantages.

First, the unit always operates in a complete energy transfer mode. Consider the switching action: When Q1 turns off, flyback current will flow in the output circuitry. The transformer voltages will be reversed and the drive winding P2, negative. Consequently, Q1 will remain turned off until all the energy stored in the magnetic field has been transferred to the output capacitors and load.

At that time, the voltage across all windings will decay toward zero. Now C2, which would have been charged during the flyback period, will track the positive-going change of voltage on P2 and take the base of Q1 positive. Once again, by regenerative action, augmented by the current drive through R1, Q1 will turn on. Consequently, a new "on" period will be initiated immediately after the stored energy has been transferred to the output capacitors and load. Complete energy transfer will take place irrespective of the loading or input voltage.

The transformer design is simplified, as there will be no DC component to consider in the design process, and the full flux capability of the core can be exploited with confidence. There is a further protection mechanism should the core begin to saturate for any reason. This saturation effect will be recognized by the increase in current in R4, and the "on" pulse will be terminated earlier. As a result of this action, there will be an increase in the frequency of operation such that saturation no longer occurs. This allows the designer to confidently utilize the full flux excursion ability of the core without the need for excessive flux margin to prevent saturation.

A typical plot of frequency against load for this type of converter is shown in Fig. 2.6.1. Note that at very light loads, very high operating frequencies are possible. To prevent excessive dissipation in the switching transistor and snubber components, this high-frequency mode should be avoided by using the power unit for applications where the minimum load is not less than 10%. Alternatively, dummy load resistors may be applied.

The normal snubber arrangements and voltage clamps described in Part 1, Chap. 18 and Figs. 2.3.1 and 2.3.2 would be used. A transformer designed for the fixed-frequency flyback converter (Sec. 2.2) will be found to operate quite satisfactorily in this variable-frequency unit. However, some improvement in efficiency may be realized by making use of the extra flux capability and reducing the primary turns accordingly. To give good regenerative action, the drive voltage generated by P2 should be at least 4 V.

In the final design, extra circuitry will often be used to improve the overall performance; for example, a positive bias may be applied in series with the drive winding P2 to speed up the turn-on action, which would tend to be rather slow in the example shown. A square-wave voltage bias may be applied to the base of Q2 by a capacitor (shown dashed in Fig. 2.6.4) or a resistor to improve the switching action under light load conditions. This decreases the minimum loading requirements by reducing the switching frequency under light loads and thus reducing the current at which squegging occurs.

6.6 SQUEGGING

In this application, "squegging" refers to a condition in which a number of pulses are generated, followed by a quiescent period, on a repetitive basis. The cause of "squegging" is that for correct switching action at light loads, a very narrow minimum "on" period is necessary. However, as a result of the transistor storage time, under light load conditions, this minimum "on" period will have more energy content than is required to maintain the output voltage constant. Hence there will be a progressive increase in output voltage as a number of pulses are generated. Because progressive control is lost at some point, the control circuit has no option but to turn off the switching transistor completely, and a quiet period now follows until the output voltage recovers to its correct value. With good drive design, giving minimum storage time, this "squegging" action will not occur except at loads below 2% or 3%. In any event, it is a nondamaging condition.

6.7 SUMMARY OF THE MAJOR PARAMETERS FOR SELF-OSCILLATING FLYBACK CONVERTERS

The component count is clearly very low, giving good reliability at economic cost.

The converter transformer may be designed to operate very near the maximum flux density limit, as the power transistor switches off at a well-defined current level. Any tendency to saturate is recognized by the control circuit, and the "on" pulse is terminated. (The frequency automatically adjusts to a higher value at which the core will not saturate.) This self-protecting ability leaves the designer free to use the maximum flux range, if desired, giving a more efficient power transformer with fewer primary turns.

The unit always operates in the complete energy transfer mode, so that by using current-mode control, automatic protection for overloads is provided and performance improved. (The current control mode is explained more fully in Part 2, Chap. 7 and Part 3, Chap. 10.)

The complete energy transfer mode (discontinuous mode) avoids the "right-half-plane zero" stability problems. (See Part 3, Chap. 9.)

Additional windings on the main transformer will provide isolated auxiliary supplies for the control circuit, or additional outputs.

Input-to-output isolation may be provided by optocouplers or control transformers in the control feedback path.

A possible disadvantage of the technique is that the frequency will change with variations of load or input voltage. This should not be a problem so long as adequate input and output filtering is provided and the supply is located or screened so that magnetic radiation from the wound components will not interfere with the performance of adjacent equipment.

This type of supply has been used very successfully in video display units and has replaced fixed-frequency or synchronized power units in many applications.

6.8 PROBLEMS

1. What are the major advantages of the self-oscillating off-line flyback converter?

2. What is the major disadvantage of the self-oscillating technique?

3. The flyback self-oscillating converter operates in a complete energy transfer mode. Why is this an advantage?

CHAPTER 7

APPLYING CURRENT-MODE CONTROL TO FLYBACK CONVERTERS

7.1 INTRODUCTION

In the flyback converter, the primary inductance of the transformer (more correctly a multiple-winding inductor) is generally much lower than the inductance of its counterpart in the forward converter. Hence, the rate of change of current during the primary conduction phase (the "on" period) is large, giving a large triangular shape to the primary current pulse. This triangular waveform is ideal for the application of current-mode control, providing good noise immunity and well-defined switching levels to the current comparator.

With current-mode control there are two control loops in operation. The first, fast-acting inside loop controls the peak primary current, while the second, much slower outside loop adjusts the current control loop to define the output voltage. The overall effect of these two control loops is that the power supply responds as a voltage-controlled current source.

There are a number of advantages to be gained from current-mode control. First of all, the system responds as if the primary is a high-impedance current source and the effective inductance of the converter transformer is removed from the output filter equivalent circuit. This results in a simple first-order transfer function. Hence the control circuit may have a good high-frequency response, improving the input transient performance. Line ripple rejection and loop stability are improved. A second major advantage is that primary current limiting is automatically provided without additional components. (See Part 3, Chap. 10.)

7.2 POWER LIMITING AND CURRENT-MODE CONTROL AS APPLIED TO THE SELF-OSCILLATING FLYBACK CONVERTER

The self-oscillating complete energy transfer flyback converter responds particularly well to the application of current-mode control. This will be explained by reference to the circuit shown in Fig. 2.6.4.

The voltage across R4 (which sets the maximum collector current) cannot exceed 0.6 V under any conditions, because at this point Q2 will turn on, turning off the power device Q1. This will occur irrespective of the condition of the voltage control circuit, because the control circuit cannot take the voltage at the base of Q2 negative. Consequently, the peak input current is defined by R4 and Q2 V_{be} and cannot be exceeded. Thus R4 and Q2 provide automatic primary power limiting. The action of the remainder of the control circuit can only reduce the limit still further.

Although a very simple circuit is used, the power limiting action is excellent. The "on" period is terminated when the primary current has reached a defined peak level on a pulse-by-pulse basis. This current-defining loop also sets a maximum limit on the transmitted power, $\frac{1}{2}L_p \cdot I_p^2 \cdot f$.

It should be noted that because the limiting action is a constant power limit, as the output voltage goes toward zero under overload conditions, the output current will increase. If this is unacceptable, then additional circuitry will be required to reduce the power limit as the load goes toward short-circuit, or to turn the supply off for overload conditions.

The second, much slower voltage control loop (R11, R12, V1, OC1a, and OC1b) responds to output voltage changes and adjusts the bias on Q2 so as to reduce the peak value of primary current required to turn Q2 on. This level is adjusted in response to the voltage control circuit to maintain the output voltage constant.

7.3 VOLTAGE CONTROL LOOP

Figure 2.7.1 shows the collector and emitter current waveforms of Q1 under steady-state voltage-controlled conditions. The emitter current waveform shows a DC offset as a result of the base drive current component I_b. An analogue voltage of the emitter current will be developed across R4. The voltage across R5 (V_{R5}) shows the effect of the snubber current in R5 imposed on the voltage across R4, resulting in a rapid increase in voltage toward the end of the conduction period. The voltage waveform on Q2 shows the DC bias introduced by the control current from OC1b and R6 developed across R3 and imposed on the voltage waveform of R5.

When Q1 turns on, current will build up in the primary of the transformer, as shown in Fig. 2.7.1. A "sawtooth" voltage waveform will be developed across R4. This waveform will be applied to the base of Q2 via R5 and R3. A control current flowing in R6 and OC1 will bias the base of Q2 more positive, and the "on" pulse will be terminated when the voltage on the base of Q2 reaches 0.6 V. This turns Q2 on and Q1 off.

As soon as Q1 starts to turn off, the voltage on its collector will rise, and a current will flow in the snubber components D2, C5, and R5, generating a further increase in the voltage across R5 because R5 has a higher resistance than R4. This increase is applied to the base of Q2, providing regenerative turn-on action of Q2 and hence turn-off of Q1.

If the optocoupler OC1b is not conducting, Q2 will still turn on when the voltage across R4 has reached 0.6 V, removing the base drive from the main transistor Q1 and turning it off. Consequently, irrespective of the state of the control circuit, the maximum primary current is defined without the need for further current limiting circuits.

However, the shunt regulator U1 will conduct when the required output volt-

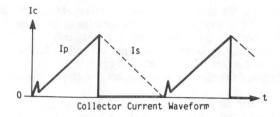

Collector Current Waveform

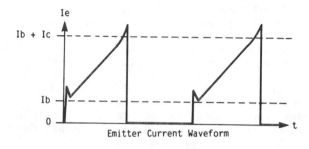

Emitter Current Waveform

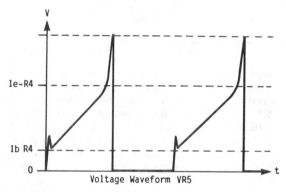

Voltage Waveform VR5

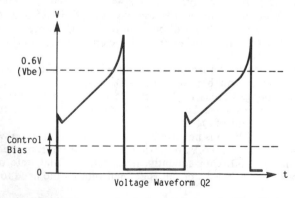

Voltage Waveform Q2

FIG. 2.7.1 Current and voltage waveforms of self-oscillating converter.

age of 12 V is obtained, driving current through the optocoupler diode OC1a. In response to the optodiode current, the optotransistor OC1b will conduct, driving current into the resistor network R3, R5, and R4. The current from OC1b will add a fixed positive bias to the triangular waveform developed across R4 by the emitter current. Consequently, Q2 will turn off with a lower current amplitude in R4. The optotransistor OC1b can be considered a constant-current (high-impedance) source. Hence it will not interfere with the shape of the triangular waveform apart from adding the DC bias. The loop stabilizing components R10 and C7 make the response of the voltage control loop relatively slow compared with the current control loop.

So long as the required output power is less than the limiting value, the voltage control loop will reset the bias on the base of Q2 so as to maintain the output voltage constant.

Resistor R6 provides a fixed offset bias to Q2 for final adjustment of the power limiting value. It also reduces the peak current when the input voltage is increased because the auxiliary voltage on R6 tracks the input voltage. Finally, it also introduces some compensation in the variable-frequency system to improve input ripple rejection.

7.4 INPUT RIPPLE REJECTION

For constant-frequency, complete energy transfer (discontinuous-mode) flyback converters, the current-mode control also provides automatic input ripple rejection.

If the converter input voltage changes, the slope of the collector and hence the emitter voltage will also change. For example, if the collector input voltage starts to rise as a result of the normal rising edge of the input ripple voltage, then the slope of the collector and hence the emitter current will also increase. As a result, the peak current level will be reached in a shorter time and the "on" pulse width will be automatically reduced without the need for any control signal change. Since the peak primary current remains constant, the transferred power and the output voltage will also remain constant (irrespective of the input voltage changes). Consequently, without the need for any action from the control circuit, input transient voltage changes and ripple voltages are rejected from the output.

This effect may be further demonstrated by considering the input energy for each cycle. As this unit operates in a complete energy transfer mode, the energy at the end of an "on" period will be

$$\frac{1}{2} \cdot L_p \cdot I_p^2$$

where L_p = primary inductance
$\quad\ \ I_p$ = peak primary current

Since L_p and the peak current I_p remain constant, the transferred power is constant. This action is very fast, as it responds on a pulse-by-pulse basis, giving good input transient rejection.

A more rigorous examination reveals a degrading second-order effect in the variable-frequency self-oscillating system, caused by an increase in the operating frequency as the input voltage rises. This frequency change reduces the ripple rejection. However, since the increase in frequency results in only a small increase in input power, the effect is quite small. In the circuit shown in Fig. 2.6.4, the effect is compensated by an increase in the drive current component in R4 and an increase in the current in R6, caused by the change in the base drive and auxiliary voltages which track the input change. This compensation is optimized when the forced drive ratio is approximately 1:10.

7.5 USING FIELD-EFFECT TRANSISTORS IN VARIABLE-FREQUENCY FLYBACK CONVERTERS

At the time of writing, FETs were available with voltage ratings of up to typically 800 V; consequently, their use for flyback off-line converters was somewhat limited.

The maximum rectified DC header voltage V_{cc} for 220-V units, and also for dual input voltage units in which voltage doubler techniques are used, will be approximately 380 V DC. As the flyback voltage stress will normally be at least twice this value, the margin of safety when 800-V power FETs are used is hardly sufficient.

However, when higher-voltage competitively priced FET devices become available, there could be significant advantages to be gained from using FETs and operating the flyback unit at higher frequency. This could result in a reduction in the size of the wound components and output capacitors. A more suitable circuit which reduces the voltage stress for lower-voltage power FETs is shown in Fig. 2.5.1.

7.6 PROBLEMS

1. Why do flyback converters lend themselves particularly well to current-mode control?
2. Why does current-mode control provide a simple first-order transfer function in the complete energy transfer mode?
3. Why is the input ripple rejection extremely good in the current-mode control topology?

CHAPTER 8
DIRECT-OFF-LINE SINGLE-ENDED FORWARD CONVERTERS

8.1 INTRODUCTION

Figure 2.8.1 shows the power stage of a typical single-ended forward converter. For clarity, the drive and control circuits have been simplified, and the input rectification omitted. Since the output inductor L_s carries a large DC current component, the term "choke" will be used to describe this component.

Although the general appearance of the power stage is similar to that of the flyback unit, the mode of operation is entirely different. The secondary winding S1 is phased so that energy will be transferred to the output circuits when the power transistor is "on." The power transformer T1 operates as a true transformer with a low output resistance, and therefore a choke L_s is required to limit the current flow in the output rectifier D1, the output capacitor C_o, and the load.

8.2 OPERATING PRINCIPLES

Under steady-state conditions, the circuit operates as follows.

8.2.1 Choke Current

When transistor Q1 turns on, the supply voltage V_{cc} is applied to the primary winding P1, and a secondary voltage V_s will be developed and applied to output rectifier D1 and choke L_s. Neglecting diode drops and losses, the voltage across the choke L_s will be V_s less the output voltage V_{out} (assuming the output capacitor C_o is large so that the output voltage can be considered constant). The current in L_s will be increasing linearly as defined by the following equation:

$$\frac{di}{dt} = \frac{V_s - V_{out}}{L_s}$$

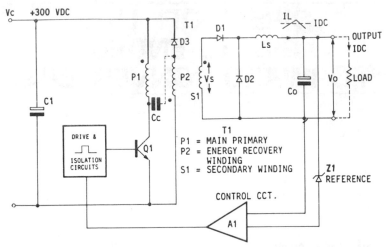

FIG. 2.8.1 Forward (buck-derived) converter with energy recovery winding, showing interwinding capacitance C_c.

At the end of an "on" period, Q1 will turn off, and the secondary voltages will reverse by the normal flyback action of T1. Choke current I_L will continue to flow in the forward direction under the forcing action of L_s, bringing diode D2 into conduction. (This diode is often referred to as the flywheel diode, as it allows the current to continue circulating in the loop D2, L_s, C_o, and load.) The voltage across the choke L_s is now reversed, with a value equal to the output voltage (neglecting diode voltage drops). The current in L_s will now decrease as defined by the following equation:

$$-\frac{di}{dt} = \frac{V_{\text{out}}}{L_s}$$

Using the same design criterion that was applied to the flyback converter transformer, the volt-seconds applied to the choke L_s in the forward and reverse directions must be equal for steady-state conditions. Hence, when the on and off periods are equal, the output voltage will be half V_s (once again neglecting diode drops and losses).

The mean value of the inductor current is the required output current I_{dc}. It should be noted at this stage that the absolute value of inductor L_s has not been defined, as it does not change the action of the circuit in principle; its value simply controls the peak-to-peak ripple current.

8.2.2 Output Voltage

When the ratio of the "on" time to the "off" time is reduced from the 50% duty factor, the output voltage will fall until forward and reverse volt-seconds equality is once again obtained. The output voltage is defined by the following equation:

$$V_{\text{out}} = \frac{V_s \times t_{\text{on}}}{t_{\text{on}} + t_{\text{off}}}$$

where V_s = secondary voltage, peak V
$\quad t_{\text{on}}$ = time that Q1 is conducting, μs
$\quad t_{\text{off}}$ = time that Q1 is "off," μs

Note: The ratio $t_{\text{on}}/(t_{\text{on}} + t_{\text{off}})$ is called the duty ratio.

It should be noted that when the input voltage and duty cycle are fixed, the output voltage is independent of load current (to the first order), and this topology provides an inherently low output resistance.

8.3 LIMITING FACTORS FOR THE VALUE OF THE OUTPUT CHOKE

8.3.1 Minimum Choke Inductance and Critical Load Current

The minimum value of L_s is normally controlled by the need to maintain continuous conduction at minimum load current. Figure 2.8.2 shows continuous-mode

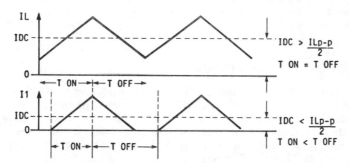

FIG. 2.8.2 Secondary current waveforms, showing incomplete energy transfer (coninuous-mode operation) and complete energy transfer (discontinuous-mode operation).

conduction in the top waveform, and discontinuous-mode conduction in the lower waveform. It should be noted that if the input and output voltages remain constant, the slope of the current waveform does not change as the load current is reduced.

As the load current I_{dc} decreases, a critical value is reached where the minimum value of the ripple current in the choke will just touch zero. At this point the critical load current is equal to the mean ripple current in the choke and is defined as follows:

$$I_{\text{dc}} = \frac{I_{L(\text{p-p})}}{2}$$

where I_{dc} = output (load) current

$I_{L(p-p)}$ = peak-to-peak choke current

At load currents below this critical value, L_s will go into a discontinuous current mode of operation. However, this is not an ultimate minimum limit to the load current, since the output voltage can still be maintained constant by reducing the mark space ratio. Still, at the critical current, there is a sudden change in the transfer function. At currents higher than critical, the mark space ratio will remain nearly constant, irrespective of load current (continuous-mode operation). Below the critical value, the mark space ratio must be adjusted for both load variations and input voltage variations (discontinuous-mode operation).

Although the control circuit can be designed to be stable for these two conditions, the stability criteria must be carefully checked. There will be two poles in the transfer function for the continuous mode, and only one pole for the discontinuous mode. (See Part 3, Chap. 8.)

A second limiting factor to the value of L1 may come into play for multiple-output applications. If the control loop is closed to the main output line, and the current on this line is taken below the critical value, then the mark space ratio will be reduced to maintain the output voltage on this line constant. The remaining auxiliary outputs, which are assumed to have constant loads on them, will respond to this variation in duty ratio, and their voltages will fall. This is the reverse of what might be expected. It is usually the need to maintain the auxiliary output voltages constant which controls the minimum value of L_s in multiple-output units. Even then, a current in excess of the critical value must be maintained on the main output if the auxiliary voltages are to be maintained reasonably constant.

8.3.2 Maximum Choke Inductance

The maximum value of L_s is usually limited by considerations of efficiency, size, and cost. Large inductors carrying DC currents are expensive. From a performance standpoint, large values of L_s will limit the maximum rate of change (slew rate) of the output current for large transient load changes. The output capacitor C_o is usually much too small to maintain the output voltage constant for large load changes in this type of converter.

8.4 MULTIPLE OUTPUTS

Extra windings on the main transformer can provide additional auxiliary outputs. Once again, the value of secondary voltage will be chosen so that the volt-seconds on the output chokes in the forward and reverse directions will equate to zero under steady-state conditions. Therefore, if the voltage of the main output line is stabilized, the voltage of the auxiliary lines will also be stabilized, provided that the load conditions remain reasonably constant. If the load on any output falls below the critical current for its particular choke, then the output voltage on this line will begin to rise. Eventually, under zero load conditions, it will be equal to the peak voltage on the transformer secondary. (For a 50% duty cycle, this would be twice the normal output voltage.)

Hence, whereas in the flyback converter transformer the secondary voltages are all clamped by the main output and are thus well defined, in the forward converter the output voltages for loads below critical can be very high. Therefore, in the forward converter it is essential that the critical value of inductor currents is lower than the minimum load presented to the output. If the load is required to go to zero or near zero, it will be necessary to provide dummy loading or voltage clamping to prevent excessive output voltages. This problem can be very much reduced by using coupled output inductors for multiple-output applications. (See Refs. 15 and 48.)

8.5 ENERGY RECOVERY WINDING (P2)

During the "on" period of transistor Q1, energy will be transferred to the output circuit. At the same time, the primary of the transformer will take a magnetizing current component and store energy in the magnetic field of the core.

When Q1 turns off, this stored energy will result in damagingly large flyback voltages on the collector of the switching transistor Q1 unless a clamping, or energy recovery, action is provided.

Note: During the flyback ("off") period, the output diodes will be reverse-biased and will not provide any clamping action, unlike the flyback converter.

In this example an "energy recovery winding" P2 and diode D3 are connected so that the stored energy will be returned to the supply line during the flyback period. Note that during the flyback period, the voltage across the flyback winding P2 is being clamped at V_c by D3, with the finish of the winding being positive. Hence, the voltage on the transistor collector will be twice V_c (assuming that the primary and flyback turns are equal), because the start of the primary is already connected to the supply voltage V_c.

To prevent excessive leakage inductance between P1 and P2, and hence excessive voltage overshoot on the transistor collector, it is conventional to bifilar-wind the energy recovery winding P2 with the main primary P1. In this arrangement it is important that diode D3 be placed in the top end of the energy recovery winding. The reason for this is that the interwinding capacitance C_c (which can be considerable with a bifilar winding) will appear as a parasitic capacitance between the collector of Q1 and the junction of P2 and D3, as shown by parasitic capacitor C_c in Fig. 2.8.1. When connected in this way, this capacitance is isolated from the collector by diode D3 during the turn-on of Q1. Hence D3 blocks any current flow in C_c during the turn-on transient of Q1. (Note that the finishes of both windings P1 and P2 go negative together, and there is no change of voltage across C_c.) Further, C_c provides additional clamping action on the collector of transistor Q1 during the flyback period, when any tendency for voltage overshoot will result in a current flow through C_c and back to the supply line via D3. Very often this parasitic winding capacitance will be supplemented by an extra real external capacitor in position C_c to improve this clamping action. However, add extra capacitance with care, as a capacitor with too large a value will result in line ripple frequency modulation in the output voltage.

As a result of the high-voltage stress across the bifilar winding in high-voltage off-line applications, special insulation would normally be required. However, if an additional clamping capacitor C_c is fitted, the energy recovery winding may be wound on a separate (nonbifilar) insulated layer. This reduces the voltage stress without compromising the clamping action. Alternatively, low-loss energy snubber systems, as shown in Part 1, Chap. 18, may be used.

8.6 ADVANTAGES

Some of the advantages of the forward converter, when compared with the flyback, are as follows:

1. The transformer copper losses in the forward converter tend to be somewhat lower, since the peak currents in primary and secondary will tend to be lower than in the flyback case (the inductance is higher, as no air gap is required). Although this may result in a smaller temperature rise in the transformer, in most cases the improvement is not sufficient to allow a smaller core to be used.

2. The reduction in secondary ripple currents is dramatic. The action of the output inductor and flywheel diode maintains a reasonably constant current in the output load and reservoir capacitors.

   Since the energy stored in the output inductor is available to the load, the reservoir capacitor can be made quite small, its main function being to reduce output ripple voltages. Furthermore, the ripple current rating for this capacitor will be much lower than that required for the flyback case.

3. The peak current in the primary switching device is lower, for the same reason as in point 1.

4. Because of the reduction in ripple current, the output ripple voltages will tend to be lower.

8.7 DISADVANTAGES

Some of the disadvantages are

1. Increased cost is incurred because of the extra output inductor and flywheel diode.

2. Under light loading conditions, when L_s reverts to the discontinuous mode, excessive output voltages will be produced, particularly on auxiliary outputs, unless minimum loads are specified or provided by ballast resistors.

   In other respects, the performance to be expected from the forward converter is very similar to that of the flyback.

8.8 PROBLEMS

1. From what class of converters is the transformer-coupled forward converter derived?

2. During what phase of operation is the energy transferred to the secondary circuit in the forward converter?

3. Why is an output choke required in the forward converter topology?

4. Why is the utilization of the primary switching device often much greater in the forward converter than in the flyback converter?

5. A core gap is not normally required in a forward converter transformer. Why is this?

6. Why is an energy recovery winding required in the forward converter?

7. Why is a minimum load required for correct operation of a forward converter?

CHAPTER 9
TRANSFORMER DESIGN FOR FORWARD CONVERTERS

9.1 INTRODUCTION

The design of switchmode transformers for forward converters may be approached in many different ways. The designer should choose a method that he or she is comfortable with.

In the following example a nonrigorous design approach is used, starting with the primary turns calculation. Manufacturers' published nomograms are utilized for the selection of core size and optimum induction. The resulting transformer is suitable for prototype evaluation and will not be far from optimum. It should be remembered that for multiple-output applications, exact voltage results are not always going to be possible for all outputs, as windings can only be applied in increments of one, or in some cases one-half, turn. Further, the core size will often be a compromise selection.

This design approach for the forward converter is very similar to the method used for the previous flyback example, except that the primary turns will be calculated for maximum "on" pulse width and nominal DC voltage. This will result in slightly more primary turns.

The reason for this selection is that in the forward converter, the output inductor will limit the rate of change of output current when a transient load is applied. To compensate for this, the control amplifier will take the input pulse width to maximum so as to force an increase in inductor current as rapidly as possible. Under these transient conditions, the high primary voltage and maximum pulse width will be applied to the transformer primary at the same time. Although this only occurs for a short period, core saturation can occur unless the transformer is designed for this condition.

The control circuit will be designed so that at maximum line input voltages, the pulse width and the slew rate of the control circuit will be limited. This prevents maximum pulse width and line voltage from coinciding. This must be checked in the final design.

In the forward converter, it is undesirable to store energy in the core, as this energy must be returned to the supply line during the flyback period. A small gap may be required to ensure that the flux returns to a low residual level during the flyback period and to allow the maximum flux density excursion. The air gap will be kept as small as possible (two or three thousandths of an inch will usually be sufficient). An examination of Fig. 2.2.1a and b shows that the residual flux will be much lower, even with a small gap in the core; also, a small gap will stabilize the magnetic parameters.

9.2 TRANSFORMER DESIGN EXAMPLE

9.2.1 Step 1, Selecting Core Size

The core size is selected on the basis of transmitted power, and the manufacturers' recommendations make a good starting point. A typical core selection nomogram is shown in Fig. 2.2.2. In the following example, a transformer will be designed for a 100-W application at 30 kHz. The output voltages and currents will be as follows:

$$+ 5 \text{ V at } 10 \text{ A} = 50 \text{ W}$$

$$+ 12 \text{ V at } 2 \text{ A} = 24 \text{ W}$$

$$- 12 \text{ V at } 2 \text{ A} = 24 \text{ W}$$

$$\text{Total power} = 98 \text{ W}$$

$$\text{Input voltage } 90\text{--}130 \text{ V or } 180\text{--}260 \text{ V}, 47\text{--}60 \text{ Hz}$$

Allowing approximately 3% increase in size for each auxiliary line to provide extra insulation and window space, a core will be selected for a transmitted power of 104 W. From Fig. 2.2.2, E core; E42-15 would be a suitable selection.

Core parameters:

$$\text{Effective core area } A_e = 181 \text{ mm}$$

9.2.2 Step 2, Select Optimum Induction

The optimum induction B_{opt} is chosen so as to make the core and copper losses approximately equal. This gives minimum overall loss and maximum efficiency, provided that core saturation is avoided.

From Fig. 2.9.1, at 100 W and a frequency of 30 kHz, an optimum peak flux density B_{opt} of approximately 150 mT is recommended for push-pull operation. Remember, in the push-pull case, the differential excitation (ΔB) will be twice this peak value, giving a flux density swing of 300 mT p–p. (See Fig. 2.9.2a.)

In Fig. 2.9.1, the recommended optimum flux density assumes the total peak-to-peak excursion. Hence, in the single-ended forward converter, a peak flux density of 300 mT would be indicated for maximum efficiency. Note that in the single-ended forward converter, only the forward quadrant of the *BH* characteristic is used (see Fig. 2.9.2b). To avoid saturation, it is necessary to allow some margin of safety for residual flux, the effects of high temperatures, and transient conditions. A total excursion of 250 mT is chosen in this example; this is less than optimum, and the design is said to be "saturation-limited."

The core losses will now be somewhat lower than the copper losses. This will often be the case with single-ended converters unless the core has been designed for single-ended applications.

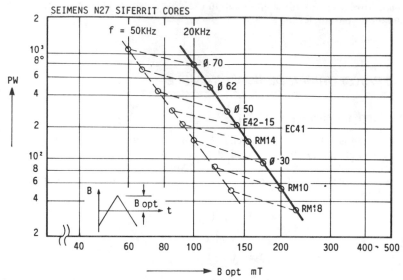

FIG. 2.9.1 Optimum working flux density swing for N27 ferrite material as a function of output power, with frequency and core size as parameters. (*Courtesy of Siemens AG.*)

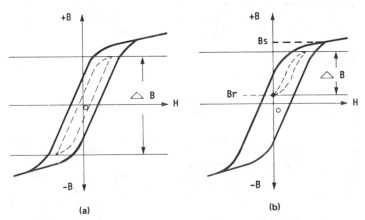

FIG. 2.9.2 (*a*) *B/H* loop showing the extended working range of *B/H* for push-pull operation. (*b*) First quadrant, showing limited *B/H* loop range for single-ended forward and flyback operation.

9.2.3 Step 3, Calculate Primary Turns

Output power   100 W
Selected core   E42-15
Frequency   30 kHz
Flux density   250 mT

The drive voltage waveform is square, so the primary turns will be calculated using the volt-seconds approach. The maximum pulse width is assumed to be 50% of the period.

$$\text{Total period } t_p = \frac{1}{f} = \frac{1}{30 \times 10^3} = 33 \ \mu s$$

Therefore

$$t_{\text{on(max)}} = \frac{T_p}{2} = 16.5 \ \mu s$$

Calculate Primary Voltage (V_{cc}). The primary voltage will be calculated for nominal input and full-load operation. The input rectifier network will be configured for dual-range 110/220-V operation, so a voltage doubler will be used at 110 V ac input.

The approximate conversion factors are

$$V_{cc} = V_{\text{rms}} \times 1.3 \times 1.9$$

(See Part 1, Chap. 6.) Hence, at 110 V rms line input, the primary DC voltage V_{cc} will be

$$110 \times 1.3 \times 1.9 = 272 \text{ V}$$

Therefore, the minimum primary turns will be

$$N_{\text{min}} = \frac{V_{cc} \times t_{\text{on}}}{\hat{B} \times A_e}$$

where V_{cc} = primary DC voltage, V
   t_{on} = maximum "on" time, μs
   $\hat{B}$ = maximum flux density, T
   A_e = effective core area, mm^2

Therefore

$$N = \frac{272 \times 16.5}{0.25 \times 181} = 100 \text{ turns}$$

9.2.4 Step 4, Calculate Secondary Turns

The secondary turns will be calculated for the lowest output voltage, in this case 5 V. The output filter and transformer secondary are shown in Fig. 2.9.3.

From Fig. 2.9.3, for the continuous-current-mode operation,

$$V_{\text{out}} = \frac{V_s \times t_{\text{on}}}{t_{\text{on}} + t_{\text{off}}}$$

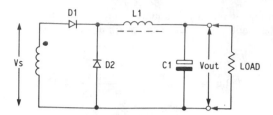

FIG. 2.9.3 Output filter of single-ended (buck-derived) forward converter.

But at maximum pulse width, $t_{on} = t_{off}$. Hence

$$V_s = 2V_{out} = 10 \text{ V}$$

Therefore, the minimum secondary voltage will be 10 V. Allowing for a 1-V drop in diode and inductor, V_s becomes 11 V. This must be available at the minimum line input of 90 V and maximum "on" period of 16.5 μs.

At 90 V in the DC primary voltage, V_{cc} will be

$$V_{cc} = 90 \times 1.3 \times 1.9 = 222 \text{ V}$$

With $N_{min} = 100$ turns on the primary, the volts per turn (V_{pt}) will be

$$V_{pt} = \frac{V_{cc}}{N} = \frac{222}{100} = 2.22 \text{ V/turn}$$

and the minimum secondary turns will be

$$\frac{V_s}{V_{pt}} = \frac{11}{2.22} = 4.95 \text{ turns}$$

The secondary turns are rounded up to 5 turns. The primary turns will be adjusted at this point as follows:

$$\frac{V_{cc}}{N_p} = \frac{V_s}{N_s}$$

Therefore

$$N_p = \frac{222 \times 5}{11} = 101 \text{ turns}$$

In a similar way, the remaining 12-V output turns may be calculated. Once again, diode and choke winding loss of 1 V is estimated; therefore

$$V_s \text{ (12 V)} = 2 \times V_{out} + V_{drop} = 2 \times 12 + 1 = 25 \text{ V}$$

and secondary turns for the 12-V outputs will be

$$\frac{V_{cc}}{N_p} = \frac{V_s(12\ \text{V})}{N_s(12\ \text{V})} : N_s(12) = \frac{25 \times 101}{222} = 11.4 \text{ turns}$$

and this will be rounded up to 11.5 turns. (See Fig. 2.9.5a and b.)

9.2.5 Multiple-Output Applications

Figure 2.9.4 shows a typical multiple-output forward converter secondary, in which all outputs share a common return line. The negative output is developed by reversing D5 and D6. Note that the phasing of the secondary is such that D3 and D5 conduct at the same time during the "on" period of Q1.

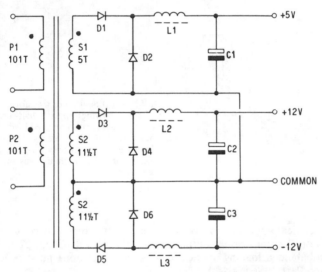

FIG. 2.9.4 Transformer and output circuit of typical multiple-output forward converter.

9.2.6 Special Case Half Turns

In this particular example there are two equal and opposite polarity 12-V outputs, and a center-tapped 23-turn secondary winding is used. This arrangement of the positive and negative 12-V outputs is a special case which allows half turns to be used with E cores without causing flux imbalance in the legs of the E core. The mmf from the two half turns (effectively one half on each leg of the E core) will cancel, and the core flux distribution will not be distorted, as shown in Fig. 2.9.5a and b.

In many applications, this half turn is not practicable and the secondary will be rounded to the nearest integer, producing an error in the output voltage. How-

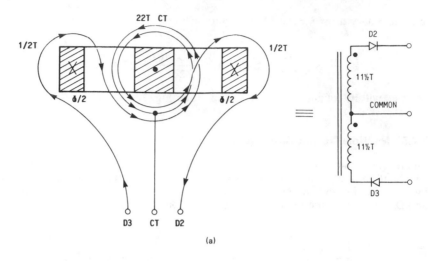

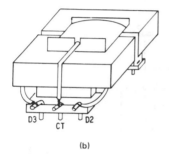

FIG. 2.9.5 (a) Section of transformer core, and schematic showing special half-turn arrangement, for dual-output, balanced-load forward and flyback operation. (Suitable for balanced secondary load specifications only.) (b) Practical implementation of balanced half turns on an E core transformer.

ever, for critical applications, the voltage can be corrected by using saturating inductors on the transformer output leads. (See Part 1, Chap. 22.)

For unbalanced loading applications, special methods must be used for half turns. (See Part 3, Sec. 4.14.)

9.2.7   Step 5, Selecting Transformer Wire Gauge

The wire types and sizes must now be selected and the transformer buildup established.

Part 3, Sec. 4.15 provides information for making the wire size choice. If a bifilar winding is to be used for energy recovery, then wire with suitable insulation ratings must be used.

In high-voltage off-line applications, a high voltage exists between the main collector and the energy recovery windings, and this can be a source of failure. (See Part 2, Chap. 8.) It is recommended that one of the alternative energy recovery techniques which does not require a bifilar-wound energy recovery winding be considered.

9.3 SELECTING POWER TRANSISTORS

A power switching transistor that will have good gain and saturation characteristics at the maximum primary current will be chosen. This current can be calculated as follows.

At minimum line input, $V_{cc} = 222$ V (assume secondary and transformer efficiency to be 75%); then input power is

$$\frac{P_{out}}{P_{in}} = \frac{75\%}{100\%} : P_{in} = \frac{98 \times 100}{75} = 130 \text{ W}$$

Collector current I_c may now be calculated.

At 90 V input, the "on" period will be the maximum of 50%; therefore, the mean primary current during the "on" period is

$$I_{mean} = \frac{P_{in} \times 2}{V_{cc}} = \frac{130 \times 2}{222} = 1.2 \text{ A}$$

Allowing for a 20% ripple current (which would be the case where the critical output inductor current is 10% of full load, as in Fig. 2.9.6), then I_p maximum would be 1.32 A.

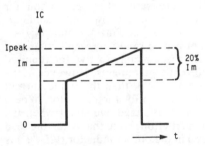

FIG. 2.9.6 Primary current waveform of a continuous-mode forward converter, showing a magnetization current of 20%.

In practice, a transistor with a current rating of at least twice this value will normally be selected to ensure reasonable current gain and efficient switching. It should be remembered that under transient conditions, the current will be greater than 1.3 A, as there will be an overshoot in the current flowing in the various output inductors. Further, there is a primary magnetizing current to consider, and the transistor must not come out of saturation for full load and transient current conditions.

9.3.1 Transistor Voltage Rating

During the turn-off transient, the collector voltage will fly back to at least twice the maximum supply voltage, and as a result of the leakage inductance, there will normally be an overshoot on top of this value.

Hence, at maximum line input and zero load, V_{cc} maximum will be

$$V_{in(rms)} \times \sqrt{2} \times 2$$

Therefore $\quad V_{cc} = 130 \times \sqrt{2} \times 2 = 367 \text{ V DC (max)}$

Allowing 5 V for losses in the rectifiers and transformers, the DC voltage V_{cc} will be 362 V.

The flyback voltage will be twice this value, plus an allowance for inductive overshoot (say 10%). Hence

$$V_{fb} = 2 \times 362 + 10\% \times 362 = 760 \text{ V}$$

With the correct drive waveform and snubber networks, the collector current will have dropped to zero before this high-voltage condition is reached; therefore, the transistor will be chosen for a V_{cex} rating of 760 V minimum.

To prevent secondary breakdown, the designer must ensure that the collector current has reached zero before a collector voltage of V_{cex} is approached. This is achieved by using a suitable snubber network. (See Part 1, Chaps. 17 and 18.)

9.4 FINAL DESIGN NOTES

To complete the design, refer to Part 1, Chap. 6 for input filters, rectifiers, and storage capacitors.

If simple primary overload protection is to be used, it is necessary that all auxiliary output lines be capable of taking the full VA of the unit without damage, since a single output may be overloaded. More satisfactory protection for multiple-output units will be given if each output line has its own individual current limit. (For suitable current-limiting transformers, see Part 3, Chap. 14.)

In forward converters, the output capacitor can be relatively small compared with the one used in the flyback case, as it is chosen mainly to give the required output ripple voltage rather than for ripple current requirements. However, if the series inductance is very small (fast-response systems), then the ripple current requirements may still be the major selection criterion for the capacitor. Where large transient loads are to be accommodated, the voltage overshoot when the load is suddenly removed may be the selection criterion. Note that when the load is suddenly dropped to zero, the energy stored in the output inductor ($\frac{1}{2}L \cdot I^2$) is dumped into the output capacitor, causing a voltage overshoot.

9.5 TRANSFORMER SATURATION

With this transformer design, it is possible for the transformer to saturate under transient conditions, unless steps are taken in the design of the control circuit to prevent it. For example, consider the condition when the input voltage is high and the load current is very small; a sudden increase in load would normally result in the drive circuit going to maximum pulse width. This condition would exist for a number of cycles until the current in the output inductor had risen to the required value. Under these conditions, at the flux density levels chosen, the transformer would saturate. To prevent this, a lower flux level may be used. However, this is not the method recommended here, as it results in an increase in primary turns and a reduction in transformer efficiency.

For this example, it has been assumed that the drive circuit will have a "pulse-by-pulse'' primary current limit, which will recognize the onset of transformer saturation and limit any further increase in drive pulse width. This will give the maximum response time and prevent failure resulting from transformer saturation.

9.6 CONCLUSIONS

The major parameters of a single-ended forward converter have been discussed and a practical transformer design examined.

Although the peak currents in the transformer are generally lower than in the flyback case, the requirements for extra primary turns tends to eliminate any improvement in the transformer efficiency.

There is a considerable reduction in ripple current for the output rectifiers and capacitors, which makes the forward technique more suitable than the flyback for low-voltage, high-current applications. The reduction in component size is somewhat offset by the need for an extra output inductor and rectifier diode.

A critical minimum load condition exists for all auxiliaries and main outputs. Operation at currents below this critical value leads to discontinuous inductor operation, which can appear as a loss of voltage regulation, particularly for the auxiliary outputs. Under open-circuit conditions, output voltages exceeding the nominal by more than 2 to 1 are possible unless dummy loads or other clamping action is provided. These requirements must be considered in the application and design of this type of converter.

Where linear regulators are used for low-current outputs, the poor voltage regulation at light loads is not normally a problem, as the excess voltage will be dropped across the series regulator and the dissipation will be small for the light loading condition. The voltage rating of the regulator may become a limiting factor, however.

CHAPTER 10

DIAGONAL HALF-BRIDGE FORWARD CONVERTERS

10.1 INTRODUCTION

The diagonal half-bridge forward converter, otherwise known as the two-transistor forward converter, has a primary power switch arrangement that is similar to its counterpart in the flyback converter. Figure 2.10.1 shows the general circuit of the power system.

This arrangement is particularly suitable for field-effect transistor (FET) operation, as the energy recovery diodes D1 and D2 provide hard clamping of the switching devices to the supply line, preventing any overshoot during the flyback action. The voltage across the power switches will not exceed the supply voltage by more than two diode drops, and therefore the voltage stress will be only half of what it would have been in the single-transistor, single-ended converter.

In this example, provision has been made for dual voltage line input selection. The input is user selectable from a nominal 110- to 220-V operation by removing the link LK1.

Voltage doubler action is used for 110-V operation; consequently, the DC voltage V_{cc} will be approximately 300 V for both 110- and 220-V operation. The maximum voltage for high line off-load operation will be approximately 380 V; consequently, power FETs with a voltage rating of 400 V may be used with confidence. (Some FETs have a voltage derating applied for high-temperature operation, and this should be considered.)

10.2 OPERATING PRINCIPLES

FET 1 and FET 2 (the power switches) turn on, or off, simultaneously. When the devices are switched on, the primary supply voltage V_{cc} will be applied across the transformer primary, and the starts of all windings will go positive.

Under steady-state conditions, a current will have been established in the output choke L1 by previous cycles, and this current will be circulating by flywheel action in the choke L1, capacitor C1, and load, returning via the flywheel diode D4.

When the secondary emf is established (by turning on the power FETs), the current in the secondary of the transformer and the rectifier diode D3 will build

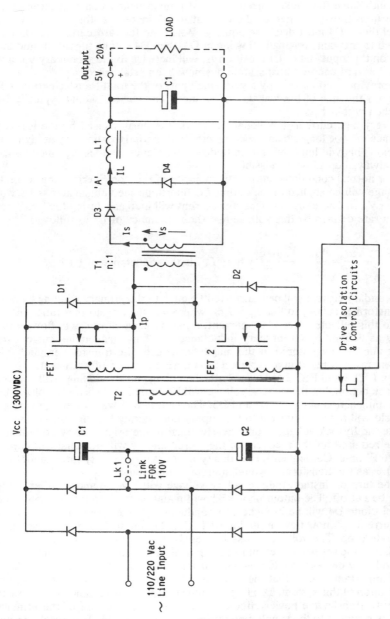

FIG. 2.10.1 Diagonal half-bridge (dual-FET) forward converter.

up rapidly, limited only by the leakage inductance in the transformer and secondary circuit. Since the choke current I_L must remain nearly constant during this short turn-on transient, then as the current in D3 increases, the current in the flywheel diode D4 must decrease equally. When the forward current in D3 has increased to the value originally flowing in D4, then D4 will be turned off and the voltage on the input end of L1 (position A) will increase to the secondary voltage V_s. The forward energy transfer state has now been established.

The previous actions occupy a very small part of the total transfer period, depending on the size of the leakage inductance. The current would typically be established within 1 μs.

For very high current, low-voltage outputs, the delay caused by the leakage inductance may be longer than the complete "on" period (particularly at high frequencies). This will limit the transmitted power. Hence, the leakage inductance should always be as low as possible.

Under normal conditions, during the majority of the "on" period, the secondary voltage will be applied to the output LC filter, and the voltage across L1 will be $(V_s - V_{out})$. Therefore, the inductor current will increase during the "on" period at a rate defined by this voltage and the inductance of L1 as follows:

$$\frac{dI}{dt} = \frac{V_s - V_{out}}{L_{L1}} \qquad \text{where } L_{L1} \text{ is the inductance of L1}$$

This secondary current will be transferred through to the primary winding by normal transformer action, so that $I_p = I_s/n$, where n is the transformer ratio. In addition to this reflected secondary current, a magnetizing current will flow in the primary as defined by the primary inductance L_p. This magnetizing current results in energy being stored in the magnetic field of the transformer, and this stored energy will result in a flyback action during the turn-off transient.

When FET 1 and FET 2 are turned off, the voltage on all windings will reverse by flyback action, but the flyback voltage will be limited to the supply voltage by the clamping action of diodes D1 and D2. The energy that was stored in the magnetic field will now be returned to the supply lines during the turn-off period.

Since the flyback voltage is now nearly equal to the original forward voltage, the time required for the recovery of the stored energy will be equal to the previous "on" time. Consequently, for this type of circuit, the duty ratio cannot exceed 50%, as the transformer would staircase into saturation.

At the turn-off instant, the secondary voltage will reverse and rectifier diode D3 will be cut off. The output choke L1 will maintain the current constant, and flywheel diode D4 will be brought into conduction. Under the forcing action of L1, a current will now flow in the loop L1, C1, D4, and node A will go negative by a diode drop. The voltage across L1 equals the output voltage (plus a diode drop), but in the reverse direction to the original "on"-state voltage. The current in L1 will now decrease to its original starting value, and the cycle is complete.

It is important to note that the leakage inductance plays an important role in the operation of this system. Too large a value of leakage inductance results in an inability to transfer the power effectively, as a large proportion of the primary current is returned to the supply line during the "off" period. This results in unproductive power losses in the switching devices and energy recovery diodes.

The reverse recovery time of diode D4 is particularly important, since during the turn-on transient, current will flow from D3 into the output inductor and also into the cathode of D4 during its reverse recovery period. This will reflect through to the primary switches as a current overshoot during the turn-on transient.

The action of this converter has been described in some detail in order to highlight the importance of the transformer leakage inductance and the need for fast recovery diodes. These effects become particularly important for high-frequency operation, where the advantages of power FETs are better utilized.

It should be remembered that the leakage inductance is not solely contained within the transformer itself; it is made up of all the external circuitry. The various current loops should be maintained at the minimum inductance by using short, thick wiring, which should be twisted where possible or run as tightly coupled pairs.

The energy recovery diodes D1 and D2 should be fast high-voltage types, and a low-ESR capacitor should be fitted across the supply lines as close as possible to the switching elements. The ESR and ESL of the output capacitor C1 are not so critically important to the function of the converter, since this capacitor is isolated from the power switches by the inductor L1.

The main function of C1 is to reduce output ripple voltages and provide some energy storage. It is often more cost-effective to use an additional *LC* filter to reduce noise, so as to avoid the use of expensive low-ESR electrolytic capacitors in this position. (See Part 1, Chap. 20.) The transformer design for the half-bridge converter is shown in Chap. 11.

CHAPTER 11
TRANSFORMER DESIGN FOR DIAGONAL HALF-BRIDGE FORWARD CONVERTERS

11.1 GENERAL CONSIDERATIONS

The transformer for the converter shown in Fig. 2.10.1 follows the same general principles used for the single-ended forward converter. The major difference is that an energy recovery winding will not be required.

11.1.1 Step 1, Selecting Core Size

Unfortunately, there are no fundamental equations for the selection of core sizes. The choice depends on a number of variables, including the type of material, the shape and design of the core, the location of the core, the type of cooling provided, and the allowable temperature rise. This in turn depends on the type of materials and insulation used in the transformer design, the equipment environmental operating temperatures, the frequency of operation, and the type of ventilation provided (i.e., forced air, convected air, or conduction cooled).

Most manufacturers provide graphs or nomograms with recommended core sizes for particular conditions of operation. These graphs can provide a good starting point for the design and will be used in this example.

A typical core selection graph is shown in Fig. 2.11.1.

11.1.2 Example Specification

Consider a requirement for a 100-W off-line forward power supply to deliver a single output of 5 V, 20 A. The switching frequency is to be 50 kHz.

Reference to Fig. 2.11.1 shows that core type EC41 should be suitable, since the operating point P1 is well inside the permitted operating range.

The supply voltage V_{cc} is to be derived from the 110-V nominal line via a voltage doubler so that dual input voltage operation will be possible by removing a link.

In the diagonal half-bridge circuit, the duty ratio cannot exceed 50%, as the same primary winding is used for both forward and flyback conditions and the

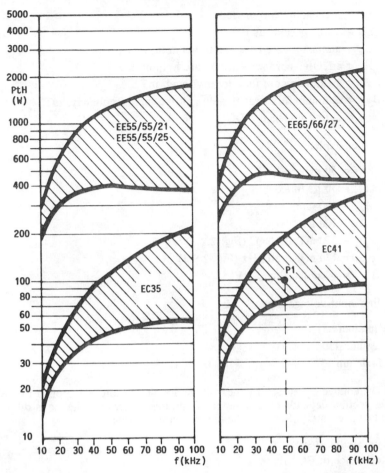

FIG. 2.11.1 Core size selection chart for forward converters, showing throughput power as a function of frequency with core size as a parameter. (*Courtesy of Mullard Ltd.*)

voltage applied for both conditions is the same, V_{cc}. Hence, since the operating frequency is 50 kHz, the maximum "on" period will be 10 μs.

In this example, a voltage-controlled feedback loop will be used and very fast transient response provided; consequently, during transient conditions, maximum voltage and pulse width can coincide. To prevent transformer saturation, the worst-case conditions will be used for the transformer primary design.

It will be assumed that the following parameters apply to the transformer design:

V_{ac} = line input voltage, V ac

V_{cc} = rectified DC converter voltage, V DC

f = converter frequency, kHz

t_{on} = maximum "on" period, μs

t_{off} = "off" period, μs

A_{cp} = minimum core cross-sectional area, mm^2

A_e = effective core area (total core in mm^2)

B' = maximum core flux density level, mT

B_{opt} = optimum flux density value at operating frequency, mT

V_{out} = output voltage, V DC

For this example,

f = 50 kHz

t_{on} = 10 μs (max)

A_{cp} = 100 mm^2

A_e = 120 mm^2

B_s = 350 mT at 100°C (Fig. 2.2.3)

B_{opt} = 170 mT at 50 kHz (Fig. 2.9.1)

V_{out} = output voltage, 5 V

The line input voltage range V_{ac} is

Minimum 85

Nominal 110

Maximum 137

The DC voltage developed by the voltage doubler circuit and applied to the converter section depends on a number of variable factors and hence is difficult to calculate accurately. Some of the major factors will be

Line source resistance

Resistance of EMI filter

Hot resistance inrush thermistors (where used)

Rectifier drop

Size of doubler capacitance

Load current

Line frequency

Part 1, Chap. 6 shows a graphical method of establishing the approximate DC voltage and gives recommendations for capacitor selection. The final value of DC voltage V_{cc} is probably best measured in the prototype model, as it is unlikely that all the variables will be known. For this example a simple empirical approximation for the minimum DC voltage will be used. For the voltage doubler connection,

$$V_{cc} = V_{ac} \times 1.3 \times 1.9$$

Hence

$$V_{cc} \begin{cases} \text{minimum} = 209 \text{ V} \\ \text{nominal} = 272 \text{ V} \\ \text{maximum} = 338 \text{ V} \end{cases}$$

From Fig. 2.9.1, the recommended flux density B_{opt} (peak) for the EC41 core at 50 kHz is 85 mT. This value is applicable for optimum design in the push-pull case. In push-pull converters the total change in flux density is from $+B_{opt}$ to $-B_{opt}$, and the core loss is dependent on the total change ΔB. For the same core loss in the single-ended converter, $B_{opt} = \Delta B = 2 \times 85 = 170$ mT because the flux density excursion is all in one quadrant (see Chap. 9). This flux density will apply at nominal input voltage for the effective core area A_e. (It determines the core losses and so applies to nominal core size and normal input voltage.)

If this flux density is to apply at the nominal line voltage (110 V ac), then at maximum line voltage (137 V ac) and maximum pulse width the maximum flux density $\hat{B}$ will be

$$\hat{B} = \frac{B_{opt} \times V_{cc}(\text{max})}{V_{cc} \text{ (nominal)}} = \frac{170 \times 380}{222} = 290 \text{ mT}$$

This is less than the saturation value for the core (Fig. 2.2.3) and hence is acceptable.

11.1.3 Step 2, Primary Turns

Since the primary waveform is a square wave, the minimum primary turns may be calculated using Faraday's law:

$$N_{min} = \frac{V \times t_{on}}{\hat{B} \times A_e}$$

where V_{cc} (max) = maximum DC supply voltage (380 V)
   t_{on} = maximum "on" period, μs (10 μs)
   $\hat{B}$ = maximum flux density, T (0.29 T)
   A_e = effective core area, mm^2 (120 mm^2)

Therefore $N_{min} = \dfrac{380 \times 10}{0.29 \times 120} = 109$ turns

Note: The effective core area A_e is used here rather than the minimum area, as the flux level has been chosen for core loss considerations rather than maximum flux density.

11.1.4 Step 3, Calculating Secondary Turns

The required output voltage is 5 V. For the LC filter used in this example, the output voltage is related to the transformer secondary V_s by the following equation:

$$V_s = \frac{V_{out} \times (t_{on} + t_{off})}{t_{on}}$$

where V_s = secondary voltage
   V_{out} = required output voltage (5 V)

In this example the secondary voltage will be calculated for the maximum duty ratio of 50%, which will occur at minimum line input voltage. Hence $t_{on} = t_{off}$, and the secondary voltage V_s may be calculated:

$$V_s = \frac{5 \times (10 + 10)}{10} = 10 \text{ V}$$

Allowing an extra volt for diode voltage drop and wiring and inductor resistance losses, the secondary voltage V_s will be 11 V.

In this example, the maximum "on" period has been used, and the calculated secondary voltage is the minimum usable value. This must be available at minimum line voltage; hence the minimum secondary turns must be calculated for these conditions.

At a line voltage of 85 V, V_{cc} will be 209 V.

Allowing a voltage drop of 2 V each for FET1 and FET2, the voltage applied to the primary V_p will be

$$V_p = 209 - 4 = 205 \text{ V}$$

Consequently, the minimum secondary turns will be

$$\frac{N_p \times V_s}{V_p} = \frac{109 \times 11}{205} = 5.8 \text{ turns}$$

The turns will be rounded up to 6 turns and the primary turns increased proportionally, resulting in a lower flux density and core loss. Alternatively, the primary turns can stay at the same number; the control circuit will then reduce the pulse width to give the required output. This results in a larger pulse current amplitude but lower drop-out voltage.

The choice is with the designer.

11.2 DESIGN NOTES

Under normal operating conditions, the pulse width required for 5 V out will be considerably less than the maximum value of 50%. Consequently, the transformer will normally be operating with a smaller flux excursion, and the core losses will be optimized.

Under transient conditions, the maximum flux condition can occur. Consider the unit operating at maximum line input and minimum load. The input voltage V_{cc} will be approximately 380 V. If the load is now suddenly applied to the output, the current in the output choke L1 cannot change immediately, and the output voltage will fall. Since the control amplifier is designed to give maximum transient response, it will rapidly increase the input pulse width to maximum (50% or 10 μs in this example).

The volt-seconds applied to the transformer primary will now be at maximum, and a maximum flux density condition will occur for a number of cycles as follows:

$$\hat{B} = \frac{V_{cc} \times t_{\text{on}}}{N_p \times A_{cp}} = \frac{380 \times 10}{109 \times 100} = 348 \text{ mT} \qquad \text{(very close to saturation)}$$

where $\hat{B}$ = maximum flux density in the smallest area of the core
A_{cp} = minimum core area (for this type of core the center leg is smaller than A_e and the minimum core area is considered to check that no part of the core will saturate)

When the output current has built up to the required load value, the pulse width will return to its original value.

Hence, the requirement to sustain the high stress voltage together with maximum pulse width conditions (to give fast transient response) requires a larger number of primary turns with increased copper losses. However, fewer turns may be used if steps are taken to prevent transformer saturation under transient conditions. Suitable methods would be

1. Reduced control-loop slew rate.
2. Primary current limiting (or control, which will recognize the onset of transformer saturation and reduce the "on" period).
3. Provision of a pulse-width end stop which is inversely proportional to applied voltage.

Although all these methods result in reduced transient performance, the results will often be acceptable for most applications, and improved transformer efficiency may then be achieved by using fewer turns.

Suitable drive and control circuits will be found in Part 1, Chaps. 15 and 16, and the design of output chokes and filters in Part 3, Chaps. 1, 2, and 3.

To minimize the energy stored in the primary inductance, the transformer core is not normally gapped in the forward converter. A small gap will sometimes be introduced to reduce the effects of partial core saturation; however, this gap will rarely exceed 0.1 mm (0.004 in).

CHAPTER 12
HALF-BRIDGE PUSH-PULL DUTY-RATIO-CONTROLLED CONVERTERS

12.1 INTRODUCTION

The half-bridge converter is the preferred topology for direct-off-line switchmode power supplies because of the reduced voltage stress on the primary switching devices. Further, the need for large input storage and filter capacitors, together with the common requirement for input voltage doubling circuits, supplies one-half of the bridge as a natural part of the input circuit topology.

12.2 OPERATING PRINCIPLES

Figure 2.12.1a shows the general arrangement of power sections for the half-bridge push-pull converter. The switching transistors Q1 and Q2 form only one side of the bridge-connected circuit, the remaining half being formed by the two capacitors C1 and C2. The major difference between this and the full bridge is that the primary of the transformer will see only half the supply voltage, and hence the current in the winding and switching transistors will be twice that in the full-bridge case.

Assume steady-state conditions with capacitors C1 and C2 equally charged so that the voltage at the center point, node A, will be half the supply voltage V_{cc}.

When the top transistor Q1 turns on, a voltage of half V_{cc} will be applied across the primary winding T1p with the start going positive. A reflected load current and magnetization current will now build up in the transformer primary and Q1.

After a time defined by the control circuit, Q1 will be turned off.

Now, as a result of the primary and leakage inductance, current will continue to flow into the start of the primary winding, being supplied from the snubber capacitors, C3 and C4. The junction of Q1 and Q2 will swing negative, and if the energy stored in the primary leakage inductance is sufficiently large, diode D2 will eventually be brought into conduction to clamp any further negative excursion and return the remaining flyback energy to the supply.

The voltage at the center of Q1 and Q2 will eventually return to its original central value with a damped oscillatory action. Damping is provided by the snubber resistor R1.

After a period defined by the control circuit, Q2 will turn on, taking the start

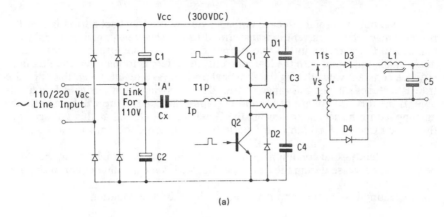

(a)

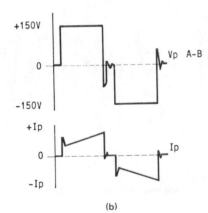

(b)

FIG. 2.12.1 (*a*) Power section of half-bridge push-pull forward converter. (*b*) Collector voltage and current waveforms for half-bridge forward converter.

of the primary winding negative. Load and magnetizing currents will now flow in Q2 and into the transformer primary winding finish so that the former process will repeat, but with the primary current in the opposite direction. The difference is that at the end of an "on" period, the junction of Q1 and Q2 will go positive, bringing D1 into conduction and returning the leakage inductance energy to the supply line. The junction of Q1 and Q2 will eventually return to the central voltage, with a damped oscillatory action. The cycle of operations is now complete and will continue. This waveform is shown in Fig. 2.12.1*b*.

As C1 and C2 are the main reservoir capacitors for the input filter, their value is very large. Consequently, the voltage at the center point of C1 and C2, node A, will not change significantly during a cycle of operations.

The secondary circuit operates as follows: When Q1 is "on," the start of all windings will go positive, and diode D3 will conduct. Current will flow in L1 and into the external load and capacitor C5.

When transistor Q1 turns off, the voltage on all transformer windings will fall toward zero, but current will continue to flow in the secondary diodes as a result of the forcing action of choke L1. When the secondary voltage has dropped to zero, diodes D3 and D4 will share the inductor current nearly equally, acting as flywheel diodes and clamping the secondary voltage at zero.

A small but important effect to consider during the flywheel period is that the primary magnetizing current is transformed to the secondary, giving a slight imbalance to the currents in the two output diodes. Although this current is normally small compared with the load current, the effect is to maintain the flux density at a constant value during the flywheel period. As a result, when the opposite transistor turns on, the full range of flux density swing from $-B$ to $+B$ is available. If the forward voltage of D3 and D4 is not matched, there will be a net voltage applied to the secondary during the flywheel period, and as this voltage is in the same direction for both "off" periods, it will drive the core toward saturation.

Under steady-state conditions, the current will increase in L1 during the "on" period and decrease during the "off" period, with a mean value equal to the output current.

Neglecting losses, the output voltage is given by the equation

$$V_{\text{out}} = \frac{V_{cc} \times D}{n}$$

where V_{cc} = primary voltage
　　　n = turns ratio N_p/N_s
　　N_p = primary turns
　　N_s = secondary turns
　　　D = duty ratio $[t_{\text{on}}/(t_{\text{on}} + t_{\text{off}})]$
　　t_{on} = "on" time
　　t_{off} = "off" time

Hence, by using suitable control circuits to adjust the duty ratio, the output voltage can be controlled and maintained constant for variations in supply or load.

12.3　SYSTEM ADVANTAGES

The half-bridge technique is often used, as it has a number of advantages, particularly for high-voltage operation.

One major advantage is that transistors Q1 and Q2 will not be subjected to a voltage in excess of the supply voltage (plus a diode drop). Diodes D1 and D2 behave as energy recovery components and clamp the collectors to the supply line, eliminating any tendency for voltage overshoot. Consequently, the transistors work under well-defined voltage stress conditions.

The frequency doubling effect in the output biphase rectifiers provides two energy pulses for each cycle of operation, reducing the energy storage requirements for L1 and C5.

The series arrangement for input capacitors C1 and C2 lends itself to a simple voltage doubling approach when 110-V operation is required.

The transformer primary winding P1 and core flux density swing are fully utilized for both half cycles of operation. This provides better utilization of the transformer windings and core than the conventional push-pull converter, where one half winding is unused during each half cycle. (A bridge rectifier in the output will provide the same action for the output winding, but this is usually reserved for high-voltage outputs for diode efficiency reasons.)

Finally, energy recovery windings are not required on the primary. This action is provided by D1 and D2 as a natural result of the topology.

12.4 PROBLEM AREAS

The designer must guard against a number of possible problems with this type of converter.

A major difficulty is staircase saturation of the transformer core. If the average volt-seconds applied to the primary winding for all positive-going pulses is not exactly equal to that for all negative-going pulses, the transformer flux density will increase with each cycle (staircase) into saturation. The same effect will occur if the secondary diode voltages are unbalanced. As storage times and saturation voltages are rarely equal in the two transistors or diodes, this effect is almost inevitable unless active steps are taken to prevent it. A small gap in the transformer core will improve the tolerance to this effect, but will not eliminate it.

Fortunately, there is a natural compensation effect as the transformer approaches saturation. The collector current on one transistor will tend to increase toward the end of an "on" period as the core starts to saturate. This results in a shorter storage time and hence a shorter period on that particular transistor, and some natural balancing action occurs.

However, where very fast switching transistors or power FETs with low storage times are used, there may be insufficient storage time for such natural corrective action.

To prevent staircase saturation, current-mode control can be used. For 110-V voltage doubler input rectifier connections, a DC path exists in the primary, and no special DC restoration circuits are required. However, for 220-V bridge operation, when current-mode control is used, a DC path through the primary must be provided, and special restoration circuits must be used. (See Part 3, Sec. 10.10.)

At lower power, a workable alternative is to have transistors Q1 and Q2 selected for near-equivalent storage times. Output diodes D3 and D4 should be selected for equivalent forward voltage drop at the working current. The transistor types and drive topology should be such that a reasonable storage time exists during the turn-off period.

Finally, the slew rate of the control amplifier must be slow so that large variations in pulse width cannot occur between cycles (otherwise the transistor that is operating near saturation will, of course, immediately saturate). The transient response of such a design will be degraded, but this may not be important. It depends on the application.

Although the partial saturation of the core caused by the staircase-saturation effect may not be a major problem under steady-state conditions, severe problems can arise during transient loading.

Assume that the power supply has been operating at a relatively light load and that steady-state operating conditions have been established. The natural tendency to staircase saturation will have resulted in the transformer operating very near a saturated condition for one transistor or the other. A sudden increase in output current demand will result in the output voltage falling initially, as the inductor current cannot change instantaneously. The control circuit will respond to the drop in output voltage by increasing the drive pulse width to maximum. The transformer will immediately saturate for one half cycle, with possible catastrophic results on the switching device.

Consequently, some other control mechanism, perhaps primary current limit or amplifier slew rate limitation, will need to be brought into action to prevent catastrophic failure. Both of these would impose a severe limitation on transient response time. More effective methods of dealing with staircase saturation (current-mode control methods) are discussed in Part 3, Chap. 10.

12.5 CURRENT-MODE CONTROL AND SUBHARMONIC RIPPLE

A coupling capacitor C_x is often fitted to prevent a DC path through the transformer winding when the supply is linked for 110-V operation. This capacitor is intended to prevent staircase saturation by blocking DC current in the transformer primary. Unfortunately, it can also introduce an undesirable effect, characterized by alternate cycles being high-voltage, narrow pulse width and low-voltage, wide pulse width as a result of the capacitor C_x developing a DC bias. This unbalanced operation results in alternate power cycles being of different amplitude and introduces subharmonic ripple into the output voltage. Even when this capacitor C_x is not fitted, this problem can still occur when the input link is removed for 220-V operation, as C2 and C3 now provide DC blocking. Special DC restoration techniques must be used if automatic balancing is to be provided, or current-mode control is applied to the half-bridge converter. (See Part 3, Sec. 10.10.)

12.6 CROSS-CONDUCTION PREVENTION

Cross conduction can be a major problem in the half-bridge arrangement. Cross conduction occurs when both Q1 and Q2 are "on" at the same instant, usually as a result of excessive storage time in the "off"-going transistor. This fault applies a short circuit to the supply lines, usually with disastrous results.

Two methods are suggested to stop this effect. The simple approach is to apply a fixed end stop to the drive pulse width so that the conduction angle can never be wide enough to allow cross conduction to take place. The problem with this approach is that the storage time is variable, depending upon transistor type, operating temperature, and loading. Consequently, to be safe, a wide margin must be provided, and hence the range of control and the utility factor of the transformer, transistors, and diodes will be reduced. (The power must be transferred during a relatively narrow conduction period.)

An alternative approach which does not suffer from these limitations is the active "cross-coupled inhibit" or "overlap protection" circuit. (See Part 1, Chap. 19.)

In this arrangement, if, say, Q1 is "on" for any reason, then the drive to Q2 is inhibited until Q1 comes out of saturation, and conversely for Q2. This automatic inhibit action has the advantages of accommodating variations in the storage time and always allowing a full conduction angle to be utilized.

12.7 SNUBBER COMPONENTS (HALF-BRIDGE)

Components C3, C4, and R1 are often referred to as snubber components; they assist the turn-off action of the high-voltage transistors Q1 and Q2 so as to reduce secondary breakdown stress. As the transistors turn off, the transformer inductance maintains a current flow, and the snubber components provide an alternative path for this current, preventing excessive voltage stress during the turn-off action.

The conventional diode, capacitor, snubber circuit should not be used in the half-bridge connection, as it provides a low-impedance cross-conduction path during the turn-on transient of Q1 and Q2.

12.8 SOFT START

When the converter is first switched on, the drive pulses should be progressively increased to allow a slow buildup in output current and voltage. This is known as soft start. If this soft-start action is not provided, there will be a large inrush current on initial switch-on, with an overshoot in output voltage; also, the transformer may saturate as a result of flux doubling effects. (See Part 3, Chap. 7.) The soft-start action should always be invoked following a shutdown of the converter—for example, after an overvoltage or overload protection shutdown. (See Part 1, Sec. 9.2.)

12.9 TRANSFORMER DESIGN

Core Size

The transformer size will be selected to meet the power requirements and temperature rise for the selected operating frequency. This information is available from manufacturers' data, and typical graphs for transmissible power are shown in Figs. 2.2.2 and 3.4.3.

Example

Assume that a conservative design is to be made for an operating frequency of 30 kHz and a temperature rise of 40°C at 100 W output.

A single output of 5 V, 20 A is to be provided. The efficiency of such a system would be of the order of 70%, giving a transmissible power of approximately 140 W (assuming that the majority of the losses will be in the transformer and output circuitry). From Fig. 2.2.2, a suitable choice would be the EC41 (FX3730) or similar.

12.10 OPTIMUM FLUX DENSITY

The choice of optimum flux density B_{opt} will be a matter for careful consideration. Unlike with the flyback converter, both quadrants of the B/H loop will be used, and the available induction excursion is more than double that of the flyback case. Consequently, core losses are to be considered more carefully as these may exceed the copper losses if the full induction excursion is used. For the most efficient design, the copper and core losses should be approximately equal.

Figure 2.12.2 shows the temperature rise for the 41-mm core plotted against total transformer loss. If we assume a permitted temperature rise of 40°C, the permitted power loss in the transformer is 2.6 W (note that the hot-spot temperature is somewhat higher than the average overall temperature of the core). Consequently, if this power loss is to be split equally between core and winding, then the maximum core loss will be 1.3 W.

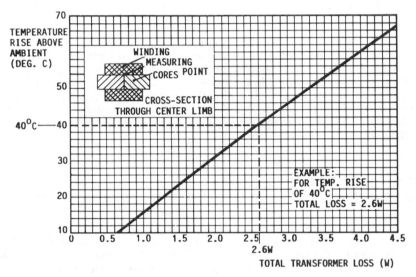

FIG. 2.12.2   Temperature rise of an FX 3730 transformer as a function of total internal dissipation, in free air conditions. (*Courtesy of Mullard Ltd.*)

From Fig. 2.12.3 it can be seen that at 30 kHz the FX 3730 cores have a loss of 1.3 W at a total flux Φ of approximately 19 μWb. The area of the center pole is 106 mm², so the peak flux density $\hat{B}$ in the center pole will be

$$\hat{B} = \frac{\Phi}{A_{cp}} = \frac{19 \ \mu\text{Wb}}{106 \ \text{mm}^2} = 180 \ \text{mT}$$

Note:

$$1 \ \text{T} = 1 \ \text{Wb/m}^2$$

A second consideration for the selection of peak flux density is the possibility of core saturation under transient loading conditions at maximum input voltage.

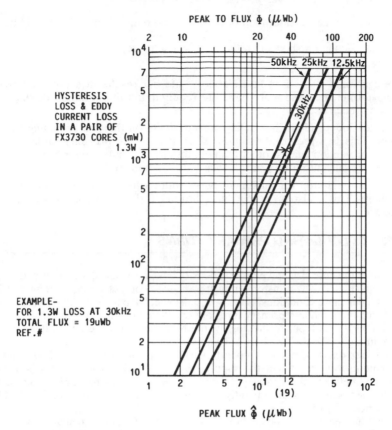

FIG. 2.12.3 Hysteresis and eddy-current loss in a pair of FX 3730 cores as a function of total flux Φ at 100°C, with frequency as a parameter. (*Courtesy of Mullard Ltd.*)

12.11 TRANSIENT CONDITIONS

When the converter operates under closed-loop conditions, as the input voltage increases, the pulse width will normally decrease in the same ratio so as to maintain the output voltage constant. Under these conditions, the peak flux density of the core remains constant at the designed value, in this case 180 mT. However, under transient conditions, it is possible for the pulse width to increase to maximum irrespective of the supply voltage. This can occur at maximum input voltage. The transformer was designed to operate at 180 mT at minimum voltage and maximum pulse width. Hence the increase in flux density at maximum input voltage will follow the same ratio as the increase in voltage (in this case 50%). In this example, the flux density would abruptly increase from 180 mT to 310 mT. From Fig. 2.2.3, it will be seen that this is still below the saturation limit, so as long as the transformer is made to operate symmetrically about zero flux, the transient loading will not cause saturation of the core, and reliable operation would be expected.

If this calculation shows that the core will saturate, then one of the following changes is recommended:

1. Design for a lower flux level. This is safe but results in a lower-efficiency transformer, because more turns are required and the optimum fluxing level has not been used.

2. Provide independent, fast-acting current limits on the two switching transistors. This is a preferred solution, since it not only prevents saturation but gives protection against other fault conditions. A similar action would be provided by current-mode control. (See Part 3, Chap. 10.)

3. Have an end stop for maximum pulse width which is proportional to input voltage. This is also an acceptable solution but degrades the transient performance.

12.12 CALCULATING PRIMARY TURNS

Once optimum core size and peak flux density have been selected, the primary turns may be calculated. The transformer must provide full output voltage at minimum line input. Under these conditions, the power pulse will have its maximum width of 16.5 μs. Hence the minimum primary turns are calculated for this condition.

With 90 V rms input to the voltage doubling network, the DC voltage will be approximately 222 V. (See Part 1, Chap. 6.)

Consider one half cycle of operation. The capacitors C1 and C2 will have a center point voltage of half the supply, that is, 111 V. When Q1 turns on, the difference between the center point voltage and V_{cc} will be applied across the transformer primary. Consequently, the primary will see a voltage V_p of 111 V for a period of 16.5 μs.

The turns required for a peak flux density of 180 mT can be calculated as follows:

$$N_{mpp} = \frac{V_p \times t_{on}}{\Delta B \times A_{cp}}$$

where V_p = primary voltage $V_{cc}/2$, V DC
 t_{on} = "on" time, μs
 ΔB = total flux density change during the "on" period,
  (Note: $\Delta B = 2\hat{B}$, see Appendix 1)
 A_{cp} = minimum core area, mm^2
 N_{mpp} = minimum primary turns (push-pull operation)
 $\hat{B}$ = peak flux density (with respect to zero), T

12.13 CALCULATE MINIMUM PRIMARY TURNS

With

$$V_p = V_{cc}/2 = 111\text{V}$$
$$t_{on} = 16.5 \text{ μs}$$
$$B_{opt} = 180 \text{ mT (optimum flux density at 30 kHz)}$$

$$\Delta B = 2 \cdot \hat{B} = 0.36 \text{ T (see Appendix 1)}$$
$$A_{cp} = 106 \text{ mm}^2$$

Therefore
$$N_{mpp} = \frac{111 \times 16.5}{0.36 \times 106} = 48 \text{ turns}$$

Therefore, the minimum primary turns required in this example would be 48 turns.

12.14 CALCULATE SECONDARY TURNS

The required output is 5 V. Allowing a voltage drop in the rectifier diode, inductor, and transformer winding of 1 V (a typical figure), the transformer secondary voltage will be 6 V. (This assumes that the maximum pulse width is 50%, giving a near-square-wave output.) Since there are 2.3 V per turn on the primary winding, the secondary turns will be

$$\frac{6}{2.3} = 2.6 \text{ turns}$$

Half turns are not convenient, as they can cause saturation of one leg of the transformer unless special techniques are used. (See Part 3, Sec. 4.23.) We have the option of rounding up to 3 turns or down to 2 turns. If the turns are rounded downward, it would become necessary to reduce the number of primary turns to maintain the correct output voltage, as the pulse width cannot be increased beyond 50%. This reduction in turns would result in an increase in the flux density of the core, and saturation could occur under transient conditions. Therefore, the choice is made to increase the secondary turns to 3. The primary turns and maximum flux density will then remain as before, and the pulse width will be reduced to give the correct output voltage. Hence, at minimum input voltage, the pulse width will be less than 16.5 μs, and it is now possible to have a fixed end stop on the pulse width and provide a "dead band" to prevent cross conduction. (Cross conduction occurs when both power transistors are "on" at the same time, presenting a short circuit across the supply lines and usually resulting in rapid catastrophic failure. See Part 1, Chap. 19.)

We now have the basic information for the winding of the power transformer. The selection of wire sizes, shapes, and topology is described in Part 3, Chap. 4.

The design approach outlined here is by no means rigorous and is intended for general guidance only. More comprehensive information is available in Part 3, Chaps. 4 and 5, and the designer is urged to study these sections and references 1 and 2.

12.15 CONTROL AND DRIVE CIRCUITS

The control and drive circuits used for this type of converter are legion. They range from fully integrated control circuits, available from a number of manufacturers, to the fully discrete designs favored by many power supply engineers. A discussion of suitable drive circuits will be found in Part 1, Chaps. 15 and 16.

For reliable operation, the drive and control circuits must provide the following basic functions:

1. *Soft start.* This reduces inrush current and turn-on stress, and helps to prevent output voltage overshoot during the turn-on action. In push-pull applications, it also prevents saturation of the transformer core by flux doubling effects. (See Part 3, Chap. 7.)

2. *Flux centering.* This circuit differentially controls the pulses supplied to the upper and lower transistors in push-pull applications to maintain the mean flux in the core at zero. Various methods may be used, and these are discussed in Part 3, Chap. 6.

3. *Cross-conduction inhibit.* Cross conduction occurs when both power transistors are "on" at the same instant. This can occur even though the drive to each device does not exceed 50%, as the storage time in the power devices can cause the power pulses to overlap. Fixed end stops or an active cross-conduction limit may be used. (See Part 1, Chap. 19.)

4. *Current limiting.* Current limiting may be applied to input or output and must maintain control down to a short-circuit condition. For switchmode systems, a constant-current limit is recommended, since this will prevent lockout with nonlinear loads. (See Part 1, Chaps. 13 and 14.)

5. *Overvoltage protection.* On higher-current units, it is generally acceptable to provide overvoltage protection by converter shutdown, rather than by SCR crowbar techniques. (See Part 1, Chap. 11.)

6. *Voltage control and isolation.* Stable closed-loop control of the output voltage must be maintained for all conditions of operation. Where outputs are to be isolated from inputs, a suitable isolation device must be used. These include optical couplers, transformers, and magnetic amplifiers.

   Note: Where output performance is not critical, it is possible to provide control on the primary circuit by using an auxiliary winding on the transformer for both voltage control and auxiliary requirements.

7. *Primary overpower limiting.* A number of possible failure mechanisms can be prevented by providing independent current limits on the primary power transistors. Problems associated with cross conduction and transformer saturation can be avoided in this way. It is sometimes a requirement of the specification that a short circuit on a transformer secondary not cause catastrophic failure. (See Part 1, Chap. 13.)

8. *Input undervoltage protection.* The control circuit should recognize the state of the input voltage and allow operation only when this is high enough for satisfactory performance. Hysteresis should be provided on this guard circuit to prevent squegging at the critical turn-on voltage. (See Part 1, Chap. 8.)

9. *Ancillary functions.* Electronic inhibit (usually TTL compatible), synchronization, power good signals, power-up sequencing, input RFI filtering, inrush current limiting, and output filter design are all covered in their various sections. (See Index.)

12.16 FLUX DOUBLING EFFECT

The difference in the operating mode for the single-ended transformer and the push-pull balanced transformer is not always fully appreciated.

For the single-ended forward or flyback converter, only one quadrant of the

B/H loop is used, there is a remnant flux B_r, and the remaining range of induction is often quite small. (Figure 2.9.2a and b shows the effect well.)

In the push-pull transformer, it is normally assumed that the full B/H loop may be used and that B will be incremented from $-B_{max}$ to $+B_{max}$ each cycle, so that the total change ΔB during an "on" period will be $(2 \times \hat{B})$. Often this value of $2\hat{B}$ is used in the calculation of primary turns, resulting in half the number of turns being required on the push-pull transformer compared with the single-ended case. Hence

$$N_{mpp} = \frac{V_{cc} \times t_{on}}{2\hat{B} \times A_e}$$

where N_{mpp} = minimum primary turns for push-pull operation
V_{cc} = primary DC voltage
t_{on} = maximum "on" time, μs
$\hat{B}$ = maximum flux density, mT
A_e = effective core area, mm^2

However, the designer should be careful when applying this approach, as $2\hat{B}$ may not always be valid for ΔB. For example, consider the condition when the converter is first switched on. The core flux density will be sitting between $-B_r$ and $+B_r$ (the remanence values). Hence, for the first half cycle, a flux change of $2\hat{B}$ may take the core into saturation (the so-called "flux doubling effect"). Thus the full voltage and pulse width cannot be applied for the first few cycles of operation, and a soft-start action is required (that is, the pulse width must increase slowly over the first few cycles). Further, if under steady-state conditions staircase-saturation effects are allowed to take the core toward saturation, the full $2\hat{B}$ range may not be available in both directions. (This is important when rapid changes in pulse width are demanded under transient conditions.) Hence there is a real need for symmetry correction in push-pull converters.

For these reasons it is common practice to reduce the range of ΔB and use more turns than the above equation would indicate in order to provide a good working margin for any start-up and asymmetry problems. The margin required depends on how well the above effects have been controlled.

12.17 PROBLEMS

1. Why is the term "half bridge" used to describe the circuit shown in Fig. 2.12.1a?
2. Describe the operating principle of the half bridge.
3. What is the function of diodes D1 and D2 in the half-bridge circuit shown in Fig. 2.12.1a?
4. In a duty-ratio-controlled half-bridge forward converter with full-wave rectification, what prevents core restoration during the period when Q1 and Q2 are both "off"?
5. What is the intended function of the series capacitor C_x in the primary circuit?

6. Why does the series capacitor C_x in Fig. 2.12.1 often cause subharmonic ripple in half-bridge applications?

7. What can be done to reduce the effects of core staircase saturation in the half-bridge circuit?

8. Why is the problem of staircase saturation more noticeable when power FETs are used?

9. Describe a method of control which eliminates staircase-saturation problems.

10. Why is it bad practice to fit a capacitor in series with the primary winding when current-mode control is being used?

11. What is meant by "flux doubling," and what is its cause?

12. What is normally done to prevent core flux doubling during the initial turn-on of a push-pull converter?

13. What is meant by "cross conduction" in a half-bridge circuit?

14. Give two methods commonly used to prevent cross-conduction problems.

15. Diodes are not used in the snubber circuits applied to the half-bridge circuit. Why is this?

16. Why is the optimum flux density swing in high-frequency push-pull converters often considerably less than the peak core capability?

CHAPTER 13
BRIDGE CONVERTERS

13.1 INTRODUCTION

The full-bridge push-pull converter requires four power transistors and extra drive components. This tends to make it more expensive than the flyback or half-bridge converter, and so it is normally reserved for higher-power applications.

The technique has a number of useful features; in particular, a single primary winding is required on the main transformer, and this is driven to the full supply voltage in both directions. This, together with full-wave output rectification, provides an excellent utility factor for the transformer core and windings, and highly efficient transformer designs are possible.

A second advantage is that the power switches operate under extremely well defined conditions. The maximum stress voltage will not exceed the supply line voltage under any conditions. Positive clamping by four energy recovery diodes eliminates any voltage transients that normally would have been generated by the leakage inductances.

To its disadvantage, four switching transistors are required, and since two transistors operate in series, the effective saturated "on"-state power loss is somewhat greater than in the two-transistor push-pull case. However, in high-voltage off-line switching systems, these losses are acceptably small.

Finally, the topology provides flyback energy recovery via the four recovery diodes without needing an energy recovery winding.

13.2 OPERATING PRINCIPLES

13.2.1 General Conditions

Figure 2.13.1 shows the power section of a typical off-line bridge converter. Diagonal pairs of switching devices are operated simultaneously and in alternate sequence. For example, Q1 and Q3 would both be "on" at the same time, followed by Q2 and Q4. In a pulse-width-controlled system, there will be a period when all four devices will be "off." It should be noted that when Q2 and Q4 are "on," the voltage across the primary winding has been reversed from that when Q1 and Q3 were "on."

In this example, a proportional base drive circuit has been used; this makes the base drive current proportional to the collector current at all times. This tech-

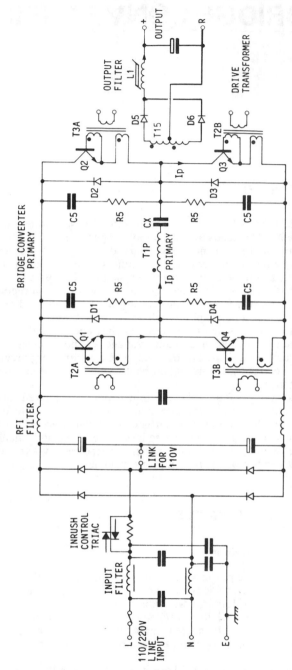

FIG. 2.13.1 Full-bridge forward push-pull converter, showing inrush limiting circuit and input filter.

nique is particularly suitable for high-power applications and is more fully described in Part 1, Chap. 16.

During the "off" period, under steady-state conditions, a current will have been established in L1, and the output rectifier diodes D5 and D6 will be acting as flywheel diodes. Under the forcing action of L1, both diodes conduct an equal share of the inductor current during this "off" period (except for a small magnetizing current). Provided that balanced diodes are used, the voltage across the secondary windings will be zero, and hence the primary voltage will also be zero (after a short period of damped oscillatory conditions caused by the primary leakage inductance). Typical collector voltage waveforms are shown in Fig. 2.13.2.

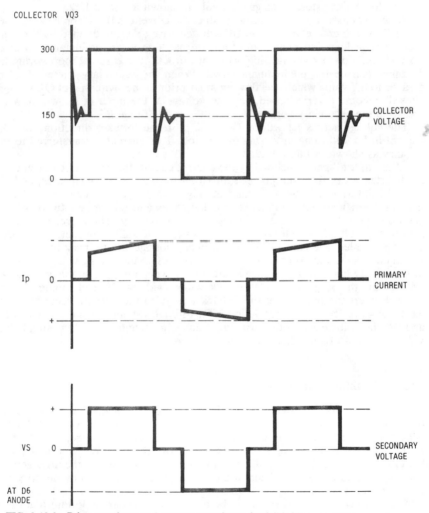

FIG. 2.13.2 Primary voltage and current waveforms for full-bridge converter.

13.2.2 Cycle of Operations

Consider a cycle of operation under steady-state conditions in Fig. 2.13.1. Assume that the drive circuit initiates a turn-on pulse for Q1 and Q3. These two devices will start to turn on. Collector current will now flow via the primary winding of T1P through the primary of the proportional drive transformers T2A and T2B. By positive regenerative feedback, the turn-on action of the two transistors is enhanced, and this results in rapid switching to the fully "on" state, with Q1 and Q3 fully saturated.

As soon as Q1 and Q3 turn on, current will start to build up in the primary winding of T1P at a rate defined by the primary leakage inductance. This current is made up of the reflected load current and a small proportion of magnetizing current for the transformer magnetic field, as shown in Fig. 2.13.2.

Simultaneously, during the turn-on edge, the current in the secondary rectifier diode D5 will increase and that in D6 will decrease at a rate defined by the total secondary leakage inductance and the external loop wiring inductance through D5 and D6. For low-voltage, high-current outputs, the external loop wiring inductance can be the predominant effect. When the secondary current has increased to the value which was flowing in L1 prior to the switch-on of Q1 and Q3, D6 will become reverse-biased, and the voltage at the input of L1 will now increase to the secondary voltage V_s less the drop in diode D5.

The voltage across L1 will be $(V_s - V_{out})$ in the forward direction, and the current in L1 will ramp up during this period. This current is transferred to the primary as shown in Fig. 2.13.2.

After an "on" period defined by the control circuit, the base drive current will be diverted away from the power transistors by the drive transformer, and Q1 and Q3 will turn off. However, a magnetizing current has now been established in the transformer primary, and leakage inductances and the ampere-turns will remain constant, the current transferring to the secondary. Therefore, by flyback action, the voltages on all windings will reverse. If sufficient energy has been stored in the leakage inductances, the primary voltage will fly back to a point at which diodes D2 and D4 conduct, and the excess flyback energy will be returned to the supply lines. If the leakage inductance is very low, the snubber capacitor C5, R5, and the output diodes D5 and D6 will provide effective clamping. D5 and D6 will divert the majority of the flyback energy to the output. Because of the hard clamping action of primary diodes D1 through D4 and secondary diodes D5 and D6, the voltage across the switching transistors cannot exceed the supply line voltage by more than a diode drop at any time.

13.2.3 Snubber Components

During the turn-off transient, the snubber components R5, C5 will reduce the turn-off stresses on the power devices by providing an alternative path for the collector current during the turn-off transient. It is possible to replace the four snubber networks by a single *RC* network across the transformer primary, but better common-mode control of the primary voltage is given by the arrangement shown when all power transistors are off. This action is more fully described in Part 1, Chap. 17.

The flywheel action provided by the output diodes is an important feature of this type of push-pull circuit. Figure 2.9.2*a* and *b* shows the working range for the core flux swing in both push-pull and single-ended operation. The range is much wider in the push-pull case, as the core will not restore to zero, even when all transistors are turned off, because of the flywheel conduction of D5 and D6 as follows.

Because D5 and D6 remain conducting during the "off" period, the voltage across the secondary, and hence across all windings, is zero when the switching transistors are turned off. As a result, the core will not restore to B_r during the "off" period but will be held at $+\hat{B}$ or $-\hat{B}$. Hence, when the following diagonal pair of input transistors are turned "on", the full flux density range of $2\hat{B}$ (from $-\hat{B}$ to $+\hat{B}$) is available for use, allowing the transformer to be designed for a lower number of primary turns. The secondary voltage waveform is shown in Fig. 2.13.2.

This secondary diode clamping effect is lost when the load falls below the magnetization current (as referred to the secondary). However, this would not normally be a problem, as under these conditions the "on" pulse would be very short and ΔB small.

13.2.4 Transient Flux Doubling Effect

Under transient loading conditions, a problem can sometimes occur if the full range of the B/H characteristic is used. If the supply has been running under light load, the pulse width will be narrow, and the core will be working near $B = 0$. If a sudden increase in load drives the unit to full pulse width, only half the range of ΔB is now available for this transient change, and the core may saturate. Be careful to consider transient conditions and either allow sufficient flux density margin to cope with this condition or limit the control slew rate to allow the core to establish a new working condition. (This effect is sometimes referred to as "flux doubling.")

13.3 TRANSFORMER DESIGN (FULL BRIDGE)

The design approach for the push-pull transformer is relatively straightforward. The single primary winding is used for both half cycles of operation, providing extremely good utility of core and windings.

To minimize magnetization currents, the maximum primary inductance consistent with minimum turns is required. Consequently, a high-permeability material will be selected, and the core will not be gapped. (A small core gap is sometimes introduced if there is a chance of a DC current component in the transformer, as the onset of saturation is more controllable with a core gap.)

13.4 TRANSFORMER DESIGN EXAMPLE

Assume that a transformer is to be designed on a ferrite core to meet the following requirements:

Input voltage	90–137 or 180–264 (by link change)
Frequency	40 kHz
Output power	500 W
Output voltage	5 V
Output current	100 A

13.4.1 Step 1, Select Core Size

Assume an initial efficiency for the transformer and secondary rectifier circuit of 75%. The transmitted power for the transformer will then be 500/0.75 = 666 W.
From Fig. 2.13.3, for push-pull operation at this power level, an EE 55-55-21

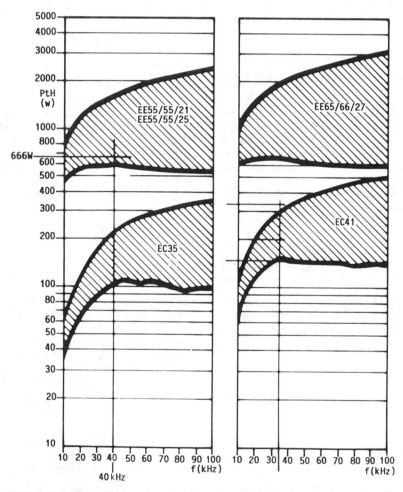

FIG. 2.13.3 Core selection chart for balanced push-pull operation, showing throughput power as a function of frequency with core size as a parameter. (*Courtesy of Mullard Ltd.*)

core is indicated for a temperature rise of 40°C under convected air-cooling conditions. Hence this core will be used in the following example.

13.4.2 Step 2, Select Optimum Flux Density

For push-pull operation, the full *B/H* loop may be used. (See Fig. 2.9.2.) A large flux density excursion gives fewer primary turns and lower copper losses, but increased core losses.

Normally, it is assumed that minimum loss (maximum efficiency) will be found near the point where the copper and core losses are equal, and this would be the normal design aim in the selection of working flux density.

Figure 2.13.4 shows how core losses increase for A16 ferrite core material as the number of turns is reduced and the peak flux density is increased from 25 to 200 m T. (At the same time, of course, the copper losses will decrease, but this is not shown here.)

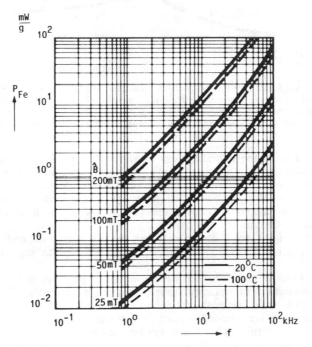

FIG. 2.13.4 Core loss per gram of A16 ferrite as a function of frequency, with peak flux density as a parameter. (Note: Graph is plotted for peak flux density $\hat{B}$; flux density sweep ΔB is $2 \times \hat{B}$.) (*Courtesy of Mullard Ltd.*)

Figure 2.13.5 shows, for a pair of EE55-55-21 cores in A16 ferrite at 40 kHz, how the core, copper, and total losses change as the number of turns is changed and the peak flux density is increased toward 200 mT. It will be seen that a minimum total loss occurs near 70 mT. (For each turns example, optimum use of the core window area and wire gauge is assumed.)

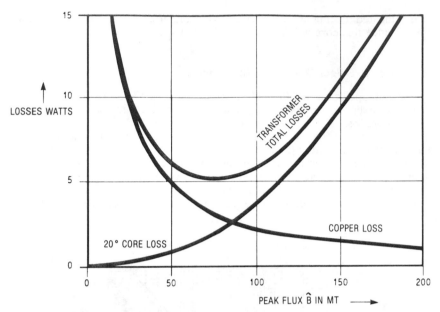

FIG. 2.13.5 A16 ferrite core loss, copper loss, and total loss for a pair of EE55/55/21 cores, when wound for optimum performance in a typical switchmode transformer. Loss is shown as a function of the peak flux density. Note that the minimum total loss occurs when the transformer induction (turns) is optimized so that the core loss is 44% of the total loss.

In this example, the minimum loss (maximum efficiency) occurs when the core loss is 44% of the total loss at 70 mT. However, the minimum loss condition has a relatively wide base, and the choice of peak flux density for optimum efficiency is not critical in the 50- to 100-mT range. The normal assumed optimum choice (where the copper and core losses are equal) would be 80 mT, which is not very far from optimum.

For each design there is an optimal flux density swing, depending upon the operating frequency, the core loss, the topology, and the winding utilization factors.

Figure 2.13.6a, b, and c shows the manufacturers' peak and optimum flux density recommendations for optimum transformer designs using the EE55-55-21 and other cores in forward and push-pull applications. From Fig. 2.13.6c (at 40 kHz), the manufacturer's recommended peak flux density is 100 mT, which is not far from optimum, and this higher value will be used in this example to reduce the number of turns.

13.4.3   Step 3, Calculate Primary Voltage (V_{cc})

As the peak flux density was chosen for near optimum efficiency and is well clear of saturation, the design approach used here will be to calculate the primary turns for the maximum "on" period (50% duty ratio) with the input voltage at minimum. This will occur at the minimum line input of 90 V rms, and the DC voltage for the voltage doubler connection will be

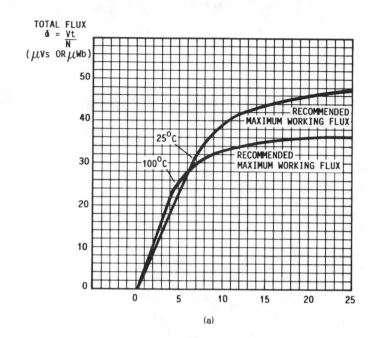

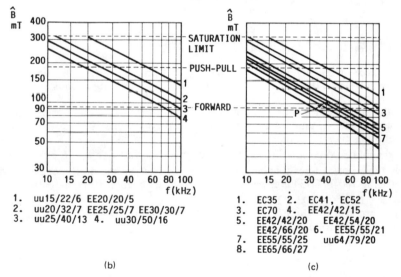

FIG. 2.13.6 (a) Magnetization curve for N27 ferrite material at 25°C and 100°C. (*Courtesy of Siemens AG.*) (b), (c) Optimum peak flux density as a function of frequency, with core size as a parameter. (*Courtesy of Mullard Ltd.*)

$$V_{cc} = V_{in} \times 1.3 \times 1.9$$

where V_{in} = ac input voltage, rms

Note: The derivation of the DC voltage and ripple component is more fully covered in Part 1, Chap. 6. Hence, at 90 V input, using the voltage doubler connection,

$$V_{cc} = 90 \times 1.3 \times 1.9 = 222 \text{ V DC}$$

13.4.4 Step 4, Calculate Maximum "On" Period

If cross conduction (two series transistors "on" at the same time) is to be avoided, then the maximum "on" time cannot exceed 50% of the total period. Hence

$$t_{on(max)} = 50\% \ P$$

At 40 kHz,

$$t_{on} = \frac{1}{2} \times \frac{1}{40,000} = 12.5 \ \mu s$$

13.4.5 Step 5, Calculate Primary Turns

The voltage waveform applied to the transformer primary during an "on" period is rectangular, and the turns can be calculated using the volt-seconds (Faraday's law) approach.

In the push-pull transformer, both quadrants of the BH loop are used, and the flux density swing, under steady-state balanced conditions, will go from $-B$ to $+B$ for a positive half cycle.

It should be noted that Fig. 2.13.4 shows the peak flux density $\hat{B}$, but assumes losses for a peak-to-peak swing ΔB of $2 \times \hat{B}$. For optimum efficiency, $\hat{B}$ was selected at 100 mT. Therefore, the peak-to-peak change (flux density swing) $\Delta B = 2 \times \hat{B}$, or $= 200$ mT for this example.

The core area for the EE55-55-21 is 354 mm², and the primary turns may now be calculated as follows:

$$N_p = \frac{V_{cc} \times t_{on}}{\Delta B \times A_m}$$

where V_{cc} = minimum DC header voltage

t_{on} = maximum "on" period, μs

ΔB = total flux density swing, T

A_m = minimum pole area, mm²

Hence

$$N_p = \frac{222 \times 12.5}{0.2 \times 354} = 39 \text{ turns (or } 5.7 \text{ V/turn)}$$

13.4.6 Step 6, Calculate Secondary Turns

When the bridge converter is operating at full conduction angle (maximum output), the primary waveform tends to a square wave. Consequently, the rectified output tends to DC, and the output voltage will be the secondary voltage less rectifier, choke, and wiring losses.

Allowing 1 V for all losses, the transformer secondary voltage V_s will be 6 V. Therefore, the turns for each half of the secondary winding will be

$$\frac{V_s}{\text{V/turn}} = \frac{6}{5.7} \simeq 1 \text{ turn}$$

Note: The secondary is normalized to one turn and the primary to 37 turns, giving a peak flux density slightly higher than 100 mT in this example.

The secondary voltage V_s used for this calculation is the voltage produced at the minimum line input of 90 V. At this voltage the pulse width is at maximum. At higher input voltages, the pulse width will be reduced by the control circuit to maintain output voltage regulation.

To minimize the copper losses and leakage inductance, it is important to choose the optimum gauge and shape of transformer wire and to arrange the makeup for minimum leakage inductance between primary and secondary windings. A split-layer winding technique will be used in this example.

The high-current secondary winding should use a copper strip spanning the full width of the bobbin (less creepage distance). Suitable winding techniques and methods of optimizing wire shapes and gauges are more fully covered in Part 3, Chap. 4.

13.5 STAIRCASE SATURATION

There will, inevitably, be some imbalance in the forward and reverse volt-seconds conditions applied to the transformer. This may be caused by differences in storage times in the transistors or by some imbalance in the forward voltage of the output rectifier diodes. However it is caused, this imbalance results in the flux density in the transformer core staircasing toward saturation with each cycle of operation.

Restoration of the core during the "off" period cannot occur, because during this period the secondary is effectively short-circuited by the clamping action of diodes D5 and D6, which will both be conducting, under the forcing action of L1 (provided that the load is above the critical current value for L1).

When the core reaches saturation, there is a compensation effect, as the transistors which are conducting the higher current on the saturating cycle will have their storage times reduced, and so a measure of balance will be restored. However, a problem still exists for transient operation, and this is discussed in the following section.

13.6 TRANSIENT SATURATION EFFECTS

Assume that the power supply has been operating for a period under light loading conditions, staircase saturation has occurred, and one pair of transistors is operating near the saturated point. If a transient increase in load is now applied, the control circuit will demand a rapid increase in pulse width to compensate for losses and to increase the current flow in L1. The core will immediately saturate in one direction, and one pair of transistors will take an excessive current, with possible catastrophic results.

If the power transistors have independent fast-acting current limits, then the "on" pulse will be terminated before excessive current can flow, and failure of the power devices can be avoided. This is not an ideal solution, since the transient response will now be degraded.

Alternatively, the slew rate of the control amplifier can be reduced so that the increase in pulse width is, say, less than 0.2 μs on each cycle. Under these conditions, the storage self-compensation effect of the power transistors will normally be able to prevent excessive saturation. However, the transient response will be very much degraded, once again. Nevertheless, these two techniques are commonly used.

13.7 FORCED FLUX DENSITY BALANCING

A much better solution to the staircase saturation problem may be applied to the push-pull bridge circuit shown in Fig. 2.13.1.

If two identical current transformers are fitted in the emitters of Q3 and Q4, the peak values of the currents flowing in alternate pairs of transistors (and, therefore, in the primary winding) can be compared alternately on each half cycle.

If any unbalance in the two currents is detected, it acts upon the ramp comparator to adjust differentially the width of the drive pulses to the power transistors. This can maintain the transformer's average working flux density near the center of the B/H characteristic, detecting any DC offset and adjusting the drive pulses differentially to maintain balance.

It should be noted that this technique can work only if there is a DC path through the transformer winding. A capacitor is sometimes fitted in series with the primary winding to block any DC current; DC transformer saturation is thus avoided. However, under unbalanced conditions this capacitor will take up a net charge, and so alternate primary voltage pulses will not have the same voltage amplitude. This results in a loss of efficiency and subharmonic ripple in the output filter; further, maintaining balanced transformer currents and maintaining capacitor charge are divergent requirements, leading to runaway condition. Therefore, this DC blocking arrangement using a capacitor is not recommended.

The series capacitor C_x must *not* be fitted if the forced current balancing system is used, as it will eliminate the detectable DC component, and the circuit will not be able to operate. This is more fully explained in Part 3, Sec. 6.3.

If the transformer's working point can be maintained close to its center point, full advantage can be taken of the working flux density range, giving improved transient capability without the possibility of transformer saturation and power device failure.

When current-mode control is used for the primary pulse-width modulation, flux balancing happens automatically, provided C_x is not fitted. See Part 3, Chap. 10.

13.8 PROBLEMS

1. Why is the full-bridge converter usually reserved for high-power applications?
2. What is the major advantage of the full-bridge converter?
3. Why is the proportional drive circuit favored in the bridge converter?
4. Why is it particularly important to prevent staircase saturation in the full-bridge converter?
5. What measures are usually taken to prevent staircase saturation in the bridge converter

CHAPTER 14
LOW-POWER
SELF-OSCILLATING
AUXILIARY CONVERTERS

14.1 INTRODUCTION

Many of the larger power converters require a small amount of auxiliary power for the supply of the control and drive circuits. Often the auxiliary requirements are derived from 60-Hz line transformers. This is not always very efficient, as the size of the transformer will often be determined by the need to meet VDE and UL creepage distance specifications rather than by the power needs. As a result, the transformer will often be larger in size than is required to meet the power requirements alone.

In applications in which an uninterruptible power supply (UPS) is required, the backup supply may be a battery-derived DC and a 60-Hz input may not be available. Hence, in this type of system, a 60-Hz transformer cannot be used.

One solution is to use a low-power, high-frequency converter to supply the auxiliary needs.

Very efficient low-cost converters can be produced using self-oscillating techniques. Some suitable examples are considered in this chapter.

14.2 GENERAL OPERATING PRINCIPLES

In self-oscillating converters, the switching action is maintained by positive feedback from a winding on the main transformer. The frequency is controlled either by saturation of the main or a subsidiary drive transformer, or in some cases by a drive clamping action which responds to the increase in magnetization current during the "on" period.

In these simple systems, the frequency is subject to variations caused by changes in the magnetic properties of the core, loading, or applied voltage.

14.3 OPERATING PRINCIPLE, SINGLE-TRANSFORMER CONVERTERS

Figure 2.14.1 shows a single-transistor version of a self-oscillating converter. This converter operates in a flyback mode and is more useful for low-power, constant-load applications, such as auxiliary supplies for the control circuits of a large converter. (In this example, an output of 12 V, 150 mA is developed.)

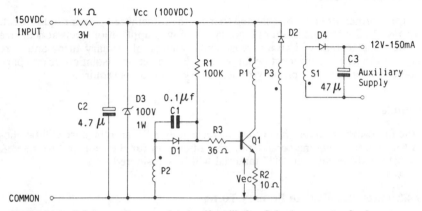

FIG. 2.14.1 Primary voltage-regulated self-oscillating flyback converter for low-power auxiliary supplies.

When first switched on, current flowing in R1 initiates the turn-on of Q1. As Q1 turns on, positive regenerative feedback is developed by drive winding P2 and applied to Q1 via C1 and D1 so that Q1 turns on rapidly. The current in the collector, and hence the emitter, of Q1 will now ramp up linearly at a rate defined by the primary inductance and the applied voltage.

As the emitter current increases, the voltage across the emitter resistor R2 (V_{ec}) will also increase until it approaches the value generated by the feedback winding P2. At this point, the base current to Q1 will be "pinched off," and Q1 will start to turn off. By normal flyback action, the voltages on all windings will now reverse, and regenerative turn-off will be applied to the base of Q1 by the drive winding P2 and capacitor C1.

This "off" state will now continue until all the energy that was stored in the transformer during the "on" period is transferred to the output circuit. At this point, the voltage across all windings will begin to fall toward zero. Now, as a result of the charge that has been developed across C1 by the current in R1 as the drive winding P2 returns to zero, the base of Q1 will, once again, be taken positive, and Q1 will be turned on again to repeat the cycle.

The frequency of operation is controlled by the primary inductance, the value of R2, the reflected load current and voltage, and the selected feedback voltage on P2.

To minimize the frequency change resulting from load variations, the turn-off time must be maintained nearly constant. This is achieved by storing sufficient energy during the "on" period to keep the energy recovery diode D2 in conduction during the complete flyback period. By this means, the flyback voltage is maintained constant. This requires that the flyback energy considerably exceed the load requirements, so that the spare energy will be returned to the supply line during the complete flyback period, maintaining D2 in conduction. The transformer inductance will be selected to obtain this condition by adjusting the core gap size. In low-power, constant-load applications, a zener diode D3 in the supply line will stabilize the supply voltage and ensure a fixed frequency and a stabilized output voltage.

14.4 TRANSFORMER DESIGN

14.4.1 Step 1, Select Core Size

The transformer size may be selected to meet the transmitted power requirement. (See Fig. 2.2.2.) But more often, for very low power applications, a practical size will be selected that will give a reasonable number of primary turns and wire gauge. Further, the core must be large enough to meet any isolation or creepage distance requirements if primary-to-secondary isolation is required.

Example

In the following example, the output power is only 3 W, and the core will be chosen for practical winding considerations rather than for the temperature rise. A core size EF16 in Siemens N27 material will be considered.

14.4.2 Step 2, Calculate Primary Turns

Assume the following specifications:

Frequency	30 kHz ($\frac{1}{2}$ period t = 16.5 μs)
Core area A_e	20.1 mm^2
Supply voltage V_{cc}	100 V
Flux density swing ΔB	250 mT

$$N_p = \frac{V_{cc} \cdot t}{\Delta B \cdot A_e} = \frac{100 \times 16.6 \times 10^{-6}}{250 \times 10^{-3} \times 20.1 \times 10^{-6}} = 330 \text{ turns}$$

The flyback voltage will be equal to the forward voltage, as bifilar windings of equal turns are used in this example.

14.4.3 Step 3, Calculate Feedback and Secondary Turns

The feedback winding should be selected to generate approximately 3 V to ensure an adequate feedback factor for fast switching action of Q1.

$$N_{fb} = \frac{N_p \times V_{fb}}{V_{cc}} = \frac{330 \times 3}{100} = 9.9 \text{ turns (use 10 turns)}$$

The output voltage is to be 12 V, and allowing 0.6 V for diode losses, the secondary voltage will be 12.6 V.
  Hence

$$N_s = \frac{N_p \times V_s}{V_{cc}} = \frac{330 \times 12.6}{100} = 42 \text{ turns}$$

14.4.4 Step 4, Calculate Primary Current

The output power is 3 W; therefore, assuming 70% efficiency, the input power will be 4.3 W.

Hence the mean input current at $V_{cc} = 100$ V will be

$$I_m = \frac{W}{V_{cc}} = \frac{4.3}{100} = 43 \text{ mA}$$

For a complete energy transfer system, the primary current will be triangular, and the peak current can be calculated. The flyback voltage is the same as the forward voltage, so the "on" and "off" periods will also be equal. (See Fig. 2.14.2.)

From Fig. 2.14.2,

$$I_{peak} = 4 \times I_{mean} = 172 \text{ mA}$$

The actual collector current should exceed this calculated mean current by at least 50% to ensure that D2 will be maintained in conduction during the complete flyback period. This defines the flyback voltage and gives a smart switching action. High efficiency will still be maintained, as the spare flyback energy will be recovered by energy recovery diode D2 and returned to the supply. Hence, in this example, the primary current will be increased by 50%, requiring a lower inductance:

$$I_p = 1.5 \times I_{peak} = 258 \text{ mA}$$

14.4.5 Step 5, Establish Core Gap (Empirical Method)

Place a temporary gap of 0.010 in in the transformer core. Connect the transformer into the circuit, and operate the unit with a dummy load at the required power. Adjust the transformer gap for the required period (collector current slope and transmitted power). Note:

$$\frac{\Delta i_c}{\Delta t} = \frac{258}{16.5} = 15.5 \text{ mA/}\mu\text{s}$$

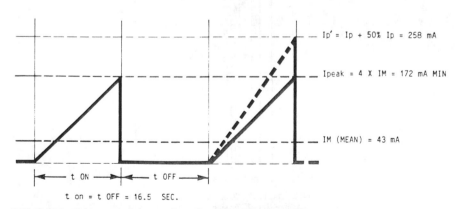

FIG. 2.14.2 Primary current waveform for self-oscillating auxiliary converter.

Resistor R1 is selected to give reliable starting and resistor R3 to give the required base drive current. These values depend upon the gain of Q1; those shown in Fig. 2.14.1 are typical for a small 5-W unit.

R2 must be selected to pinch off the drive when the collector current reaches 253 mA. (At this point the base current will have dropped to zero.) Hence

$$R2 = \frac{V_{fb} - V_{be}}{I_p} = \frac{3 - 0.6}{258 \times 10^{-3}} = 9.3 \ \Omega$$

Finally, R2 may be adjusted to give the required operating frequency.

14.4.6 Establish Core Gap (by Calculation and Published Data)

From Fig. 2.14.2, the required inductance may be calculated from the slope of the current and the value of the applied voltage, as follows:

$$L_p = \frac{V_{cc} \times \Delta t}{\Delta I_p} = \frac{100 \times 16.5}{258} = 6.4 \text{ mH}$$

The required A_L factor (nH/turn^2) may now be calculated:

$$A_L = \frac{L}{n^2} = \frac{6.4 \times 10^{-3}}{330^2} = 59 \text{ nH/turn}$$

From Fig. 2.14.3 (the published data for the E16 core) the required gap is obtained by entering the figure with $A_L = 59$ nH
  A gap of 0.6 mm (0.023 in) is indicated.

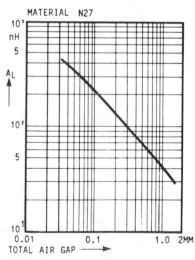

FIG. 2.14.3 A_L factor as a function of core gap size for E16 size N27 ferrite cores. (*Courtesy of Siemens AG.*)

CHAPTER 15

SINGLE-TRANSFORMER TWO-TRANSISTOR SELF-OSCILLATING CONVERTERS

15.1 INTRODUCTION

Figure 2.15.1 shows the circuit of a very basic current-gain-limited two-transistor saturating transformer converter. This is sometimes referred to as a DC transformer. In this type of converter, the main transformer is driven to saturation, giving a large core loss; hence the transformer is not very efficient. Also, the maximum collector current in the switching transistors is gain-dependent and not well specified. Hence, this type of converter is more suitable for low-power applications, typically from 1 to 25 W.

Because of the full-conduction square-wave push-pull operation, the rectified output is nearly DC, and hence the topology is capable of much higher secondary currents than the flyback converter considered in Chap. 14. The primary transistors see a voltage stress of at least twice the supply voltage, so the circuit is preferred for lower-input-voltage applications.

The major advantages are simplicity, low cost, and small size. Where the performance is acceptable, these simple converters provide very cost-effective solutions to the smaller auxiliary power needs.

15.2 OPERATING PRINCIPLES (GAIN-LIMITED SWITCHING)

The circuit shown in Fig. 2.15.1 operates as follows.

On initial switch-on, a start-up current flows in R1 to the bases of Q1 and Q2. This will initiate a turn-on action on the transistor with the lowest V_{be} or highest gain; assume Q1 in this example.

As Q1 turns on, the finish of all windings goes negative (starts positive), and positive regenerative feedback from drive winding P2 will take the base of Q1 more positive and the base of Q2 more negative. Hence, Q1 will switch rapidly into a fully saturated "on" state.

The supply voltage V_{cc} is now applied across the left-hand side of the main

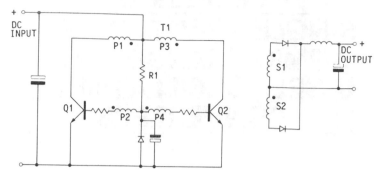

FIG. 2.15.1 Low-voltage, saturating-core, single-transformer, push-pull self-oscillating converter (DC transformer).

primary winding P1, and magnetization current plus any reflected load current will flow in the collector of Q1.

During the "on" period, the flux density in the core will increase toward saturation—say, point S1 in Fig. 2.15.2*a*. After a period defined by the core size, saturation flux value, and number of primary turns, the core will reach its saturated state.

At S1, H is increasing rapidly, giving a large increase in collector current. This continues until a gain limitation in Q1 prevents any further collector current increase. At this point the magnetizing field H will not be further increased, and the required rate of change of B (dB/dt) will not be sustained. As a result, the voltage across the primary winding must collapse, and the voltage on the collector of Q1 will rise toward the supply voltage.

As the voltage across the primary falls, the base drive voltage on P2 will also collapse, and Q1 will turn off.

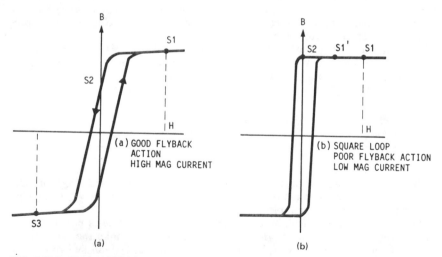

FIG. 2.15.2 Typical B/H loops for ferrite cores with B_r/B_s ratios (squareness ratios) of (*a*) less than 70% and (*b*) greater than 85%.

Because of the finite slope at the top of the *BH* characteristic (the post-saturation permeability), there will be a flyback action in the transformer from S1 to S2, and since *dB/dt* is now negative, the voltages on all windings will reverse. This will further turn Q1 off and start to turn Q2 on.

By regenerative action, Q2 will now be turned on and Q1 off, and the cycle will repeat on the second half of the primary winding. The flux in the core will now change from S2 toward S3, where the turn-off action will be repeated.

Note: The start of the turn-off action was initiated by transistor gain-current limiting. This limiting action is not well defined, as the collector current will vary with device type and temperature. With high-gain transistors, a very large collector current would flow before the transistor would become gain-limited. This results in poor switching efficiency and increased EMI problems. Further, the ill-defined collector current may even result in failure of the switching device under some conditions if it has not been carefully selected for the correct (low) current gain and rating.

15.3 DEFINING THE SWITCHING CURRENT

Figure 2.15.3 shows a more satisfactory method of limiting the termination current. Emitter resistor R1 or R2 develops an increasing voltage drop as the primary current rises. The voltages on the transistor base will track this voltage, and when it reaches the clamp value of the zener diode D1 or D2, the base drive current will be "pinched off," and turn-off action will commence.

In this circuit, the peak collector current is defined by the choice of emitter resistors and clamp zener voltage, and turn-off termination current no longer depends upon the gain of the transistors.

The full action of Fig. 2.15.3 is as follows. On switch-on, current flows through R4 and starts to turn on, say, Q1. Positive regenerative feedback from P2 will assist the turn-on action, and Q1 will saturate into a fully "on" state.

A drive current loop is now set up from P2 through R3, Q1 base-emitter, and R1, returning via D2, which acts as a diode in the forward direction. The drive current value is large so that the transistors are fully saturated throughout the "on" period.

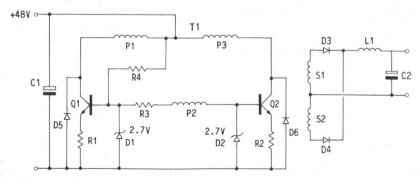

FIG. 2.15.3 Single-transformer, nonsaturating, push-pull self-oscillating converter with primary current limiting.

With Q1 "on," the current in the collector will increase with time, as defined by the primary inductance, the reflected secondary inductance L1, and the load.

As the current increases, the voltage across R1 and hence on the base of Q1 will also increase until zener diode D1 is brought into conduction.

The value of R1 is chosen so that drive pinch-off occurs at a controlled maximum collector current, which will be just into core saturation at, say, point S1' on Fig. 2.15.2b. Since at the clamp current H cannot be incremented further, the voltages across all windings must fall toward zero, and only a very small flyback action is required to finally turn Q1 off and Q2 on. The cycle of operations will continue for Q2. The collector termination current no longer depends on the gain of the transistors.

Collector diodes D5 and D6 provide a path for reverse current when the reflected load current is lower than the magnetization current. These diodes operate in a cross-connected fashion; for example, when Q1 turns off, magnetization current will cause the collector of Q1 to fly positive. At the same time, the collector of Q2 will fly negative until D6 conducts. D6 now provides clamping action for Q1 at a collector voltage of twice V_{cc}. To reduce the effects of leakage inductance, P1 and P3 would be bifilar-wound.

To reduce RF1 and secondary breakdown stress, snubber networks may be required in addition to D5 and D6.

To ensure good switching action, with sufficient regenerative feedback, the drive voltage from P2 should be at least 4 V in the circuit shown.

15.4 CHOOSING CORE MATERIALS

The efficiency and performance of these saturating single-transformer converters is critically dependent upon the choice of transformer material. For high-frequency operation, low-loss square-loop ferrite materials are usually chosen. Square-loop, tape-wound toroids may also be used, but these tend to be somewhat expensive at the higher frequencies, as very thin laminated tape core material must be used to reduce losses. Some of the more recently developed amorphous square-loop materials show excellent properties for this application.

With some square-loop materials, the postsaturation permeability (the slope of the B/H characteristic in the saturation region) will be very low (B_r is high). This is shown for the H5B2 material in Fig. 2.15.4b. Hence, the change in flux density between S1 and S2 is very small, and the flyback action at turn-off is weak. This gives a lazy switching action during the changeover phase. A more energetic switching action will be found with the H7A or similar core materials, as shown in Fig. 2.15.4a, but the core losses will often be higher with such materials, since they have a low B_r (remanence) value. With toroidal cores it would seem that the designer must make a compromise choice between a very square loop low-loss core (giving low core loss but lazy switching) or a lower-permeability, higher-loss core which will give good switching but higher magnetizing currents (lower inductance) and larger losses. With other core topologies, the problem can be solved by introducing an air gap. This gives energetic switching without increasing core loss. Suitable square-loop metal tape-wound materials include square Permalloy, Mumetal, and the various HCR materials. Some of the less square amorphous materials can also be used. Choose square-loop materials with care, as some are so square they will not oscillate, because the flyback action is too weak. (B_r/B_s ratios should be less than 80%.)

For high-frequency applications, the square-loop ferrites are preferred. Suit-

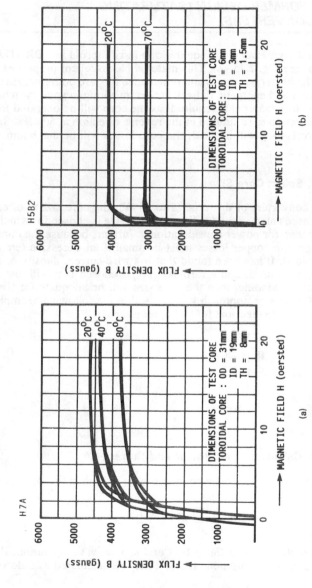

FIG. 2.15.4 (*a*) Magnetization curves for TDK H7A ferrite at 20, 40, and 80°C. This material shows a low residual flux density and low loss, making it more suitable for self-oscillating converter applications. (*b*) Magnetization curves for TDK H5B2 square-loop ferrite material. (*Courtesy of TDK Ltd.*)

able types include Fair-Rite #83, Siemens T26, N27, Philips 3C8, TDK H7C1, H5B2, H7A, and many more.

15.5 TRANSFORMER DESIGN (SATURATING-CORE-TYPE CONVERTERS)

In the following design example, a square-loop ferrite toroid in TDK H7A material will be used (see Fig. 2.15.4a). This material has excellent properties for this application, as the hysteresis losses and remanence flux are both low. Hence, this core gives low loss and energetic flyback action, resulting in good switching performance. A low-loss material is required, as the core will be operated from saturation in one direction to saturation in the reverse direction at 50 kHz. Since the complete hysteresis loop will be used, the core losses will be maximum.

15.5.1 Step 1, Select Core Size

For low-power converters of this type (5 to 25 W), the core size is often larger than the power needs alone would indicate, as the size is chosen for winding convenience rather than for power considerations. Further, because the core loss is intrinsically large, the copper losses must be small if an excessive temperature rise is to be avoided. It has been found that if a wire current density of approximately 150 A/cm^2 is used, and a toroidal core is chosen that will just allow a single-layer primary winding, then the core size will be adequate for the power requirements. This design approach will be used in the following example.

Consider a design to meet the following requirements:

Output power	10 W
Input voltage	48 V DC
Operating frequency	50 kHz
Output voltage	12 V
Output current	830 mA

15.5.2 Step 2, Calculate Input Power and Current

The output power is 10 W. Assume efficiency $\eta = 70\%$; then the input power will be

$$P_{in} = \frac{P_{out} \times 100}{\eta} = \frac{10 \times 100}{70} = 14.3 \text{ W}$$

With full-wave rectification at the output and square-wave operation, the input current will be almost a continuous DC and may be calculated as follows:

$$I_{in} = \frac{P_{in}}{V_{in}} = \frac{14.3}{48} = 0.3 \text{ A}$$

15.5.3 Step 3, Select Wire Gauge

At 150 A/cm^2, from Table 3.1.1, a 24 gauge wire is required for 0.3 A. Hence, the wire diameter is 0.057 cm.

Note: Two wires will be used for the bifilar primary winding, taking up a space of 0.114 cm per bifilar turn.

15.5.4 Step 4, Calculate Primary Turns

In this type of self-oscillating converter, the duration of the "on" period, and hence the frequency, is set by core saturation. Since this is a push-pull circuit, the total flux density change ΔB from $-B$ to $+B$ must occur for each half cycle. From Fig. 2.15.4a, for the H7A material at 80°C, the saturation flux density is 3500 G. The swing will be 2 × 3500 = 7000 G (700 mT).

A core size is now required that will just allow a single-layer bifilar winding of 24 AWG wires. However, there are two variables to be resolved to obtain the core size:

1. The primary turns required to saturate the core at 48 V in the 10-μs (50-kHz) half period (which are inversely proportional to the core area).
2. The number of turns of bifilar 24 AWG that can be wound on the core in a single layer (which is proportional to the circumference of the center hole; see Fig. 2.15.5).

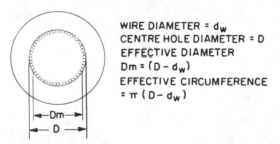

WIRE DIAMETER = d_w
CENTRE HOLE DIAMETER = D
EFFECTIVE DIAMETER
$Dm = (D - d_w)$
EFFECTIVE CIRCUMFERENCE
$= \pi (D - d_w)$

FIG. 2.15.5 Effective inner circumference of toroid with single-layer winding.

The relationship between center hole circumference and core area is not known in this example and will be different for various core designs, so a graphical solution will be used. The graph is developed as follows.

Consider Table 2.15.1. It will be assumed that a suitable core will be found in the range from T6-12-3 through T14.5-20-7.5.

The area of each core is given in the table, so the turns required for saturation of each core at the operating frequency and applied voltage may be calculated as follows:

$$N_p = \frac{V_{cc} \times t_{on}}{\Delta B_s \times A_e}$$

TABLE 2.15.1 Toroidal Core Dimensions

Core type	A Dimensions	B mm inch	C mm inch	Core factor l_e/A_e, cm^{-1}	Effective length l_e, cm	Effective area A_e, cm^2	Effective volume V_e, cm^3	Weight, g
T2-4-1	4.0±0.2 0.157	2.0±0.2 0.079	1.0±0.15 0.039	90.6	0.871	9.61×10^{-3}	$8.37\times10\times^{-3}$	0.045
T3-6-1.5	6.0±0.3 0.236	3.0±0.25 0.118	1.5±0.2 0.059	60.4	1.31	21.6×10^{-3}	28.3×10^{-3}	0.15
T4-8-2	8.0±0.3 0.315	4.0±0.25 0.157	2.0±0.2 0.079	45.3	1.74	38.4×10^{-3}	67.0×10^{-3}	0.36
T5-10-2.5	10.0±0.4 0.394	5.0±0.3 0.197	2.5±0.25 0.098	36.3	2.18	60.1×10^{-3}	0.131	0.71
T6-12-3	12.0±0.4 0.472	6.0±0.3 0.236	3.0±0.25 0.118	30.2	2.61	86.5×10^{-3}	0.226	1.2
T7-14-3.5	14.0 ± 0.4 0.551	7.0 ± 0.3 0.276	3.5 ± 0.25 0.138	25.9	3.05	0.118	0.359	1.9
T8-16-4	16.0 ± 0.4 0.630	8.0 ± 0.3 0.315	4.0 ± 0.3 0.157	22.7	3.48	0.154	0.536	2.9
T9-18-4.5	18.0 ± 0.4 0.709	9.0 ± 0.3 0.354	4.5 ± 0.3 0.177	20.1	3.92	0.195	0.763	4.1
T10-20-5	20.0 ± 0.4 0.787	10.0 ± 0.3 0.394	5.0 ± 0.3 0.197	18.1	4.36	0.240	1.05	5.7
T14.5-20-7.5	20.0 ± 0.4 0.787	14.5 ± 0.4 0.571	7.5 ± 0.3 0.295	26.1	5.33	0.204	1.09	5.4
T16-28-13	28.0 ± 0.4 1.10	16.0 ± 0.4 0.630	13.0 ± 0.3 0.512	8.64	6.56	0.760	4.99	26

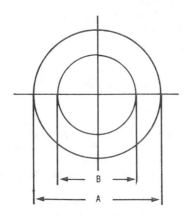

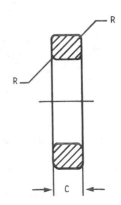

where N_p = number of primary turns
V_{cc} = supply voltage, V
t_{on} = "on" time (½ cycle), μs
ΔB_s = flux density change, T (from $-B_{sat}$ to $+B_{sat}$)
A_e = effective core area, mm^2

The turns required to saturate each core in the range T6-12-3 through T14.5-20-7.5 are calculated as shown above and entered in Table 2.15.2.

The size of the center hole is given in Table 2.15.1, and the number of turns that will fit each core in a single layer, as shown in Fig. 2.15.5, can be calculated as follows.

Consider the toroid center hole with closely packed windings. Figure 2.15.5 shows that the diameter of the center hole is related to the number of turns (assuming that the turns are large and the wire diameter is small compared with the center hole diameter) by the following formula:

$$N_w = \frac{\pi(D - d_w)}{d_w}$$

where D = diameter of center hole, mm
d_w = wire diameter, mm
N_w = number of primary wires (that will just fit on the core in a single layer)

Note: The primary is to be wound as a bifilar winding, and primary "turns" N_p refers to a pair of wires which will be split finally to form each side of the center-tapped primary. Hence, each turn consists of two wires in parallel, so $N_p = N_w/2$.

The number of bifilar turns N_p which will just fit on each core in a single layer for the range T6-12-3 through T14.5-20-7.5 is calculated and also entered in Table 2.15.2.

A graph is now drawn (Fig. 2.15.6) of turns required for saturation at 50 kHz against core area (plot A) and maximum bifilar turns that will just fit on the core in a single layer against core area (plot B). The intersection of these two lines gives optimum core area, from which the core size can be established.

TABLE 2.15.2 Windings for Ferrite Toroidal Cores

Core type	Diameter of center hole D, mm	Cross-sectional area of core, mm^2	Maximum turns for a single-layer winding of 24 AWG N_p N_p' turns	Minimum turns for core saturation at 50 kHz Np turns
T6-12-3	6	8.6	15	79
T7-14-3.5	7	11.8	18	58
T8-16-4	8	15.4	22	44
T9-18-4.5	9	19.5	25	35
T10-20-5	10	24.0	28	28.5
T14.5-20-7.5	14.5	20.4	42	33

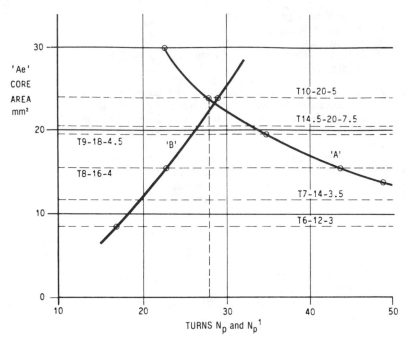

FIG. 2.15.6 Graphical method for finding optimum toroidal core size for a single-layer winding.

It will be seen that the ideal area is between cores T10 and T14, and so the nearest larger core T10-20-5 is chosen. The core area is entered on the graph (horizontal dotted lines). A vertical line from the intersection of the core area and the 'A' line gives the number of primary turns required to saturate the core (28 in this example), and the intersection of the core area and the B line shows that 29 turns of 24 AWG can be fitted in a single layer on this core.

Clearly, the matching point will vary with different core topologies and wire sizes, and although a larger core can always be used, because core saturation occurs, the total power loss will increase (and efficiency fall) with the larger core.

For this example, the T10-20-5 is chosen and a primary winding P1, P3 of 28 turns is used. Two 24-gauge wires will be wound together to form a bifilar winding.

The turns per volt are

$$T_{pv} = \frac{V_{cc}}{N_p} = \frac{48}{28} = 1.71 \text{ turns/V}$$

Selecting a feedback voltage of approximately 5 V provides adequate regenerative feedback, resulting in fast switching action. In this example, 3 turns would be used for the feedback winding P2. The secondary winding will be designed for the required output voltage by normal transformer action; in this case 8 turns are used.

The value of the transistor base feed resistor R3 will be selected in accordance with the maximum collector current and transistor gain. The emitter resistors R1 and R2 are chosen to ensure that transistor drive "clamp-off" will not occur under normal full-load conditions, but will occur before excessive collector current flows. A current margin of about 30% above full load is a good compromise. Remember, the emitter current is the sum of the transformed load current, the magnetization current, and the base drive current. In this example a value of 2.2 Ω is used.

Cross conduction (that is, conduction caused by both power transistors being "on" at the same time) is eliminated in this circuit arrangement, as it is the turn-off action of one transistor that initiates the turn-on action of the second. Consequently, variations in storage time are automatically accommodated without the possibility of cross conduction. This is a useful feature of the self-oscillating converter.

The frequency of this type of simple self-oscillating converter will change with input voltage, core temperature (because of the change in saturated flux level at high temperatures), and load. Increasing the load results in a voltage drop in the transformer resistance and transistors, so that the effective transformer primary voltage is lower; hence the frequency will fall with increasing load.

15.6 PROBLEMS

1. Explain the major advantages of the self-oscillating converter.
2. Give a typical application for a small self-oscillating converter.
3. Why is the frequency stability of the self-oscillating converter rather poor?
4. Why does the single-transformer self-oscillating converter tend to be limited to low-power applications?
5. Explain why ungapped very square loop magnetic materials (such as square amorphous material) are considered unsuitable for self-oscillating single-transformer converters.

CHAPTER 16
TWO-TRANSFORMER SELF-OSCILLATING CONVERTERS

16.1 INTRODUCTION

In the two-transformer self-oscillating converter, the switching action is controlled by saturation of a small drive transformer rather than the main power transformer, resulting in improved performance and better power transformer efficiency.

As the main power transformer is no longer driven to saturation, more optimum transformer design is possible, and the shape of the B/H loop is not so critical. Further, better transistor switching action is obtained, because the collector termination current is lower and the power transistors turn off under more controlled conditions. The switching frequency is also more constant, as the drive transformer is not loaded by the output and the voltage across the drive transformer is not a direct function of the input voltage. Hence, the operating frequency is less sensitive to supply and load changes.

With all these advantages over the simple saturating converter, the two-transformer converter is preferred in many higher-power applications. It is often used for low-cost DC transformers. (See Chap. 17.)

There are two major disadvantages of the simple self-oscillating two-transformer square-wave converter which tend to limit its applications:

1. Soft start is difficult to provide because the pulse width cannot be easily reduced during the initial turn-on action.
2. Outputs are unregulated because of the square-wave (100% duty ratio) output.

16.2 OPERATING PRINCIPLES

A basic two-transformer circuit is shown in Fig. 2.16.1. This operates as follows.

At switch-on, a current flows in R1 to initiate turn-on of one of the two drive transistors, say Q1. (It depends on the two devices' gains and their discrete emitter-base voltages.) As Q1 turns on, regenerative feedback is provided by proportional current drive from a separate drive transformer T2. The primary T2P1

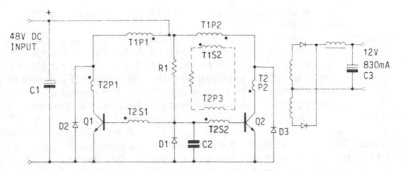

FIG. 2.16.1   Two-transformer self-oscillating converter.

couples with the collector current of Q1 to provide the base drive to Q1 via the secondary winding T2S1. The phasing of the windings is such that an increase in current in the collector of Q1 will cause an increase in base current in Q1. Since T2 is a current transformer, the base drive current is defined by the collector current and the turns ratios of T2. (A ratio of ⅕, for a forced beta of 5, is used in this example.)

With Q1 in its "on" state, the voltage V_s across the secondary of the drive transformer T2S1 will be the base-emitter drop of Q1 plus the diode drop of D1 (approximately 1.3 V total). The voltage across the drive winding to Q2, T2S2, will be the same value, but in the opposite direction, taking the base of Q2 0.7 V negative.

After a period of time defined by the core area and secondary voltage V_s, the drive transformer T2 will saturate, and the drive voltage to Q1 will fall to zero. Q1 will then turn off. This is arranged to happen before the main transformer T1 saturates.

With Q1 off, the collector current will fall to zero, and the voltages on T2 will reverse by flyback action. Q2 will be brought into conduction, and Q1 will turn hard off. The same cycle of operations will now occur on transistor Q2, with the collector current for Q2 flowing in the second half of the drive transformer T2P2 and the main transformer T1P2, reversing the phase.

Since the forward voltage of the base-emitter junction of Q1 and D1 or Q2 and D1 is allowed to define the voltage across the secondary of T2, the timing voltage for T2 is fixed and independent of the supply voltage. Consequently, the operating frequency is largely independent of supply and load changes. However, changes in temperature will affect both diode and base voltages and the saturation flux density of the drive transformer T2. Hence, the frequency is still sensitive to temperature changes. In many applications, small changes in frequency are not important.

Clamp diodes D2 and D3 are fitted across the collector-emitter junctions of the switching transistors to provide a path for the reverse magnetization current, which would otherwise flow from base to collector in Q1 and Q2 during turn-on, when the reflected load current is less than the magnetization current. Although this will occur only under very light loading conditions, with proportional drive it can cause problems and warrants further examination.

Consider the conditions immediately before the turn-off of Q1: A transformer magnetization current will have been established in the primary of the power transformer T1P1 and the collector of Q1, flowing from right to left in the winding. When Q1 turns off, this magnetization current will try to continue in the same direction in the winding, taking the collector of Q1 positive and the collector of Q2 negative (flyback action).

Since the magnetization current cannot instantly drop to zero, and with Q1 off cannot flow in the collector of Q1, it will commutate from the collector of Q1 to the collector of Q2, to flow from right to left in T1P2 from the clamp diode D3. If D3 were not fitted, the current would flow in the base-collector junction of Q2, and since it is now flowing in the reverse direction to the normal collector current, it diverts the base drive current away from the base-emitter junction into the base-collector junction. (The transistor is reverse-biased.)

This reverse collector current would also flow in the primary of the drive transformer T2P2, but in the reverse direction to that required for positive feedback. This reverse current would prevent or at least delay the turn-on of Q2 and is very undesirable. However, by providing a low-resistance alternative path through the clamp diode D3, the majority of this undesirable current is diverted away from Q1. In some cases a further series blocking diode will be required in the collectors of Q1 and Q2.

In any event, under light loading conditions, until the magnetization current has fallen to less than the reflected load current, there will not be a forward collector current or a regenerative base drive action. If this light loading condition is to be a normal mode of operation, an extra voltage-controlled drive winding will be required to maintain the drive until normal forward current is established. This extra winding T2P3 and T1S2 is shown as a dashed-line detail in Fig. 2.16.1.

If the load current always exceeds the magnetization current, then this refinement is not necessary and diodes D2 and D3, are redundant (except as small snubber components if required).

Once again, in this topology, cross conduction is eliminated, since it is the turn-off action of one transistor that initiates the turn-on action of its partner. It is interesting to note that staircase saturation of T1 cannot occur because the drive transformer T2 equates forward and reverse volt-seconds in the longer term. Even variations in the storage times of the two transistors will be accommodated.

If a drive core with a square B/H loop and a high remanence is used, "flux doubling" on initial switch-on is eliminated. Both cores retain a magnetic "memory" of the previous direction of operation. When the system is turned off, the residual flux remains in the drive core as a "memory" of the last operating pulse direction. If, on the next switch-on, the first half cycle of operation is in the same direction, then the drive core will saturate more rapidly, and this pulse will be shortened; hence saturation of the main transformer will not occur. For this reason the residual flux level B_r of the drive transformer material should be higher than that of the main transformer.

With this good control of the state of the flux density in the main core, the working flux excursion of the main transformer can be selected with confidence, for optimum efficiency.

It can be seen from the preceding discussion that although the self-oscillating circuits appear to be extremely simple, they operate in a fairly sophisticated manner, providing very efficient converter action if correctly designed. It is this type of converter (or DC transformer, as it is known) that is used in tandem with the primary series buck switching regulator (Chap. 18) to provide the extremely cost effective multiple-output supplies.

To its disadvantage, two transformers are required for this type of converter. However, the drive transformer T2 is very small, as it delivers little power.

16.3 SATURATED DRIVE TRANSFORMER DESIGN

The drive transformer T2 is essentially a saturating current transformer.

Having decided on the operating power, select the power transistors and operating current. From the transistor data, find the forced beta required to ensure good saturation and switching action. (Assume that a forced beta of 5:1 is chosen in this example; the actual value depends on the transistor parameters.)

This means that for every turn on the primary (collector) winding of T2P1, 5 turns must be provided on the secondary base drive winding T2S1. Consequently, it will be seen that the base winding can be incremented only in steps of 5 turns.

The secondary voltage of T2S1 is defined at a diode drop D1 plus V_{be} of Q1 (approximately 1.3 V), and this voltage, secondary turns T2S1, and core size set the frequency. A core size must now be selected so that the correct operating frequency will be obtained with 5 turns (or an increment of 5 turns) on the base drive winding T2S1. (The smaller the core, the larger the number of turns for the same frequency.)

16.4 SELECTING CORE SIZE AND MATERIAL

Assume that the operating frequency is to be 50 kHz, that the required forced beta is 5, and that a single turn will be used on the collector winding. The secondary will have 5 turns, and the voltage seen by the secondary winding will be 1.3 V (V_{be} plus a diode drop).

The area of core required for a half period (10 μs) can be calculated as follows:

$$A_e = \frac{V_s \times t}{\Delta B \times N_s}$$

where V_s = secondary voltage (1.3 V)
  t = half period (10 μs)
  ΔB = flux density change ($-B$ to $+B$)
  N_s = secondary turns (5 in this case)

If the TDK H7A material is chosen, then from Fig. 2.15.4a, the saturating flux density at 40°C is 0.42 T, and B = 0.84 T peak-to-peak. Hence

$$A_e = \frac{1.4 \times 10}{0.74 \times 5} = 3.78 \text{ mm}^2$$

From Table 2.15.1, the area required will be nearly satisfied by core T 4-8-2, and this is selected for T2 in this example.

This core is now bifilar-wound with two wires of 5 turns to form the secondary windings T2S1 and T2S2. A wire to each transistor collector is passed through the toroid from opposite directions to give a single primary turn for each transistor, and the drive transformer is complete.

16.5 MAIN POWER TRANSFORMER DESIGN

The design of the main transformer follows the same procedure used for any driven push-pull converter. Typical designs are covered in Part 3, Chap. 4.

16.6 PROBLEMS

1. What is the major advantage of the two-transformer self-oscillating converter?
2. Why is the two-transformer converter more suitable for high-power applications?
3. What is the functional name that is often applied to the square-wave self-oscillating converter?
4. What are the major disadvantages of the two-transformer self-oscillating converter?
5. By what process is staircase saturation eliminated in the two-transformer self-oscillating converter?
6. By what process is flux doubling eliminated in the two-transformer self-oscillating converter?

CHAPTER 17
THE DC-TO-DC TRANSFORMER CONCEPT

17.1 INTRODUCTION

Throughout the previous designs, the high-frequency transformer has been considered an integral part of the converter topology. However, it can be shown that the transformer does not change the fundamental form of the converter. The basic building blocks remain the buck and boost regulators. In all the forward converter topologies, the main function of the transformer was to provide input-to-output galvanic isolation and voltage transformation.

It would be useful at this stage to introduce the concept of the ideal DC-to-DC transformer (a transformer that would provide a DC output for a DC input). Although this may seem a strange concept at first, consider the conventional ideal transformer model. The ideal transformer would pass all frequencies from DC upward in both directions, with no power loss. It would provide any required voltage or current ratio, and would have infinite galvanic isolation between input and output. It would also provide the same performance for all windings. Clearly such an ideal device does not exist.

17.2 BASIC PRINCIPLES OF THE DC-TO-DC TRANSFORMER CONCEPT

The reason that the DC-to-DC transformer concept seems strange should be recognized as the unquestioned acceptance of a nonideal device. The practical transformer can transform a DC input to a DC output only for a very short time, because it has a limited inductance and the core soon saturates. However, this practical limitation can be overcome by reversing all terminals of the transformer before saturation occurs.

The rotary DC converter and the synchronous vibrator (shown in Fig. 2.17.1, but now obsolete) did just this, physically reversing the transformer input and output terminals with a commutator or an electromechanical synchronous relay. This simple vibrator will be examined in more detail.

In Fig. 2.17.1, switches S1(a) and S1(b) are on the same armature and are thus synchronized to each other. This synchronous vibrator circuit (apart from its mechanical limitations) provides a nearly ideal DC-to-DC transformer. It is inter-

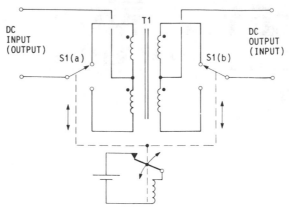

FIG. 2.17.1   DC-to-DC transformer (mechanical synchronous vibrator type).

esting to note that it is symmetrical and hence fully bidirectional; it can transform DC from input to output or vice versa. This is entirely due to the bidirectional nature of the mechanical switches. This circuit provides a good model for the modern DC-to-DC transformer concept.

In the modern semiconductor equivalent, the bidirectional property is lost as a result of replacing the mechanical switches by unidirectional semiconductors. In the square-wave DC-to-DC converter, the primary switching S1(a) is provided by two unidirectional transistors or power FETs, and the secondary synchronous switching S1(b) by diode rectification. However, the basic concept is the same as that of the synchronous vibrator (keep reversing the input and output of the transformer before saturation can occur, and thus provide DC-to-DC transformation).

17.3   DC-TO-DC TRANSFORMER EXAMPLE

In this example, a push-pull converter will be used to demonstrate the principle. Figure 2.17.2 shows a typical self-oscillating square-wave converter. This type of converter is characterized by full conduction angle operation; that is, no pulse-width modulation is applied, and each transistor is conducting for a full 50% of the period. Consequently, in this simple form, the converter will not provide regulation and the DC output voltages will change in sympathy with the input.

When operated in this mode, the converter is a true DC-to-DC transformer. Its main functions are to provide DC voltage and current step-up or step-down, galvanic isolation, and multiple outputs. The outputs may be isolated, common, or inverted as required.

Being self-oscillating, the converter has the advantage of very low cost, since very few drive components are required. The full conduction angle results in a near DC input and output, so that very little filtering will be required.

The output impedance of the DC-to-DC transformer reflects the input impedance and can be very low. The transformer utility and the efficiency are very high as a result of the full conduction angle operation. For multiple-output applica-

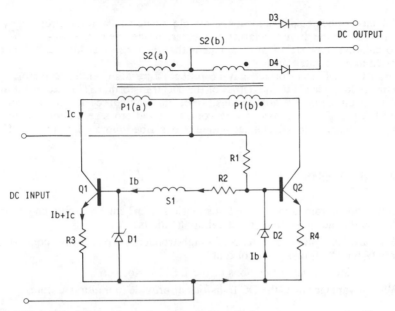

FIG. 2.17.2 DC transformer (transistor, self-oscillating, square-wave, push-pull converter with biphase rectification).

tions, the tracking between independent outputs will be very good, typically better than 2%.

The simple self-oscillating converter shown in Fig. 2.17.2 is a modification of the Royer circuit, which has been described more fully in Chaps. 14, 15, and 16. Although this converter normally depends on saturation of the transformer to initiate commutation, in this example the collector current at commutation is specified by R3 and R4 and the zener voltages D1 and D2, and the transformer does not saturate; hence a measure of overload protection is provided. This is a considerable improvement on the gain-limited switching action of the original Royer circuit.

There are many variations of this simple DC-to-DC converter, but the principle of operation will be similar. The main attractions are the extremely low cost and efficient operation. The frequency variations which occur with these simple systems (because of input voltage and load changes) are sometimes considered a disadvantage. However, since the input and output should be DC, the frequency of operation should not matter, and with efficient screening should be acceptable in all cases.

The DC-to-DC transformer (self-oscillating square-wave converter) is unlikely to be used on its own because it does not provide regulation. Hence it is normally supplemented with other regulating circuits to form a regulated system.

For low-power applications, up to, say, 10 W, this type of converter is often used with linear three-terminal output regulators to provide fully regulated multiple outputs from a single DC input.

These low-power regulated converters are very often manufactured as small encapsulated units, and may be used to generate extra local supply voltages. They will often be mounted directly on the users' printed circuit boards.

For larger-power applications, the simple single-transformer circuit may be used to provide the drive to a larger converter, giving rise to the two-transformer self-oscillating converter. The advantage is that the larger transformer will not be operated in a saturated mode.

The DC-to-DC transformer may be preceded by a switching regulator, with the control loop closed to the output of the DC transformer. This combination is ideal for multiple-output applications, since the low source impedance for the DC-to-DC transformer provides very good load and cross regulation for the additional auxiliary outputs. This technique is described more fully in Chap. 18.

17.4 PROBLEMS

1. In the "ideal transformer," the transmission of DC current would be possible. Why is this not possible in practical applications?
2. What are the basic requirements of a high-frequency square-wave converter to satisfy the DC transformer concept?
3. Give the ideal transfer functions of the DC transformer.
4. What advantages does the DC transformer provide in multiple-output applications?

CHAPTER 18

MULTIPLE-OUTPUT COMPOUND REGULATING SYSTEMS

18.1 INTRODUCTION

It has been shown[9,10,14,15,16,17] that it is possible to combine DC-to-DC transformers (nonregulated DC-to-DC converter circuits) with buck- or boost-derived regulators to produce regulated single- or multiple-output converter combinations.

Using integrated magnetics and various modeling techniques,[16] a wide range of possible combinations have been demonstrated. In many cases, these combinations are not clearly distinguishable as combinations of converters. Various advantages and disadvantages are claimed for each combination; for example, in the Ćuk converter it has been shown that by combining output and input chokes it is possible to suppress the input and output ripple currents to near zero. Hence combinations of converters can provide some very useful properties.

The power supply designer who is fully conversant with the various combinations and is able to select the most appropriate for a particular application has indeed very powerful design tools.

Full coverage of the many techniques is beyond the scope of this book, but one particularly useful topology is covered in the following sections. For more information, the interested engineer should study the many excellent papers and books referred to in the Bibliography.

18.2 BUCK REGULATOR, CASCADED WITH A DC-TO-DC TRANSFORMER

The following section considers just one of the more popular and straightforward arrangements of a buck regulator cascaded with a square-wave free-running voltage-fed DC-to-DC converter (referred to as a DC transformer). The general block schematic for a FET version of this converter is shown in Fig. 2.18.1. This combination is particularly useful for multiple-output off-line switchmode supplies.

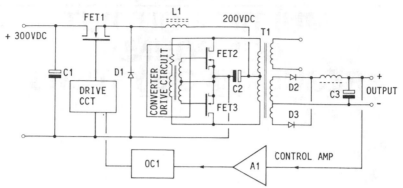

FIG. 2.18.1 Voltage-regulated DC-to-DC transformer (DC converter), consisting of a combination of a primary buck switching regulator and a DC-to-DC transformer.

18.3 OPERATING PRINCIPLES

In simple terms, in the example shown in Fig. 2.18.1, the input buck regulator FET1, L1, and D1 reduces the high-voltage (300-V) input to a more manageable preregulated 200-V DC to the push-pull DC transformer FET2, FET3, and T1. The control loop to the buck regulator is closed to the main output of the DC transformer, to maintain this output and hence the other auxiliary outputs near constant.

In push-pull operation, the DC transformer FETs are subjected to a voltage stress of at least twice the preregulated supply voltage. The buck regulator, by reducing this voltage and removing input variations, considerably reduces the stress on the converter FETs and improves the reliability.

The voltage control loop is closed to the output which is to be most highly regulated, in this case the 5-V output, by amplifier A1 and optical coupler OC1. Thus, the buck regulator maintains the DC supply to the DC transformer section at a level that will maintain the 5-V output constant. Since the input to the converter section is now almost entirely isolated from ac supply variations, and a much lower voltage stress is applied to the converter switching components, there will be a reduction in output ripple and an improved reliability.

Further, as a result of the closed-loop control, the DC transformer volts per turn will be maintained constant (to the first order), and any other outputs wound on the same transformer will be semistabilized.

For input voltage transients, the natural filtering provided by the large input capacitor C1 and the buck regulator and filter L1, C2 gives good noise immunity. A small low-pass filter on the output further eliminates switching and rectifier recovery noise from the output. Since the DC transformer runs under full duty ratio (square-wave) conditions, the rectified output is nearly DC, and only a small high-frequency filter is required in the output circuit. This becomes particularly cost-effective where a large number of outputs are required.

In some applications, the converter operating frequency will be synchronized to the buck regulator to prevent low-frequency intermodulation components which tend to generate extra output ripple.

A power FET is used in the buck regulator section so that this can be operated at a high frequency without introducing excessive switching loss. This arrangement means that several power pulses will be provided from the buck regulator for each half cycle of the converter section, reducing intermodulation ripple.

This buck-derived combination is subject to low-frequency instability at light loads when bipolar switches are used in the buck regulator position. This instability is believed to be caused by modulation of the storage time of the bipolar transistor by direct positive feedback from the DC transformer section. The effect is difficult to compensate, because it is outside of the normal control loop. The negligible storage time of a FET switch in the buck regulator position eliminates this problem.

It should be noted that the DC transformer is voltage-fed, capacitor C2 being large enough to maintain the voltage nearly constant over a cycle. This provides a low-impedance ripple-free preregulated DC input to the DC transformer and permits the use of secondary duty ratio control if additional secondary regulation is required. It also reduces cross-regulation effects. Without additional secondary regulation, output regulation on the auxiliary outputs of ±5% would be expected when the loop is closed to the 5-V output, or ±2% when the loop is closed to a higher-voltage, lower-current output.

18.4 BUCK REGULATOR SECTION

Figure 2.18.2 shows in block schematic form the basic elements of the buck regulator section. In general, this is a current-mode-controlled system, very similar to the buck regulator described in Chap. 20.

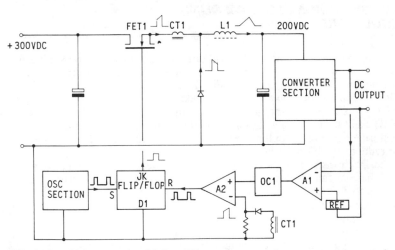

FIG. 2.18.2 Voltage-regulated DC-to-DC transformer, using current-mode control on the buck regulator section with the voltage control loop closed to the secondary.

Briefly, the buck regulator power FET is turned on in response to a clock signal from the oscillator section. It is turned off in response to the current in L1 as sensed by the current transformer CT1. The current switching level is defined by the much slower voltage control loop from the output via A1 and OC1.

The clock signal from the oscillator section sets the bistable switch D1 to turn on the series power switch FET1, which then delivers current to the series inductor L1. During the "on" period, the current in inductor L1 will increase, and this increase is translated by the current transformer CT1 to a ramp voltage on the input of the voltage comparator A2. When the amplitude of the ramp voltage has reached the reference value set up on the noninverting input by the control amplifier A1, a reset signal is generated, and the driver bistable switch D1 is reset to the "off" state. To permit operation at pulse widths beyond 50%, the current feedback would have slope compensation to prevent subharmonic instability. (See Part 3, Sec. 10.5)

It will be noted that the current-mode control defines the peak current in L1, making the buck regulator a voltage-controlled, constant-current source. This high-impedance source is converted to a low-impedance voltage source by the capacitor C2 at the input of the DC transformer.

The voltage control amplifier A1 responds to changes in output voltage and adjusts the reference level for the ramp comparator A2. By this means the current in L1 and C1 is adjusted so as to maintain the output voltage of the DC transformer constant.

18.5 DC TRANSFORMER SECTION

In the simple circuit examples shown in Figs. 2.18.1 and 2.18.2, the DC transformer would be a self-oscillating square-wave bipolar or FET converter, and suitable circuits are covered in Chaps. 14, 15, 16, and 17.

18.6 SYNCHRONIZED COMPOUND REGULATORS

If the DC transformer and buck regulator frequencies are to be synchronized, a driven converter design is required. The same oscillator will drive the buck regulator and the DC converter. (See Fig. 2.18.3.)

It is recommended that the buck regulator operate at twice the frequency of the DC transformer, so that a power pulse is delivered to L1 for each "on" state of FET2 and FET3. This will assist in maintaining balanced conditions for the DC transformer, reducing the tendency for staircase saturation in the transformer. It also reduces the output ripple.

The DC transformer should operate with a nearly pure square wave, so that the biphase or full-wave-rectified outputs are nearly pure DC. Very small output LC filters will then be required to remove noise spikes resulting from commutation of the output rectifiers. These filter components can be quite small, as they are not required to store significant amounts of energy.

With driven DC converters using bipolar transistors, cross-conduction problems can occur near full 50% conduction as a result of the storage time in the transistors during the turn-off edge. This problem can be eliminated without introducing a "dead time" by using dynamic cross-coupled inhibit drive circuits. (See Part 1, Chap. 19.)

Power FETs have an advantage for this type of square-wave DC converter because of the negligible storage times. No special precautions against cross conduction are required when FETs are used.

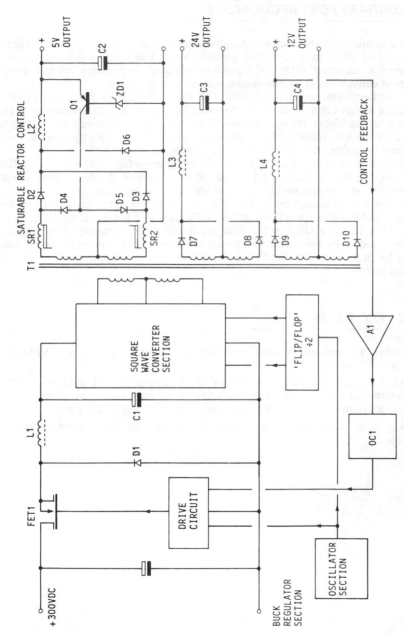

FIG. 2.18.3 Multiple-output compound regulator/converter, with secondary saturable reactor post-regulation on the 5-V output.

18.7 COMPOUND REGULATORS WITH SECONDARY POST REGULATORS

Where improved performance is required, additional switching or linear regulators may be added to the auxiliary outputs. When this is done, it is important to ensure that the input to the DC transformer is a low-impedance voltage source; an input capacitor will be provided as shown in Fig. 2.18.2.

A further useful technique is shown in Fig. 2.18.3. Here the control loop is closed to a high-voltage output, in this case the +12-V line, to provide good regulation on the higher-voltage outputs. However, extra regulation has been provided for the low-voltage, high-current output, +5 V at 40 A, by using a highly efficient saturable reactor regulator. (See Chap. 20.)

Using this approach, the regulation of the higher-voltage auxiliary lines will be very good, and the tendency for loss of regulation in the low-voltage, high-current line is compensated by the saturable reactor regulator.

In this hybrid arrangement, the +12- and +5-V outputs are fully regulated, and the additional outputs (−12 and +24 V) also have good regulation as a result of the tight coupling to the fully regulated +12-V winding. The very low losses incurred by the saturable reactor regulator, buck regulator, and square-wave DC transformer make it possible to provide overall system efficiencies in excess of 70%.

18.8 PROBLEMS

1. Explain the meaning of the term "compound regulating systems" as used in Chap. 18.
2. What are the particular advantages of combining a buck regulator with a square-wave DC-to-DC self-oscillating converter (DC transformer) in a multiple-output direct-off-line switchmode power supply system?
3. Why is the DC transformer particularly suitable for multiple-output applications?

CHAPTER 19
DUTY-RATIO-CONTROLLED PUSH-PULL CONVERTERS

19.1 INTRODUCTION

The push-pull converter is not generally favored for off-line applications, because the power switches operate at collector stress voltages of at least twice the supply voltage. Also, the primary utility factor for the main transformer is not as good as in the half-bridge or full-bridge converter, because a center-tapped primary is used, and only half of the winding is active at each half cycle.

However, at low input voltages, the push-pull technique has some advantages over the half or full bridge, as only one switching device is in series with the supply and primary winding at any instant. The full supply voltage is applied to the active half winding, and for the same output power, the switching losses are lower.

19.2 OPERATING PRINCIPLES

The power section of a typical low-voltage DC-to-DC push-pull converter is shown in Fig. 2.19.1. This operates in the following way.

The two power switches Q1 and Q2 will be turned on for alternate half cycles by the drive circuit. When Q1 turns on, the primary voltage is impressed across one-half of the primary winding P1a. Under this condition, the starts of all windings will go positive, and the collector of Q2 will be stressed to a voltage of twice the supply voltage by transformer action from Q1. The starts of the secondary winding S1a will go positive, and a current flows in the secondary windings, output rectifier D6, and inductor L1.

The primary current will now consist of the reflected load current plus a small component of primary magnetizing current defined by the primary inductance. During the "on" period, the primary current will increase as the magnetization component increases in both the primary inductance and the secondary filter choke L1. At the end of an "on" period, defined by the control circuit, Q1 will turn off.

As a result of the energy stored in the primary and leakage inductances, the collector voltage of Q1 will fly positive, and by transformer action, the collector of Q2 will be taken negative. When the voltage on Q2 goes below zero, the en-

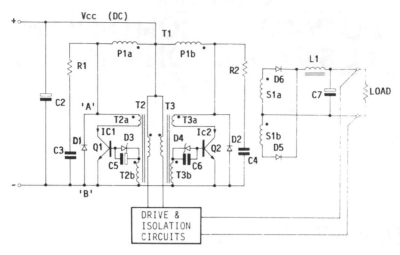

FIG. 2.19.1 Push-pull converter with duty ratio control and proportional base drive circuit.

ergy recovery diode D2 will be brought into conduction to return part of the flyback energy to the supply line. At the same time, the output rectifier diode D5 will be conducting some of the flyback energy to the output circuit (depending upon the value and distribution of the primary and secondary leakage inductances).

With "duty ratio control," there now follows a quiescent period during which both switching devices will be off. During this period, the output choke L1 will maintain a current flow in the rectifier diodes D5 and D6, the output load, and capacitor C1. This current will return via the center tap of the secondary winding and both rectifier diodes. If the current in L1 exceeds the reflected magnetizing current in magnitude (the normal case), then both output rectifiers will be conducting nearly equally throughout the quiescent period, with the same forward voltage drop. Hence, the net voltage across the secondary winding will be zero (the rectifier forward voltages are equal but opposite across the secondary winding). Consequently, during the quiescent period, the flux density in the core will not change; that is, the core will not be restored toward zero while Q1 and Q2 are off. This is an important property of this type of circuit, as it permits full use of the *B/H* characteristic.

After a quiescent period defined by the control circuit, transistor Q2 will be turned on, and the previous process will repeat for Q2, completing a cycle of operations.

For the continuous conduction mode, the load current is not permitted to fall below the critical value,* and L1 will conduct current at all times. The output voltage will be defined by the primary voltage, the turns ratio, and the duty ratio as in the following equation:

$$V_{\text{out}} = \frac{V_{cc} \times N_s \times t_{\text{on}}}{N_p \times (t_{\text{on}} + t_{\text{off}})}$$

*See Sec. 8.3 and Fig. 2.20.1*b*.

where V_{out} = output voltage, DC
$\quad\quad V_{cc}$ = supply voltage, DC
$\quad\quad N_s$ = secondary turns
$\quad\quad N_p$ = primary turns
$\quad\quad t_{on}$ = on time, Q1 or Q2
$\quad\quad t_{off}$ = off time (Q1 and Q2 off)

The control circuit will adjust the duty ratio, $t_{on} / (t_{on} + t_{off})$ to maintain the output voltage constant.

Under normal loading conditions, the load current exceeds the magnetization current as referred to the secondary, and the primary and secondary waveforms will be as shown in Fig. 2.19.2. For steady-state conditions, there will be a bal-

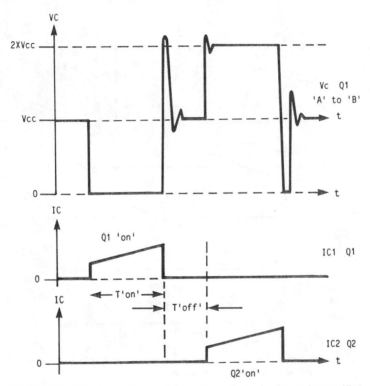

FIG. 2.19.2 Collector, voltage, and current waveforms for duty-ratio-controlled converter.

anced excursion of flux density between the negative and positive quadrants, as shown in Fig. 2.19.3.

When the load current is less than the reflected magnetizing current, the energy recovery diodes will continue to conduct for a longer part of the "off" period, and the current and voltage waveforms will be as shown in Fig. 2.19.4. It should be noted that the collector current is reversed for a short period at the beginning of the "on" period.

A problem can develop for this condition of operation. If recovery current is

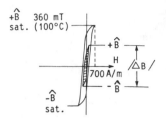

FIG. 2.19.3 Flux density excursion for balanced push-pull converter action.

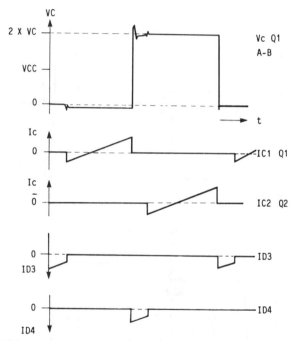

FIG. 2.19.4 Voltage and current waveforms at light loads for duty-ratio-controlled push-pull converter.

still flowing in say D1 when Q1 turns on, the collector voltage on Q1 will be negative, and the base drive current will be diverted from the base-emitter junction to the base-collector junction. This drive current will be in the reverse direction to the normal collector current of Q1. Although this is not in itself damaging, it can be a problem for proportional drive systems, since this is the wrong direction for the proportional drive transformer, and there will not be regenerative feedback from the primary of the drive transformer for this condition.

With the drive transformer positioned as shown in Fig. 2.19.1, there will, in fact, be negative feedback from the drive transformer. Hence the "drive and isolation circuit" must be more powerful and maintain its active drive condition until the correct forward current has been established in Q1. This effect must also be considered for light loading conditions in full-conduction-angle square-wave converters. Under some conditions, extra drive components will be required for this mode of operation.

19.3 SNUBBER COMPONENTS

To assist the turn-off action of Q1 and Q2, snubber components R1, R2, C3, and C4 have been fitted. Now the inductively maintained collector current is diverted into the snubber components R1 and C3 during the turn-off edge, reducing the rate of change of voltage on the transistor collectors. This allows Q1 and Q2 to turn off under lower stress conditions. Resistor R1 is chosen to restore the working point for capacitor C3 during the minimum "on" period.

This snubber action is more fully explained in Part 1, Chap. 18.

19.4 STAIRCASE SATURATION IN PUSH-PULL CONVERTERS

The push-pull converter is particularly vulnerable to staircase saturation effects. A direct DC path exists through the main transformer primary and primary switches Q1 and Q2.

If the average volt-seconds impressed across the transformer when Q1 is "on" is not exactly equal to that impressed by Q2, then there will be a net DC polarizing component. With high-permeability materials, the core will quickly staircase into saturation with successive cycles.

An imbalance can be caused by variations in the saturation voltage and storage times of Q1 and Q2, by differences in the forward voltage drop of the output rectifiers D1 and D2, or by differences in the winding resistance between the two halves of the primary or secondary windings. Consequently, some DC bias toward saturation is almost inevitable. The effects of this DC bias can be reduced to barely acceptable limits by careful matching of the drive and output components and by introducing an air gap in the core. However, an air gap will also reduce the permeability, increasing magnetization currents.

When bipolar transistors are used, a natural correcting effect occurs. The transistor which is driving the core toward the saturated region will experience a rapidly increasing collector current toward the end of the "on" period. This will tend to reduce its storage time, limiting any further excursion into the saturated state. Although this compensating effect may prevent failure under steady-state conditions, it puts a severe limitation on the transient performance of the unit. Rapid variations in pulse width cannot be accommodated by this natural balancing effect. Hence, the control circuit must have a limited slew rate, reducing the transient performance.

When power FETs are used in the positions Q1 and Q2, the natural balancing effect will be absent, as these devices have negligible storage times, and the pulse-width error is not self-correcting. Dynamic flux-balancing techniques or current-mode control should be used with these devices. (See Part 3, Chap. 6.)

19.5 FLUX DENSITY BALANCING

When a push-pull transformer operates with a nonsymmetrical flux excursion, there are two detectable primary current changes before saturation occurs. There will be a DC offset in the current in each half of the primary winding; also, as a result of the curvature of the B/H characteristic, the magnetizing current components will not be symmetrical.

With suitable control circuitry, the difference between alternate primary currents pulses can be detected and used to differentially adjust the drive pulse width to maintain the mean flux density in the core near zero. This will ensure that the core operates in a balanced condition.

This technique is almost essential for high-power duty-ratio-controlled push-pull converters. Alternatively, current-mode control should be used. (See Part 3, Chap. 10.)

19.6 PUSH-PULL TRANSFORMER DESIGN (GENERAL CONSIDERATIONS)

It will be clear from the preceding section that the design of push-pull transformers presents some rather unique problems. Figure 2.19.1 shows that a center tap is required on the primary winding, and that each half of this winding is active only for alternate power pulses. This means that the utility factor of the primary winding copper is only 50%. The unused part of the winding is occupying space on the bobbin and increasing the primary leakage inductance.

To prevent excessive voltage overshoot on the collectors of Q1 and Q2 during the turn-off transient, very close coupling is required between the two halves of the primary winding.

Note: The flyback voltage clamping and energy recovery action are provided to the transistor being turned off by the diode at the opposite end of the primary winding.

To minimize leakage inductance, the two halves of the primary would normally be bifilar-wound. However, when the transformer is wound in this way, a large ac voltage appears between adjacent turns at the collector ends of the windings. This introduces considerable undamped capacitance between the supply lines and the collector connections; also, adjacent transformer turns are subjected to high stress voltages, with the risk of breakdown between the windings.

These limitations become particularly severe in off-line applications with their inherent high-voltage conditions. This is probably the major reason why the push-pull technique is more favored for low-voltage DC-to-DC converters.

19.7 FLUX DOUBLING

(See Part 3, Chap. 7.) Generally, greater utility would be expected from the transformer core in push-pull applications, as both quadrants of the B/H characteristic should be available for flux excursion. Under steady-state, balanced conditions, this is indeed the case. (See Fig. 2.19.3.)

However, when such systems are first switched on, or during some transient conditions, it should be remembered that the flux density in the core may start at zero rather than at $+\hat{B}$ or $-\hat{B}$. Consequently, the available flux excursion ΔB at the instant of turn-on will be only half that normally available under steady-state conditions. (This is the so-called "flux doubling" effect.) If core saturation at switch-on is to be avoided, then the drive and control circuitry must recognize these start conditions and prevent the application of wide drive pulses until the normal working conditions of the core are restored. (This is known as "volt-second clamping.")

These procedures (soft start, slew rate control, and volt-second clamping) are more fully explained in Part 1, Chaps. 7, 8, and 9. The following example assumes that these techniques will be implemented so that full advantage can be taken of the larger flux density excursion available in the push-pull transformer.

19.8 PUSH-PULL TRANSFORMER DESIGN EXAMPLE

19.8.1 Step 1, Establish Full Specification

Assume that a transformer is required to meet the following specification for a 200-W DC-to-DC converter:

Input voltage, nominal	48 V DC; range 42–52 V DC
Operating frequency	40 kHz
Output voltage	5 V
Output current	50 A
Operating temperature range	0–55°C
Efficiency	75%

19.8.2 Step 2, Select Core Size

To demonstrate a different design approach, in this example the core size and flux density will be chosen to give a *selected temperature rise* and optimum efficiency.

The maximum permitted temperature rise for the transformer is to be 40°C, giving a maximum surface temperature of 95°C at an ambient of 55°C and an approximate hot spot temperature of 105°C at maximum ambient temperature.

From Fig. 2.19.5, for balanced push-pull operation and a throughput power of 300 W at 40 kHz, a core type EE42/42/20, with a power throughput ranging from 220 W to 700 W, should be suitable. [The actual power rating of the core depends on the permitted temperature rise, duty ratio, winding design, wire types (plain or Litz wire), insulation requirements, and make-up of the windings—hence the wide range shown in Fig. 2.19.5.]

Since this example is a low-voltage DC-to-DC converter, the input-to-output voltage stress and insulation requirements are not severe, and the creepage distance allowance is only 2 mm. The transformer will be vacuum impregnated to eliminate air voids, giving good thermal conduction properties. Hence a thermal resistance toward the lower value indicated in Table 2.19.1 will be assumed for the finished product.

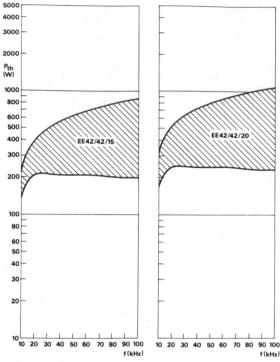

FIG. 2.19.5 Core selector chart for balanced push-pull operation, showing throughput power as a function of frequency, with core size as a parameter. (Courtesy of Mullard Ltd.)

The range of transformer thermal resistance for the EE42/42/20 (from Table 2.19.1) is 10 to 11.5°C/W. (The loss figures are not precise, as they depend on the materials used, methods of makeup, and finish.) In this example, a value of 10.5°C/W will be assumed.

19.8.3 Step 3, Transformer Power Losses

If the thermal resistance is 10.5°C/W and the maximum permitted temperature rise is 40°C, the total internal dissipation of the transformer may be calculated:

$$P_{id} = \frac{T_r}{R_{th}} = \frac{40}{10.5} = 3.8 \text{ W}$$

where P_{id} = total internal transformer dissipation, W
R_{th} = thermal resistance, °C/W
T_r = temperature rise, °C

For optimum efficiency, the core loss P_c should be 44% of the total losses P_{id} (see Fig. 2.13.5); that is, $3.8 \times 44/100 = 1.67$ W. The optimum flux density swing to

TABLE 2.19.1 EC Transformer Parameters and Core Loss

Catalogue number*	Core type	$A_{cp(min)}$, mm²	A_e, mm²	V_e, mm³ × 10³	l_e, mm	B_{CF}, mm	H_{CF}, mm	l_{ave}, mm	R_{th}, °C/W	$R_{th(z)}$,[†] °C/W	Weight,[§] g
4322 020 52500	EC35/17/10	66,5	84,3	6,53	77,4	21,4	4,6	53	17,4	20,0	36
4322 020 52510	EC41/19/12	100	121	10,8	89,3	24,4	5,5	62	15,5	17,0	52
4322 020 52520	EC52/24/14	134	180	18,8	105	28,2	7,5	70	10,3	11,9	100
4322 020 52530	EC70/34/17	201	279	40,1	144	41,3	11,5	96	7,1	7,8	252
4312 020 34070	EE20/20/5	23,5	31,2	1,34	42,8	10,5	3,0	38	35,4	—	7,2
4312 020 34020	EE25/25/7	52,0	55,0	3,16	57,5	—	—	—	30,0	—	16
4312 020 34550	EE30/30/7	46,0	59,7	4,00	66,9	16,3	4,8	56	23,4	—	22
4312 020 34110	EE42/42/15	172	182	17,6	97,0	26,2	6,8	93	10,4	12,2	88
4312 020 34120	EE42/42/20	227	236	23,1	98,0	26,2	6,8	103	10,0	11,5	116
4312 020 34170	EE42/54/20	227	236	28,8	122	35,3	6,8	103	8,3	9,8	130
4312 020 34190	EE42/66/20	227	236	34,5	146	50,0	6,8	103	7,3	8,1	—
4312 020 34100	EE55/55/21	341	354	43,7	123	32,5	7,7	116	6,7	7,4	216
3122 134 90210	EE55/55/25	407	420	52,0	123	32,5	7,7	124	6,2	6,8	—
4312 020 34380	EE65/66/27	517	532	78,2	147	38,6	10,2	150	5,3	6,1	—
3122 134 90690	UU15/22/6	30,0	30,0	1,44	48,0	10,0	4,0	45	33,3	—	—
3122 134 90300	UU20/32/7	52,2	56,0	3,80	68,0	14,5	5,5	57	24,2	—	—
3122 134 90460	UU25/40/13	100	100	8,60	86,0	19,0	7,0	75	15,7	—	—
3122 134 90760	UU30/50/16	157	157	17,4	111	26,0	9,0	104	10,2	—	—
3122 134 91390	UU64/79/20	289	290	61,0	210	—	—	110/(98‡)	5,4	6,2	—

*Core material 3C8
†For IEC class 2 insulation.
‡Wound on both legs.
§Core weight per pair.

2.155

give this optimum core loss at 40 kHz can be established in several ways, two of which are shown in the next section.

19.8.4 Step 4, Select Optimum Flux Density (Swing)

From Fig. 2.13.4, the optimum peak flux may be obtained directly from the material loss if the weight of the total core, the working frequency, and the total watt loss per gram for the ferrite material are known. This is the most universal approach.

Note: The loss graphs normally assume balanced push-pull operation, so the peak loss shown for $\hat{B}$ applies to a flux density swing ΔB of twice the peak value shown. It is important to remember this when calculating losses for single-ended operation, where the loss-producing flux swing is less than half the $\hat{B}$ peak-to-peak loss shown in the loss graph in Fig. 2.13.4.

The total weight of a pair of 42/21/20 half cores (a 42/42/20 core) is 116 g. At a total core loss of 1.67 W, the watt loss per gram will be $1.67/116 = 14.4$ mW/g. Entering Fig. 2.13.4 with this value at 40 kHz yields a peak flux density $\hat{B}$ of approximately 100 mT (a swing ΔB of 200 mT p–p).

Figure 2.19.6 shows an alternative method for establishing the optimum flux level for the pair of 42/21/20 half cores. In this nomogram, the total flux Φ is given. Entering this figure at $P_c = 1.67$ W yields a result for push-pull peak-to-peak flux (top of figure) of $\Phi = 50$ μWb.

Note: The area of the core A_e is 240 mm^2, and the peak-to-peak flux density ΔB may be calculated:

$$\Delta B = \frac{\Phi}{A_e} = \frac{50}{240} = 208 \text{ mT (A very similar result.)}$$

The corresponding copper losses should be 56% of P_{id} or 2.13 W. It remains to be seen if it is possible to obtain a copper loss of this value (or less).

19.8.5 Step 5, Calculate Secondary Turns

In this example, only one secondary is required, and the number of secondary turns will be an integer; hence the design approach will be to calculate secondary turns first, rounding to the nearest whole number if required.

The optimum turns per volt may be calculated as follows:

$$N/V = \frac{t_{on}}{\Delta B \times A_e}$$

where N/V = optimum turns per volt

$\quad\quad t_{on}$ = maximum "on" time, μs

$\quad\quad \Delta B$ = optimum flux density swing, T

$\quad\quad A_e$ = area of core, mm^2

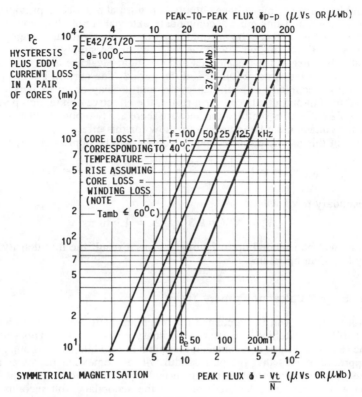

FIG. 2.19.6 Hysteresis and eddy-current losses for a pair of E42/21/20 cores as a function of peak flux φ, with frequency as a parameter. (Courtesy of Mullard Ltd.)

For this calculation, the value of the flux density swing ΔB is twice the peak value $\hat{B}$, because both quadrants of the B/H loop are used for push-pull operation.

The maximum "on" time for each transistor is 50% of the period, or 12.5 μs at 40 kHz. The area of the core is 240 mm², and the optimum flux density swing is 200 mT. Hence the turns per volt may be calculated as follows:

$$N/V = \frac{12.5}{0.2 \times 240} = 0.26 \text{ turns/V}$$

Before the secondary turns can be calculated, it is necessary to obtain the secondary voltage V_s. The absolute value of the secondary voltage cannot be obtained until the design is complete and the secondary resistance has been calculated, so some approximations must be made at this point.

A typical efficiency for a low-voltage converter of this type would be 75%, and the losses at an output of 5 V, 50 A would be 83 W. If 75% of this loss is in the secondary circuits, then at 50 A output the effective voltage drop for rectifier diodes, choke wiring, and transformer resistance would be 1.2 V (a typical overall value when Schottky diodes are used).

With high-current, low-voltage outputs, it is normal practice to provide remote voltage sensing, and this must accommodate external supply line voltage losses of up to 0.5 V total. Hence the maximum power supply terminal voltage could be as high as 5.5 V to maintain 5 V at the load end of the supply leads.

Finally, because of the transformer and secondary leakage inductance, the effective "on" period is not a full 50% of the period. The current builds up rather slowly in the secondary at each half cycle, and a delay of up to 1 μs could be expected for full output current. As a result, the effective conduction period will be only 46% (92% after rectification because there are two power pulses), and the secondary voltage must be increased to allow for this effect.

Hence in this example the secondary voltage will be approximately

$$(V_{out} + \text{losses}) \times \frac{t_{on} + t_{off}}{t_{on}} = (5.5 + 1.2) \times \frac{100}{92} = 7.3 \text{ V}$$

The secondary turns will be

$$V_s \times N/V = 7.3 \times 0.26 = 1.9 \text{ turns}$$

The turns will be rounded up to 2 turns, and as a result the flux density will be slightly less than optimum.

19.8.6 Step 6, Calculate Primary Turns

The primary turns will be calculated for minimum input voltage with an allowance of 10% for switching transistor and primary resistance loss. This will ensure that the required output is obtained at minimum input voltage. (Although this design approach means that the pulse width will be somewhat narrower and the peak secondary voltage higher, under normal input conditions the total flux excursion will remain as calculated, because the secondary volt-seconds will be unchanged.)

The minimum supply voltage V_{cc} is 42 V. Allowing 10% margin, the effective primary voltage V_p will be 38 V.

The turns per volt were modified by the rounding-up process to $2/7.3 = 0.274$ turns/V.

Hence the primary turns N_p will be

$$N_p = \frac{V_p}{N/V}$$

Therefore $\qquad N_p = 38 \times 0.274 = 10.4 \text{ turns}$

In this example the turns will be rounded to 11 turns, giving slightly less output voltage margin.

The primary and secondary turns have now been established, and it remains to select suitable wire gauges for the primary and copper strip size for the secondary to remain within the copper loss design requirement of 2.13 W.

19.8.7 Step 7, Select Wire Sizes and Winding Topology

For high-current outputs of this nature, the secondary will normally be wound from a copper strip and the primary from multifilament parallel insulated magnet wire or litz wire.

To minimize leakage inductance and skin and proximity effects and to optimize the F_r ratio (AC/DC resistance ratio), the correct gauge of wire and strip thickness must be used.

The winding makeup (that is, the number of layers and how they are configured with respect to one another) is also important to ensure minimum leakage inductance and proximity effects. This is more fully covered in Part 3, App. 4.B.

A split primary winding will be used as indicated in Fig. 3.4.8c. If desired, the design may be further checked and the temperature rise predicted from total calculated losses.

More often, at this stage, the prototype transformer would be wound using the wire gauges and makeup recommended in Part 3, Chap. 4. The temperature rise will be measured in the working environment, where the final result depends upon many intangible factors, including ventilation, conduction, radiation, and the effect of neighboring components, as well as on the transformer design.

19.9 PROBLEMS

1. Why is the push-pull converter not favored for high-voltage applications?
2. When proportional drive techniques are used with the push-pull converter, it is necessary to provide a minimum load. Why is this?
3. Why is the push-pull converter particularly vulnerable to staircase saturation effects?
4. What techniques are recommended to prevent staircase saturation in the push-pull converter?

CHAPTER 20
DC-TO-DC SWITCHING REGULATORS

20.1 INTRODUCTION

The following range of DC-to-DC converters, in which the input and output share a common return line, are often referred to as "three-terminal switching regulators."

Functionally, switching regulators have much in common with three-terminal linear regulators, taking unregulated DC inputs and providing regulated DC outputs. They will often replace linear regulators when higher efficiencies are required. Switching regulators are characterized by the use of a choke rather than a transformer between the input and output lines.

The switching regulator differs from its linear counterpart in that switching rather than linear techniques are used for regulation, resulting in higher efficiencies and wider voltage ranges. Further, unlike the linear regulator, in which the output voltage must always be less than the supply, the switching regulator can provide outputs which are equal to, lower than, higher than, or of reversed polarity to the input.

The four building blocks described below form the basis of all DC-to-DC converters, since the introduction of a transformer does not change the basic transfer functions. Further, it can be shown[16] that the inverting and Ćuk regulators are combinations of the basic buck and boost regulators. Hence all converters are derived from combinations of these two basic types and have performance characteristics that are linked to their root type. The four switching regulators of major interest to the power supply designer are as follows:

Type 1, Buck Regulators. See Fig. 2.20.1a. In buck regulators, the output voltage will be of the same polarity but always lower than the input voltage. One supply line must be common to both input and output. This may be either the positive or negative line, depending on the regulator design.

Type 2, Boost Regulators. See Fig. 2.20.2a. In boost regulators, the output voltage will be of the same polarity but always higher than the input voltage. One supply line must be common to both input and output. This may be either the positive or negative line, depending on the design.

The boost regulator has a right-half-plane zero in the transfer function. (See Part 3, Chap. 9.)

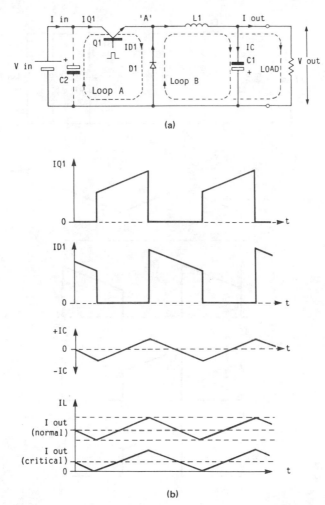

FIG. 2.20.1 (*a*) Basic power circuit of a buck switching regulator. (*b*) Current waveforms for buck regulator circuit.

Type 3, Inverting Regulators. See Fig. 2.20.3. Derived from a combination of the buck and boost regulators, this type is otherwise known as a "buck-boost" regulator. In this type, the output voltage is of opposite polarity to the input, but its value may be higher, equal, or lower than that of the input. One supply line must be common to both input and output, and either polarity is possible by design.

The inverting regulator carries the right-half-plane zero of the boost regulator through to its transfer function.

Type 4, The Ćuk Regulator. This is a relatively new class of boost-buck-derived regulators, in which the output voltage will be reversed but may be equal, higher, or

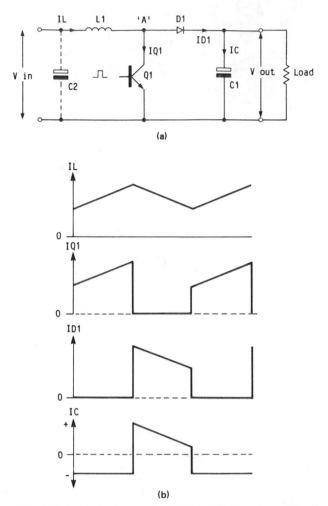

FIG. 2.20.2 (*a*) Basic circuit of DC-to-DC "boost regulator."
(*b*) Current waveforms for boost regulator.

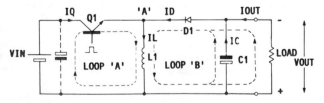

FIG. 2.20.3 Basic circuit of DC-to-DC "inverting regulator" (buck-boost).

lower than the input. Again one supply line must be common to both input and output, and either polarity may be provided by design.

This regulator, being derived from a combination of the boost and buck regulators, also carries the right-half-plane zero through to its transfer function.

Although work by Middlebrook, Ćuk, Bloom, Severns, and others[9,10,16] has shown that these topologies are related, for the purpose of the following discussion, each type will be considered separately. This will provide a working feel for the performance of the basic types. The specialist will of course be familiar with the many variations and combinations of the basic types, covered more fully in the references.

To ease the following explanation, the continuous conduction operating mode will be assumed. Hence the load currents are always above the critical value shown in Fig. 2.20.1b.

20.2 OPERATING PRINCIPLES

20.2.1 Type 1, Buck Regulators

Figure 2.20.1a shows the general arrangement of the power stages of a typical buck regulator.

Switching device Q1 will be turned on and off by a square-wave drive circuit with a controlled on-to-off ratio (duty ratio control).

When Q1 is "on," the voltage at point A will rise to the supply voltage V_{in}. Under steady-state conditions, a forward voltage of $V_{in} - V_{out}$ is now impressed across the series inductor L1, and the current in this inductor will increase linearly during the period Q1 is "on." The current waveform is shown in Fig. 2.20.1b.

When Q1 turns off, the inductor will try to maintain the forward current constant, and the voltage at point A will fly negative (by normal flyback action) until diode D1 is brought into conduction. The current in L1 will now continue to circulate in the same direction as before around loop B and the load. However, since the voltage now impressed across L1 has reversed (being V_{out} plus a diode drop in the reverse direction), the current in L1 will now decrease linearly to its original value during the "off" period.

To maintain steady-state conditions, the input volt-seconds applied to the inductor in the forward direction (when Q1 is "on") must be equal to the reverse output volt-seconds applied (when Q1 is "off"). Hence, the output voltage will be defined by the input voltage and the ratio of the "on" to "off" periods.

By inspection, when Q1 is "on," the volt-seconds appled to L1 is

$$(V_{in} - V_{out}) \times t_{on}$$

When Q1 is "off," the diode is conducting, and the volt-seconds applied to L1 is

$$V_{out} \times t_{off}$$

Hence, to meet the volt-seconds equality on L1 (neglecting losses),

$$(V_{in} - V_{out}) \times t_{on} = V_{out} \times t_{off}$$

Therefore

$$V_{out} = V_{in} \times \left(\frac{t_{on}}{t_{on} + t_{off}} \right)$$

where V_{out} = output voltage
t_{off} = "off" time, Q1
V_{in} = input voltage
t_{on} = "on" time, Q1

With the ratio

$$\frac{t_{on}}{t_{on} + t_{off}}$$

defined as the duty ratio D, the preceding equation will simplify to

$$V_{out} = V_{in} \times D$$

Since D cannot be greater than unity, it will be clear from this equation that accounting for losses, the output voltage must always be less than the applied input voltage in the buck regulator.

Since the power loss in L1 and D1 will be very small, and since Q1 operates in either a low-loss saturated "on" or high-resistance "off" state, extremely efficient DC regulation is obtained.

From the preceding equation, it will be seen further that there are no output-current-related factors in the transfer function; hence the output resistance of the buck regulator is extremely low. In the first order (neglecting losses), the duty ratio is not required to change to provide more output current. However, this is true only when the load current does not fall below a critical minimum value, shown by the lower waveform in Fig. 2.20.1b. This critical minimum load current is defined by the value of the inductance L1.

Although operation below the critical minimum load current is permitted, some degradation of performance would be expected, because the duty ratio is then required to change in response to load variations. Further, the output resistance of the power section in this low-current discontinuous-mode region is high, and the transfer function has changed. If stable operation is to be maintained in this area, the control circuit must be able to compensate for these transfer function changes.

This type of buck regulator is often used to provide additional secondary regulation in a multiple-output converter. When it is used in this way, it would be common practice to synchronize the switching regulator repetition rate to that of the converter to eliminate low-frequency intermodulation effects (beat frequency) in the output ripple.

To enable common drive and control circuitry to be used for the switching regulators in a multiple-output application, it is better that all the converter outputs share a common return line where possible.

It should be noted that the input current to the buck regulator is discontinuous (pulsating), and input filtering would normally be required. Since the buck regulator has an effective negative input resistance slope (when the input voltage rises, the input current falls), the designer should be careful to avoid instability caused by input filter resonance. It is important that the supply impedance presented to Q1 is low, so that when Q1 turns on and demands a large current, the input voltage is maintained nearly constant. Normally a large low-ESR capacitor will be fitted as close as possible to the collector of Q1, and input inductors may require some resistive damping.

20.2.2 Type 2, Boost Regulators

Figure 2.20.2a shows the general arrangement of the power sections of a boost regulator. The operation is as follows.

When Q1 turns on, the supply voltage will be impressed across the series inductor L1. Under steady-state conditions, the current in L1 will increase linearly in the forward direction. Rectifier D1 will be reverse-biased and not conducting. At the same time (under steady-state conditions), current will be flowing from the output capacitor C1 into the load. Hence, C1 will be discharging. Figure 2.20.2b shows the current waveforms.

When Q1 turns off the current in L1 will continue to flow in the same direction and will take point A positive. When the voltage at point A exceeds the output voltage on capacitor C1, rectifier diode D1 will conduct, and the inductor current will be transferred to the output capacitor and load. Since the output voltage exceeds the supply voltage, L1 will now be reverse-biased, and the current in L1 will decay linearly back toward its original value during the "off" period of Q1.

Unlike in the buck regulator, the current into the output capacitor C1 from rectifier diode D1 is always discontinuous, and a much larger output capacitor will be required if the output ripple voltage is to be as low as in the buck regulator. The ripple current in C1 is also much larger.

To the boost regulator's advantage, the input current is now continuous (although there will be a ripple component depending on the value of the inductance L1); hence less input filtering is required, and the tendency for input filter instability is eliminated.

As with the buck regulator, for steady-state conditions, the forward and reverse volt-seconds across L1 must equate. The output voltage V_{out} is controlled by the duty ratio of the power switch and the supply voltage, as follows.

By inspection (to meet the volt-seconds equality on L1),

$$V_{in} \times t_{on} = (V_{out} - V_{in}) \times t_{off}$$

Therefore

$$V_{out} = V_{in} \times \left(\frac{t_{off} + t_{on}}{t_{off}} \right)$$

But

$$\frac{t_{off} + t_{on}}{t_{off}} = \frac{1}{1 - D}$$

Hence

$$V_{out} = \frac{V_{in}}{1 - D}$$

It will be seen from this equation that the output voltage is not related to load current (neglecting losses), and the output resistance is very low. Again, this is true only when the load current is not below the critical value shown in Fig. 2.20.1b. At loads lower than the critical value, the duty ratio must be reduced to maintain a constant output voltage.

It should also be noted that in the boost regulator, a sudden increase in load current requires that the "on" period be increased initially (to increase the current in L1 and to make up for voltage losses). However, increasing the "on" time reduces the "off" time, and the output voltage will initially fall (the reverse of what is required) until the current in L1 increases to the new load level. This introduces an extra 180° phase shift during the transient. This is the cause of the right-half-plane zero in the transfer function. (See Part 3, Chap. 9.)

The value of L1 is usually chosen to ensure that the critical current is below the minimum load requirement. Further, L1 must not saturate at maximum load and maximum "on" period.

20.2.3   Type 3, The Inverting-type Switching Regulator

Figure 2.20.3a shows the power circuit of a typical inverting (buck-boost) regulator which operates as discussed below.

When Q1 is "on," current will build up linearly in inductor L1 (loop A). Diode D1 is reverse-biased and blocks under steady-state conditions. When Q1 turns off, the current in L1 will continue in the same direction, taking point A negative (by normal flyback action). When the voltage at point A goes more negative than the output voltage, diode D1 is brought into conduction, transferring the inductor current into the output capacitor C1 and load (loop B).

During the "off" period, the voltage across L1 is reversed, and the current will decrease linearly toward its original value. The output voltage depends on the supply voltage and duty cycle (t_{on}/t_{off}), and this is adjusted to maintain the required output. The current waveforms are the same as those for the boost regulator, shown in Fig. 2.20.2b.

As previously, the forward and reverse volt-seconds on L1 must equate for steady-state conditions, and to meet this volt-seconds equality (neglecting polarity),

$$V_{in} \times t_{on} = V_{out} \times t_{off}$$

Therefore

$$V_{out} = V_{in} \times \left(\frac{t_{on}}{t_{off}}\right)$$

The ratio t_{on}/t_{off} is defined as the duty cycle d'. Hence

$$V_{out} = V_{in} \times d'$$

Note that the output voltage is of reversed polarity but may be greater or less than V_{in}, depending on the duty cycle.

In the inverting regulator, both input and output currents are discontinuous, and considerable filtering will be required on both input and output.

20.2.4   Type 4, The Ćuk Regulator

Since its introduction in 1977 by Slobodan Ćuk,[9] the "boost-buck" regulator shown in Fig. 2.20.4a, which Ćuk described as the "optimum topology converter," has attracted a great deal of interest. It was the first major new switching

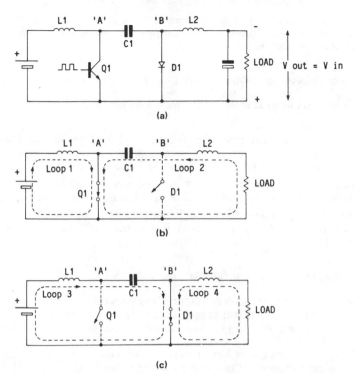

FIG. 2.20.4 (*a*) Basic circuit of Ćuk (boost-buck) regulator. (*b*) Storage phase. (*c*) Transfer phase.

regulator topology to appear for some years. Subsequent analysis and duality circuit modeling show that this circuit is the dual of the type 3 inverting (buck-boost) regulator.[16]

For the Ćuk regulator, some of the major features of interest to the power supply engineer are as follows:

1. Both input and output currents are continuous; further, the input and output ripple currents and voltages may be suppressed to zero by correct coupling of L1 and L2.
2. Although in the basic topology the polarity of the output voltage is reversed, the output voltage may have values which are equal to, higher than, or lower than those of the input voltage. In fact, the supply may range through the output voltage value, and the output can be maintained constant.
3. By replacing diodes by active switching devices, it is possible to reverse the direction of energy transfer, a useful attribute for vehicle or machine control. (Although this bidirectional ability also applies to the previous regulator types, in the Ćuk topology the transfer function is not changed by reversing the current flow.)
4. As with other topologies, DC isolation of input and output (galvanic isolation) can be provided by introducing transformers without compromising the other major features.

5. A major advantage in some applications is that a failure of a diode or switching device will normally result in zero output (an advantage for crowbar-type overvoltage protection requirements).

20.2.5 Possible Problem Areas (Ćuk Regulators)

The designer should be cautious of the following possible limitations of the Ćuk regulator:

1. Internal resonances can cause discontinuities in the transfer function or poor input ripple rejection at some frequencies. Check open-loop response to ensure that this will not be a problem for the intended application.
2. In coupled inductor and transformer isolated versions, output voltage reversal can occur during initial "switch-on" as a result of inrush current coupling.
3. As with any boost-derived topology, loop stability problems can occur as a result of the right-half-plane zero in the transfer function. (See Part 3, Chap. 9.)

20.2.6 Operating Principle (Ćuk Regulators)

The operation cycle of the Ćuk regulator is more complex than those of the previous three types, and excellent papers on this topology have been published.[9,10] However, in keeping with the previous simple volt-seconds approach, the following explanation is offered.

The operating principle is best understood by considering Fig. 2.20.4 under steady-state operation with the following initial conditions:

Input and output voltages are equal but of opposite polarity.

Q1 is operating with a 50% duty cycle (equal "on" and "off" times).

The regulator is loaded at a current in excess of the critical value, so that L1 and L2 are continuously conducting (continuous-mode operation).

The two inductors are identical, so that L1=L2. Current is flowing into L1 from left to right (forward current direction) and into L2 from right to left (reverse current direction).

Note: Because the output and input voltages are equal, although of reversed polarity, and losses are assumed to be zero, the mean input current must equal the mean output current.

Let the voltage on C1 be $V_{in}+V_{out}$ ($2 \times V_{in}$ in this example); also, let C1 be very large, so that its voltage does not change significantly during a cycle. Finally, let C1 terminal A be positive with respect to B.

Consider first the circuit shown in Fig. 2.20.4b with the operating conditions as specified above.

When Q1 turns on, point A will go to zero, and the current in inductor L1 will continue to flow into Q1 around loop 1. The voltage across L1 (V_{L1}) will be the supply voltage acting in the forward direction (from left to right), and the current in L1 will be increasing linearly.

Hence during the "on" period of Q2, the voltage applied to L2 (V_{L2}), neglecting polarity, is

$$V_{L2} = V_{in}$$

At the same time as the A terminal of C1 is taken to zero volts, the B terminal will be taken to $-(2 \times V_{in})$ (the initial starting voltage across C1). Diode D1 will be reverse-biased and hence open-circuit. The current in L2 will continue to flow from right to left into C1 and Q1, around loop 2, under the forcing action of L2. Hence, C1 will be discharging toward zero.

During this "on" period of Q1, the voltage applied to L2 (V_{L2}), neglecting polarity, is

$$V_{L2} = (2 \times V_{in}) - (V_{out})$$

Since $V_{out} = V_{in}$, $\qquad\qquad\qquad V_{L2} = V_{in}$

Thus the volt-seconds applied to the output inductor are equal to the volt-seconds applied to the input inductor. Since the voltage is applied in the forward direction in one case and the reverse direction in the other, the output current in L2 will increase at the same rate but in the opposite direction as that in L1.

It is interesting to note at this stage that if L1 and L2 were to be wound on the same core with the same number of turns and phased correctly, the ripple current components would exactly cancel to zero, and the effective input and output currents would be a steady DC. This can in fact be done, although it is not considered further here.[30]

Consider Fig. 2.20.4c. When Q1 turns off, the forcing action of L2 will drive D1 into conduction (loop 4), taking the voltage at point B to zero (neglecting diode drops). At the same time, point A will have been taken to $2V_{in}$ (the voltage across the capacitor), so that the voltage across L1 will have reversed, with an effective value of V_{in}, but in the opposite direction to that in the "on" state; however, under the forcing action of L1, the current in L1 will still continue in the forward direction, but will be decreasing. With Q1 "off," the current path for L1 will be from the supply into C1 and through D1 (loop 3).

Hence the voltage applied to L1 during the "off" period (V_{L1}) is

$$V_{L1} = V_{in}$$

Current is now flowing into terminal A of capacitor C1, replacing the previously removed charge. Since D1 is conducting, the voltage across L2 for the "off" period is simply V_{out} (neglecting diode drops), and the voltage polarity is such that the current in L2 will be decreasing.

Hence the voltage applied to L2 during the "off" period (V_{L2}) is

$$V_{L2} = V_{out}$$

But $\qquad\qquad\qquad V_{out} = V_{in}$

Hence $\qquad\qquad\qquad V_{L2} = V_{in}$

The volt-seconds applied to L1 and L2 are the same during the "on" and "off" period.

It has been shown that the magnitude of the voltages and volt-seconds applied to L1 and L2 for both "on" and "off" periods are the same, although the polarities are reversed. Also, the forward and reverse volt-seconds equate for both inductors. Hence, with equal "on" and "off" periods, the initial assumed conditions for steady-state operation are satisfied.

In general, to satisfy the "on" and "off" volt-seconds equality in, say, L1 (neglecting polarity),

$$V_{in} \times t_{on} = (V_{C1} - V_{in}) \times t_{off}$$

But

$$V_{C1} = V_{in} + V_{out}$$

Hence

$$V_{in} \times t_{on} = V_{out} \times t_{off}$$

Therefore

$$V_{out} = V_{in} \times \left(\frac{t_{on}}{t_{off}}\right)$$

But

$$\frac{t_{on}}{t_{off}} = d'$$

Hence

$$V_{out} = V_{in} \times d'$$

Applying this approach to L2 yields the same result. Hence for a 50% duty cycle (where $t_{on} = t_{off}$), $V_{out} = V_{in}$ but the polarity is reversed.

20.3 CONTROL AND DRIVE CIRCUITS

There are many suitable control and drive circuits in both discrete and integrated circuit form. Many of the single-ended control circuits used for the forward and flyback converters can be used for the Ćuk regulator.

Although many switching regulator control circuits use duty ratio control quite successfully, the more recent current-mode control techniques can be applied. These will yield advantages similar to those found in conventional transformer converters.

Research and development work is still in progress on this interesting and useful range of regulators using transformers and chokes in integrated magnetic assemblies. Full coverage of these techniques is beyond the scope of this book. For more information on this subject, the interested engineer should study papers by Slobodan Ćuk, R. D. Middlebrook., Rudolf P. Severns., Gordon E. Bloom., and others, some of which are covered in References 9, 10, 16, and 20.

20.4 INDUCTOR DESIGN FOR SWITCHING REGULATORS

It will be clear from the preceding that the inductors (or chokes) play a critical part in the performance of the regulators.

These inductors carry a large component of DC current, as well as sustaining a large high-frequency ac stress. The inductor must not saturate for any normal condition, and for good efficiency the winding and core losses must be small.

The choice of inductance is usually a compromise. Theoretically, the inductance can have any value. Large values are expensive and lossy, and give poor transient load response. However, they also have low ripple currents and give continuous conduction at light loads. Small values of inductance give high ripple currents, increasing switching losses and output ripple. Discontinuous conduction will occur at light loads, which changes the transfer function and can lead to instability. However, the transient response is good, and efficiency is high, size small, and cost low. The choice is a compromise between minimum size and cost, and acceptable ripple current values, consistent with continuous-mode conduction.

Although there are no set rules, one arbitrary choice that is often applied is to choose an inductance such that the critical current (the current at which discontinuous conduction occurs) is below the minimum specified load current. (See Fig. 2.20.1a.) However, if the minimum load current is zero, this cannot be done. Hence, although this criterion may be a useful guide, a well-designed regulator must work without instability well below the critical current value.

It should be noted that although operation at currents below the critical value is permitted, the load and line regulation will be degraded. Also, with the ripple regulator circuit (Sec. 20.7), the frequency can be very low at light loads.

A practical approach used by the author is to chose the inductance value to be as small as possible consistent with acceptable ripple current. Values of ripple current between 10% and 30% of I_{DC} load (maximum) are used depending on the ripple and transient response requirements. Remember, the smaller the inductance, the lower the cost and the better the transient load response.

20.5 INDUCTOR DESIGN EXAMPLE

Calculate the inductance required for a 10-A, 5-V type 1 buck regulator operating at 40 kHz with an input voltage from 10 to 30 V, when the ripple current is not to exceed 20% of I_{DC} (2 A).

Procedure: Maximum ripple current will occur when the input voltage is maximum—that is, when the voltage applied across the inductor is maximum.

1. Calculate the "on" time when the input is 30 V.

$$t_{on} = \frac{V_{out} \times t_p}{V_{in}}$$

where t_p = total period ($t_{on} + t_{off}$)

Hence

$$t_{on} = \frac{5 \times 25}{30} = 4.166 \ \mu s$$

2. Select the peak-to-peak ripple current. This is by choice 20% of I_{DC}, or 2 A in this example.
3. Calculate the voltage across the inductor V_L.

$$V_L = V_{in} - V_{out} = 25 \ V$$

4. Calculate the inductance.

$$V_L = -L \times dI/dt$$

Therefore

$$L = V_L \times \frac{\Delta t}{\Delta I} = \frac{25 \times 4.166 \times 10^{-6}}{2} = 52 \ \mu\text{H}$$

The critical load will be $\frac{1}{2} I_{\text{p--p}} = 1$ A, or 10% in this example.

The design of these inductors (chokes) is covered more fully in Part 3, Chaps. 1, 2, and 3.

20.6 GENERAL PERFORMANCE PARAMETERS

Where input-to-output galvanic isolation is not essential, switching regulators can provide extremely efficient voltage conversion and regulation. In multiple-output switchmode power supply applications, independent fully regulated secondary outputs can be provided by these regulators. The performance of the overall power unit can then be extremely good.

The user should specify the range of load currents for which full performance is required. This range should be as small as is realistic (the lower the minimum current, the larger the inductor and the cost). In particular, full performance continuous conduction at very light loads should not be demanded, since this may require a very large series inductor, which would be expensive and bulky and would introduce considerable power loss.

20.7 THE RIPPLE REGULATOR

A control technique which tends to be reserved for the buck-type switching regulator is the so-called "ripple regulator."[17] This is worthy of consideration here, as it provides excellent performance at very low cost.

The "ripple regulator" is best understood by considering the circuit of the buck regulator shown in Fig. 2.20.5a.

A high-gain comparator amplifier A1 compares a fraction of the output voltage V_{out} with the reference V_R; when the output fraction is higher than the reference, the series power switch Q1 will be turned off. A small hysteresis voltage (typically 40 mV) is provided by positive feedback resistor R1, so that Q1 will stay "off" until the voltage has fallen by 40 mV, at which point Q1 will turn on again and the cycle will be repeated.

By this action the output voltage is made to ramp up and down about its mean DC value between the upper and lower limits of the hysteresis range (Fig. 2.20.5b).

The time taken for the voltage to ramp up to the higher limit is defined by the value of the inductor, output capacitor, and supply voltage.

The time taken for the voltage to ramp down to the lower limit is defined by the output capacitor and load current. Since both periods are variable, the operating frequency is variable.

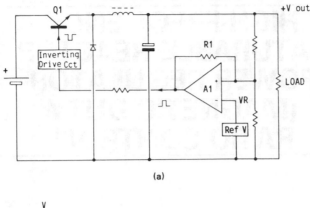

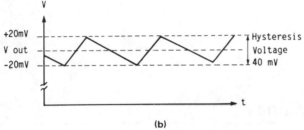

(b)

FIG. 2.20.5 (*a*) Ripple-controlled switching buck regulator circuit (ripple regulator). (*b*) Output ripple voltage of ripple regulator.

The output ripple is always constant at the hysteresis value (40 mV in this example), irrespective of load. If the load is very small, the frequency can be very low.

20.8 PROBLEMS

1. The term "switching regulator" is used to describe a particular type of DC-to-DC converter. In what way do these converters differ from the more conventional transformer-coupled converters?

2. What is the major advantage of the switching regulator over the more familiar three-terminal linear regulator?

3. Explain the major transfer properties of the following regulators: buck regulators, boost regulators, inverting regulators, and the Ćuk regulator.

4. In what way does the Ćuk regulator differ from the three previous types of regulators?

5. Why is the boost regulator particularly prone to loop stability problems?

6. Describe some of the limitations of the coupled inductor integrated magnetic Ćuk regulator.

7. Why is a minimum load necessary on a switching regulator?

8. Why is the output choke in a buck regulator relatively large?

CHAPTER 21
HIGH-FREQUENCY SATURABLE REACTOR POWER REGULATOR (MAGNETIC DUTY RATIO CONTROL)

21.1 INTRODUCTION

The saturable reactor regulator, otherwise referred to as the saturable-core magnetic regulator or magnetic pulse-width modulator, as applied to high-frequency switchmode power supplies, is a relatively new development of the very well established line frequency magnetic amplifier power control technique. However, in high-frequency applications, the mode of operation is quite different.

As a result of the recent advances in magnetic materials, particularly the low-loss square-loop field-annealed amorphous alloys, these techniques are now finding interesting reapplications in high-frequency switching regulators.

The major attraction of this method of control is that large currents at low voltages can be efficiently regulated. The power loss in the saturable reactor is mainly limited to a small resistive loss in the winding. The core loss can usually be neglected, as it is independent of the load current being controlled and is usually small compared with the transmitted power. A further advantage is the inherent high reliability of the saturable reactor and its ability to provide independent isolated secondary regulation in multiple-output applications.

21.2 OPERATING PRINCIPLES

In simple terms, the saturable reactor is used in high-frequency switchmode supplies as a flux-saturation-controlled power switch, providing regulation by secondary pulse-width control techniques.

The method of operation is best explained by considering the conventional buck regulator circuit shown in Fig. 2.21.1. This figure shows the output LC filter and rectifiers (such as would be found on the secondary of a typical single-ended

2.174

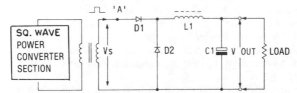

FIG. 2.21.1 Typical secondary output rectifier and filter circuit of a duty-cycle-controlled forward converter.

forward converter). In this type of filter arrangement, the output voltage is related to the transformer secondary voltage by the following equation:

$$V_{out} = \frac{V_s \times t_{on}}{t_{on} + t_{off}}$$

where V_{out} = output voltage

V_s = transformer secondary peak voltage

t_{on} = "on" time, μs (when voltage at point A is high and positive)

t_{off} = "off" time, μs (when voltage at point A is negative)

The ratio $t_{on}/(t_{on} + t_{off})$ is often referred to as the duty ratio D. It can be seen from the above equation that adjusting the duty ratio or input voltage will control the output voltage.

In many of the previous techniques, the *width* of the power pulses on the primary of the main transformer will be adjusted by the primary power switches to provide dynamic control of the output voltage. In some multiple-output buck-regulator converters, the input *voltage* to a square-wave converter (DC transformer) will be controlled so as to provide a fixed output voltage. In some multiple-output applications, additional regulation will be required on auxiliary outputs, and linear regulators will often be used even though high dissipation will occur.

Clearly, if some form of pulse-width control is used to provide the required regulation on the secondary outputs, it should be possible to obtain higher efficiency. Pulse-width control can be introduced at several positions in the secondary loop. In Fig. 2.21.2, for example, a switch S1 has been introduced in the secondary circuit at point A.

This switch may be operated in synchronism with the applied power pulses to further reduce the pulse width applied to the output filter L1, C1. This reduction

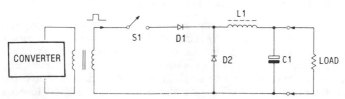

FIG. 2.21.2 Output filter, showing duty cycle (pulse width) secondary control switch S1 in series with the rectifier diode D1.

in pulse width may be achieved by either switching off the power pulse early, thus removing the trailing edge of the power pulse, or switching on the power pulse late, thus removing the leading edge of the power pulse. In either case, the effective pulse width to the output filter will be reduced, and by applying dynamic control to this switch, output regulation can be obtained but at a lower output voltage.

If the switch and other components are very low loss, then the efficiency of this method of power control will be high. To achieve this, the switch must be nearly perfect. That is, it needs near-zero "on"-state resistance, very high "off"-state resistance, and low switching losses.

Clearly, the switch in Fig. 2.21.2 could be replaced with any device which provides suitable switching action. Some examples would be power FETs, SCRs, triacs, or transistors (e.g., Bisyn, a power FET designed for synchronous switching). However, all these devices have relatively high losses at high output currents. In the following section is a magnetic reactor (choke) which will be considered "on" when saturated and "off" when not saturated.

21.3   THE SATURABLE REACTOR POWER REGULATOR PRINCIPLE

Consider what would be required from a magnetic reactor to make it behave like a good magnetic on-off switch. It must have a high effective inductive reactance in its "off" state (nonsaturated state), and low effective inductive reactance in its "on" state (saturated state). Also, it must be able to switch rapidly between these two states with low loss. These properties can be obtained from the saturable magnetic reactor, if the correct core material is used.

Consider the B/H characteristic for a hypothetical, near ideal, square-loop magnetic material, as shown in Fig. 2.21.3.

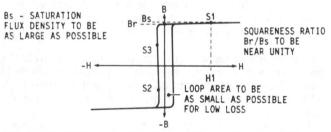

FIG. 2.21.3 B/H loop of an "ideal" saturable core for pulse-width modulation.

This near-square magnetization characteristic has the following properties:

1. In the nonsaturated state (points S2 and S3), the characteristic is vertical, showing negligible change in H (current change) for the full excursion of ΔB (applied volt-seconds); that is, the permeability is very high. An inductor wound on such a core would have nearly infinite inductance. Hence, provided that the core is not allowed to saturate, negligible current would pass. (The reactor would be in its "off" state and would make a good low-loss "off" switch.)

2. Now consider the core in a saturated state (at, say, point S1 on the characteristic). The B/H characteristic in this area is nearly horizontal, so that a negligible change in B will result in a large change in H; that is, the permeability is near zero, and the inductance is near zero. The impedance will be very close to the resistance of the winding, only a few milliohms. In this state the reactor presents very little impedance to the current flow. (It is now in its "on" state and makes an efficient "on" switch.)

Since the area of the ideal B/H loop is negligible, very little energy is lost as the B/H loop is traversed. Consequently, this ideal core may be switched between the "on" and "off" states at high frequency, and the losses will be very small.

It now remains to see how the core may be switched between these two states.

21.4 THE SATURABLE REACTOR POWER REGULATOR APPLICATION

Consider a reactor wound on a core of ideal square-loop material and fitted in series with output rectifier diode D1 (position A in Fig. 2.21.1). This gives the circuit shown in Fig. 2.21.4.

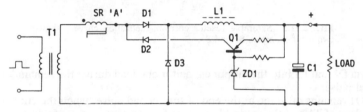

FIG. 2.21.4 Single-winding saturable reactor regulator with simple voltage-controlled reset transistor Q1.

In the circuit shown in Fig. 2.21.4, assume that the core is unsaturated at a point S3 on the B/H characteristic shown in Fig. 2.21.5. When the start of the secondary winding of T1 goes positive, D1 will conduct, and a voltage will be impressed across the winding SR of the saturable reactor. The applied volt-seconds increases the flux density B from S3 toward positive saturation, as shown by the dashed line. Provided that the "on" period is short, the change in flux density ΔB is small, and the core will not saturate; B moves to, say, a point level with S2. Hence, only a small magnetization and core loss current will flow into the output. (The area of the B/H loop has been exaggerated for clarity.)

If Q1 is conducting during the "off" state (secondary voltage of T1 reversed), then the core will be reset to point S3, and for the next power pulse, the same small B/H loop will be followed. Hence, the only current allowed through to the output load will be the magnetization and loss current of the reactor (which is very small compared with the load).

If, on the other hand, the core was not reset after the first forward voltage pulse (Q1 turned off), then the core will have reset to a point on the $H = 0$ line level with S2. Now the second forward voltage pulse will take the core from point S2 into saturation, along the dashed line to, say, point S1 on the B/H characteristic. The impedance of the reactor will now be very low, and a large current can

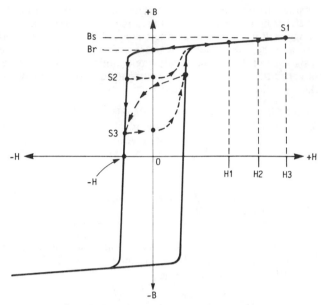

FIG. 2.21.5 Saturable reactor core magnetization curves, showing two reset examples S2 and S3.

flow via D1 and L1 into the output capacitor and load during the remainder of this second pulse.

If Q1 remains off, then at the end of the second pulse, when the current in the secondary of T1 and the saturable reactor falls to zero, the core will return to its remanence value (B_r on the B/H characteristic). The rectifier diode D1 prevents any reversal of current in the reactor winding, and there will not be a reverse reset action.

Hence, at the beginning of the third and subsequent "on" periods, only a very small increment in B (applied volt-seconds), from B_r to B_s, is required to saturate the core. Hence only a very short forward volt-seconds stress is required to take the core back into saturation at S1. With the ideal core material, the slope (permeability) of the characteristic in the saturated area (point S1) will be near zero, and the inductance is negligible.

Hence, after initial switch-on, only two forward polarization pulses were required to make the saturable reactor look like a magnetic switch in its "on" state. All the time it remains in this state, it presents very little impedance to the flow of forward (output) current. It introduces only a slight winding resistance and gives a short delay to the leading edge of the applied pulse, as the core is incremented from B_r to B_s with each pulse. Hence, this ideal saturable reactor may be made to behave like a magnetic on-off switch by either resetting or not resetting the flux density level before the beginning of each "on" pulse.

In practice, the core will normally be reset between pulses to some intermediate point between S3 and B_r on the B/H characteristic. Consequently, when the next forward voltage pulse is applied, there will be a delay in the current flow while the core is taken from its nonsaturated point (say S2 on the characteristic) into a saturated state. At saturation the core switches to its low-impedance "on" state, and the remainder of the power pulse will be allowed through to the output circuit.

In this type of circuit, it is only possible to reduce the effective pulse width and hence reduce the output voltage. The forward current pulse is made narrower by presetting the flux density further down the *B/H* characteristic. By using Q1 to give a controlled reset during the "off" period (when the secondary voltage of T1 is reversed), the output voltage can be controlled to some lower value. (The reset increases the delay applied to leading edge of the power pulse, as shown in Fig. 2.21.6.)

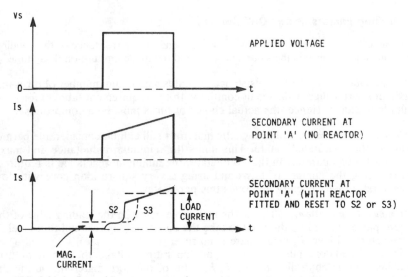

FIG. 2.21.6 Secondary current waveforms with saturable reactor fitted.

For a given core size, the time taken to bring the core from, say, S3 on Fig. 2.21.5 to saturation (the leading-edge delay) will be defined by the number of turns, the applied voltage, and the required flux density increment (the change in ΔB from the reset value to the saturated value), as defined by Faraday's law:

$$t_d = \frac{N \times \Delta B \times A_e}{V_s}$$

where t_d = delay time, μs
  N = turns
  ΔB = preset flux density change S3 to B_s, T
  A_e = area of core, mm^2
  V_s = secondary voltage

21.5 SATURABLE REACTOR QUALITY FACTORS

The effectiveness of the saturable reactor as a power switch will be determined by several factors as follows:

Controlling Factors as an "Off" Switch.

The magnetization current can be considered a leakage current in the "off" state of the switch. The reactor's quality as an "off" switch—that is, its maximum impedance—will be defined by its maximum inductance. This in turn depends on the permeability of the core in the unsaturated state and the number of turns. Increasing the number of turns will, of course, increase the inductance and reduce the magnetization current. However, large numbers of turns will increase the copper losses and minimum turn-on delay, degrading the performance as an "on" switch.

Controlling Factors as an "On" Switch.

The reactor's quality as an "on" switch will depend on the resistance of the winding, the residual inductance in the saturated state, and the minimum turn-on delay time.

1. Minimum Resistance.   A low resistance indicates a minimum number of turns and maximum wire gauge, which is in conflict with the requirement (above) for maximum inductance. Hence, the actual choice of turns must be a compromise.

2. Minimum Inductance.   Some magnetic materials still exhibit considerable permeability in the "saturated" state. This limits the minimum inductance and maximum let-through current. In the "on" state, the minimum reactance is required, and reducing the number of turns and using a very square loop core with low postsaturation permeability improves this parameter.

3. Minimum Turn-On Delay.   This is the inevitable delay on the leading edge of the current pulse, caused by the need to take the core from the residual flux level B_r to the saturated level B_s when there is no reset (the "on" state).

For a defined core, number of turns, and operating voltage, the minimum turn-on delay time is controlled by the B_r/B_s ratio of the core material. During the "off" period, provided that there is no reset current, the flux density in the core will return to its residual value B_r. When the next power pulse is applied, there is an inevitable delay in reactor conduction as the flux density increments from B_r to B_s at point S1 on Fig. 2.21.5. This minimum undesirable delay may be calculated from the previous equation if B_{sat} and B_r are known.

Because of the delay, the maximum width of the current pulse let-through to the output filter will be narrower than the applied secondary voltage pulse, and some of the control range will be lost. It should be noticed from the previous equation that the delay time is proportional to the number of turns, while inductance is proportional to the number of turns squared.

Although it is possible to reduce the turn-on delay by prebiasing the core into its saturated state with a control winding, generally it is not economic to use the control power in this way.

Once again, the ideal solution is a very square loop material with a B_r/B_s ratio close to 1.

4. Power Loss.   At a fixed frequency, the power loss depends on two factors, copper loss and core loss.

First, the copper loss is controlled by the size of the wire and the number of turns, and this loss decreases as the core gets larger. Second, the core loss gets larger as the core size and control range get larger. Hence, as in normal transformers, the core size is a compromise choice. However, because of the high cost of the core material, the smallest practical core size is often selected, even if this does not give the smallest overall power loss. (The copper loss will be large.)

21.6   SELECTING SUITABLE CORE MATERIALS

The ideal core material would match the ideal B/H characteristic shown in Fig. 2.21.3 as closely as possible. That is, it would exhibit high permeability in the nonsaturated state (values from 10,000 to 200,000 are possible), and the saturated permeability and hysteresis losses would be very low.

To minimize the turn-on delay, the squareness ratio B_r/B_s should be as high as possible (values between 0.85 and 0.95 are realizable). The hysteresis and eddy-current losses should be small to minimize core loss and allow high-frequency operation—remember, in this application, the flux density swing is large. Square-loop materials are now available which have usable square-loop magnetic parameters and acceptable losses up to 100 kHz. Unfortunately, most manufacturers do not quote the saturated permeability of their cores at this time, and no standards are available; hence some research in this area is necessary.

For low-frequency operation, up to, say, 30 kHz, suitable materials will be found in the grain-oriented, cold-rolled, field-annealed Permalloys and Mumetals. These materials are nickel-iron alloys available in tape-wound toroidal form for maximum permeability and best squareness ratios. Magnetic field annealing will improve the squareness ratio.

For higher-frequency applications, up to, say, 75 kHz, some of the more recent amorphous nickel-cobalt alloys are more suitable. At the time of writing, efficient operation much above 50 kHz is not possible with these materials because of their excessive core loss and modified pulse magnetization characteristics. However, rapid improvements are being made.

For higher frequencies, square-loop ferrite materials are more suitable.

Some of the core materials found most suitable for this application are listed in Table 2.21.1.

21.7   CONTROLLING THE SATURABLE REACTOR

As explained in Sec. 21.2, to control the saturable reactor in switching regulator applications, it is necessary to reset the core during the "off" period to a defined position on the B/H characteristic prior to the next forward power pulse.

The reset (volt-seconds), may be applied by a separate control winding (transductor or magnetic amplifier operation; see Fig. 2.21.7) or by using the same primary power winding and applying a reset voltage in the opposite direction to the previous power pulse during the "off" period (saturable reactor operation; see Fig. 2.21.5). Although the circuits for these modes of operation are quite different, as far as the core is concerned the action is identical.

In Fig. 2.21.7, the reset is provided during the "off" period of D1—that is, when the drive winding has gone negative—by the reset transistor Q1, which applies a voltage to a separate reset winding via D2. The reset current flows from the positive output, through Q1, the SR reset winding, and D2, to the transformer secondary. (This secondary is negative during the reset period.)

The reset magnetizing ampere-turns will be equal and opposite to the previous forward magnetizing ampere-turns. (This magnetizing current is shown on the leading edge of the waveform in Fig. 2.21.6 and is considered leakage current as far as the perfect magnetic switch is concerned.) The advantage of having a separate reset winding is that the reset current can be reduced by increasing the turns. However, remember: forward and reverse volt-seconds/turn must be equal. More reset turns require more reset volts or more reset time.

TABLE 2.21.1 Properties of Typical Square-Loop Magnetic Materials

Trade name	Squareness ratio B_r/B_s	Saturation flux density, T	Core loss, W/kg (0.002 in at 300 mT)		Curie temp, °C	Frequency
			50 Hz	35 kHz		
Ultraperm Z	0.91	0.8	0.01	60	400	High
Permax Z	0.94	1.25	0.06	180	520	Low
Square Permalloy 80	0.9	0.7	0.01	70		High
Mumetal	0.6	0.7	0.03	180	350	Middle
H.C.R.	0.97	1.54	0.15	60 (5 kHz)		Low
Sq. Metglass	0.5	1.6	0.1	150		Low
Vitrovac 6025	0.9	0.55	0.003	50	85*	High
Sq. Ferrite	0.9	0.4	0.01	60		High
Fair-Rite 83						

Note: Square Metglass and Vitrovac 6025 are amorphous material.
*Maximum long-term operating temperature. (Amorphous properties degrade slowly above this temperature.)

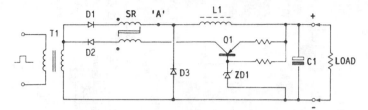

FIG. 2.21.7 Two-winding saturable reactor regulator (transductor type) applied to buck regulator output circuit.

The saturable reactor control circuit shown in Fig. 2.21.4 operates as follows.

As the output voltage tries to exceed the zener diode ZD1 voltage, Q1 will turn on, increasing the reset of the core. The core is reset to a position on the B/H curve that provides the correct delay on the leading edge of the next power pulse to maintain the output voltage constant.

Since the pulse width can only be reduced, the required output voltage must be obtained at a pulse width which is narrower than the normal secondary forward "on" period.

In Fig. 2.21.4, the reset voltage is applied directly to the main winding of the saturable reactor. A major advantage of this arrangement is that the reset current from Q1 automatically provides the preloading of the output necessary to exactly reabsorb the reactor "leakage" current component from the previous forward "on" period.

In this example, reset current flows via Q1 and diode D2 into the main winding when the transformer secondary voltage is negative. Diode D1 is reverse-biased (turned off) during this reset action.

The current waveforms at point A with and without the saturable reactors are shown in Fig. 2.21.6. The excursions on the B/H loop for two different pulse-width conditions are shown in Fig. 2.21.5.

It is important to notice at this point that the current taken during the resetting of the core is given by the value of the negative magnetization force $-H$ required to take the core from B_r to a point, say, S3 on the B/H characteristic. For a particular reactor at a fixed input voltage, this current is entirely controlled by the core parameters, not the control circuit or load. The previous forward current (and hence the load power) does not affect the value of the reset current, as the core always returns to the same remanence value B_r when the forward current has fallen to zero. (When the current is zero, H must be zero.)

Hence, irrespective of the forward current or the position on the B/H loop at which the core was saturated during forward conduction (i.e., H1, H2, H3, etc.), the reset current remains the same during the following reset period. Consequently, very large powers can be controlled with small reset currents, giving high efficiency control. Remember, after the core has saturated, the inductance is near zero, and no further energy is being stored in the core as the output current increases. (In other words the incremental $\frac{1}{2}LI^2$ tends to zero as L tends to zero in the saturated state.)

21.8 CURRENT LIMITING THE SATURABLE REACTOR REGULATOR

Figure 2.21.8 shows a simple current-limiting circuit which operates as follows. When the output current is such that the voltage across RI exceeds 0.6 V, tran-

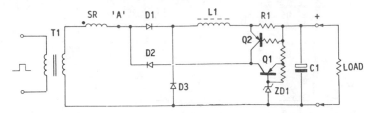

FIG. 2.21.8   Saturable reactor buck regulator with current-limiting circuit R1 and Q2.

sistor Q2 will conduct and provide a reset current via D2 when the secondary voltage goes negative, thus limiting the maximum current. When this type of current limiting is provided, the saturable reactor must be designed to withstand the maximum forward volt-seconds without saturating (after being fully reset on the previous half cycle). Remember, this applies even when the output voltage is zero (e.g., when a short circuit is applied to the output). If the voltage drop in the reset circuit exceeds the forward drop in D1, L1, it may be necessary to tap the reset diode into the SR winding to ensure a full reset when the output is short-circuited.

For high-current applications, the loss in the current-sensing resistor R1 may be unacceptable; in this case, a current transformer should be used in series with D1. The DC current transformer shown in Part 3, Sec. 14.9 is particularly suitable in this application, and would be placed in the DC path, in series with L1.

21.9 PUSH-PULL SATURABLE REACTOR SECONDARY POWER CONTROL CIRCUIT

The discussion so far has been limited to single-ended systems. In such systems, the same time (volt-seconds) is required to reset the core during the "off" period as was applied to the core to set it during the "on" period. Therefore, if control is to be maintained under short-circuit conditions, the duty ratio cannot exceed 50% unless a high-voltage reset circuit is provided or a reset tapping point is provided on the SR winding.

In the push-pull system shown in Fig. 2.21.9, two saturable reactors are used to provide two forward power pulses per cycle. These current pulses are routed via diodes D1 and D2 into the output *LC* filter. Consequently, even for full square-wave input (100% duty ratio), separate alternate saturable reactor paths are in operation for each half cycle, and there is a 50% "off" period for the reset of each reactor providing the required reset volt-seconds.

A single control transistor Q1 may be used to reset both reactors, as the reset current will be automatically routed to the correct reactor by the gating action of diodes D3 and D4 (one diode or the other being reverse-biased during the reset).

Current limiting is provided by Q2, which turns on when the voltage across RI exceeds 0.6 V as a result of an overload current. Once again, current transformers may be used if preferred, but would be positioned in series with D1 or D2 (or in the DC path to L1 in the case of the DC current transformer).

The push-pull technique is recommended for higher output currents, as it will considerably reduce the output ripple filtering requirements.

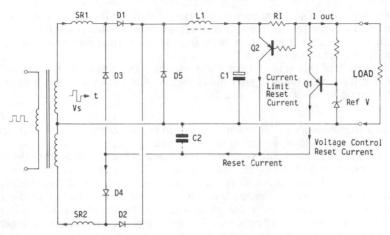

FIG. 2.21.9 Push-pull saturable reactor secondary regulator circuit.

21.10 SOME ADVANTAGES OF THE SATURABLE REACTOR REGULATOR

For low-voltage, high-current secondary outputs, the saturable reactor control is particularly valuable. The "on"-state impedance may be very close to the resistance of the copper winding (a few milliohms in high-current applications). Consequently, the voltage drop across the reactor element will be very low in the "on" state. In the "off" state, with the right core material, the inductance and hence the reactance can be very high, and leakage current (magnetization current) is low. Consequently, very efficient duty ratio power control is possible. The reliability of the saturable reactor is very high, as it may be considered a passive component.

In a multiple-output application, the secondary saturable reactor regulator provides high-efficiency, fully independent voltage and current limit control. Further, all outputs may be isolated if required. The saturable reactor is indeed a powerful control tool.

21.11 SOME LIMITING FACTORS IN SATURABLE REACTOR REGULATORS

The saturable reactor is not a perfect switch. Several obvious limitations, such as maximum and minimum "off" and "on" reactance, have already been mentioned. Some of the less obvious but important limitations will now be considered.

1. Parasitic Reset.   When the voltage applied to the reactor reverses during the reset period, the main rectifier diode D1 must block (turn off). During this blocking period, there will be a reverse recovery current flowing in the diode. This reverse current flows in the reactor winding and applies an unwanted reset action to the core; hence, the core is reset to a point beyond the normal remanence value B_r (even when reset is not required).

As a result of this spurious reset action, the minimum turn-on delay is increased, reducing the range of control. Therefore, fast diodes with low recovered charge should be chosen for D1 and D2.

2. Postsaturation Permeability.   The permeability in the saturated state is never zero. At best it will be at least that of an air-cored coil.

At very large currents, the saturated inductance of the core may limit the current let-through to such an extent that full output cannot be obtained. If this is a problem, then a larger core with fewer turns should be used. Remember, L is proportional to N^2, but B is proportional to N/A_e and a net reduction in the saturated inductance is obtained at the same working flux density by using a larger core with fewer turns. Ferrite cores, with their larger postsaturation permeability, are more prone to this problem.

21.12   THE CASE FOR CONSTANT-VOLTAGE OR CONSTANT-CURRENT RESET (HIGH-FREQUENCY INSTABILITY CONSIDERATIONS)

At high frequencies the area of the B/H loop increases, giving an increased core loss and a general degradation of the desirable magnetic properties.

In particular, some materials show a modification of the B/H loop to a pronounced S-shaped characteristic. This S shape can lead to instability if constant-current resetting is used in the control circuit. This effect is best understood by considering Fig. 2.21.10.

If constant-current reset is used, then the magnetizing force H is the controlled parameter. As H is being incremented from zero to H2 during reset, an indeterminate range is entered between H1 and H2, in which $+B$ will "flip" to $-B$ because of the negative slope of the B/H characteristic between H2 and H1, and progressive control is lost.

In practice, a very large compliance voltage from the constant reset circuit would be required to change B from $+B1$ to $-B2$ rapidly. Therefore, some measure of control is normally retained even with constant-current reset, since most real constant-current circuits have a limited compliance voltage and the control circuit reverts to a voltage-limited state during this part of the reset. However, this may not be a well-defined action.

For this reason, better stability will be obtained if controlled volt-seconds reset is used rather than constant-current reset (particularly at light loads). Voltage reset will increment the flux density B rather than the magnetization force H. This is more controllable in the negative-slope region shown in Fig. 2.21.10.

For voltage reset, a fast-slew-rate, low-output-impedance voltage-controlled amplifier is preferred. The decoupling capacitor shown in Fig. 2.21.8 converts the current control of Q1 to virtual voltage control at high frequencies, but degrades the transient response and is a compromise solution only.

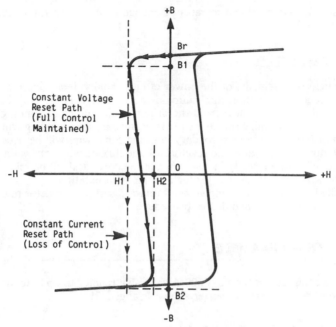

FIG. 2.21.10 High-frequency pulse magnetization B/H loop, showing S-shaped B/H characteristic.

21.13 SATURABLE REACTOR DESIGN

The most difficult design decision is the selection of core material and core size. As previously discussed, this depends on the application, frequency, and required performance. However, once the core selection has been made, the rest of the design procedure is relatively straightforward.

21.13.1 Core Material

The choice of core material is normally a compromise between cost and performance. At low frequencies, there are many suitable materials, and the controlling factors will be squareness ratio, saturation flux density, cost, and core losses. Also, at low frequencies, core losses are less important, giving a wider selection. At medium frequencies, up to, say, 35 kHz, the core loss starts to be the predominant factor, and Permalloys, square ferrites, or amorphous materials will be chosen. At high frequencies, above 50 kHz, the core loss tends to become excessive, and the parameters of the cores rapidly degrade. Ferrite materials are probably the best choice. (The author's experience is limited to frequencies below 50 kHz at this time.)

For very high frequency operation, more than 75 kHz, better results may be found using sine-wave converters and true magnetic amplifier techniques. Sine-wave operation extends the useful frequency range of the material. (Many of the

/core losses are proportional to rates of change of induction rather than to frequency per se.)

21.13.2   Core Size

The practical requirements of the power circuit usually determine the core size. In low-voltage, high-current applications, the reactor winding may be three or four turns of large-gauge wire, and the practical difficulties of winding this wire on the core will determine the core size. In many cases the winding will be a continuation of the transformer secondary, and the same wire will be used. To minimize the turns on the saturable reactor, a large flux density excursion, typically 300 to 500 mT, will normally be used; hence the core loss will be relatively large compared with the copper loss for high-frequency operation.

A typical example of the single-ended forward saturable reactor regulator will be used to demonstrate one design procedure.

21.14   DESIGN EXAMPLE

Consider a requirement for a 5-V, 20-A saturable reactor to operate at 35 kHz in the single-ended forward converter shown in Fig. 2.21.4.

21.14.1   Step 1, Select Core Material

From Table 2.21.1, suitable materials would be Permalloy, square ferrite, or Vitrovac 6025. In this example, assume that cost is less important than performance, so that the best overall material, Vitrovac 6025 amorphous material, will be used.

21.14.2   Step 2, Calculate the Minimum Secondary Voltage Required from the Converter Transformer

The maximum "on" time is 50% of the total period, or 14.3 μs at 35 kHz. When the SR is fitted, there will be an unavoidable minimum delay on the leading edge of the "on" pulse, as a result of the time required to take the core from B_r to B_{sat}, even when the reset current is zero. Previous experience with the 6025 material using fast diodes indicates that this delay will typically be 1.3 μs. (The actual value can be calculated when the turns, core size, and secondary voltage have been established.) Therefore, the usable "on" period will be approximately 13 μs. The minimum secondary voltage required from the converter transformer, to develop the required output voltage can now be calculated:

$$V_s = \frac{V_{out}(t_{on} + t_{off})}{t_{on}} = \frac{5(13 + 15.6)}{13} = 11 \text{ V}$$

21.14.3   Step 3, Select Core Size and Turns

In this example, it will be assumed that the transformer secondary has been brought out as a flying lead, and that this wire is to be wound on to the saturable reactor core to form the winding. The core size is to be such that the winding will just fill the center hole. Further, it will be assumed that the wire size on the transformer secondary was selected for a current density of 310 A/cm^2 and that 10 wires of 19 AWG were used. Assuming a packing factor of 80%, the area required for each turn will be 19.5 mm^2.

The next step is an iterative process to find the optimum core size. The larger the core size, the smaller the number of turns required, but the larger the center hole size.

Consider a standard Series 2 toroid, size 25-15-10. From the manufacturer's data, this toroid has a core area of 50 mm^2 and a center hole area of 176.6 mm^2. The turns required on this core (if the flux density change is to be 500 mT and the core is to control to full pulse width) may be calculated as follows:

$$N = \frac{V_s \times t_{on}}{\Delta B \times A_e} = \frac{11 \times 14.3}{0.5 \times 50} = 6 \text{ turns}$$

where V_s = secondary voltage
t_{on} = time that forward voltage is applied, μs
ΔB = flux density change, T
A_e = core area, mm^2

The area required for six turns of 10 × 19 AWG at 80% packing density is 117 mm^2, and this will just fit the core center hole size.

At higher input voltages, the flux density excursion will be larger, but as the core can support a total change of 1.8 T (+ $\hat{B}$ to − $\hat{B}$), there is an adequate flux density margin.

21-14-4 Step 4. Calculate Temperature Rise

The temperature rise depends on the core and winding losses and the effective surface area of the wound core. The core loss of Vitrovac 6025 at 35 kHz and 500 mT is approximately 150 W/kg. The weight of the 25 × 15 × 10 core is 17 g, so the core loss is 2.5 W. The copper loss is more difficult to predict, as an allowance must be made for the increase in effective resistance of the wire as a result of skin effect. With a multiple-filament winding of this type, the F_r ratio (ratio of DC resistance to effective ac resistance) is approximately 1.2, giving a winding resistance of 0.0012 Ω and a copper loss (I^2R loss) of 0.48 W. Hence, the total loss is approximately 3 W. The surface area of the wound core is approximately 40 cm^2, and from Fig. 3.1.9, the temperature rise will be 55°C. Since much of the heat will be conducted away by the thick connection leads, the actual rise will normally be less than this.

21.15 PROBLEMS

1. Explain the basic principle of the saturable reactor regulator.
2. What are the desirable core properties for saturable reactors?
3. How does the saturable reactor delay the transmission of the leading edge of a secondary current pulse?
4. How is the saturation delay period adjusted?

5. Why is the saturable reactor particularly suitable for controlling high-current outputs?

6. Why is constant-voltage reset preferred to constant-current reset in high-frequency saturable reactor regulators?

7. Why are fast secondary rectifier diodes recommended for saturable reactor regulators?

CHAPTER 22
CONSTANT-CURRENT POWER SUPPLIES

22.1 INTRODUCTION

Most engineers will be very familiar with the general performance parameters of constant-voltage power supplies. They will recognize that these power supplies have a limited power capability, normally with fixed output voltages and some form of current- or power-limited protection. For example, a 10-V 10-A power supply would be expected to deliver from zero to 10 A at a constant output voltage of 10 V. Should the load current try to exceed 10 A, the supply would be expected to limit the current, with either a constant or a foldback characteristic. The well-known output characteristics of one such supply are shown in Fig. 2.22.1.

22.2 CONSTANT-VOLTAGE SUPPLIES

From Fig. 2.22.1, the output characteristics of the constant-voltage supply will be recognized. The normal working range for the constant-voltage supply will be for load resistances from infinity (open circuit) to 1 Ω. In this range, the load current is 10 A or less. The voltage is maintained constant at 10 V in this "working range."

At load resistances of less than 1 Ω, the current-limited area of operation will be entered. In a constant-voltage supply, this is recognized as an overload condition. The output voltage will be decreasing toward zero as the load resistance moves toward zero (a short circuit). The output current is limited to some safe maximum value, but since this is normally considered a nonworking area, the characteristics of the current limit are not very closely specified.

22.3 CONSTANT-CURRENT SUPPLIES

The constant-current supply is not so well known, and therefore the concept can be a little more difficult to grasp. In the constant-current supply, the previous

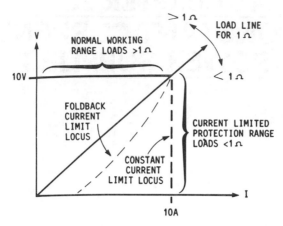

FIG. 2.22.1 Output characteristics of a constant-voltage power supply, showing constant-current and reentrant-current protection locus.

constant-voltage characteristics are reversed. Figure 2.22.2 shows the output characteristics of a typical constant-current supply.

It should be noted that the controlled parameter (vertical scale) is now the output current, and the dependent variable is the compliance voltage. The normal "working range" is now from zero ohms (short circuit) to 1 Ω, and in this load range the output *current* is maintained constant.

At load resistances in excess of 1 Ω, a compliance-voltage-limited protection area is entered. For the constant-current supply, this voltage-limited area would be considered an overvoltage condition. Since this is normally a nonworking protection area, the output *voltage* may not be well specified in this area.

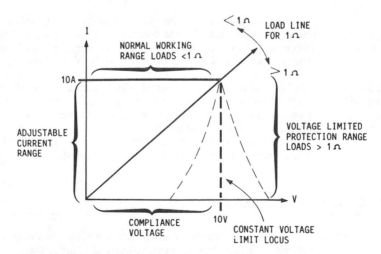

FIG. 2.22.2 Output characteristics of a constant-current power supply, showing constant-voltage compliance limits.

22.4 COMPLIANCE VOLTAGE

The terms used to describe the operation of a constant-current supply are somewhat less familiar than those used for the constant-voltage supply. For a variable constant-current unit, the output current may be adjusted, normally from near zero to some maximum value (simply described as the constant-current range).

To maintain the load current constant, the output terminal voltage must change in response to load resistance changes. The terminal voltage range over which the output current will be maintained constant is called the "compliance voltage." This compliance voltage usually has a defined maximum value.

In the example shown in Fig. 2.22.2, the compliance voltage is 10 V, and a constant current of 10 A will be maintained into a load resistance ranging from zero to 1 Ω.

Constant-current supplies have limited applications. They will be used where currents must be maintained constant over a limited range of variations of the load resistance. Typical examples would be deflection and focusing coils for electron microscopes and gas spectrometry.

Figure 2.22.3 shows the basic circuit for a constant-current linear supply. In this example, a voltage-controlled current source is shown. This is an important concept, not previously introduced. Just as constant-current supplies can be configured from voltage-controlled current sources, so can constant-voltage supplies be configured from current-controlled voltage sources. This concept has important implications for current sharing, when constant-voltage supplies are to be operated in parallel.

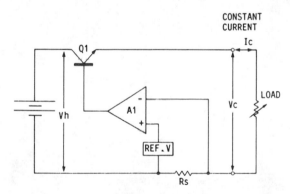

FIG. 2.22.3 Example of a constant-current linear supply (basic circuit).

In this example, the load current returns to the supply via the low-value series resistor R_s. The current analogue voltage developed across this resistor is compared with the internal reference voltage by amplifier A1, and the series regulator transistor Q1 is adjusted to maintain the voltage across R_s constant. Thus the current in R_s will be maintained constant, and provided that the amplifier input current is negligible, the load current will also be maintained constant, irrespective of load resistance, within the "compliance voltage range."

It should be noted from Fig. 2.22.3 that as the load resistance increases, the voltage across the output terminals V_c increases to approach the supply voltage V_h. When $V_t = V_h$, Q1 will have been fully saturated and has no further control.

Beyond this point, the current must start to fall, and the output voltage will be defined by the characteristics of the header supply V_h, which is not regulated and hence is not well specified in this example.

Although it is possible to reconfigure a constant-voltage supply to give constant-current performance, this is not recommended. To provide maximum efficiency and high performance, the constant-current supply will have a very low reference voltage (typically less than 100 mV), the internal current shunt must be highly stable, and internal current paths must be well defined.

22.5 PROBLEMS

1. How do the general performance parameters of a constant-current power supply differ from those of a constant-voltage power supply?
2. What is the meaning of the term "compliance voltage" in a constant-current supply?
3. What would be considered an overload condition for a constant-current supply? Compare this with a constant-voltage supply.
4. Why is the output ripple and noise voltage a meaningless parameter for a constant-current supply?
5. How should output ripple and noise be defined in a constant-current supply?

CHAPTER 23
VARIABLE LINEAR POWER SUPPLIES

23.1 INTRODUCTION

The variable linear power supply, although perhaps somewhat out of place in a switching power supply book, has been included here for several reasons.

First of all, when very low output noise is required, the linear regulator is still the best technique available. Also, the "cascaded" linear system described here is a very useful and somewhat neglected technique. Finally, the high dissipation and low efficiency of the dissipative linear regulator serve to illustrate the advantages of the switchmode variable supply, described in the next chapter.

In this section, we review the basic concepts of a linear variable supply for laboratory applications. The same general principles will apply to fixed-voltage linear regulators, except that for the latter the losses would normally be much lower.

To its advantage, the linear regulator has inherently low noise levels, usually measured in microvolts rather than the more familiar millivolts of switchmode systems. For applications in which the minimum electrical noise levels are essential (for example, sensitive communications equipment and research and development activities), the advantage of the very low noise levels of the dissipative linear regulator often outweighs the wish for maximum efficiency.

The transient response of a well-designed linear system may be of the order of 20 μs for full recovery, rather than the 500 μs for the typical switchmode regulator.

The major disadvantage of the linear regulator is that it must dissipate as heat the power difference between the used output power (volt-amperes) and the internally generated volt-amperes. This dissipation can be very large. It is at a maximum at high output currents and low output voltages.

In the example to be considered here (a 60-V, 2-A variable supply), the unregulated header voltage will be 65 V minimum. When the variable output voltage is set to zero at 2 A load (an output short circuit), a normal series regulator dissipation would be 130 W minimum. If this energy is all concentrated in series linear regulator transistors, then expensive heat sinks and transistors will be required.

The following section describes a method of secondary preregulation which allows the majority of the unwanted energy to be dissipated in passive resistors rather than in the series regulator transistors. Dissipating the energy in resistors has major advantages. It should be remembered that good-quality wirewound resistors are much more efficient at dissipating the unwanted energy, since they may run at much higher surface temperatures than semiconductor devices can. Hence smaller air flow can efficiently carry away the excess heat. Resistors are also much lower in cost than extra regulation transistors. The dissipative resistors may be positioned external to the main power supply case, allowing much smaller units to be built without an excessive internal temperature rise. Finally, less expensive linear regulator transistors and smaller heat sinks may be used.

23.2   BASIC OPERATION (POWER SECTION)

Figure 2.23.1 shows the basic block diagram of the power section of the linear supply. R1 and Q2 form a preregulator to the main linear regulator transistor Q1.

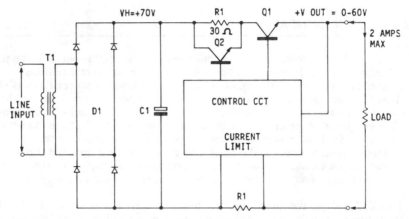

FIG. 2.23.1   Power circuit topology of a basic "piggyback" type linear variable-voltage power supply.

The unregulated DC header voltage V_H is developed from the line input, using a standard 60-Hz isolation and voltage transformer T1 and bridge rectifier D1. In this example, the supply is required to provide an output of 60 V at 2 A maximum, and to provide a margin for regulation and loss, the header voltage is 65 V minimum.

The unregulated header voltage must be large enough to allow for losses in the linear regulator, input voltage variations, and the input ripple voltage—in this example, a minimum of 65 V at the lowest line input of 105 V. At the nominal line input of 115 V, the voltage on C1 will be 70 V DC. Full-wave bridge rectification by D1 makes for a low ripple voltage on C1 and good transformer utility.

23.2.1   Preregulator Operation

The nominal 70-V DC nonregulated header voltage V_H is applied to a network of resistor R1 in parallel with transistor Q2, in series with the normal linear regulator transistor Q1, as shown in Fig. 2.23.1.

When the power supply has been set to give a low output voltage (for example, 0 V at 2 A), transistor Q2 is turned off, and the majority of the applied header voltage will appear across R1. Hence, the maximum dissipation will appear in the resistor, and Q1 is relieved of the high-stress dissipative conditions that would normally apply in a conventional series regulator.

When the output voltage is set to a high value (for example, 60 V, 2 A), transistor Q2 will be turned hard on, applying a short circuit across R1. The difference between the header voltage and the required output voltage will now appear across transistor Q1, but since this voltage is now only 10 V, the dissipation in Q1 is only 20 W.

Between these two extreme conditions, Q2 will conduct current in such a way that the transistors and resistor take up various proportions of the total stress. By the correct selection of resistor values, Q1 working voltage, and drive design, the maximum stress on Q1 and Q2 can be limited to less than 41 W rather than the 140-W stress that would have occurred in the conventional series regulator.

23.3 DRIVE CIRCUIT

Figure 2.23.2 shows the basic elements of the drive circuit for the cascaded power sections.

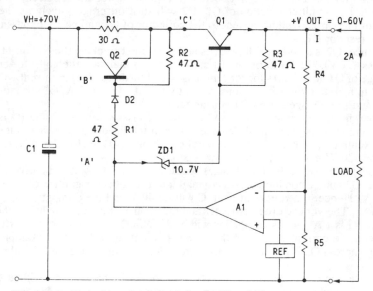

FIG. 2.23.2 Basic drive circuit for "piggyback" variable power supply.

The operation is best understood by considering four extreme operating conditions. These will be

1. Low-output-voltage, high-current conditions
2. High-output-voltage, high-current conditions
3. Intermediate-voltage, high-current conditions (say, 30 V, 2 A)
4. Low-output-voltage, intermediate-current conditions

First, consider the condition in which the output voltage has been set at zero and the load current is at a maximum of 2 A. Assume nominal line voltage and a header voltage of 70 V. Under this condition, transistor Q2 is turned off, and the output current path is via R1 and transistor Q1 to the load. Since the current is 2 A, the voltage drop across R1 will be 60 V and its dissipation will be 120 W. Transistor Q1 holds off the remainder of the voltage, i.e., 10 V, and the dissipation in Q1 is only 20 W. Q1 will maintain the output voltage near zero under the control of amplifier A1.

The drive conditions for Q1 and Q2 are as follows: Q1 is conducting, and a base drive current is flowing in the zener diode ZD1 and the base-emitter of Q1. The voltage at point A is thus $V_{be(Q1)} + V_{ZD1}$ (in this example, 11.4 V). The voltage at point B is at least a diode drop less, or 10.7 V in this example. Since the voltage at point C, the collector voltage of Q1, is 10 V, Q2 will be close to conduction but still "off." Hence, the assumed initial conditions are satisfied.

Consider now the second condition, in which the output voltage is 60 V and the load current is increased from zero to 2 A.

With the output voltage set to 60 V, as the load current is increased from zero toward 2 A, the voltage across R1 will increase and the collector-emitter voltage of Q1 will decrease.

When the collector-emitter voltage of Q1 drops below 10 V, diode D2 and the base-emitter junction of Q2 will become forward-biased, and a current will flow into the base-emitter junction of Q2. Q2 will turn on progressively as the load current increases, feeding current to the collector of Q1 and partly bypassing the series resistor R1.

Hence, Q2 is a voltage follower, its emitter output tracking the voltage at point A (less 1.4 V). Q2 will turn on just sufficiently to maintain the collector-emitter voltage of Q1 at 10 V. As the load increases still further, a larger current will flow in Q2 to maintain these conditions, so that when the load is 2 A, Q2 is conducting fully and the voltage across R1 is near zero.

Therefore, for an output of 60 V at 2 A, the dissipation in Q2 and R1 is near zero, and Q1 dissipates the difference between the header and output voltage (in this example, 2 A at 10 V). Hence the dissipation in Q1 is 20 W, relatively small for the 120-W maximum output power conditions.

Consider now the third (midrange) operating condition. With an output of 30 V at 2 A, Q2 will be conducting so as to maintain the collector-emitter voltage of Q1 at 10 V. Hence, the voltage at point C will be 40 V and the dissipation in Q1 will be 20 W. The voltage drop across R1 and Q2 is 30 V. For this condition, the current in R1 is 1 A, and its dissipation will be 30 W. Q2 must conduct the difference between the current in R1 and the output current, or 1 A in this example. The voltage across Q2 is the same as that across R1, and the power in Q2 will also be 30 W.

Hence, the power distribution in Q1, Q2, and R1 in this example will be 20, 30, and 30 W, respectively, a fairly even distribution.

Consider the final example, when the output voltage is zero and the current is to be reduced below 2 A. Let the initial conditions be the same as in the first example. The current is 2 A, the voltage across R1 is 60 V, the voltage on the collector of Q1 (point C) is 10 V, and Q2 is just turned off.

As the current is reduced, the voltage across R1 is reduced and the voltage across Q1 is increased, taking point C more positive. Hence, for these output voltage conditions, Q2 will remain "off" for all currents of 2 A or less. When the load current is 1 A, the voltage drop across R1 will be 30 V, and the dissipation in Q1 will be 40 W.

23.4 MAXIMUM TRANSISTOR DISSIPATION

It is clear from the preceding that the distribution of losses depends on loading and output voltage, and the maximum dissipation condition is not easily seen. However, the maximum loss in Q1 may be established as follows:

The power loss in Q1 is given by the following equation:

$$P_{Q1} = (V_H - I \cdot R1) \times I$$

or
$$P_{Q1} = V_H \cdot I - R1 \cdot I^2$$

where P_{Q1} = power in transistor Q1
V_H = header voltage (70 V)
R1 = series resistor value (30 Ω)
I = output current

Therefore, in this example,

$$P_{Q1} = (70 \times I) - (30 \times I^2)$$

This equation will have a maximum value when the first differential is zero. Differentiating,

$$\frac{d(P_{Q1})}{dI} = 70 - (60 \times I) \qquad \text{which} = 0 \text{ when } I = 1.166 \text{ A}$$

Hence the maximum dissipation in Q1 will be given when the load current is 1.166 A.

From the preceding equation, the loss in Q1 when $I = 1.166$ A is 40.83 W.

Therefore, the majority of the power dissipation requirements are dealt with by resistor R1. The regulator transistors are subjected to considerably less stress than would be the case if the complete supply voltage and current were to be presented to a single series device, when the dissipation could be 140 W maximum.

23.5 DISTRIBUTION OF POWER LOSSES

Figure 2.23.3 shows how the power losses are distributed between the two power transistors Q1 and Q2 and the series resistor R1 over the output voltage range for the maximum output current of 2 A.

Note that the peak power conditions for transistors Q1 and Q2 occur at different voltages and that both devices can be mounted on the same heat sink. This need be rated only for the worst-case combination, which never exceeds 41 W. This is considerably lower than the 140 W that would have been dissipated if the preregulator had not been used.

23.6 VOLTAGE CONTROL AND CURRENT LIMIT CIRCUIT

Laboratory variable supplies are usually designed to provide constant-voltage or constant-current performance with automatic crossover between the two modes.

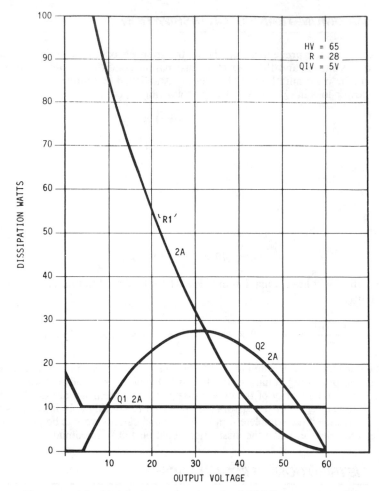

FIG. 2.23.3   Distribution of power loss in "piggyback" linear power supply.

Figure 2.23.4 shows a typical output characteristic with the supply set for 30 V and 1 A. Load lines for 60 Ω, 15 Ω, and the critical value 30 Ω are shown.

It will be seen that the mode of operation depends on the values of output voltage and current which have been selected on the supply controls, and on the load resistance applied to the output terminals. When the load resistance is higher than the critical value R_x, then the power supply will be operating in its constant-voltage mode, range A, and when the resistance is lower than the critical value, the supply will be in its constant-current mode, range B.

When the supply is used in the constant-voltage mode (the usual operating mode), the adjustable current control is used to set the overload current limit, normally for protection of the supply and external load. When the supply is operating in the constant-current mode, the voltage control sets the compliance voltage. (This voltage setting defines the external load resistance at which the

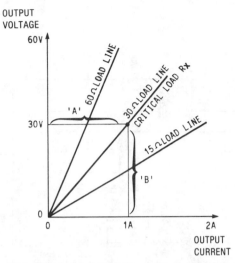

FIG. 2.23.4　Load lines for constant-voltage/constant-current "piggyback" linear power supply.

unit reverts from the constant-current to the constant-voltage mode.) The control circuit must give a well-defined performance for both modes of operation, with automatic and stable transition from one mode to the other (hence, automatic crossover).

23.7 CONTROL CIRCUIT

A suitable control circuit is shown in Fig. 2.23.5. Amplifier A1 provides the voltage control, and amplifier A2, the constant-current control.

The method of operation is best explained by considering the conditions when a 60-Ω load is applied to the output with the controls set as specified above.

Under these conditions, the supply is in the constant-voltage mode, and amplifier A1 will compare the internal reference voltage (TL 431) with the voltage developed across the divider network R15, RV2. Amplifier A1 will respond to any tendency for the output voltage to change by adjusting the drive current through diode D5 and transistors Q4 and Q3 to the power regulators, so as to maintain the output voltage constant at, say, 30 V.

At the same time, the current control amplifier A2 will compare the voltage developed across the current shunt R2 (a function of the output current) with the current reference voltage on RV1. In this example, the current is less than 1 A and the voltage across R2 is smaller than the current reference voltage developed across RV1, and so pin 2 of amplifier A2 will be low. Hence, the A2 amplifier output will be high, with output diode D4 reverse-biased and not conducting. Consequently, for this particular loading condition, the voltage control amplifier A1 defines the conditions of operation, and the supply is operating in range A.

Now assume that the load resistance is changed from 60 Ω to 15 Ω. The unit

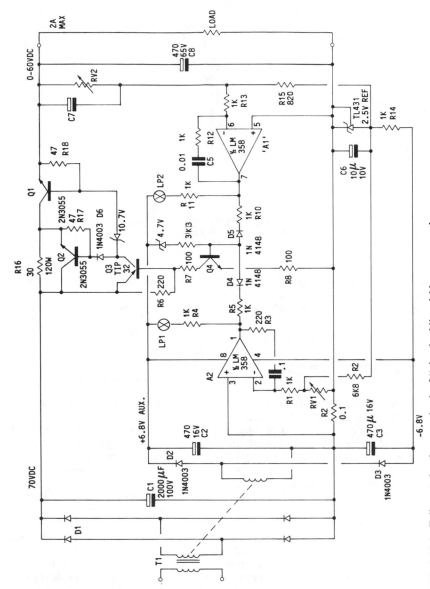

FIG. 2.23.5 Full control and power circuit of "piggyback" variable power supply.

will now be operating in a constant-current mode, delivering 1 A and developing 15 V across the 15-Ω load. The action is as follows.

Amplifier A2 compares the voltage generated across the series shunt resistor R2 (the voltage analogue of the output current) with the current reference voltage from the current control potentiometer RV1.

Amplifier A2 responds to any tendency for the current in the shunt R2 to change by adjusting the drive current through diode D4, transistors Q4 and Q3, and the output transistors so as to maintain the output current constant.

In constant-current operation, the output voltage is lower than the voltage defined by the voltage control amplifier A1. For this condition of operation, A1 will have its inverting input (pin 6) low, and its output voltage will be high. Diode D5 is reverse-biased, and the power supply is operating entirely under the control of the current control amplifier A2.

The transition from constant-voltage to constant-current control takes place at the critical load resistance R_x (30 Ω in this example). This transition is very sharp because of the high DC gain of the control amplifiers. Only a 1- or 2-mV change is necessary to turn the amplifiers from off to on. The gating diodes D5 and D4 ensure that only one amplifier is in control at any time.

Both the constant-voltage and constant-current boundaries of Fig. 2.20.4 may be changed by adjusting the appropriate variable control. The value of the critical load resistance R_x will, of course, change accordingly.

Two light-emitting diodes (LEDs) LP1 and LP2 indicate the operating mode. When the supply is operating very close to the critical crossover point, both indicator lights will be on, showing an indeterminate state of operation. This area should be avoided by adjusting the appropriate control to bring the unit into a defined mode.

23.8 PROBLEMS

1. The linear-regulator-type power supply can be very inefficient when used as a variable supply. Why is this?

2. What is the advantage of the cascade (piggyback) variable linear regulator technique?

3. What are the major advantages of the linear variable regulator compared with a switching regulator?

CHAPTER 24
SWITCHMODE VARIABLE POWER SUPPLIES

24.1 INTRODUCTION

There are many types of variable linear laboratory power supplies. Typically, they will have output voltages adjustable from zero to some maximum value and can be used in either constant-voltage or constant-current mode. However, whatever the form these supplies take, conventional variable linear supplies will have three things in common: they are large, heavy, and inefficient.

At low output voltages, the majority of the rated power must be wastefully dissipated within the power supply, and conversion efficiencies will be very low.

The use of switching regulator techniques can eliminate the extremely lossy operation and will reduce size and weight.

By using the "flyback" or buck-boost-derived switchmode converter technique, a further major advantage can be obtained. A single switchmode supply can be made to provide the voltages and currents of up to three separate variable linear supplies.

With dissipative linear supplies, it is normal practice to provide several different models to cover a range of voltages and currents, even when the rated output power is the same. For example, a 300-W linear unit may be rated at, say, 10 V, 30 A; 30 V, 10 A; or 60 V, 5 A, in three separate units. When the 30-V supply is set for an output of 10 V, it will still only deliver a maximum of 10 A; it is then operating at only 100 W output, the remainder being dissipated in the supply.

Consequently, to cover a range of currents at the same power level, several different linear variable power supplies are required.

Using the flyback switchmode technique, it is possible, in a single unit, to provide nearly constant output power over the majority of the output voltage and current range. Consequently, a single switchmode supply designed for a 300-W capability will cover the range of the previous three linear supplies. The typical output characteristic for such a supply is shown in Fig. 2.24.1.

A switchmode variable supply of this type is much more versatile than the earlier variable linear supplies and will be considerably smaller and lighter. The efficiency will be maintained near 70% over the complete current and voltage range.

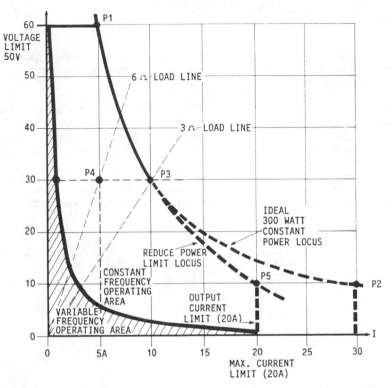

FIG. 2.24.1 Output characteristic and load lines for constant-power-rating, variable switchmode power supply (VSMPS).

24.2 *VARIABLE SWITCHMODE TECHNIQUES*

There are a number of difficulties in the design of variable switchmode supplies. Probably the most obvious is the potential for relatively large conducted and radiated electrical noise. Laboratory-grade power supplies are often used for development applications which will be sensitive to electrical noise and ripple. Hence, good ripple rejection and noise filtering must be provided; filtering cannot be too good in this application. Faraday screens for switching sections and filters are essential, and metal or screened cabinets are used to minimize EMI problems.

The requirement of operating down to zero output voltage also becomes a problem in switchmode supplies, since this demands very narrow power pulses, and it becomes difficult to control the required output power. This limitation can be overcome by using fast FET switching and a different method of control for low output voltages and currents.

24.3 SPECIAL PROPERTIES OF FLYBACK CONVERTERS

An investigation of the flyback technique (see Chap. 1) reveals a very useful property of flyback converters: The energy storage cycle and the energy transfer cycle may be considered entirely independent operations.

Consider the simple diagonal half-bridge flyback power section shown in Fig. 2.24.2. During the period when both FETs are "on," energy is being stored in the transformer magnetic field. Since the secondary is not conducting when the FETs are "on," the transformer may be considered a single-winding inductor in which energy is being stored. At the end of an "on" period, energy of $\frac{1}{2} L_p \cdot I_p^2$ is stored in the core.

At the end of the storage cycle (when the FETs turn off), the primary stops conducting, and the transformer acts as a single-winding inductor, with the secondary winding and output circuit now setting the operating conditions.

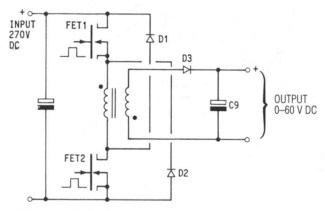

FIG. 2.24.2 Basic diagonal half-bridge power section of a typical flyback variable SMPS.

Since these two actions are independent, there is no direct connection between the operating conditions of the primary and those of the secondary (except that the power must be conserved). The secondary output voltage and current are not defined by the input, as would be the case with a normal transformer, and the output power may be taken as high-voltage, low-current or low-voltage, high-current, or in fact any combination of the two, provided that the power conservation requirements are satisfied.

A less obvious feature is that provided that the complete energy transfer mode is maintained (discontinuous-mode operation), the secondary turns also have no direct bearing on the output voltage and current.

In simple terms, for a defined primary operating condition, the complete energy transfer flyback converter transformer can be considered a constant-energy source.

24.4 OPERATING PRINCIPLES

Consider once again the flyback power section shown in Fig. 2.24.2. Assume energy has been stored in the main transformer T1 during the "on" period. It will now be shown that this energy may be transferred to the load as any combination of current and voltage, provided that the power conservation criteria are satisfied.

The first requirement that must be satisfied (if the core is to continue to operate at a defined flux density under steady-state conditions) is that the equality of the forward and flyback volt-seconds (per turn) must also be satisfied. In terms of flux density, this means that the forward and flyback ΔB must equate, to prevent core saturation.

It can be demonstrated that if a load is applied to the output of a fixed-frequency, defined-duty-ratio flyback converter, then the output voltage will adjust over a few cycles, until the conditions of volt-second and power equality are satisfied.

This self-adjusting action holds for a wide range of loads and turns ratios. Limiting values are found where the secondary turns are so low (or output voltages so high) that the flyback voltage as referred to the primary exceeds the supply voltage. At this point the flyback energy will be returned to the supply through D1 and D2, and energy transfer is lost. A further limiting value will be found where the secondary turns are so large (or output voltages so low) that the converter moves into the incomplete energy transfer mode (continuous mode) where the effective DC current component can cause core saturation.

It should also be noted that with a particular fixed load, the load voltage can be adjusted by adjusting the primary input volt-seconds (duty ratio or input voltage). Of course, if the input voltage and frequency are constant, then only the duty ratio needs to be adjusted.

Accepting for a moment the concept that the flyback converter can be made to behave as a constant energy transfer system as described above, providing a wide range of output voltages and currents within the rated power range, a constant-power envelope of voltage against current may be drawn for the maximum 300-W capability of this particular system.

Figure 2.24.1 shows the constant-power envelope. Initial arbitrary limiting values of voltage and current are shown at 60 V and 30 A. In this example these limits are selected to suit the choice of output component ratings. The constant-power envelope is swept out between points P1 and P2 by maintaining the input conditions constant and simply varying the resistance of the load. The plot assumes that complete energy transfer is maintained throughout the range, and since the pulse width and input voltage are maintained constant, the input power will be constant.

The previous assumptions may now be further tested and perhaps better understood by examining a particular operating point, say point P3 on the characteristic shown in Fig. 2.24.1.

This point is on the maximum output power characteristic, at a voltage of 30 V at 10 A, a power of 300 W. Consequently, the input power will need to be adjusted to this power plus some allowance for losses. With the control circuit set for 30-V constant-voltage operation, a second point P4 at lower current will also be considered. This represents an increase in load resistance from 3 to 6 Ω.

When the load resistance is increased to 6 Ω, the control circuit will recognize and respond to the tendency for an increase in output voltage, and will adjust the input power (pulse width) to maintain the voltage across the load nearly constant at 30 V. Hence a new working point P4 will be established at a lower power level of 150 W.

Hence, by selecting a range of loads and output voltages, a family of curves may be established which will sweep out the complete area bounded by the constant-power limit and the maximum voltage and current limits.

It has been assumed, to this point, that the storage period ("on" time) and energy transfer periods (flyback time) have been independent and self-adjusting, and that the complete energy transfer mode has been maintained throughout the different output voltages. To achieve this over a wide range of loads requires that the "on" and flyback periods be small compared with the total period. At the minimum output voltage, the flyback period must not exceed the "off" period. It is difficult to achieve this with fixed-frequency operation.

24.5 PRACTICAL LIMITING FACTORS

1. Minimum Secondary Turns.   As previously mentioned, with the circuit shown in Fig. 2.24.2, the flyback voltage as referred to the primary cannot exceed the input voltage V_{cc}, as diodes D1 and D2 would then conduct the flyback energy back to the supply and it would not be transferred to the output. Assuming that V_{cc} is 300 V, the secondary turns for maximum 60 V output must be at least 20% of the primary turns.

2. Maximum "On" Period.   If the lower-voltage operating point P5 is now considered, the output voltage is only 10 V. Since the secondary turns are constant, if a stable working flux density is to be maintained in the transformer core, the flyback volt-seconds must be constant, and the flyback time must now be six times longer than it was at 60 V.

If the output power and frequency are to remain constant, the maximum "on" period can now be only one-seventh of the total period, and the peak primary current will need to be more than three times greater in amplitude than it would be for the more conventional 50% duty ratio. Hence, the primary efficiency will be degraded, and large power devices would be required. Increasing the number of turns makes this even worse.

24.6 PRACTICAL DESIGN COMPROMISES

In the interest of component economy and optimum efficiency, some design compromises are usually made.

By allowing the unit to revert to incomplete energy transfer at the lower output voltages, and reducing the operating frequency at the lower output powers, the "on" period can be increased, reducing the peak current. Also, by using a control circuit that limits the maximum current, the stress on the switching devices, output rectifiers, and output capacitors is reduced.

However, these techniques also reduce the maximum output power at the lower output voltages. The lower dashed line in Fig. 2.24.1 shows the modified power output characteristics of the compromise example used here. The maximum output current is reduced to 20 A to limit rectifier diode size and output capacitor ripple current requirements.

In the cross-hatched area of Fig. 2.24.1, where the transferred power is 25 W or less, constant-frequency operation would require extremely narrow "on" periods. Better control can be obtained by reducing the frequency at low power. Hence. in the following example, the repetition rate will be reduced at low loads and/or low voltages.

24.7 INITIAL CONDITIONS

The following compromise parameters will be used for this design:

1. Complete energy transfer will be maintained down to 30 V output, giving a full 300 W output to this voltage. Below 30 V, the unit will revert to incomplete energy transfer, resulting in some reduction in the maximum output power, but also a corresponding reduction in the maximum primary current.

2. At 10 V the maximum output current will be limited to 20 A to reduce the stress on the output rectifiers and secondary winding, reducing the maximum power to 200 W.

3. At output powers below 25 W, the minimum pulse width will be fixed, and the repetition rate will be reduced to control the output, giving better control at light loads.

In this supply, power FETs provide considerable advantages. With these devices used as the main switching element, switching times are very short and storage times negligible. Consequently, good switching action can be maintained for very narrow conduction angles.

The transformer design is complicated by the fact that the secondary must deliver both high currents at low voltages and low currents at high voltages. The output rectifiers and capacitors must also be rated for both of these conditions. The modified power characteristic at lower output voltages reduces these problems.

24.8 THE DIAGONAL HALF BRIDGE

Figure 2.24.2 shows the arrangement for the power components. A single-ended diagonal half-bridge, dual-FET flyback primary stage is used. In this arrangement, both switching elements, FET1 and FET2, will be either in their "on" state or in their "off" state simultaneously. The energy recovery diodes D1 and D2 will return any flyback energy stored in the leakage inductance to the supply lines and will also provide voltage clamp protection on the FET switches.

This arrangement is particularly suitable for FET operation, since it prevents voltages in excess of the supply line from appearing across the FET switching devices under any conditions.

Energy will be stored in the transformer during the "on" state of FET1 and FET2. During this period, the output rectifier D3 will be reverse-biased, and the secondary current is zero.

When the FETs are turned off, the voltage across the primary windings will reverse by flyback action, initially bringing diodes D1 and D2 into conduction. At the same time, a secondary emf will be generated which will drive current into the secondary leakage inductance and output diode D3. When the secondary current has been fully established and the energy in the leakage inductance has been returned to the supply line, the primary diodes D1 and D2 will cease conducting and the majority of the stored energy will be transferred to the output capacitor C9 and the output load.

In this example, the complete energy transfer mode is maintained down to 30 V, at which point there is a change to incomplete energy transfer (at full load). Further, the primary current has been limited to 9.5 A so that the output power at point P5 is reduced to 200 W. This still gives an output current of 20 A at 10 V, a very useful increase over the linear regulator limit, which would be only 5 A for the same conditions.

24.9 BLOCK SCHEMATIC DIAGRAM (GENERAL DESCRIPTION)

Figure 2.24.3 is a block schematic of the basic functional elements of the complete variable switchmode supply. The major functions are described below.

Block 1.   Block 1 is the input filter, voltage doubler, and rectification circuit. It converts the 115/230-V ac input to a nominal 300-V DC output to the converter section, block 3. Block 1 also contains current limiting to prevent excessive in-rush current when the system is first switched on, and the voltage doubler option for 115-V operation.

Block 2.   This is the auxiliary supply, where a small 60-Hz transformer provides the power for the various control functions. In addition, this section provides a soft-start action to the pulse-width modulator, block 5.

Block 3.   This is the main power converter. It is a dual-power-FET, diagonal half-bridge flyback converter.

Output transformer T1 delivers the required output power, via rectifier D4 and smoothing capacitor C9, to the external load. Transformer T2 provides drive to the two switching devices, FET1 and FET2. T3 is a current transformer which provides information on the peak primary current to the control circuit in block 5. Output current shunt R6 provides information on the DC secondary current to the current control amplifier A2 in block 6.

Block 4.   This is the main oscillator, operating at a nominal 22 kHz. It provides a clock signal to the pulse-width modulator in block 5. The oscillator normally runs at a fixed frequency, except at very low output powers (<25 W), when a signal from block 5 will increase the ''off'' period. This reduces the effective frequency.

Block 5.   This is the main pulse-width modulator, which controls the duty cycle to the power switches, FET1 and FET2, in response to the control input. It responds to information from one of the two control amplifiers, A1 or A2 in block 6, to maintain the output voltage or current constant. It also provides soft start, primary power limiting, and frequency reduction at loads below 25 W.

Block 6.   This contains the voltage and current control amplifiers. Secondary current limiting is provided by amplifier A2 in response to the voltage analogue of the output current developed across shunt R6 in block 3. Voltage control is pro-

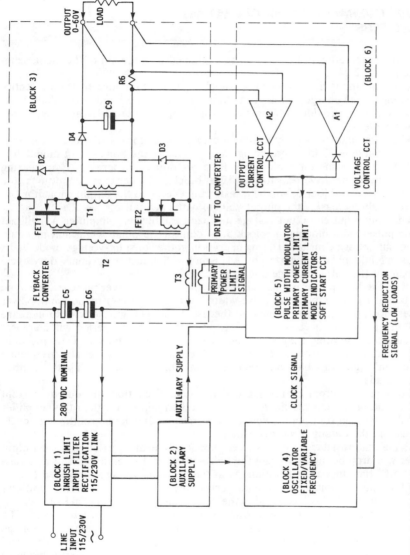

FIG. 2.24.3 Block diagram of variable switchmode power supply.

vided by amplifier A1. Block 6 also contains the mode state indicators and the current-mode-controlled pulse-width control amplifier A3, shown more fully in Fig. 2.24.6.

24.10 OVERALL SYSTEM OPERATING PRINCIPLES

The overall operating principles will be explained with the aid of the block schematic in Fig. 2.24.3.

When the unit is first switched on, block 1 will limit the inrush current, rectify the ac input, and charge up the main storage capacitors C5 and C6 in block 3. At the same time, it provides a filtered 60-Hz ac supply to the 60-Hz auxiliary transformer in block 2.

When the auxiliary supply has been established, block 4 (the oscillator unit) will be running at 22 kHz, providing clock pulses to the pulse-width modulator, block 5. At the same time, block 5 will be providing a progressively increasing pulse width to the drive transformer T2 in block 3. Under these conditions, the converter section, block 3, will be transferring energy to the output capacitor C9, progressively increasing the output voltage with each pulse.

When the required output voltage has been established, amplifier A1 will act on the pulse-width modulator, block 5, to control the pulse width and maintain the output voltage constant. During operation in the constant-voltage mode, variations in load will be recognized by A1, which will adjust the primary duty cycle to maintain the output voltage constant.

When the load power is less than 25 W, the primary power pulse will become constant at 1 μs (the minimum pulse-width limit). Any further reduction in power causes block 6 to provide a signal to the oscillator section, block 4, to increase the "off" period. Hence, for powers below 27 W, the "off" period will increase, reducing the operating frequency and hence the output power. Thus the converter reverts from a fixed-frequency variable-duty-cycle to a variable-frequency (fixed "on" time, variable "off" time) system. This provides much better control at light loads.

When larger output currents are demanded, such that the output current reaches the preset current limit set by R6 and amplifier A2, then this amplifier will take over control from the voltage amplifier A1. It will control the duty cycle to maintain the output *current* constant.

As the load resistance becomes very small, say near short circuit, the output power will drop below 27 W, and the unit will revert to a fixed "on" time, variable "off" time mode of operation, as in the previous voltage-controlled case.

If both output voltage and output current are high, then the power limit amplifier (A3 in block 6) will respond to the peak primary current signal from current transformer T3 to prevent any further increase in input or output power. The power-limited mode has now been entered. Under these conditions, the voltage and current control amplifiers A1 and A2 have no further control of the unit, and the overload mode indicator will be illuminated. This is not a normal mode of operation and is provided to protect the power unit and load.

24.11 INDIVIDUAL BLOCK FUNCTIONS

The various elements of the block schematics in Figs. 2.24.4, 2.24.5, and 2.24.6. will now be considered in more detail.

Figure 2.24.4 shows the internal circuit for the input power section, block 1; the auxiliary supply, block 2; and the power converter, block 3.

Block 1, Input Filter.   In block 1 the ac line input is taken via the supply switch SW1 and fuse FS1 to the input filter inductor L1. Inrush current limiting is provided by thermistor TH1 in series with the rectifier bridge D1. Rectifier D1 provides a DC input to the filter and storage capacitors, C5 and C6, to provide the unregulated DC header voltage to the converter section, block 3.

Block 3, Converter Section.   The converter section, block 3, contains the power switches FET1 and FET2, both of which will be "on" or "off" simultaneously, so that they act as a single switch in series with the flyback transformer primary. They provide energy to the the the switching transformer T1 during the "on" period. This stored energy will be transferred to the storage capacitor C9 by flyback action during the "off" period to provide the required output voltage and current.

High-frequency output filtering is provided by inductor L2 and capacitor C10. In the flyback converter, L2 is not intended for energy storage (as it would be in a forward converter); hence, it is quite small.

Output DC current information is provided by shunt R6. The output voltage and current meters are connected as shown, and dummy load resistor R5 prevents loss of control when the output is "off" load.

To reduce the rate of change of voltage on the power switches FET1 and FET2, and also to reduce RF1 noise, snubber components R3, R4, C7, and C8 are provided.

To prevent voltage overshoot as a result of leakage inductance, clamp diodes D2 and D3 are fitted. These diodes will prevent any voltage stress in excess of the supply voltage appearing across the two switching elements FET1 and FET2 during the flyback period. Energy stored in the primary leakage inductance will be returned to the supply lines by D2 and D3. Finally, a current transformer T3 provides information on peak primary current.

Block 2, Auxiliary Supply.   In block 2, the 60-Hz ac input to the auxiliary supply transformer T4 is taken from the output of the supply filter and inrush limiting thermistor TH1.

The secondary of T4 is rectified by the bridge rectifier D5 and smoothed by capacitor C11 to provide an unregulated 20 V DC to the control circuits. A series connection of resistor R7 and zener diodes ZD1 and ZD2 provides additional regulated 15- and 5-V auxiliary outputs to the control circuits.

Comparator amplifier IC1A gives a soft-start output signal when the auxiliary voltage on the unregulated 20-V line exceeds 19.7 V. This amplifier has an open collector output, so that capacitor C12 charges up relatively slowly via R12 when the amplifier output goes high. This provides the progressively increasing soft-start signal to the pulse-width modulator IC1C in block 5.

Blocks 4 and 5, Oscillator and Pulse-Width Modulator Circuits.   Figure 2.24.5 shows the oscillator and pulse-width control circuits.

In block 4, amplifier IC1b forms a relaxation oscillator which operates in the following way.

Assume that capacitor C15 is discharged to its low state. A reference voltage has been set up on pin 5 of the comparator amplifier IC1B by the divider network

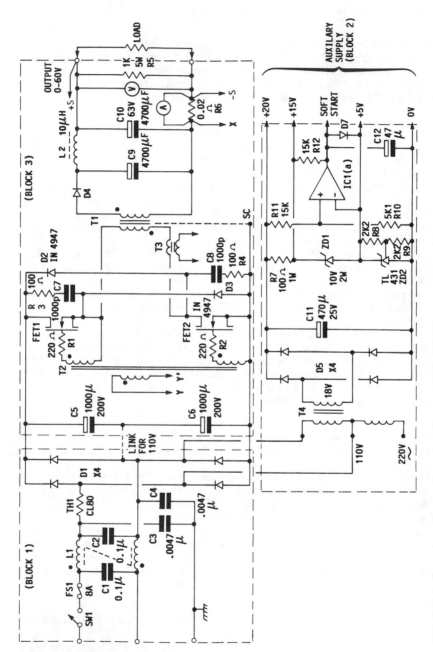

FIG. 2.24.4 Converter power section and auxiliary supply for VSMPS.

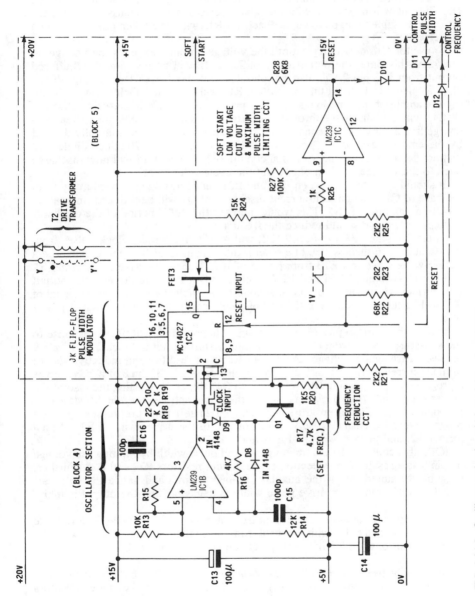

FIG. 2.24.5 Oscillator and pulse-width modulator for VSMPS.

R13, R14. Hence pin 5 will be high and pin 4 low, and the output of the amplifier (pin 2) will be high.

A noninverting buffer amplifier in IC2 provides a high output at pin 2 to diode D9. Current will now be flowing in D9 via resistor R16 to charge up capacitor C15. At the same time, positive feedback via R15 will have taken pin 5 to an even higher voltage.

Capacitor C15 now charges until the voltage on pin 4 exceeds the voltage on pin 5. At this time, the output of amplifier IC1B will go low, and the buffered output from IC2 pin 2 will also go low. This reverse-biases D9, cutting off the charge current to R16. At the same time, R15 will take pin 5 slightly lower, latching the amplifier into the low state; R15 provides a defined hysteresis voltage.

C15 will now discharge through the network D8 and the constant-current discharge transistor Q1. The constant-current value of Q1 is in turn initially defined by the emitter resistor R17 and the divider network R19, R20 (since diode D12 will not be conducting under normal conditions). Hence, C15 will now discharge back to the voltage level on pin 5, and the cycle will repeat.

It should be noted that when D12 and R21 conduct, taking current away from the base of Q1, the constant-current discharge of Q1 will be reduced, increasing the discharge time of C15. This would increase the "off" period of the oscillator and converter, and would reduce the frequency.

In block 5, the "*JK* flip-flop," IC2, will toggle on alternate clock pulses from the oscillator. In the absence of a reset signal on pin 12 of IC2, there would be a square-wave drive to FET3 from pin 15 of IC2. Hence the drive transformer T2 and power switches FET1 and FET2 would deliver a square drive to the output transformer. However, under normal conditions, the square-wave drive signal of IC2 will be terminated before the end of a 50% period by a reset signal at pin 12 of IC2.

During the start-up phase, the reset signal is provided by IC1C in response to the soft-start voltage from C12 in the auxiliary start circuit in block 2. IC1C pin 9 receives a triangular voltage signal from R23 as a result of the inductive rise of primary current in the primary of T2 after turn-on of FET3. Pin 8 of IC1C receives a progressively increasing DC voltage from C12 (block 2), starting the instant the power supply is turned on. As a result of these two actions, the "high" state output signal from IC1C is progressively delayed as soft start progresses, giving a progressively increasing pulse width from IC2 to the drive switch FET3, transformer T2, and power switches FET1 and FET2.

IC1C also controls the maximum permitted pulse width by limiting the voltage on pin 8 as a result of the selection of the divider resistors R24 and R25. Further, it should be noted that if the auxiliary voltage is low, and hence the soft-start signal on C12 is low, the drive pulse width is negligible and the output is inhibited.

Under normal conditions, as the pulse width from IC1C increases during the soft start, a second pulse-width control from the control section via D11, will take over the control of IC2. This control section is shown in Fig. 2.24.6.

Block 6, Control Amplifiers and Pulse-Width Modulator.   Figure 2.24.6 shows the voltage and current limit control amplifiers A1 and A2, together with the mode indicators and minimum pulse-width limiting circuits.

Consider the supply operating in a voltage-controlled mode. The voltage control amplifier A1 [IC2(a)] has a reference voltage of +5 V connected to pin 3. The inverting input, pin 2, is taken to the divider network RV1, R30 and monitors the output voltage at the power supply output terminals.

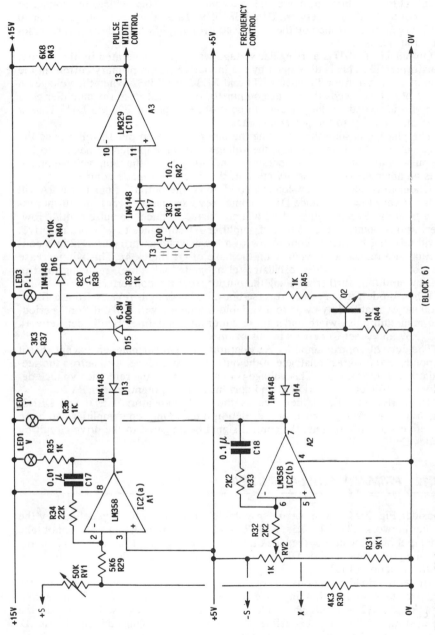

FIG. 2.24.6 Voltage and current control amplifiers for VSMPS.

In response to an increasing output voltage, pin 2 will go high, taking the output of A1 low and bringing diode D13 into conduction. This reduces the voltage at the input to the divider network D16, R38, R39. Thus, the input voltage to pin 10 of IC1D will be controlled by the voltage control amplifier. Diode D14 is reverse-biased at this time.

On pin 11 of IC1D is a triangular voltage waveform, generated by the current transformer T3. This is developed by the inductive rise in primary current in the primary of transformer T1 when FET1 and FET2 are "on." When the voltage on pin 11 of IC1D exceeds the voltage on pin 10, the output from the amplifier goes high and is routed to the pulse-width-control reset pin of IC2 via D11. Thus it turns off the drive to the power switches.

The phasing is such that reducing the input voltage on pin 10 will reduce the drive pulse width. Consequently, the voltage control amplifier A1 has control of the pulse width under normal operating conditions. Since the ramp voltage on pin 11 is an analogue of the primary current, this is current-mode control.

Under low-voltage or low-loading conditions, the output of pin 1 of IC2A will be taken very low until diode D16 becomes reverse-biased; hence any further reduction in the output voltage of A1 will not further reduce the pulse width. However, at this point, as the voltage of amplifier A1 continues to go more negative, it will take the frequency control line to a lower voltage, bringing D12 into conduction and reducing the voltage on the base of Q1 (block 4). This will increase the "off" time of the drive oscillator (at a fixed "on" time), reducing the frequency, and thus further reducing the output from the converter.

When diode D16 becomes reverse-biased, the system reverts from a fixed-frequency variable-duty-cycle to a variable-frequency with a fixed "on" period. The minimum pulse width under these conditions is defined by divider network R40, R39 and is set to be 1 μs in this example.

The current control amplifier A2 operates in a similar manner, except that it responds to the voltage analogue of the output current, developed across the secondary current shunt R6. This voltage is compared with the reference voltage developed by divider network R31 and current control potentiometer RV2.

The active amplifier is identified by the mode indicators, LED1 and LED2, which indicate voltage or current regulation mode. Only one amplifier can be in an active state at any time, the control signal being gated to the drive circuit by diodes D13 or D14.

24.12 PRIMARY POWER LIMITING

Consider Fig. 2.24.6. The maximum voltage on pin 10 of IC1D is defined by the divider network R37, D16, R38, R39, R40. (The amplifiers A1 and A2 cannot take D16 input high because of the blocking diodes D13 and D14.) A voltage clamping action on the input of diode D16 is provided by the zener diode D15 and the base-emitter junction of Q2.

Under overload conditions, the outputs of A1 and A2 are both higher than D15 clamp voltage, and the limiting condition is entered. The outputs of both amplifiers A1 and A2 will be high, diodes D13 and D14 are reverse-biased, and the peak primary current in T1 is defined by the maximum voltage on pin 10 of IC1D. The unit will be in a power-limited state.

This overload condition is indicated by mode indicator LED3, which will be turned on as a result of the clamp current from ZD15 flowing in the base-emitter of Q2 turning this transistor on.

24.13 CONCLUSIONS

This completes the basic circuit description. In general, it has been shown that the switchmode variable supply consists of a flyback converter of special design operating in a constant energy transfer mode. The energy input is adjusted by the control circuit to maintain the output voltage or current constant.

For practical reasons, the operating mode is modified to a variable-frequency system at low voltages or low loads. To meet these needs, the control circuit must provide a number of extra functions. In particular, it reverts to a variable mode of operation at light loads to maintain complete control. Also, the maximum primary current is limited, preventing primary overloading.

The use of the flyback complete energy transfer mode of operation provides a trade-off between output voltage and current to provide constant-power ability over a wide operating range.

Transformer design for the switchmode variable power supply is covered in Chap. 25.

CHAPTER 25
SWITCHMODE VARIABLE POWER SUPPLY TRANSFORMER DESIGN

25.1 DESIGN STEPS

25.1.1 General Considerations

In the flyback converter design in Chap. 24, the flyback voltage (as referred to the primary winding) cannot exceed the supply voltage, as diodes D2 and D3 would then return the flyback energy to the primary circuit, and it would not be transmitted to the output.

The need to satisfy the above requirement defines the turns ratio that must be used at the maximum output voltage of 60 V. Further, since the output voltage covers a wide range (from 0 to 60 V), and the secondary turns cannot be changed when the output voltage is changed, the secondary winding must be capable of providing the maximum output current (20 A in this example). Hence, to minimize copper losses, the secondary turns should be kept to a minimum.

Hence, the minimum secondary turns are defined by the maximum output voltage, the maximum permitted flyback voltage, and the primary turns. The minimum secondary wire gauge is defined by the maximum secondary current. Because the secondary turns and current are larger than normal, the transformer core will be larger than that normally expected for the required transmitted power.

25.1.2 Step 1, Selecting Operating Mode

In the switchmode variable supply, it is the transformer design which defines the mode of operation (complete or incomplete energy transfer). The choice depends on a number of practically oriented compromises. The factors to be considered are as follows.

If complete energy transfer is to be maintained down to very low output voltages, then the primary "on" period will need to be very short. This results in a very low primary efficiency because the peak primary current will be very large.

Conversely, if incomplete energy transfer is to be maintained at high output voltages, then the primary inductance will need to be very large. It will then be difficult to avoid core saturation at low output voltages, where the transformer will be operating with a large DC current component.

In this example a compromise choice is made, so that the mode changes from complete to incomplete at half output voltage.

These three fundamental criteria will now be used as a basis for the transformer design.

25.1.3 Step 2, Transformer Core Size

To minimize secondary turns, a core with a large cross-sectional area is required. Also, to allow sufficient window space for the high-current secondary winding, the core size will be larger than is normally expected for the required power output of 300 W.

From Fig. 2.2.2, the PM #87 core is rated for 800 W, and this will be considered. To allow the maximum range of pulse-width control, the minimum operating frequency will be used. As this must be above the audio band, 22 kHz is selected.

Further, as previously explained, the flyback voltage referred to the primary at 60 V output cannot exceed the minimum supply voltage. Consequently, to prevent saturation of the core, a duty ratio of less than 50% applies at 60 V output. (The forward and flyback volt-seconds must equate.)

At 30 V out, the operating mode will change to the continuous mode. Since the secondary turns are constant, the flyback voltage will be only half the applied voltage (as referred to the primary); hence, to maintain volt-seconds equality, the flyback period must be twice the forward period. Therefore, at 22 kHz, the "on" period will be 15 μs and the "off" period 30 μs, making up the total period of 45 μs.

By just completely using the total available period at an output of 30 V, this voltage becomes the transition point for the move from the complete to the incomplete energy transfer mode (continuous mode). Figure 2.25.1 shows the primary current waveform for the above conditions.

The minimum secondary and primary turns required to meet these conditions can now be calculated.

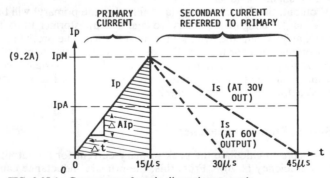

FIG. 2.25.1 Current waveform in discontinuous mode.

25.1.4 Step 3, Calculate Minimum Primary Turns

From Faraday's law,

$$N_{p(\text{min})} = \frac{V_p \cdot t}{B \cdot A_{cp}}$$

where $N_{p(\text{min})}$ = minimum primary turns
V_p = minimum primary voltage (280 V)
t = maximum "on" time, μs (15 μs)
B = optimum flux density, mT (150 mT)
A_{cp} = area of center pole, mm^2 (700 mm^2)

Therefore

$$N_{p(\text{min})} = \frac{280 \times 15}{0.15 \times 700} = 40 \text{ turns}$$

Calculate the primary volts per turn (P_v/N):

$$\frac{P_v}{N} = \frac{V_p}{N_p} = \frac{280}{40} = 7 \text{ V/turn}$$

25.1.5 Step 4, Calculate Minimum Secondary Turns

Since the secondary volts per turn must not exceed the primary volts per turn during the flyback period, the minimum secondary turns required for an output of 60 V may be calculated:

$$N_{s(\text{min})} = \frac{V_{\text{out}}}{P_v/N} = \frac{60}{7} = 8.5 \text{ turns}$$

where V_{out} = maximum output voltage

The secondary turns will be rounded up to the nearest integer, i.e., 9 turns. To maintain the correct turns ratio, the primary is adjusted to 42 turns, and the flux density will be slightly smaller.

At 30 V output, the flyback voltage (as referred to the primary) will be only half that at 60 V out, although the primary voltage and transformer turns remain the same. Hence, at 30 V out, the flyback time must be twice the preceding "on" time if a stable working point for the core flux density is to be maintained (the forward and flyback volt-seconds per turn must be the same for steady-state conditions).

The primary and secondary current waveforms, referred to the primary, for 30- and 60-V outputs are shown in Fig. 2.25.1

25.1.6 Step 5, Primary Inductance

It is now possible to calculate the peak primary current for an output of 300 W assuming an efficiency of 70%. From this, the primary inductance can be established

$$\text{Input power} = \frac{P_{\text{out}} \times 100}{\text{Eff}} = \frac{300 \times 100}{70} = 428 \text{ W}$$

where P_{out} = required output power, W

Eff = efficiency from primary to output, %

At 280 V input, the mean primary current will be

$$I_{\text{ave}} = \frac{\text{input power}}{V_{\text{in}}} = \frac{428}{280} = 1.53 \text{ A}$$

By inspection of Fig. 2.25.1, the peak primary current will be six times the mean current. Therefore

$$I_p = 6 \times I_{\text{ave}} = 9.2 \text{ A}$$

The slope of the primary current with respect to the "on" time will be

$$\frac{dI}{dt} = \frac{9.2}{15} = 0.613 \text{ A/}\mu\text{s}$$

The primary inductance may now be calculated as follows:

$$V_p = \frac{-L\,di}{dt}$$

therefore

$$|L| = \frac{V_p \times dt}{di} = \frac{280 \times 15 \times 10}{9.2} = 456 \ \mu\text{H}$$

25.1.7 Step 6, Core Air Gap Size

The A_L factor for an ungapped PM 87 core is 12 μH for a single turn. With the minimum primary turns of 42, this would give an inductance of 21 mH. It is clear that a very large air gap will be required with this core for a primary inductance of 456 μH. (See Chap. 2 for methods of establishing the gap size.)

25.1.8 Step 7, Power Transfer Limits

At voltages above 30 V, the transformer is operating in a complete energy transfer mode, and the energy in the core at the end of a primary "on" period is given by $E = \frac{1}{2}L_p \cdot I_p^2$ joules/cycle. This energy is transferred to the output at each cycle. Hence, by setting a maximum limit for the peak primary current, the maximum value of the transferred power is defined.

In the range from 60 V output to 30 V output, the maximum output power (at 70% efficiency) is given by

$$P_{\text{out}} = 70\% \text{ of } E \times \text{frequency}$$

At 30 V output, the time taken to store the required maximum power in the core is 15 μs. This storage time is the same at 60 V output, as the primary voltage, inductance, and turns remain the same. However, the flyback period at 60 V output will be only half that of the 30-V condition (15 μs instead of 30 μs). This reduction is indicated on Fig. 2.25.1, and shows that the converter will operate in the discontinuous mode (complete energy transfer) between 30 and 60 V output.

Note: Although the average secondary current at 60 V is now only half the value shown for 30 V, the output power remains constant.

At voltages below 30 V, the unit reverts to the continuous mode (incomplete energy transfer). Figure 2.25.2 shows the waveform to be expected at 10 V out. The criteria for developing this waveform are as follows:

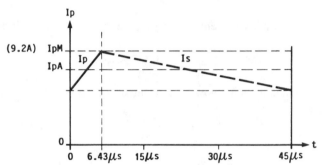

FIG. 2.25.2   Current waveform in continuous mode.

The peak primary current I_p will be limited to 9.2 A by the primary current limit.

To maintain forward and flyback volt-seconds equality, with a secondary flyback voltage of only 10 V, the flyback period must be 6 times longer than the "on" period. Hence, with a total period of 45 μs, the maximum "on" period is 6.43 μs.

Since the primary voltage, inductance, and transformer turns remain unchanged, the current slope dI_p/dt during the "on" period remains the same. The starting current (at the beginning of an "on" period) may be calculated as follows:

The slope of the primary current is

$$\frac{\Delta I_p}{\Delta t} = \frac{9.2}{15} = 0.613 \text{ A/μs}$$

Since the current at the end of a primary "on" period cannot exceed the limiting value of 9.2 A at maximum output, the current at the start of the "on" period may be calculated:

$$I_{ps} = I_{\text{max}} - \frac{\Delta I_p \times t_{\text{on}}}{\Delta t} = 9.2 - (0.613 \times 6.43) = 5.26 \text{ A}$$

With incomplete energy transfer, the energy transferred at each cycle is that value which was stored in the core at the end of an "on" period, less the energy remaining in the core at the beginning of the next "on" period.

Since all other factors remain constant, the percentage of power remaining in the core may be calculated from the ratio of the current (squared) at the end of an "on" period to the current (squared) at the beginning of the next "on" period. The energy (%) remaining in the core at the start of a period is

$$\frac{I_{ps}^2}{I_{pm}^2} \times 100\% = \frac{(5.26)^2}{(9.2)^2} \times 100 = 33\%$$

Since 33% of the energy remains in the core and the maximum output power at 10 V output is 67% of 300 W, or 200 W. This gives an output of 20 A at 10 V. The reduced power locus is shown in Fig. 2.24.1.

Hence, as a result of the change from complete to incomplete energy transfer below 30 V, the power curve is somewhat reduced if the peak primary current is limited to 9.2 A. However, a very useful 20-A output current is still available at the lower output voltages.

It is recommended that a primary current limit always be used with this type of supply. It reduces the high current stress which would normally apply to both input and output circuits, if the constant power curve is maintained at the lower output voltages.

The remainder of the transformer design, i.e., selection of wire sizes and general design parameters, will be very similar to that used in the design of the flyback transformers shown in Chap 2., except that the secondary must be rated for the higher current of 20 A. Part 3, Chap. 4 covers general design and wire selection.

25.1.9 Final Transformer Specification

Core size	= PM 87
Center pole area A_{cp}	= 700 mm²
Operating frequency	= 21 kHz
Total period	= 45 μs
Maximum "on" period	= 33.3% (15 μs)
Minimum primary volts	= 280 V DC
Optimum flux density	= 0.15 T
Primary turns	= 42
Secondary turns	= 9

25.2 VARIABLE-FREQUENCY MODE

If the minimum pulse width is limited to 1 μs, it can be shown (by the same methods as used above) that the maximum input power at the limiting input current of 9.2 A will be 57 W.

This limiting power condition will occur at very low output voltages, but the limiting output current will still be 20 A. Hence the internal losses will be large (say 30 W). Hence, if the output power falls below, say, 27 W at low output voltages, the power must be further reduced, and the system reverts to a variable-frequency mode.

At high output voltages, where complete energy transfer takes place, the maximum input power for a 1-μs period will be only 1.9 W. Hence, constant-frequency operation will be maintained down to this much lower power level at higher voltages. Figure 2.24.1 shows the area of variable-frequency operation.

25.3 PROBLEMS

1. The switchmode variable supply may replace two or three linear variable supplies in the same power range. Why is this?
2. Why is the flyback technique particularly suitable for switchmode variable supplies?
3. Why is constant-frequency operation abandoned at low output powers in the variable switchmode supply?
4. Why is the output power at low output voltages somewhat less than that at high output voltages?

P · A · R · T · 3

APPLIED DESIGN

CHAPTER 1

INDUCTORS AND CHOKES IN SWITCHMODE SUPPLIES

The following types of wound components (inductors and chokes) are covered in this chapter:

- 1.2 Simple inductors (no DC current)
- 1.3 Common-mode line-filter inductors (special dual-wound inductors which carry large but balanced line frequency currents)
- 1.7 Series-mode line-filter inductors (inductors which carry large and unbalanced line frequency currents)
- 1.8 Chokes (inductors with a large DC bias current wound on gapped ferrite cores)
- 1.12 Rod chokes (chokes wound on ferrite or iron powder rods)

The derivation of magnetic equations and the development of nomograms are shown in Appendixes 3.A, 3.B, and 3.C.

1.1 INTRODUCTION

For the purpose of this discussion, the term "inductors" will be reserved for wound components which do not carry a DC current, and the term "chokes" will be used for wound components which carry a large DC bias current, with relatively small ac ripple currents.

The design and materials used for the wound component can vary considerably depending on the application. Further, the design process tends to be iterative; a number of interactive but often divergent variables must be reconciled.

The engineer who fully masters all the theoretical and practical requirements for the optimum design of the various wound components used in switchmode supplies has a rare and valuable design skill.

The design approach used here will depend on the application. The final design often tends to be a compromise, with emphasis being placed on minimum cost, minimum size, or minimum loss. Since the optimum conditions for these three major requirements are divergent, a compromise choice will often have to be made. The designer's task is to obtain the best compromise.

In switchmode applications, inductors (no DC bias) will normally be confined to low-pass filters used in the supply line. Here, their function is to prevent the conduction of high-frequency noise back into the supply lines. For this application, high core permeability would normally be regarded as an advantage.

Chokes (inductors which carry a large DC bias current) will be found in high-frequency power output filters and continuous-mode buck-boost converter "transformers." In such applications, low permeability and a low high-frequency core loss would normally be considered advantages.

To minimize the number of turns and hence reduce copper loss, it might have been assumed that a high-permeability core material with a low core loss would be the most desirable. Unfortunately, in choke design, the large DC current component and the limited saturation flux density of real magnetic materials force the selection of a low-permeability material or the introduction of an air gap in the core. However, as a result of the low effective permeability, more turns are needed to obtain the required inductance. Hence, in choke design, the desired low copper loss and high efficiency are compromised by the need to support a large DC current.

1.2 SIMPLE INDUCTORS

In power supply applications, pure inductors (those which do not carry a DC component or a forced high-current ac component) are rare. Since the design of such inductors is relatively straightforward (the inductance may be obtained directly from the A_L value provided for the core, because no gap is required), their design will not be covered here. However, remember that with such inductors the inductance increases as N^2; therefore

$$L = N^2 \times A_L$$

1.3 COMMON-MODE LINE-FILTER INDUCTORS

Figure 3.1.1a shows a balanced line filter typical of those used to meet the conducted-mode RFI noise rejection limits in direct-off-line switchmode supplies. It shows two separate inductors, L1(a) and L1(b), which are wound on a single core to form a dual-wound common-mode line-filter inductor. Also shown is L2, which is a single-wound series-mode inductor.

Typical examples of dual-wound common-mode filter inductors are shown in Fig. 3.1.1b and c.

The common-mode filter inductor has two isolated windings with the same number of turns. The windings are connected into the circuit in such a way that the two windings are in antiphase for series-mode line frequency currents. Hence, the magnetic field that results from the normal series-mode ac (or even DC) supply currents will cancel to zero.

When the two windings are connected in this way, the only inductance presented to series-mode currents will be the leakage inductance between the two windings. Hence the low-frequency line current will not saturate the core, and a high-permeability material may be used without the need for a core air gap. Thus a large inductance can be obtained with few turns.

However, for common-mode noise (noise currents or voltages which appear

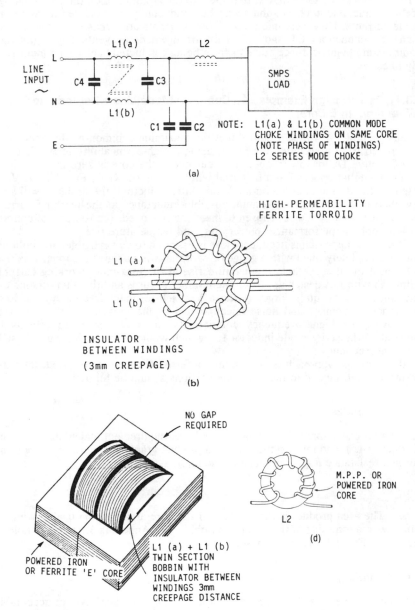

FIG. 3.1.1 (a) Line input filter for reduction of SMPS common- and differential-mode conducted noise. (b) and (c) Typical examples of common-mode line-filter inductors. (d) Typical series-mode line-filter choke using low-permeability high-loss iron dust toroidal cores.

on both lines at the same time with respect to the ground plane), the two windings are in parallel and in phase, and a very high inductance is presented to common-mode currents. Hence common-mode noise currents are bypassed to the ground plane by capacitors C1 and C2. This arrangement prevents any significant common-mode interference currents from being conducted back to the input supply lines.

1.3.1 Basic Design Example of a Common-Mode Line-Filter Inductor (Wound on an E Core)

In this example, it will be assumed that the maximum common-mode inductance is required from a specified core size, using a high-permeability ferrite E core. The effective DC or low-frequency ac current in the core is zero as a result of using two equal opposed and balanced windings, as shown in Fig. 3.1.1a. Very often, in the design of common-mode line-filter inductors, the designer will simply choose to obtain the maximum possible inductance at the working current from a particular core size, chosen to meet the size needs (consistent, of course, with acceptable performance, power loss, and temperature rise).

When this approach is used, core loss is assumed to be negligible, and bobbins will be completely filled with a gauge of wire that will just give a copper loss that will result in an acceptable temperature rise at the maximum working current. Although with this design approach the number of turns and the interwinding capacitance may be quite large, giving a low self-resonant frequency, the low-frequency inductance and noise rejection will be maximized for the core size. Moreover, the higher-frequency components can often be more effectively blocked by the series-mode inductor L2, which would normally have a high self-resonant frequency.

If this design approach is chosen, and a ferrite E core is to be used, then the design steps discussed in the following sections should be followed.

1.3.2 Core Size

Select a core size that suits the mechanical size requirement, calculate the "area product" (*AP*), and refer to the core area product graph, Fig. 3.1.2, to obtain the thermal resistance R_{th} of the finished inductor.

$$AP = A_{cp} \times A_{wb} \quad \text{cm}^4$$

Note: The area product is the product of the core area and the usable winding window area (one side of the E core on bobbin; see Appendix 3.A; an example of the use of Fig. 3.1.2 is given in Sect. 1.4).

1.3.3 Winding Dissipation

Calculate the permitted winding dissipation *W* that will just give an acceptable temperature rise ΔT. Then obtain the winding resistance R_w at the working (rms) current *I*. Assume zero core loss.

$$W = \frac{\Delta T}{R_{th}} \quad \text{W}$$

and

$$R_w = \frac{W}{I^2} \quad \Omega$$

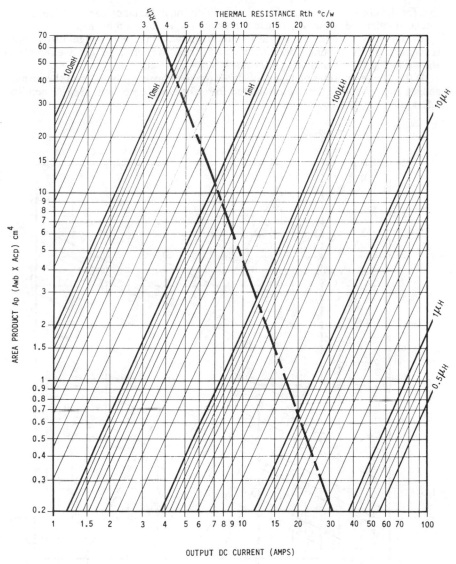

FIG. 3.1.2 Nomogram for establishing the area product (and hence the size) for chokes in ferrite material, as a function of DC load current and inductance, with thermal resistance as a parameter.

From this permitted maximum resistance (of the fully wound bobbin), the wire gauge, turns, and inductance can be established by one of the methods discussed in the following sections.

1.3.4 Establish Wire Size, Turns, and Inductance

Many manufacturers provide information on the resistance and maximum number of turns of a fully wound bobbin using various wire gauges. Also, the A_L factors for the core are often provided, from which the inductance can be calculated. Because balanced windings are used, there is no need for an air gap.

In some cases a nomogram is available, from which the wire gauge, turns, and resistance of the wound component can be read directly. (A good example is shown in Fig. 3.1.3. and an application is shown in Sec. 1.4.)

If the above information is not available for the chosen bobbin, then the turns, gauge, and resistance may be calculated from the basic core data. (See Appendix 3.B.)

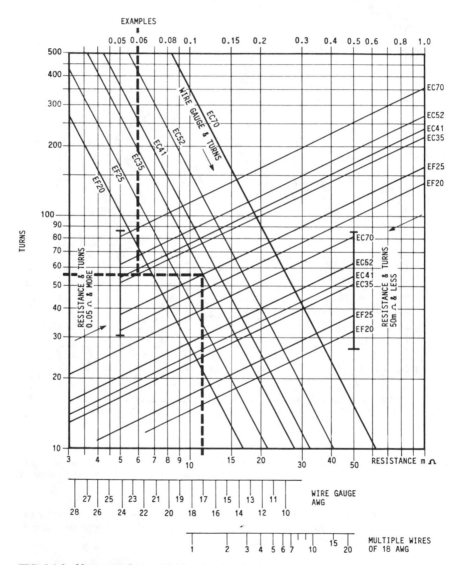

FIG. 3.1.3 Nomogram for establishing the wire size for chokes in ferrite material, as a function of turns and core size, with resistance as a parameter.

An inductor wound following the preceding simple steps provides the maximum inductance possible on the selected core size, at the maximum rated current and selected temperature rise. The finished choke will look much like the example shown in Fig. 3.1.1c.

1.4 DESIGN EXAMPLE OF A COMMON-MODE LINE-FILTER INDUCTOR (USING A FERRITE E CORE AND GRAPHICAL DESIGN METHOD)

Assume that an EC35 core is to be used to provide the maximum inductance for a common-mode line-filter inductor, with a temperature rise not to exceed 30°C at an input current of 5 A rms.

The area product for the EC35 is 0.7 (when a bobbin is used). Entering the left side of Fig. 3.1.2 with $AP = 0.7$ gives the thermal resistance 20°C/W (at the top of the nomogram). Hence the dissipation permitted for a temperature rise ΔT of 30°C will be

$$P = \frac{\Delta T}{R_{th}} = \frac{30}{20} = 1.5 \text{ W}$$

At a current of 5 A rms, the maximum resistance is related to power by

$$I^2 R = P$$

Hence

$$R = \frac{1.5}{25} = 0.06 \ \Omega$$

From the nomogram shown in Fig. 3.1.3, the maximum number of turns to give a resistance of 0.06 Ω on the EC35 core can be established as follows:

Find the required resistance (0.06 Ω) on the top horizontal scale of resistance. (An example is shown at top left.) Project down to the upper (positively sloping) "resistance and turns" lines for the EC35, as shown in the example. The intersection with the EC35 line is projected left to give the number of turns (56 in this example). From the same point, project right to the intersection with the negatively sloping "wire gauge and turns" line for the EC35 core. This point is then projected down to the lower scale as shown to give the wire gauge (approximately #17 AWG in this example).

In a common-mode inductor, the winding will be split into two equal parts. Hence the EC35 bobbin would be wound with two windings of 28 turns of #17 AWG.

Note: For resistance values of less than 50 mΩ, enter the graph from the bottom scale of resistance and project up to the lower group of "resistance and turns" lines. For example, assume that a winding of 40 mΩ is required on an EF25 core. Enter the graph from the lower 40-mΩ scale and project up to the EF25 "resistance and turns" line. Project left to give the total number of turns (34 in this example). The intersection of this turns line with the "wire gauge and turns" line for the EF25 core is then projected down to give the gauge (#17 AWG again in this example). Hence two windings of 17 turns of #17 AWG would be used.

For applications not covered by the nomogram, the winding parameters may be calculated as shown in Appendixes 3.A and 3.B.

1.5 CALCULATING INDUCTANCE (FOR COMMON-MODE INDUCTORS WOUND ON FERRITE E CORES)

In dual-winding, common-mode inductors, the series-mode line frequency or DC magnetization force will cancel out, as a result of the reverse phasing of the two windings. Hence, a high-permeability core material may be used, and a core gap is not required. Therefore, in most cases the inductance can be calculated from the published A_L values.

For the example shown above, the A_L value for the EC35 without an air gap is approximately 2000 nH. The inductance for each 28-turn winding can be calculated as follows:

In general,

$$L = N^2 A_L$$

For the above example,

$$L = 28^2 \times 2000 \times 10^{-9} = 1.57 \text{ mH}$$

For the purpose of calculating the common-mode inductance, both windings are effectively in parallel, and the inductance is that of a single winding.

This graphical design approach also gives the maximum common-mode inductance that can be obtained from this core at 5 A for a temperature rise of 30°C.

If the power supply input power P_{in} is to be used to obtain the rms current for the above calculations, remember to allow for the power factor of the capacitor rectifier input filter (typically 0.63) when calculating the rms current. Also calculate the maximum rms current at minimum input voltage. For example,

$$I_{\text{rms(max)}} = \frac{P_{in}}{V_{\text{in(min)}} \times 0.63}$$

The choice of core material depends on the noise spectrum. In most cases, the noise will be most troublesome in the lower-frequency range (from the switching frequency to, say, 800 kHz). For this application, a large inductance is required, and high-permeability iron or ferrite core material will be used. If the problem noise is in the higher-frequency spectrum, then powdered iron cores, with their inherently larger high-frequency power loss, may provide better results.

1.6 SERIES-MODE LINE-INPUT-FILTER INDUCTORS

L2 in Fig. 3.1.1a is a series-mode line-filter inductor. Even though these inductors do not carry DC in this position, the peak line frequency currents are large, and a high forcing voltage exists. Also the duration of the current pulse is very long compared with the switching frequency. Hence, the inductor can be considered to be driven from a constant-current line frequency source, and at the peak line current this has a saturating effect similar to that of DC. L2 must be designed to carry this current without saturating.

Consider a typical off-line capacitive input rectifier circuit. Because of the large input capacitor, a large current pulse flows on the peak of the applied voltage waveform as the input rectifier diodes conduct. It is essential to ensure that L2 will not saturate during this current pulse, even under full-load conditions, and the design of these input inductors should follow the same approach used for DC current choke design. (See Sec. 1.8.) To prevent saturation of L2, it may be necessary to use gapped ferrite cores or low-permeability iron powder cores.

With capacitor rectifier input circuits, it may be difficult to calculate the peak current in L2, since it depends upon a number of ill-defined variables. These include the line source impedance, circuit resistance, input capacitor ESR values, and total loop inductance. It is often better to simply measure the current and calculate the peak flux density in the core. A 30% safety margin between the peak and saturation flux densities should be provided to allow for component and line impedance variations. Part 1, Chap. 6 shows graphical methods of establishing the peak rectifier current.

If the peak rectifier current is known, the inductor design may proceed in the same way as a choke design, with the peak forced ac current taking the place of the DC current in the calculations.

1.7 CHOKES (INDUCTORS WITH DC BIAS)

Chokes (inductors which carry a large component of DC current) are to be found in some form in all switchmode supplies. Choke design can be quite complex, and a good working knowledge is most essential. The power supply engineer will need to develop considerable skill in the choice of core material, core design, core size, and winding design if the most cost-effective chokes are to be produced. The subject is very broad, and this discussion will be confined to those types of chokes most often used in high-frequency switchmode applications.

Chokes range from small ferrite beads used, for example, to profile the base drive currents of switching transistors, up to the very large high-current chokes used in power output filters. Typical examples of switchmode chokes are shown in Fig. 3.1.4.

1.7.1 Core Material

The core material will be chosen to suit the operating frequency, ratio of DC to ac current, inductance, and mechanical requirements. Where the ac component or frequency is low—for example, in series line input filters—then a laminated silicon iron or similar material may be chosen. This will have the advantage of high saturation flux density and will need fewer turns for the required inductance, resulting in lower copper losses. Where the operating frequencies and ac currents are higher, core losses will need to be considered, and gapped ferrite, Molypermalloy, or powdered iron materials may be chosen.

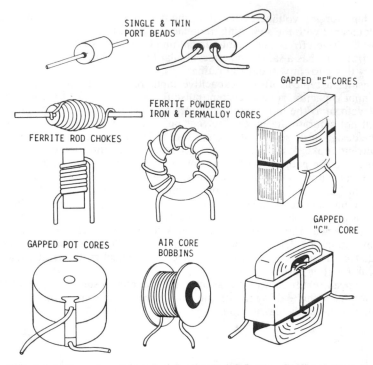

SINGLE & TWIN
PORT BEADS

GAPPED "E"CORES

FERRITE POWDERED
IRON & PERMALLOY CORES

FERRITE ROD CHOKES

GAPPED
"C" CORE

GAPPED POT CORES   AIR CORE
                          BOBBINS

FIG. 3.1.4 Examples of typical output chokes and differential-mode input chokes.

1.7.2 Core Size

Often the most difficult choice is that of core size and configuration. A bewildering range of core topologies exist, and it can be difficult to decide which would be the optimum choice for a particular application.

The area product of the core provides a good method of selecting core size. It has been shown[1,2] that the area product factor tends to be reasonably constant for all core topologies of the same general power rating, and this can be used to select the core size.

Note: The area product AP is the product of the winding window area and the core center pole area. (See Appendix 3.A.)

In general,

$$AP = A_w \cdot A_e \qquad \text{cm}^4$$

where AP = area product, cm^4
   A_w = core winding window area, cm^2
   A_e = area of center pole, cm^2

The area product is quoted by many core manufacturers, or may be easily calculated from the core dimensions.

1.7.3 Temperature Rise

In general, the temperature rise of the wound component in free air cooling conditions will depend on the total loss in the wound component and the component's surface area. The surface area is also related to the area product for "scrapless" core geometries.

Figure 3.1.2 shows the typical thermal resistance (dashed line) expected from a core whose size is defined in terms of its area product AP. The graph was derived from the measured values of thermal resistance for the EC, ETD, RM, and PQ families of cores.[1,2,15]

The actual temperature rise ΔT that may be expected from a particular core size AP will be given by the thermal resistance and total dissipation, as follows:

$$\Delta T = P \cdot R_t$$

where ΔT = temperature rise, °C
$\quad\quad P$ = total dissipation, W
$\quad\quad R_t$ = thermal resistance, °C/W

Note: In choke design, the loss P will be mainly copper loss. Core losses are small in most cases, and may be neglected.

1.7.4 Core Air Gaps

If considerable DC currents flow in the choke, the use of gapped E or C cores may be considered. The materials used here range from the various iron alloys for low-frequency operation (or even for high-frequency operation where flux excursions are very small) to gapped ferrites when higher ac currents and high-frequency operation are required.

Since chokes will normally be required to support the DC component without saturation, relatively large air gaps are used, and the effective permeability, irrespective of the material chosen, is usually very low—typically between 10 and 300.

Figure 3.1.5 shows the effect of introducing an air gap into a high-permeability ferrite core (lower B/H loops), and also the difference between a gapped ferrite and a gapped iron core (upper B/H loop). It should be noted that the value of H_{dc} (proportional to DC current) that would have caused saturation of a nongapped ferrite core at point B_{sat} results in a lower value of magnetization B_{dc} when a gap is used. There is sufficient margin to support the ac ripple component ΔB in the gapped example. Also shown is the difference in flux density and permeability between a gapped ferrite and gapped silicon iron core for the same magnetizing force H_{dc}.

Note: Ferrite material saturates at a lower flux density than iron, even when gapped. Further, to prevent saturation, a large gap must be used in the core, resulting in a lower effective permeability (slope of the B/H curve) and giving lower inductance. This lower permeability means that for a defined ΔB_{ac} (amplitude of applied ripple voltage), the ripple current ΔH_{ac} will be larger with the ferrite material even when a gap is used. This effect is clearly seen in the larger horizontal spread of ΔH in the ferrite B/H loop of Fig. 3.1.5, compared with that of the iron core.

The higher saturating flux density of the iron core permits a smaller gap, giving a larger permeability for the same DC bias conditions; hence the inductance is

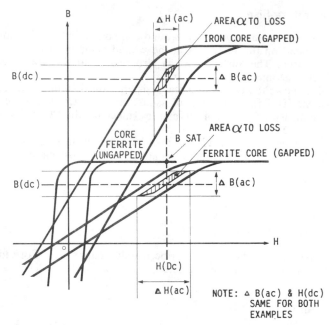

FIG. 3.1.5 Comparison of the *B/H* characteristics of ferrite and iron cored chokes, with and without air gaps, showing the change in magnetizing current swing (ripple current) as a function of a fixed flux density swing (ripple voltage) for the two materials.

greater, and the ripple current will be smaller. In Fig. 3.1.5 the ripple current is proportional to ΔH_{ac}. It will be seen that for the same applied ripple voltage ΔB, the iron core gives a smaller ripple current. Further, if ΔB is small (large DC current, low ac voltage applications), the laminated iron core loss can be acceptably low; hence laminated iron or powdered iron cores should be considered for this type of application.

A further advantage of the gapped E or C core is that the effective permeability can be optimized for the application by adjusting the gap size for the most effective performance. This cannot be done with a toroidal core.

Although magnetic radiation from the gap can sometimes be a problem, a copper shield can reduce this by up to 12 dB or more (see Sec. 4.5 in Part 1). Hence the use of a gapped core should not be rejected on the grounds of possible magnetic radiation problems.

The core size depends on the total loss and the permitted temperature rise. The copper loss depends on the DC current, turns (inductance), and wire size. The core loss, and hence the choice of core material, depends on the ac volt-seconds that the choke must withstand, that is, the flux density swing ΔB and the operating frequency.

1.8 DESIGN EXAMPLE OF A GAPPED FERRITE E-CORE CHOKE (USING AN EMPIRICAL METHOD)

In this example it will be assumed that the choke is required to support a large DC current with considerable high-frequency ripple current; hence a low-core-loss material will be used, and an air gap will be required. A typical application would be an output filter inductor for a high-frequency forward converter. In this example the DC current is 10 A mean, with a ripple current not exceeding 3 A at 100 kHz.

It will also be assumed that the maximum core size is defined by the mechanical rather than the ideal electrical needs. Hence the design approach will be to obtain the minimum ripple current (maximum inductance) from the defined core size. The following simple empirical approach may be used to obtain optimum inductance and air gap size:

1. Select core and bobbin size (as defined by the mechanical needs), and completely fill the bobbin with a gauge of wire that will give acceptable I^2R power loss and hence acceptable temperature rise (say, 40°C). Use the maximum mean DC current value for this calculation, and proceed as for the example in Sec. 1.4. (Many manufacturers provide information on wire gauges and the resistance of a fully wound bobbin; alternatively, the temperature rise may be obtained from Fig. 3.1.2 and the wire gauge and turns from Fig. 3.1.3.)

2. Assemble core and bobbin, allowing an adequate air gap (say, 20% of the diameter of the center pole). Fit the choke in the power filter position in the supply, and observe the choke ripple current waveform. Adjust the air gap under maximum load and input voltage conditions until a minimum ripple current is observed. (A clear minimum will be seen.) By this means, maximum dynamic inductance has now been obtained. Increase the air gap above this minimum by at least 10%, to allow for variations in materials and the reduction in saturation flux density level at higher temperatures. With a little practice, it is possible to get very close to the optimum value with minimum effort and time.

The empirical design is now complete. This approach results in the maximum inductance that this particular choice of core size can give for the conditions of operation (current) and the chosen temperature rise.

Note:

1. With ferrite material, the core loss is normally much less than the copper loss; hence it has been neglected in this example. With iron powder or laminated cores, the core loss may be considerable and must be added to the I^2R loss when establishing the temperature rise.

2. In some applications, the absolute value of the inductance may be critical (it may control transient performance and loop stability). In such cases a more complex design approach which yields the required inductance must be used.

1.9 DESIGN EXAMPLE OF CHOKES FOR BUCK AND BOOST CONVERTERS (BY AREA PRODUCT GRAPHICAL METHODS AND BY CALCULATION)

In the nonisolated buck and boost switchmode regulators shown in Fig. 3.1.6a and b, the output choke current is continuous and large. This results in a large DC magnetization bias. However, the ripple current is normally relatively small compared with the maximum load current, giving a low magnetization swing and hence low core loss.

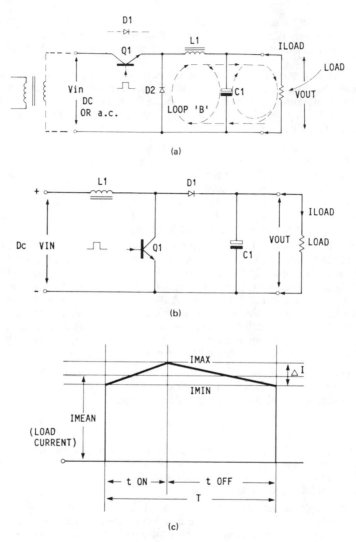

FIG. 3.1.6 (*a*) Basic buck regulator circuit. (*b*) Basic boost regulator circuit. (*c*) Output current waveform for buck regulator and input current waveform for boost regulator, with L1 in continuous-mode conduction.

In the case of the buck regulator, Fig. 3.1.6*a*, the input to the choke L1 may be a switched DC from Q1, as in a DC-to-DC switchmode regulator, or a pulse-width-modulated square wave from the secondary of a transformer, in the case of a fully isolated switchmode converter. In the second case, the switching transistor Q1 will be replaced by the rectifier diode D1.

The boost regulator circuit shown in Fig. 3.1.6*b* is normally reserved for nonisolated DC-to-DC switchmode regulators where the output voltage must be higher than the supply. The output current is discontinuous (even though the choke current is continuous), giving higher output ripple voltages. In isolated transformer-coupled applications, where higher output voltages are required, the voltage step-up is normally provided by the transformer, and a buck-type secondary regulator would be used.

In continuous-mode applications, the current in the filter choke L1 for both buck and boost circuits under steady-state conditions will have a shape similar to that shown in Fig. 3.1.6c. The current at the start of a cycle will have the same value as at the end of a cycle. The peak-to-peak ripple current I_{min} to I_{max} is normally small compared with the mean current.

In the buck application, the mean choke current is equal to the output DC load current, the supply current being discontinuous. In the boost application, the supply current is the same as the mean choke current, the output current being discontinuous.

In both types of converter, the choke inductance will have been chosen to give continuous-mode conduction from a small load (normally 10% or less) to full-load operation. Hence, the peak-to-peak ripple current would be less than 20% of the full-load current. Even in applications in which the load is fixed, the ripple current will still be made quite small, to ensure good switching efficiency and low output ripple voltages.

To minimize the choke size, the core may be operated at a flux density near saturation, since the change in flux density is small (ripple currents are small). If a ferrite core is used, the core loss will be very small, and the temperature rise will depend mainly on the winding copper losses. Hence, the core size, for a defined temperature rise, will depend on the winding copper loss at full load.

1.9.1 General Conditions for Graphical Choke Design (Using Gapped Ferrite Cores)

The following starting conditions will be assumed:

1. The choke is to be designed for a buck or boost regulator for which the ripple current does not exceed 20% of the maximum defined load current.

2. A gapped ferrite core will be used; hence the core losses will be small and can be neglected.

3. The maximum temperature rise is to be limited to 30°C above ambient.

4. To allow some margin for maximum current limiting, the flux density is not to exceed 0.25 T for normal operation.

5. A copper packing factor K_u of 0.6 is assumed (that is, 60% of the bobbin winding window area is utilized by copper). This is a typical value for a single winding using round wire. See Appendix 3.B.

6. The value of the required inductance will be established by assuming a 20% peak-to-peak ripple current and a maximum "off" period of 80%. (A typical choke current waveform for a buck regulator meeting these conditions is shown in Fig. 3.1.6c.)

With the above parameters defined at typical values, a very simple area product graphical approach may be used to establish the choke requirements.

1.9.2 The Area Product Design Nomograms

The area product AP is the product of the core winding window area and the core center pole area. This product value tends to be similar for all core topologies of similar power rating and general size. Further, the *AP* links the inductance value to the load current and copper loss, and hence the temperature rise, and will be used to establish the core size for the following choke design. The area product derivation is shown in Appendix 3.A.

Many core manufacturers provide the *AP* figures for their standard cores. Alternatively, the *AP* may be easily established from the dimensions of the core. (See Sec. 7.2 in Part 1.)

Figure 3.1.2 has been developed for gapped ferrite cores. It shows the area product (and hence the core size) as a function of the load current, with the inductance as a parameter. It assumes an ambient of 20°C, a temperature rise from ambient of 30°C, a maximum flux density $\hat{B}$ of 250 mT, and a copper packing factor of 0.6.

The *AP* values shown in Fig. 3.1.2, and those for the EC70, EC52, EC41, and EC35 cores (listed in Table 3.3B.1, page 3.62), are based on the product of core area and useful bobbin window area (rather than core window area). Where bobbin windings are to be used, this should be normal practice when using Fig. 3.1.2, as the value of *AP* can be considerably reduced by the window space used for the bobbin material.

1.10   CHOKE DESIGN EXAMPLE FOR A BUCK REGULATOR (USING A FERRITE E CORE AND GRAPHICAL AP DESIGN METHOD)

Sample Choke Specification

$$\begin{aligned}
\text{Output voltage} &= 5 \text{ V} \\
\text{Maximum output current} &= 10 \text{ A} \\
\text{Frequency} &= 25 \text{ kHz} \\
\text{Maximum ripple current} &= 20\% \, I_{\text{max}} \, (2 \text{ A}) \\
\text{Maximum temperature rise} &= 30°\text{C}
\end{aligned}$$

1.10.1   Step 1, Establish Inductance

Draw the current waveform for full load and maximum input voltage (see Fig. 3.1.6*c*). From the slope of the current waveform, establish the required inductance as follows:

Starting values:

Frequency f = 25 kHz. Hence, $T = 1/f = 40 \, \mu s$

Output voltage = 5 V

Input voltage = 25 V

Calculate "on" time t_{on}:

$$\text{Duty ratio} = \frac{t_{\text{on}}}{T} = \frac{V_{\text{out}}}{V_{\text{in}}}$$

Hence

$$t_{on} = \frac{T \times V_{out}}{V_{in}} = \frac{40 \times 10^{-6} \times 5}{25} = 8 \ \mu s$$

$$t_{off} = T - t_{on} = 40 - 8 = 32 \ \mu s$$

During the "off" period, the current falls by 20% of $I_{load(max)}$, or 2 A in this example, and the inductance may be calculated.

Consider Fig. 3.1.6a. During the "off" period, current loop B is established, and the magnitude of the voltage across the choke, $|e|$, is the output voltage plus the D2 diode voltage drop (the diode, being forward-biased, takes the input to L1 negative). During the "off" period, the current is flowing in loop B under the forcing action of choke L1. During this period the current will be decaying at a constant rate defined by

$$|e| = \frac{L \ dI}{dt} = 5.6 \ V$$

Hence

$$L = e \frac{\Delta t}{\Delta I} = \frac{5.6 \times 32 \times 10^{-6}}{2} = 89.6 \ \mu H$$

1.10.2 Step 2, Establish Area Product *AP*

Using the graph shown in Fig. 3.1.2, find the area product *AP* for the required load current and inductance. In this example, entering the bottom of the graph with the required current of 10 A and projecting upward to meet the required inductance of 90 μH yields (to the left) an *AP* of approximately 1.5. This value falls between the EC41 core ($AP = 1.46$) and the EC52 core ($AP = 2.96$). The smaller EC41 core is chosen in this example, although it will give a larger temperature rise. Note that as the core sizes only change in large increments, absolute values of *AP* and hence temperature rise are not always possible.

1.10.3 Step 3, Calculate Turns

The minimum number of turns that may be used on a core to give the required inductance without exceeding the flux density is given by Eq. (3.A.9):

$$N_{min} = \frac{L \cdot I_{pm} \cdot 10^4}{B_{max} \cdot A_e}$$

where N_{min} = minimum turns
L = inductance (90×10^{-6} H)
I_{pm} = maximum current (11 A)
B_{max} = maximum flux density (250×10^{-3} T)
A_e = center pole area (106×10^{-2} cm^2)

Hence

$$N_{\min} = \frac{90 \times 10^{-6} \times 11 \times 10^4}{250 \times 10^{-3} \times 106 \times 10^{-2}} = 37 \text{ turns}$$

1.10.4 Step 4, Establish Optimum Wire Size

For minimum loss, the wire size should be such that the winding space is just completely filled when the required number of turns and insulation have been wound on the bobbin. Information on the gauge and number of turns for a fully wound bobbin, together with the winding resistance, is often provided by the bobbin or core manufacturers.

Alternatively, the gauge may be calculated, or the nomogram shown in Fig. 3.1.3 may be used.

In this example, entering the nomogram shown in Fig. 3.1.3 at the left with 37 turns gives the wire gauge at the first intercept with the EC41 core line as #14 AWG (lower scale).

If preferred, the wire size can be calculated as follows: The cross-sectional area of a wire that will just fill the bobbin is given by

$$A_x = \left(\frac{A_w \times K_u}{N}\right)^{1/2}$$

where A_x = wire cross-sectional area, mm²
A_w = total winding window area, mm²
K_u = winding packing factor
N = turns

In this example,

$A_w = 138 \text{ mm}^2$ (EC41)
$K_u = 0.6$ (for round wire)
$N = 37$

Hence
$$A_x = \left(\frac{138 \times 0.6}{37}\right)^{1/2} = 1.5 \text{ mm}^2$$

giving a wire size of #14 AWG (which is the same as the size established by the nomogram method).

1.10.5 Step 5, Calculate Core Gap

If it is assumed that most of the reluctance will be in the air gap (the normal case), then the approximate air gap length l_g, neglecting fringe effects, will be given by the following formula:

$$l_g = \frac{\mu_0 \cdot \mu_r \cdot N^2 \cdot A_e \cdot 10^{-1}}{L}$$

where l_g = total air gap, mm
  $\mu_0 = 4\pi \times 10^{-7}$
  $\mu_r = 1$ (for air)
  N = turns
  A_e = area of center pole, cm^2
  L = inductance, H

In this example,

  $N = 37$
  $A_e = 106 \times 10^{-2}$
  $L = 90 \times 10^{-6}$

Hence

$$l_g = \frac{4\pi \times 10^{-7} \times 37^2 \times 106 \times 10^{-2} \times 0.1}{90 \times 10^{-6}} = 1.74 \text{ mm}$$

For a minimum external magnetic field, the gap should be confined to the center pole only. However, with this type of inductor, when the ripple current component is small, the gap may extend right across the core, and the radiation will not be excessive. The majority of any remaining external field may be effectively eliminated by fitting a copper screen as shown in Fig. 1.4.5.

With the EC core, the area of the center pole is less than the sum of the outer legs, and if the gap extends right across the core, then the effective leg gap will be reduced by this ratio. In any event, because of the neglected core permeability and fringe effects, some adjustment of the air gap may be necessary to obtain optimum results.

1.10.6   Step 6, Check Temperature Rise

The temperature rise will depend mainly on the total power loss (core loss plus copper loss), the surface area and emissivity, and the air flow. In the interest of simplicity, a number of second-order effects have been neglected in the design procedure. These neglected effects will only result in a small error in the final calculated temperature rise. In any event, the temperature of the choke should be checked finally in the working prototype, where the layout and thermal design will also introduce additional "difficult to determine" thermal effects.

The preceding graphical design approach assumed that the core loss using ferrite cores would be negligible. This may be verified as follows.

1.10.7   Step 7, Check Core Loss

The core loss is made up of eddy-current and hysteresis losses, both of which increase with frequency and flux excursion. The loss factor depends on the material and is provided in the material specifications. Usually the published graphs assume a symmetrical flux density excursion about zero (push-pull operation), so the indicated B_{max} is only half ΔB peak-to-peak. For first-quadrant buck and boost chokes and flyback applications, the calculated ΔB peak-to-peak should be divided by 2 to enter the graphs and obtain the core loss.

In the preceding example, the ac flux density excursion is given by

$$\Delta B_{ac} = \frac{|e| \cdot t_{off}}{N \cdot A_e}$$

where ΔB_{ac} = ac flux density swing, T
$\quad\quad |e|$ = output voltage plus diode drop, V
$\quad\quad t_{off}$ = "off" time, μs
$\quad\quad N$ = turns
$\quad\quad A_e$ = area of core, mm^2

For the above example,

$|e| = 5.6$ V

$t_{off} = 32$ μs

$N = 37$

$A_e = 71$ mm^2

Hence B_{ac} is

$$B_{ac} = \frac{5.6 \times 32}{37 \times 71} = 68 \text{ mT}$$

With a typical ferrite material, at this flux density and a frequency of 20 kHz, the core loss will be less than 4 mW/g (see Fig. 2.13.4), giving a total core loss of 104 mW with the EC41 core (a negligible loss). Hence, with ferrite material, core loss will not be significant, except for high-frequency and large-ripple-current applications.

Much greater losses will be found with powdered iron cores, and it may not be possible to neglect the core loss with these materials.

1.10.8 Step 8, Check Copper Loss

The DC resistance of the wound choke can be obtained from the bobbin manufacturer's information, or it may be calculated using the mean diameter of the wound bobbin, turns, and wire size. In any event, it should finally be measured, as winding stress and packing factors will depend on the winding technique, and will affect the overall resistance. Remember, the resistance of copper will increase approximately 0.43%/°C from its value at 20°C. This makes it 34% higher at 100°C.

The copper power loss is given by I^2R. (Since the ripple current is small, the skin effects are negligible, and the mean DC current and DC resistance can be used with little error.) Hence

$$\text{Power loss} = I^2R \quad\quad \text{W}$$

In this example,

$$I = 10 \text{ A} \quad\quad \text{and} \quad\quad I^2 = 100 \text{ A}$$

The length of the winding, and hence the resistance, may be established from the mean diameter of the bobbin and the number of turns, as follows:

$$\text{Mean diameter of EC41 bobbin } d_m = 2 \text{ cm}$$

$$\text{Total length of wire } l_t = \pi \times d_m \times N \quad \text{cm}$$

Therefore

$$l_t \pi \times 2 \times 37 = 233 \text{ cm}$$

From Table 3.1.1, the resistance of #14 AWG wire is between 83 $\mu\Omega$/cm at 20°C and 111 $\mu\Omega$/cm at 100°C, giving a total resistance between 19.3 and 25.8 mΩ.
Hence the power loss I^2R will be between 1.9 and 2.58 W.
From Table 2.19.1, page 2.155, the thermal resistance of the EC41 is 15.5°C/W,

TABLE 3.1.1 AWG Winding Data (Copper Wire, Heavy Insulation)

AWG	Diameter, copper, cm	Area, copper, cm²	Diameter, insulation, cm	Area, insulation, cm²	Ω/cm 20°C	Ω/cm 100°C	A for 450 A/cm²
10	.259	.052620	.273	.058572	.000033	.000044	23.679
11	.231	.041729	.244	.046738	.000041	.000055	18.778
12	.205	.033092	.218	.037309	.000052	.000070	14.892
13	.183	.026243	.195	.029793	.000066	.000088	11.809
14	.163	.020811	.174	.023800	.000083	.000111	9.365
15	.145	.016504	.156	.019021	.000104	.000140	7.427
16	.129	.013088	.139	.015207	.000132	.000176	5.890
17	.115	.010379	.124	.012164	.000166	.000222	4.671
18	.102	.008231	.111	.009735	.000209	.000280	3.704
19	.091	.006527	.100	.007794	.000264	.000353	2.937
20	.081	.005176	.089	.006244	.000333	.000445	2.329
21	.072	.004105	.080	.005004	.000420	.000561	1.847
22	.064	.003255	.071	.004013	.000530	.000708	1.465
23	.057	.002582	.064	.003221	.000668	.000892	1.162
24	.051	.002047	.057	.002586	.000842	.001125	.921
25	.045	.001624	.051	.002078	.001062	.001419	.731
26	.040	.001287	.046	.001671	.001339	.001789	.579
27	.036	.001021	.041	.001344	.001689	.002256	.459
28	.032	.000810	.037	.001083	.002129	.002845	.364
29	.029	.000642	.033	.000872	.002685	.003587	.289
30	.025	.000509	.030	.000704	.003386	.004523	.229
31	.023	.000404	.027	.000568	.004269	.005704	.182
32	.020	.000320	.024	.000459	.005384	.007192	.144
33	.018	.000254	.022	.000371	.006789	.009070	.114
34	.016	.000201	.020	.000300	.008560	.011437	.091
35	.014	.000160	.018	.000243	.010795	.014422	.072
36	.013	.000127	.016	.000197	.013612	.018186	.057
37	.011	.000100	.014	.000160	.017165	.022932	.045
38	.010	.000080	.013	.000130	.021644	.028917	.036
39	.009	.000063	.012	.000106	.027293	.036464	.028
40	.008	.000050	.010	.000086	.034417	.045981	.023
41	.007	.000040	.009	.000070	.043399	.057982	.018

giving a temperature rise of between 29.5 and 40°C, depending on the ambient (starting) temperature.

1.10.9   Step 9, Check Temperature Rise (Graphical Method)

The graphical area product design approach using Fig. 3.1.3 should result in a temperature rise of 30°C in free air.

A further design check may be made by obtaining the thermal resistance from Fig. 3.1.2. With this, the temperature rise may be calculated and confirmed.

Entering the graph once again with 37 turns, the second intercept with the EC41 resistance and turns core line gives a wound resistance at 100°C of 21.5 mΩ (lower scale).

The power loss I^2R is 2.15 W. The thermal resistance of the EC41, from Fig. 3.1.2, is 15°C/W. Hence the temperature rise ΔT will be

$$\Delta T = R_{th} \times P = 15 \times 2.15 = 32.2°C$$

giving a working temperature of 52°C, at an ambient temperature of 20°C.

The larger temperature rise is due to the choice of the smaller core. If this is not acceptable, choose a larger core or reduce the inductance requirements.

The area product graph in Fig. 3.1.2 assumes a wire current density of 450 A/cm, which will give a temperature rise of 30°C for a core size with an AP of 1 cm⁴. To obtain the same temperature rise with larger cores, a smaller current density should be used. (The surface-area-to-volume ratio is not so good in larger cores.) However, the packing factor tends to be better in larger cores when copper strip or rectangular conductors are used. Hence, provided that the complete window is utilized, this results in a lower current density in the larger cores, compensating somewhat for the reduced surface area ratio.

1.10.10   Step 10, Check Temperature Rise (Area Product Method)

The "scrapless" E-core geometry allows the surface area of the wound core to be related to the area product. Further, the surface area defines the rate of heat loss and hence the temperature rise.

The nomogram in Fig. 3.1.7 shows the surface area as a function of area product (top and left scales), and the temperature rise as a function of dissipation, with surface area as a parameter (lower scale and diagonal lines).

The thermal resistance values given in Fig. 3.1.2 are valid only for a temperature rise of 30°C. Figure 3.16.11 shows that the thermal resistance is a function of the temperature differential, and falls as the temperature differential increases. Hence, when the temperature rise of the wound core is other than 30°C, a more accurate figure will be obtained from Fig. 3.1.7.

Example

Predict the temperature rise of the EC41 core when the total wound component loss is 3.9 W.

The area product of the EC41 core is 1.46.

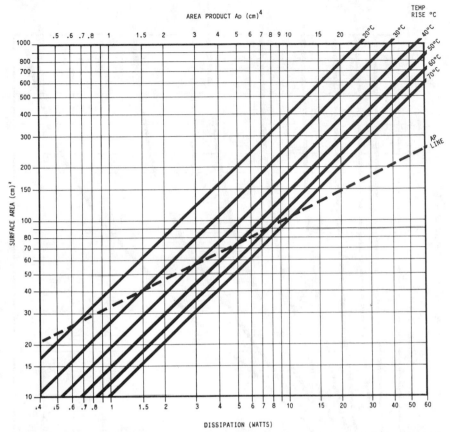

FIG. 3.1.7 Nomogram giving temperature rise as a function of area product and dissipation, with surface area as a parameter.

Enter the nomogram in Fig. 3.1.7 at the top with an *AP* of 1.46. The intersect with the *AP* line (dashed line) gives the surface area on the left scale (40 cm² in this example).

Enter the nomogram with the total dissipation (lower scale), and the intersect with the surface area line (40 cm²) gives the temperature rise on the diagonal lines (70°C in this example).

1.11 FERRITE AND IRON POWDER ROD CHOKES

For small low-inductance chokes, the use of straight open-ended ferrite or iron powder rods, bobbins, or spools or axial lead ferrite rods should be considered. In many cases, the ac current will be much smaller than the DC current (for example, in a second-stage *LC* output filter), and the high-frequency magnetic ra-

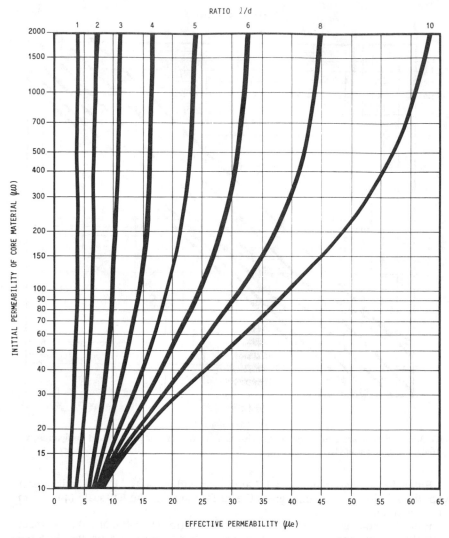

FIG. 3.1.8 Effective permeability μ_e of rod core chokes as a function of the initial material permeability, with the ratio of length to diameter as a parameter. (*By W. J. Polydoroff*)

diation from these open-ended rods, normally the most objectionable parameter, will be acceptably small.

By careful attention to minimizing the interwinding capacitance—for example, using spaced windings and insulating the wire from the ferrite rod former—the self-resonant frequency of chokes wound on open-ended rods can be made very high. An example of a wound ferrite rod, a self-resonant choke filter, is given in Part 1, Sec. 20.5. These chokes, when used in output *LC* filters with low-ESR output capacitors, can be very effective in reducing high-frequency output noise.

The intrinsically large air gap will prevent the saturation of high-permeability ferrite rods, even when the DC current is very large. The wire gauge should be chosen for acceptable dissipation and temperature rise.

The lower-cost iron powder rods are also suitable for this application. The lower permeability of these materials is not much of a disadvantage, as the large air gap swamps the initial permeability of the core material. The inductance is more dependent on the geometry of the winding than on the properties of the core.

The inductance of a "rod core" choke depends on the turns, core material permeability, and geometry. Figure 3.1.8 shows how the inductance of iron powder rod wound chokes is related to core permeability and winding geometry. It should be noted that for most practical applications, where the length-to-diameter ratio is of the order of 3:1, the initial permeability of the core material μ_0 will not significantly affect the relative permeability μ_r. Hence, the inductance is not very dependent on the core permeability. The information in Fig. 3.1.8 can also be applied to high-permeability ferrite rod chokes with little error.

Figure 3.1.9 gives the design information for single- and multiple-layer windings on rod-type chokes.

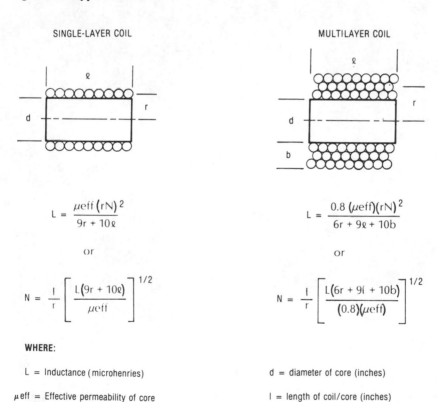

SINGLE-LAYER COIL

MULTILAYER COIL

$$L = \frac{\mu\text{eff}\,(rN)^2}{9r + 10\ell}$$

$$L = \frac{0.8\,(\mu\text{eff})(rN)^2}{6r + 9\ell + 10b}$$

or

or

$$N = \frac{1}{r}\left[\frac{L(9r + 10\ell)}{\mu\text{eff}}\right]^{1/2}$$

$$N = \frac{1}{r}\left[\frac{L(6r + 9\ell + 10b)}{(0.8)(\mu\text{eff})}\right]^{1/2}$$

WHERE:

L = Inductance (microhenries)

μ eff = Effective permeability of core

N = Number of turns

r = radius of coil (inches)

d = diameter of core (inches)

l = length of coil/core (inches)

b = coil build (inches)

FIG. 3.1.9 Methods of winding rod core chokes, and inductance calculations. (*Arnold Eng. Co.*)

1.12 PROBLEMS

1. Explain the functional difference between inductors and chokes, as defined in this chapter.
2. Why is the DC component of conducted current a problem in choke design?
3. Give a method of reducing core permeability without necessarily changing the core material.
4. By what means is the problem of core saturation eliminated in dual-wound common-mode input filter inductors?
5. Explain the conditions that would control the selection of each of the following core materials for choke construction: (*a*) iron powder; (*b*) molybdenum Permalloy (MPP); (*c*) gapped ferrite; (*d*) gapped laminated iron.

CHAPTER 2
HIGH-CURRENT CHOKES USING IRON POWDER CORES

2.1 INTRODUCTION

Iron powder cores are constructed from finely divided ferromagnetic material bonded together by a nonmagnetic material in such a way that small gaps are introduced in the magnetic path throughout the body of the core. These distributed gaps significantly lower the effective permeability and increase the energy storage capability of the core. A further advantage of the distributed gap is the reduction of the radiated magnetic field as a result of the elimination of the discontinuity associated with the discrete air gap, more common in the ferrite or silicon iron-cored chokes.

For filter applications, improved high-frequency performance is sometimes obtained from iron dust cores, because of their large high-frequency losses. The iron dust core tends to reduce radiation due to core loss, and the material can be manufactured to selectively absorb some part of the high-frequency energy. These properties, together with the relatively low interwinding capacitance possible on a single-layer toroidal winding, results in good high-frequency rejection. However, considerable core losses and hence temperature rise may occur with this material at high frequency, when the flux density swing is large, and this must be considered in the choke design. An undoubted advantage of the iron dust core is its low cost.

At high frequencies, better choke efficiency will be obtained with Molypermalloy (MPP) toroids, as these have much lower core losses at high frequencies and will give a lower temperature rise. These cores are available in a wide range of sizes, shapes, and permeability, typically ranging from 14 through 550. To their disadvantage, the cost of MPP cores is generally considerably greater than that of powdered iron. Hence, for medium-frequency switchmode output choke applications, "iron powder" cores offer a very cost-effective alternative to MPP, gapped ferrites, or laminated silicon iron.

Unlike in the ferrite core, the power loss in the powdered iron material cannot normally be neglected when establishing the total power loss and temperature rise. Using the manufacturer's core data, operating frequency, core weight, and flux density excursion, the core losses can be established. These losses must be added to the copper losses to obtain the total dissipation.

Although iron powder materials have relatively large core losses compared with ferrite and MPP materials, they have low losses compared with laminated iron. Further, the larger loss may not be a major problem, provided that the ac ripple current component is low (say, less than 10% of I_{DC}).

Hence, for many medium-frequency high-DC-current chokes, the core losses are not predominant, and the various Permalloy, iron powder, and gapped laminated cores can be extremely size- and cost-effective. The saturating flux level is much higher than it is for ferrite materials, giving more energy storage and reduced size.

Although at higher frequencies, gapped ferrites in the form of E and I or pot cores are more often used, considerable magnetic radiation can occur where outer limbs are gapped. Also, the bobbin windings used with this type of core will usually have relatively large interwinding capacitance, and the self-resonant frequency will be low. Consequently, E-core chokes will not give very efficient high-frequency noise rejection when used on their own.

If a two-stage *LC* output filter is used, good high-frequency rejection in the first stage is not so essential, as a simple, low-cost ferrite rod choke in the second stage is very effective in dealing with the higher-frequency components. In such applications, the multilayer wound E core can be very effective.

As with the gapped ferrite core previously considered, the design exercise with powdered iron tends to be an iterative process. AC and DC flux density levels are calculated to ensure good design margins, and the wire gauge is selected for minimum loss. For the best high-frequency rejection (low interwinding capacitance), single-layer windings should be used on toroidal cores.

2.2   ENERGY STORAGE CHOKES

The energy stored in the choke core is proportional to the square of the flux density and inversely proportional to the effective permeability of the material. Hence, from Eq. (3.A.15),

$$\text{Energy} \propto \frac{B^2}{\mu_{\text{eff}}}$$

The high saturation flux density of powdered iron cores (greater than 1 T), together with their low permeability (35 to 75), makes them very suitable for energy storage chokes.

2.3   CORE PERMEABILITY

The materials considered in this section are the Micrometals Mix #8 through Mix #40.* These have permeabilities in the range 35 to 75. Table 3.2.1 shows the basic properties of these materials. (Other manufacturers produce similar materials.)

Iron powder materials have a "soft" saturation characteristic. This has two advantages for choke applications. First, good overload characteristics are provided, preventing a sudden loss of inductance if the current is above the normal maximum value. Second, the chokes can be designed to "swing," that is, to pro-

---

*A trade name of Micrometals Inc.

TABLE 3.2.1   General Material Properties

Mix #	Permeability, μ_0	Temperature stability (+)	Inductance tolerance, %	Relative cost	Core color code
8	35	225 ppm/°C	+10/−5	4.0	Yellow/Red
26	75	822	+15/−7½	1.2	Yellow/White
28	22	415	+10/−5	1.7	Gray/Green
33	33	635	+10/−5	1.6	Gray/Yellow
40	60	950	+15/−7½	1.0	Green/Yellow

vide a larger inductance at low DC currents. The swinging choke maintains continuous current conduction to a lower load current. This is often an advantage in
buck regulator applications. Figure 3.2.1 shows how the initial permeability of
the iron powder material falls as the DC magnetizing force increases.

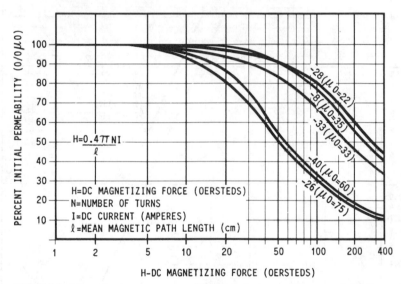

FIG. 3.2.1   Magnetization parameters of iron powder cores. (*Micrometals Inc.*)

2.4   *GAPPING IRON POWDER E CORES*

Iron powder E cores may be gapped to increase their energy storage capability,
prevent DC saturation, or obtain permeability values between the discrete values
provided by the material mix. Because of the rapid fall in effective permeability
at high values of DC magnetizing force, a small air gap can result in an increase
in effective permeability under some high-current conditions. This is particularly
true with the #26 and #40 mix materials.

2.5 METHODS USED TO DESIGN IRON POWDER E-CORE CHOKES (GRAPHICAL AREA PRODUCT METHOD)

General Conditions

In this example, in the interest of minimum cost, it has been assumed that the smallest E core consistent with a maximum temperature rise of 40°C is to be used. The required inductance and maximum DC load current are known.

With the above parameters defined, the smallest core size may be obtained using the area product nomogram in Fig. 3.2.2. [This nomogram is developed

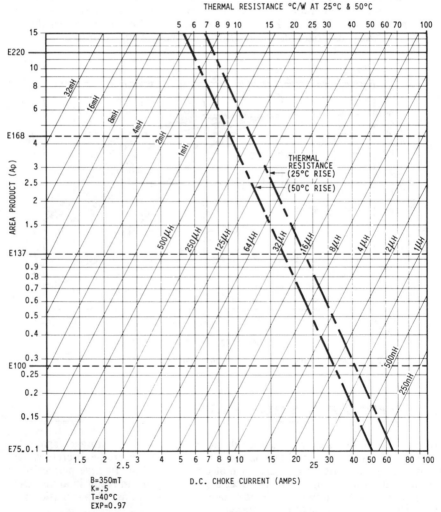

FIG. 3.2.2 Nomogram for iron powder cores, giving area product (and hence core size) as a function of DC load current and required inductance, with thermal resistance as a parameter.

from Eq. (3.A.13) and covers the current range from 1 to 100 A and inductances from 250 nH to 32 mH.]

Design Steps

In the area product graphical design approach, the following design steps should be followed:

1. To establish the core size, obtain the required area product as follows: Enter the bottom of the graph with the required DC current, project up to the required inductance, then project horizontally left, to the vertical scale, which then indicates the required area product.

   The area product translates to core size (see Table 3.2.2). Interpolate between the lines for other inductances, choose the next larger core size, or reduce the required inductance if the area product is between core sizes.

2. Calculate the number of turns needed to give the required inductance, using Eq. (3.A.9):

$$N = \frac{L \cdot I \cdot 10^4}{B \cdot A_e}$$

TABLE 3.2.2 Iron Powder E Core and Bobbin Parameters

Core type No.	Core parameters					Bobbin parameters			
	l, cm	A_e, cm^2	V, cm^3	W, cm^2	AP, cm^4	A_{wb}, cm^2	A_{pb}, cm^4	m_{lt}, cm	S_a, cm^2
E75	4.13	0.226	0.929	0.530	0.12	0.4	0.09	3.8	10.3
E100	5.08	0.403	2.05	0.810	0.32	0.62	0.25	5.1	16.5
E125	7.34	0.907	6.83	1.37	1.21	0.97	0.9	6.4	34.3
E137	7.30	0.907	6.63	1.51	1.37	1.22	1.1	7.0	36.1
E162	8.25	1.61	13.3	1.70	2.74	1.32	2.13	8.3	49.9
E168	10.3	1.84	19.0	2.87	5.28	2.32	4.3	9.2	67
E168A	10.3	2.45	25.3	2.87	7.03	2.17	5.3	10.2	73
E178	8.63	2.48	23.3	1.94	4.81	1.61	4.0	9.5	67
E220	13.1	3.46	42.3	4.07	14.08	3.33	11.5	11.9	114
E225	10.4	3.58	40.5	2.78	9.95	2.05	7.3	11.4	90
E450	20.9	12.2	279	12.7	154	10.5	128	22.8	354

Where l = magnetic path length, cm
A_e = effective core area, cm^2
V = volume of core, cm^3
W = window area of core, cm^2
AP = area product of core, cm^4
A_{wb} = window area of bobbin, cm^2
A_{pb} = area product of bobbin, cm^4
m_{lt} = mean length of turn, cm
S_a = surface area of wound core, cm^2

3. Calculate the initial permeability required from the core material, using Eq. (3.A.17).

$$\mu_r = \frac{l_e \cdot L \cdot 10^{-1}}{\mu_0 \cdot N^2 \cdot A_e}$$

4. Select the nearest core material that meets or exceeds the permeability requirements from Table 3.2.1.

5. Calculate the DC magnetizing force H_{DC} and check the percentage initial permeability from Fig. 3.2.2.

$$H = \frac{0.4\pi N \cdot I}{l_e}$$

6. Calculate the permeability of the selected core at the working magnetizing force H, and establish whether an air gap is required. (A gap is required if the permeability is too large or the core is saturating.)

7. If an air gap is required, first establish the effective length of the air gap built into the core material at the existing permeability. (This is the air gap that would exist if the distributed gap were all collected in one place.) Assume that the core material itself has zero reluctance. This calculation is quite simple; for example, if the existing permeability is 50, then the core length is 50 times longer than it would have been if it had been all air. This means that the effective air gap is only 1/50 of the actual length of the core.

Also calculate the required effective gap at the required (lower) permeability; for example, if a permeability of 40 is required, then the gap would be 1/40 of the core length. The difference between these two effective gaps is the length of the real air gap that must be added to obtain the required permeability. Hence

$$l_g = \frac{l_e}{\mu_x} - \frac{l_e}{\mu_r}$$

8. Calculate wire size and winding resistance.

9. Calculate the copper loss and check the predicted temperature rise of the finished choke from the thermal resistance shown in Fig. 3.2.2. If the ripple current exceeds 10% or the operating frequency is above 40 kHz, check the core loss and adjust the temperature rise prediction if necessary.

The following example demonstrates the use of the above equation.

2.6 EXAMPLE OF IRON POWDER E-CORE CHOKE DESIGN (USING THE GRAPHICAL AREA PRODUCT METHOD)

In this example, a 25-kHz continuous-mode buck regulator choke with the following parameters is to be designed on an iron powder E core:

$$\text{Inductance} = 1 \text{ mH}$$
$$\text{Mean DC current} = 6 \text{ A}$$
$$\text{Maximum temperature rise} = 50°C$$
$$\text{Maximum ripple current} = 10\%$$

2.6.1 Step 1, Establish Core Size

Obtain the area product from Fig. 3.2.2.

In this example, entering the graph with 6 A and projecting up to the 1-mH line gives an area product of 4.4.

From Table 3.2.2, the E168 core with an AP of 5.28 (4.3 with bobbin) meets this requirement.

2.6.2 Step 2, Calculate Turns

$$N = \frac{L \cdot I \cdot 10^4}{B \cdot A_e}$$

where N = total turns
$\quad L$ = inductance, mH
$\quad I$ = maximum DC current, A
$\quad B$ = maximum flux density, mT
$\quad A_e$ = effective core area, cm^2

Note: To limit core loss, the graph in Fig. 3.2.2 was developed using a peak flux density of 350 mT. This value is used to calculate the turns.

From Table 3.2.2, the E168 core constants are

$$\text{Area of center pole } A_e = 1.84 \text{ cm}$$

$$\text{Length of magnetic path } l_e = 10.3 \text{ cm}$$

Hence

$$N = \frac{1 \times 6 \times 10^4}{350 \times 1.84} = 93 \text{ turns}$$

2.6.3 Step 3, Calculate Required Initial Core Relative Permeability μ_r

$$\mu_r = \frac{l_e \cdot L \cdot 10^{-1}}{\mu_0 \cdot N^2 \cdot A_e}$$

where $\mu_0 = 4\pi \times 10^{-7}$
$\quad l_e$ = effective magnetic path length, cm
$\quad L$ = inductance, mH
$\quad N$ = turns
$\quad A_e$ = area of core, cm^2

Hence

$$\mu_r = \frac{10.3 \times 1 \times 10^{-1}}{4\pi \times 10^{-7} \times 93^2 \times 1.84} = 51$$

At a required permeability of 51, only the 40 and 26 core materials can be considered, as the rest have permeabilities which are already too low.

2.6.4 Step 4, Calculate the DC Magnetizing Force H_{DC}

Note: The magnetizing force is required so that the percentage initial permeability can be obtained.

$$H_{DC} = \frac{0.4\pi \, N \cdot I}{l_e} \quad \text{Oe}$$

where H_{DC} = DC magnetizing force, Oe
      N = turns (93)
      I = DC current (6 A)
      l_e = magnetic path length (10.3 cm)

Hence

$$H_{DC} = \frac{0.4\pi \times 93 \times 6}{10.3} = 68 \text{ Oe}$$

2.6.5 Step 5, Establish Effective Working Permeability and Select Core Material

From Fig. 3.2.1, at $H = 68$ Oe the percentage initial permeability is only 41% of μ-75 for the #26 mix material or 44% of μ-60 for the #40 mix material, giving

$$\mu_{r(\text{eff})} = 30.75 \text{ for the #26 material}$$
or
$$\mu_{r(\text{eff})} = 26.4 \text{ for the #40 material}$$

Hence, at $H = 68$, the effective permeability of the nongapped cores in any of the materials is lower than the required value of 51, because the core material is approaching saturation. However, if the higher-permeability material is chosen, it is possible to delay the onset of saturation by introducing an air gap. (Although the gap will reduce the permeability at low values of magnetizing force, it will increase it at high values by reducing the onset of saturation.) Hence in this example an air gap is indicated.

To obtain the highest possible permeability at full-load current, the higher-permeability (μ-75) #26 mix material is selected, and it is gapped to give a permeability of 51 at low currents.

2.6.6 Step 6, Establish Air Gap Size

The effective length of the core air gap intrinsic to the E168 μ75 core (that is, the effective length of the internal distributed gap) is l_e/μ_r (103/75 mm = 1.37 mm, this example). If the effective permeability $\mu_{r(eff)}$ is to be 51, then the effective air gap must be increased to $103/51 = 2.02$ mm. The difference is 0.65 mm (26 mil), and this is the required total air gap (in practice 13 mil in each leg). Hence

$$l_g = \frac{l_e}{\mu_x} - \frac{l_e}{\mu_r} \qquad \text{mm}$$

where l_e = effective length of magnetic path (of core), mm
$\quad l_g$ = length of air gap, mm
$\quad \mu_r$ = relative permeability (before gap is introduced)
$\quad \mu_x$ = required permeability (after gap is introduced)

In this example,

$$l_g = \frac{103}{51} - \frac{103}{75} = 0.65 \text{ mm}$$

Some adjustment of the air gap may be required to obtain the optimum inductance. Check core saturation (from page 2-29):

$$B_{DC} = \frac{\mu_0 \cdot N_p \cdot I_{DC}}{\alpha \times 10^{-3}}$$

2.6.7 Step 7, Establish Optimum Wire Size

For minimum copper loss the winding should fill the available bobbin window area with some allowance for insulation.

From Appendix 3.B, the packing factor K_u for round wire with heavy-grade insulation is 0.64.

The E168 core and bobbin have been selected. From Table 3.2.2, the window area of the bobbin $A_{w \cdot b}$ is 2.32 cm, and 93 turns are required (from step 2). Hence the area of the wire A_x is

$$A_x = \frac{A_{wb} \times K_u}{N}$$

$$= \frac{2.32 \times 0.64}{93} = 0.016 \text{ cm}^2$$

From Table 3.1.1, a #16 AWG wire is selected.

2.6.8 Step 8, Calculate Copper Loss

The total length of wire l on the fully wound bobbin can be calculated from the number of turns N and mean turn length l_m.

From Table 3.2.2, the mean turn length l_m for the E168 bobbin is 9.2 cm.

Hence

$$l = N \times l_m \qquad \text{cm}$$

Therefore, in this examples, $l = 93 \times 9.2 = 856$ cm.

The resistance per centimeter of #16 AWG copper wire RT_{cm} at 70°C (50°C rise above ambient) is 0.00015 Ω/cm (Fig. 3.4.11). Hence the total resistance of the winding R_{wT} is

$$R_{wT} = l \times RT_{cm} = 856 \times 0.00015 = 0.128 \ \Omega$$

The power P dissipated in the winding as a result of the DC current flow is

$$P = I^2 \times R = 6^2 \times 0.128 = 4.6 \text{ W}$$

2.6.9 Step 9, Calculate Temperature Rise

From Fig. 3.2.2, the thermal resistance $(R_o$ w-a) of the E168 wound assembly at a temperature rise above ambient of 50°C is given by the intersection of the horizontal E168 line with the diagonal 50°C line as 9.1°C/W (top scale). Hence the temperature rise of the wound core ΔT_a resulting from copper loss only would be approximately

$$\Delta T_a = (R_o \text{ w-a}) \times W = 9.1 \times 4.6 = 41.8°C$$

Core Loss (Iron Powder Cores).   The above example neglects the core loss and is valid so long as this loss is negligible. However, at ripple currents of more than 10% and/or frequencies higher than 40 kHz, the core loss can become significant and must be considered.

In this example, calculate the core loss for the above choke at 10% ripple current and 40-kHz operation. From Eq. (3.A.9) the flux density swing can be determined:

$$\Delta B = \frac{L \times \Delta I \times 10^4}{A_e \times N}$$

where ΔB = flux density swing, mT
$\quad\quad L$ = inductance, mH
$\quad\quad I$ = ripple current, A p-p
$\quad\quad A_e$ = effective core area, cm^2
$\quad\quad N$ = turns

Therefore
$$\Delta B = \frac{1 \times 0.6 \times 10^4}{1.84 \times 93} = 35 \text{ mT}$$

From Fig. 3.2.3, the core loss at 35 mT and 40 kHz is 50 mW/cm^3, and with a core volume of 19 cm^3, the total core loss is 950 mW. Hence, in this example, the core loss would give a 20% increase in the total loss, and the temperature rise will now be 50.3°C. Therefore some allowance must be made for core loss.

At high frequency and high ripple currents, the core loss quickly predominates and may even prevent the use of iron powder cores, as the following example shows.

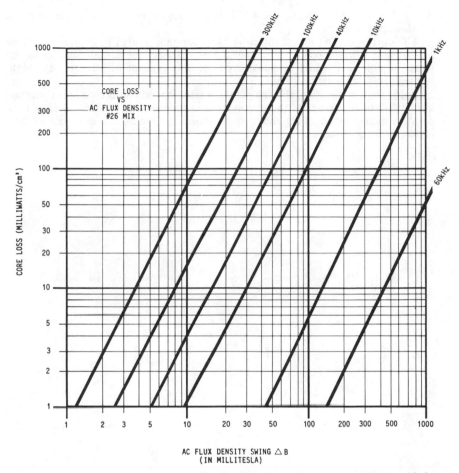

FIG. 3.2.3 Core loss as a function of ac flux density swing for iron powder #26 mix, with frequency as a parameter. (*Micrometals Inc.*)

Example

Consider the loss in the above choke example (1 mH at 6 A) if the choke were operated at 100 kHz and 20% ripple current (20% × 6 A = 1.2 A p–p ripple current).

Hence

$$\Delta B = \frac{1 \times 1.2 \times 10^4}{1.84 \times 93} = 70 \text{ mT}$$

From Fig. 3.2.3, the core loss of #26 mix at 70 mT and 100 kHz is 800 mW/cm³. The core volume (Table 3.2.2) is 19 cm³, and the total core loss is

$(0.8 \times 19) = 15.2$ W. Clearly the #26 iron powder material would not be suitable for this application.

The corresponding loss using μ-60 Molypermalloy (MPP) material would be 1.5 W, and with gapped ferrite, only 0.5 W. Hence, these materials would be more suitable and would be considered for this type of high-frequency, high-ripple-current application.

In the final application, the temperature rise depends on the complex thermal interaction of the complete thermal system in the finished product. A final temperature measurement of the choke under working conditions is essential to complete the design process.

CHAPTER 3
CHOKE DESIGN USING IRON POWDER TOROIDAL CORES

3.1 INTRODUCTION

Iron powder toroids can be very suitable for chokes in output filters of power converters provided that the DC current component is large compared with the ripple current. With iron powder, the core loss will be relatively low in the mid-frequency range, and although it will be somewhat higher than with MPP (molybdenum Permalloy) cores, there is a significant reduction in cost.

In applications in which high-frequency noise rejection is important, toroidal chokes with a single-layer winding give minimum interwinding capacitance and good noise rejection. When maximum inductance is required, a "full" multilayer winding will be used.

For the purpose of this discussion, the term "choke" will be used to describe an inductor which carries a large DC bias current and small ac ripple currents of less than 20%.

Since toroidal cores are generally supplied uncut, it is not possible to introduce an air gap into the magnetic path so as to change the permeability; hence, the design procedure must ensure that the core will not saturate under maximum current conditions, without requiring an air gap.

The area product of a toroid is often much larger than that of an E core of equivalent core area, as the window (center hole area) is relatively large. The main reason for a large window is to provide space for the winding shuttle when a winding machine is used. Unlike the bobbin-wound E core, the window space of a toroid cannot be fully utilized when machine winding is used.

It is common practice to wind single-layer windings on toroids, both for ease of winding and to get lower interwinding capacitance for improved high-frequency noise rejection. This can give very poor copper packing factors with relatively large copper loss.

A further complication unique to the toroid is the variation in the wire packing factor between the inner and outer surfaces. Because of this, small cores tend to have very poor packing factors when large-gauge wires are used.

The strict consistency of geometric ratios brought about by the "scrapless" approach to the E-core topology is not relevant to toroids; hence, there is a much greater variation in the core geometry with toroidal cores. As a result of the above limitations, together with the geometric inconsistency of the toroids, the area product design approach used for transformers and E-core chokes tends to be less generally applicable; it will not be used in the following example.

3.2 PREFERRED DESIGN APPROACH (TOROIDS)

Because the toroid window area tends to be rather large relative to the core cross-sectional area, it has been found to be better to allow the magnetic requirements to define the minimum core size, rather than the temperature rise requirements that were used for the E-core design approach.

In the following example, the need to provide a defined inductance, while preventing core saturation under maximum current conditions, will be used to define the minimum core size.

Further, it will be shown that this design approach yields a suitable core size and required number of turns directly from a single nomogram, when only the required mean current and inductance are known. The maximum temperature rise is then controlled by selecting a suitable wire gauge.

Although the nomograms used in this section were developed specifically for the Micrometals® powder cores, they may be used for any powder cores of similar geometry and permeability.

3.3 SWINGING CHOKES

Although this was not previously mentioned, chokes wound on real ferromagnetic materials will display some nonlinearity as a result of the nonlinear magnetization characteristic. The general effect is to reduce the effective small-signal dynamic inductance as the DC polarizing current increases. Where this effect is undesirable, air gaps will often be introduced to linearize the characteristic.

In some applications, a nonlinear choke characteristic (swinging choke) can be useful. For example, in buck regulators, if the inductance of the choke can be made to increase at light loads, continuous-mode converter operation can be extended to smaller minimum load currents. Conversely, at larger load currents, the inductance will be lower, giving better transient response. Chokes which display this change in effective inductance with DC bias current are known as "swinging chokes." The nonlinear B/H characteristic of nongapped iron powder toroids can be used to provide an inductance "swing" ratio of up to 2:1.

3.3.1 Types of Core Material Most Suitable for Swinging Choke Applications

For good swinging choke performance, the B/H loop of the core material should be very nonlinear. As the magnetizing force H increases (DC load current increases), the core material is taken toward saturation, and the permeability should become progressively lower. Some iron powder materials are particularly suitable for this application, as they display a rapid change in permeability.

Figure 3.2.1 shows how the permeability of the various iron powder materials falls as the DC magnetizing force H increases. The #26 and #40 materials display a rapid fall in permeability as H increases and are more suitable for swinging choke applications.

If these materials are operated at a maximum full-load magnetizing force H of 50, the permeability will be only half of the initial value, and the incremental full-load inductance will be only half of the initial (low-current) value. This change in permeability gives an inductance swing of 2:1, a useful range. Further, the core is not easily saturated at higher values of H, and the current overload safety margin is good. (A current overload of 100% will only reduce the inductance by a further 20%, and the core will not be saturated.) This provides a safe overload current margin for most applications. Further, it will be shown that choosing the magnetization H for a value of 50 gives a suitable core size, resulting in a working temperature rise not exceeding 40°C.

3.3.2 Design Example (Swinging Chokes)

The following design example is for a swinging choke with a 2:1 inductance swing ratio, wound on a toroidal powder core. A graphical design method based on the nomogram shown in Fig. 3.3.1 will be used. (The derivation of this nomogram is given in Appendix 3.C.)

Although Fig. 3.3.1 has been based on the #26 mix material, it can be used for all the iron powder materials by applying the inductance adjustment factors shown in the diagram. When this design approach is used with the low-permeability #28 and #8 materials, the major difference is that the inductance swing will be much less.

3.3.3 Choke Specification

Assume a requirement for a continuous-mode, 5-V, 10-A, 100-kHz buck regulator swinging choke. The peak-to-peak ripple current is to be 20% of the rated current at the maximum duty ratio of 48%. The inductance swing is to be 2:1, so as to give continuous choke conduction down to a minimum load of 0.5 A. To satisfy this requirement, the ripple current must not exceed 10% of rated output current. Hence at a 0.5-A load the inductance must be double that at full load, and the choke specification will be as follows:

$$V_{out} = 5 \text{ V}$$
$$I_{mean} = 10 \text{ A}$$
$$I_{L(p-p)} = 2 \text{ A (at full load)}$$
$$f = 100 \text{ kHz}$$
$$t_p = 10 \text{ } \mu s$$
$$t_{off} = 5.2 \text{ } \mu s$$
$$I_{L(p-p)} = 1 \text{ A (at minimum load, 0.5 A)}$$

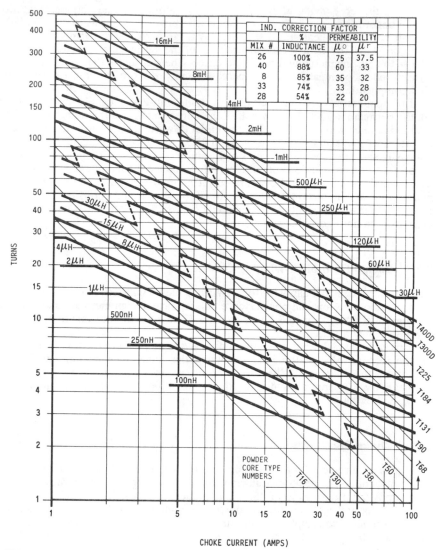

IND. CORRECTION FACTOR

MIX #	% INDUCTANCE	PERMEABILITY μo	μr
26	100%	75	37.5
40	88%	60	33
8	85%	35	32
33	74%	33	28
28	54%	22	20

CHOKE CURRENT (AMPS)

FIG. 3.3.1 Nomogram for iron powder toroidal cores, giving turns as a function of choke current, with required inductance and core size as parameters.

3.3.4 Step 1, Calculate the Inductance Required at Full load

In the buck regulator (Fig. 2.20.1*a*), during the "off" or flywheel period, the voltage *e* across the choke *L* is constant at V_{out} plus a diode drop; hence

$$|e| = L\frac{di}{dt}$$

where $|e|$ = the magnitude of the voltage across the choke during the "off"
period
di/dt = the rate of change of current during the "off" period

and L may be calculated as follows.

In this example, $\qquad e = V_{out} + 0.7 = 5.7 \text{ V}$

and

$$\frac{di}{dt} = \frac{I_{L(p-p)}}{t_{off}} = \frac{2}{5.2} \text{ A/}\mu\text{s}$$

Hence

$$L = \frac{5.7 \times 5.2 \times 10^{-6}}{2} = 14.8 \ \mu\text{H}$$

3.3.5 Step 2, Obtain Core Size and Turns from Nomogram (Fig. 3.3.1)

Enter the nomogram at the bottom scale with the required maximum load current of 10 A. Project upward to the required inductance (14.8 μH in this example). Use the 15-μH full-load inductance intersect (heavy line marked 15 μH). The nearest (lower) solid diagonal gives the required core size (T90 in this case).

The horizontal left projection from the core and inductance intersect indicates the required turns (21 in this example). Thus a T90 core wound with 21 turns will be used.

3.4 WINDING OPTIONS

Once the required turns have been determined (as shown above), there are three options normally considered for toroidal choke winding depending on the required performance.

3.4.1 Option A, Minimum Loss Winding (Full winding)

In this design option, the main aim is to minimize the power loss (mainly copper loss). Hence the maximum wire gauge that will just fill the available window space (normally limited to 55% of the total window area) is selected.

This "full winding" will give the lowest copper loss, but it is more difficult and expensive to manufacture. Further, because of larger interlayer winding capacitance, the high-frequency noise rejection is not as good as that of a single-layer winding.

3.4.2 Option B, Single-Layer Winding

In this option, a wire gauge is selected that will just give the required number of turns in a single layer (with a space between start and finish of at least one turn).

This option gives easy, low-cost windings with a low distributed capacitance and good high-frequency noise rejection. However, it has the disadvantage of poor utility of the available winding space, with higher copper loss and the largest temperature rise.

3.4.3 Option C, Winding for a Specified Temperature Rise

This option is a compromise approach in which the wire gauge is selected to give a specified temperature rise that is between the minimum value provided by option A and the maximum loss condition of the single-layer option B.

3.5 DESIGN EXAMPLE (OPTION A)

Once the inductance, turns, and core size have been selected using Fig. 3.3.1, as described in Sec. 3.4, the maximum wire size that will just fill 45% of the selected core window area is required.

3.5.1 Selecting Wire Size

Figure 3.3.2 is a nomogram showing the gauge of wire and number of turns that will just give a full winding on the selected core.

Enter the graph from the left with the required number of turns. The intersection of the number of turns with the *solid* diagonal "core line" for the selected core indicates the required wire gauge on the lower scale.

Note: The dashed lines in Fig. 3.3.2 indicate the gauge of wire to be used if a single-layer winding is to be maintained.

It should be noted that at wire gauges above #18 AWG, the option for multiple wires of 18 AWG (for ease of winding) is provided. Other wire sizes may also be used, so long as the copper cross-sectional area is maintained.

When the *solid* line is followed above the slope discontinuity on Fig. 3.3.2, the winding reverts from a single-layer to a multiple-layer winding.

Example

Using the information for the 10-A 15-μH example in Sec. 3.3.5, 21 turns on a T90 core are required. Entering Fig. 3.3.2 with 21 turns, the intersect with the T90 core *solid* line indicates a wire gauge of #13 AWG, or three wires of #18 AWG, for a full winding. (The intersect is above the slope discontinuity; hence the winding will be multiple-layer.)

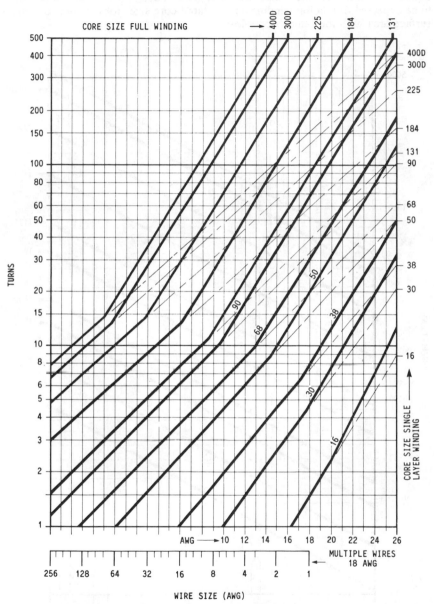

FIG. 3.3.2 Nomogram for iron powder toroids, giving wire size as a function of turns, with core size and single- or multiple-layer windings as parameters.

3.5.2   Temperature Rise, Full Winding

Figure 3.3.3 indicates the temperature rise (above ambient free air temperature) to be expected from a full winding on the stated core sizes (or for cores of similar surface area with the same copper loss).

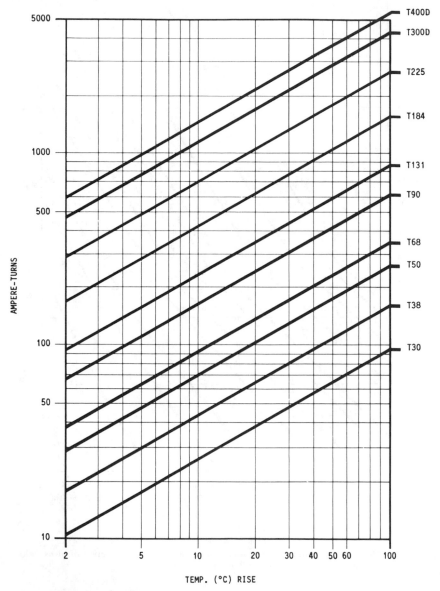

FIG. 3.3.3   Nomogram for iron powder toroids, giving temperature rise as a function of choke DC ampere-turns, with core size for single-layer windings as a parameter.

Enter the graph at the left with the required ampere-turns. The intersect with the core size indicates the temperature rise on the lower scale.

Example

Using the information from the previous example, 10 A and 21 turns gives 210 ampere-turns, on the T90 core.

Entering Fig. 3.3.3 at the left, with 210 ampere-turns, the intersect with the T90 core indicates a temperature rise of approximately 15°C (on the lower scale).

This temperature rise is a free air prediction considering copper loss and core area; it neglects core losses, which should be small. However, the temperature needs to be checked in the finished application, since core losses, together with the thermal contribution from nearby hot components, can make a considerable difference in the final temperature.

3.6 DESIGN EXAMPLE (OPTION B)

Single-layer windings have the advantage of being simple to manufacture and having low interwinding capacitance, giving good high-frequency rejection. The disadvantage is smaller wire sizes and higher copper loss, giving a higher temperature rise.

3.6.1 Selecting Wire Size

Using the same example as for the full winding, a choke of 15 μH at 10 A on a T90 core will require 21 turns.

Entering Fig. 3.3.2 with 21 turns and using the dashed T90 core line this time (for a single-layer winding) yields a smaller gauge of #15 AWG (or two wires of #18 AWG).

3.6.2 Temperature Rise

To obtain the temperature rise for a single-layer winding, enter Fig. 3.3.4 with the load current. The intersect with the selected wire gauge indicates the temperature rise on the lower scale.

In this example, 10 A and #15 AWG yields a predicted temperature rise of 22°C (47% higher than the full winding example).

3.7 DESIGN EXAMPLE (OPTION C)

This option is more difficult to design, since it requires a more iterative process. Using the preceding approaches, the temperature will be between 15 and 22°C, depending on the wire gauge. For temperature outside this range (for the same current and inductance), the same design process as in the above examples may be used, with larger or smaller cores selected to give smaller or larger tempera-

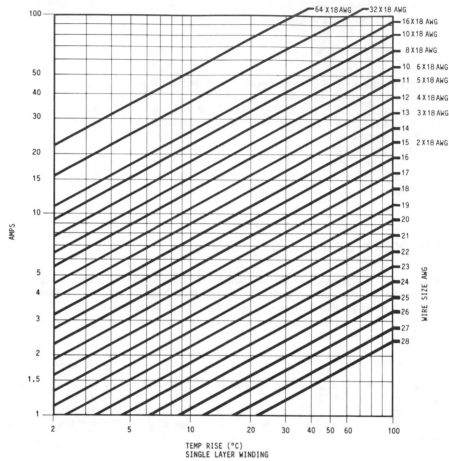

FIG. 3.3.4 Nomogram for iron powder toroids using single-layer windings, giving temperature rise as a function of DC load current, with wire size as a parameter.

ture rises. Several iterations may be required to get acceptable results. Alternatively, the process may be reversed by starting with the required temperature rise, using Fig. 3.3.4 for single-layer windings or Fig. 3.3.3 for full windings.

Note: Changing the number of turns from that indicated in Fig. 3.3.1 will change the inductance and "swing" ratio. (The swing ratio is given by the percentage permeability graph, Fig. 3.2.1.)

3.8 CORE LOSS

In all the above examples, the core loss has been neglected; hence the temperature rise prediction is valid only when the core loss is in fact negligible. At high frequency and high ripple currents, the core loss may be significant and should be considered.

For the previous example, the flux density swing at full load can be determined from Eq. (3A.9) as follows:

$$\Delta B = \frac{L \times I_{ac} \times 10^4}{A_e \times N}$$

where ΔB = flux density swing, T
  L = inductance, H
  $I_{L(p-p)}$ = ripple current, A p–p
  A_e = core area, cm^2
  N = turns

In the above example,

$$L = 15 \ \mu H$$
$$I_{L(p-p)} = 2 \ A$$
$$A_e = 0.422 \ cm^2$$
$$N = 21$$

$$\Delta B = \frac{15 \times 10^{-6} \times 2 \times 10^4}{0.422 \times 21} = 33.8 \ mT$$

From Fig. 3.2.3 (p. 3.39), at a frequency of 100 kHz and a flux density swing of 33.8 mT, the core loss is 140 mW/cm^3. The core volume is 2.68 cm^3, giving a total loss of 375 mW. Although small, this compares with the copper loss and should be considered when calculating the temperature rise. At lower frequencies the core loss can be neglected.

3.9   TOTAL DISSIPATION AND TEMPERATURE RISE

When the core loss is large and cannot be neglected, the temperature rise can be obtained from Fig. 3.3.5, using the total dissipation figure.

Add the copper and core losses to obtain the total dissipation. Enter the nomogram (Fig. 3.3.5) with the total dissipation on the lower scale. The intersect with the surface area (left) or core size (right) gives the predicted temperature rise on the diagonal lines.

3.9.1   Calculate Winding Resistance

Using the example for option A, the full winding (Sec. 3.5), calculate the copper resistance from the wire gauge, number of turns, and total winding length. For the full winding the following parameters were obtained:

$$\text{Turns } N = 21$$
$$\text{Wire gauge} = 3 \text{ wires of } \#18 \text{ AWG}$$
$$M_{lt} = 3 \text{ cm}$$
$$RT_{cm} \text{ (resistance of } \#18 \text{ AWG)} = 0.00024 \ \Omega/cm \text{ at } 50°C \text{ (p. 3.23)}$$
$$\text{Load current} = 10 \text{ A}$$

The total copper resistance of each wire is given by

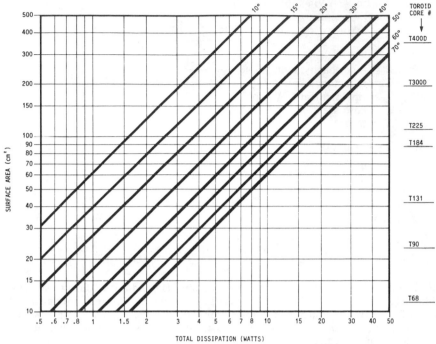

FIG. 3.3.5 Nomogram for toroidal cores, giving temperature rise as a function of total dissipation and core surface area, with the surface area of typical cores as a parameter.

$$R_{total} = N \times M_{lt} \times RT_{cm}$$

Hence, with three wires in parallel,

$$R_{total} = \frac{21 \times 3 \times 0.00024}{3} = 0.005 \ \Omega$$

$$\text{Copper dissipation} = I^2R = 0.5 \text{ W}$$

Add the core loss Sec. 3.8 (375 mW).

Total dissipation = 0.5 + 0.375 = 0.875 W

From Fig. 3.3.5, the temperature rise for the T90 wound core at full load, including the core loss, will be approximately 22°C.

3.10 LINEAR (TOROIDAL) CHOKE DESIGN

Although the design approach detailed in Secs. 3.2 through 3.7 specifically applies to "swinging chokes," it can also be used for linear choke design. Linear chokes require the core permeability to remain constant throughout the current range so that the inductance will remain constant. It is not possible to completely eliminate the inductance swing, as all materials display some curvature in the B/H characteristic. However, the #28 and #8 materials, when operated at a magnetizing force of 50 Oe, will have a maximum swing of less than 10%, as shown in Fig. 3.2.1.

Hence, for linear choke design, choose #8 or #28 material and adjust the inductance shown in Fig. 3.3.1 by the correction factor shown for the selected material. Otherwise proceed as for "swinging choke" design.

APPENDIX 3.A
DERIVATION OF AREA PRODUCT EQUATIONS
for Energy Storage Chokes

The area product (AP) is the product of the core winding window area and the center pole cross-sectional area. As such, the dimensions are in cm^4. This factor can be used by core manufacturers and designers to indicate the power-handling ability of transformer and choke cores.[1,2]

Chokes (inductors which carry a large DC current component) wound on ferrite cores will have relatively large air gaps to prevent saturation of the core. As a result, the majority of the reluctance and stored energy will be concentrated in the air gap, not in the core. On this basis the following approximations can be used in the derivation of equations relating the area product (AP) to the inductance and current ratings of chokes wound on gapped ferrite cores:

1. The reluctance of the core may be considered negligible compared with the air gap; hence all the energy will be assumed to be concentrated in the air gap.
2. With a large air gap, the permeability μ_r is nearly constant, provided that saturation is avoided; hence the B/H characteristic is assumed to be linear.
3. A uniform distribution of flux density in the air gap is assumed.
4. A uniform distribution of field intensity H in the air gap is assumed.

3.A.1. BASIC MAGNETIC EQUATIONS SI UNITS

1. In accordance with Faraday's law of induction, the emf induced in a winding is

$$e = N \cdot \frac{d\Phi}{dt} = \frac{L \cdot dI}{dt} \qquad (3.A.1)$$

Also,

$$\Phi = A_g \cdot B$$

Hence

3.54

$$e = N \cdot A_g \cdot \frac{dB}{dt}$$

For this example (assuming that the gap is concentrated in a single place), $A_g = A_p$, where A_p is the area of the core pole. (Fringe effects are neglected.)

2. Ampere's law of electromotive force (as applied to the field in the air gap) is

$$\text{mmf} = \int H \cdot d \, l_g = H_{l_g} = NI \qquad (3.A.2)$$

3. The magnetic field relationship is

$$B = \mu_r \cdot \mu_0 \cdot H$$

But $\mu_r = 1$ (in the air gap). Hence

$$H = \frac{B}{\mu_0} \qquad (3.A.3)$$

From Eq. (3.A.1),

$$e = L \cdot \frac{di}{dt}$$

Hence

$$e \cdot dt = L \cdot di$$

Multiplying both sides by I and integrating,

$$J = \int e \cdot I \cdot dt = \int L \cdot I \cdot di \qquad (3.A.4)$$

Hence

4.

$$J = \frac{1}{2} \cdot L \cdot I^2$$

From Eq. (3.A.2),

5.

$$I = \frac{H \cdot l_g}{N} \qquad (3.A.5)$$

Multiplying Eq. (3.A.5) by Eq. (3.A.1),

$$E \cdot I = N \cdot A_g \cdot \frac{dB}{dt} \cdot H \cdot \frac{l_g}{N}$$

Hence

$$E \cdot I \cdot dt = A_g \cdot l_g \cdot H \cdot dB$$

Integrating both sides,

6.

$$\int E \cdot I \cdot dt = J = A_g \cdot l_g \cdot \int H \cdot dB \qquad (3.A.6)$$

From Eq. (3.A.3),

$$H = \frac{B}{\mu_0}$$

Substituting for H in Eq. (3.A.6), with μ_0 constant,

$$J = A_g \cdot l_g \cdot \int \frac{B \cdot dB}{\mu_0} = \frac{A_g \cdot l_g}{\mu_0} \cdot \int B \cdot dB$$

Hence

$$J = \frac{A_g \cdot l_g \cdot B^2}{2 \cdot \mu_0} = \frac{1}{2} \cdot A_g \cdot l_g \cdot \frac{(B)}{(\mu_0)} \cdot B$$

But from Eq. (3.A.3),

$$\frac{B}{\mu_0} = H$$

Hence

7.

$$J = \frac{1}{2} \cdot B \cdot H \cdot A_g \cdot l_g \qquad \text{J} \qquad (3.A.7)$$

Equating Eqs. (3.A.4) and (3.A.7),

8.

$$\frac{1}{2} \cdot L \cdot I^2 = \frac{1}{2} \cdot B \cdot H \cdot A_g \cdot l_g \qquad (3.A.8)$$

Hence the circuit energy equals the magnetic energy stored in the gap.
From Eq. (3.A.2),

$$N \cdot I = H \cdot l_g$$

Substitute in Eq. (3.A.8):

$$\frac{1}{2} \cdot L \cdot I^2 = \frac{1}{2} \cdot B \cdot A_g \cdot N \cdot I$$

Simplify:

$$L \cdot I = \dot{B} \cdot A_g \cdot N$$

Hence

9.

$$N = \frac{L \cdot I}{B \cdot A_g} \tag{3.A.9}$$

where I = peak current.

Now consider the winding: The ampere-turns is the current density in the wire I_a multiplied by the usable window area A_w, modified by the packing factor K_u.

$$N \cdot I_{rms} = I_a \cdot A_w \cdot K_u$$

where I_{rms} = rms (heating) current.

Hence

10.

$$N = \frac{A_w \cdot I_a \cdot K_u}{I_{rms}} \tag{3.A.10}$$

Equating N in Eqs. (3.A.9) and (3.A.10),

$$\frac{A_w \cdot I_a \cdot K_u}{I_{rms}} = \frac{L \cdot I}{B \cdot A_g}$$

The area product $AP = A_w \times A_g$; hence, solving for AP and converting dimensions to centimeters,

11.

$$AP = A_w \cdot A_g = \frac{L \cdot I \cdot I_{rms} \cdot 10^4}{I_a \cdot K_u \cdot B} \tag{3.A.11}$$

In continuous-current choke applications, the peak current amplitude I is very close to the rms current I_{rms}; hence, $I \times I_{rms} \approx I^2$. Further, for most practical applications, the current density I_a, packing factor K_u, and peak flux density $\hat{B}$ can be considered constant. Hence, Eq. (3.A.11) relates the area product to the stored energy, $\frac{1}{2}(L \cdot I^2)K$.

The nomogram shown in Fig. 3.1.2 was developed assuming that the current density was such as to give a temperature rise of 30°C in free air cooling conditions. Further, the packing factor K_u is assumed to be constant at 0.6 (nominal for a single winding of round wire), and the peak flux density is assumed to be 0.25 T.

It has been shown[1,2] that a choke with an area product of 1 will have a temperature rise of approximately 30°C in free air at a winding current density I_a of 450 A/cm². With larger cores the current density must be decreased, as the ratio of volume to surface area decreases with size.

Hence with larger cores the current density I_a should be reduced as follows:

12.

$$I_a = 450 \, AP^{-0.0.125} \quad A/cm^2 \quad\quad (3.A.12)$$

Substituting for I, $b \cdot K_u$, and I_a in Eq. (3.A.11) gives for a 30°C temperature rise the special case for AP of

13.

$$AP = \left(\frac{L \times I^2 \times 10^4}{450 \times 0.6 \times 0.25} \right)^{1.143} \quad cm^4 \quad\quad (3.A.13)$$

A family of curves of AP developed from Eq. (3.A.13) for various values of inductance L plotted against load current is shown in Fig. 3.1.2.

3.A.2. FURTHER USEFUL DERIVATIONS

Energy Storage

From Eq. (3.A.7),

$$J = \frac{1}{2} \cdot B \cdot H \cdot l_g \cdot A_g$$

From Eq. (3.A.3),

14.

$$H = \frac{B}{\mu_r \cdot \mu_0}$$

Substituting Eq. (3.A.3) into Eq. (3.A.7),

$$J = \frac{B^2 \cdot A_g \cdot l_g}{2 \cdot \mu_0 \cdot \mu_r} \quad joules \quad\quad (3.A.14)$$

But $A_g \cdot l_g$ = volume of air gap (or volume of core and air gap where l_m is total mean length of magnetic path, A_e is effective area of core, and μ_r is permeability of core plus gap). Hence, in general, the energy density is

15.

$$J/m^3 = \frac{B^2}{2 \cdot \mu_0 \cdot \mu_r} \quad joules/m^3 \quad\quad (3.A.15)$$

Or for the air gap only (since $\mu_r = 1$),

$$J/m^3 = \frac{B^2}{2 \cdot \mu_0} \quad joules/m^3$$

3.A.3. INDUCTANCE

From Eq. (3.A.8),

$$\frac{1}{2} \cdot L \cdot I^2 = \frac{1}{2} \cdot B \cdot H \cdot A_g \cdot l_g$$

But from (3.A.5)

$$I = \frac{H \cdot l_g}{N}$$

Hence (substituting for I^2)

16.

$$L = B \cdot H \cdot A_g \cdot l_g \frac{N^2}{H^2 \cdot l_g^2} = \frac{B \cdot N^2 \cdot A_g}{H^2 \cdot l_g^2} \qquad (3.A.16)$$

Substitute $\mu_r \cdot \mu_0$ for B/H from Eq. (3.A.3):

17.

$$L = \frac{\mu_r \cdot \mu_0 \cdot N^2 \cdot A_g}{l_g} \qquad H \qquad (3.A.17)$$

APPENDIX 3.B
DERIVATION OF PACKING AND RESISTANCE FACTORS

The nomogram shown in Fig. 3.1.3 is developed from the basic physical parameters of the core and former.

3.B.1. MAXIMUM TURNS

For choke applications, a fully wound bobbin is normally used so as to obtain the maximum energy storage with minimum copper loss. The aim is to use the smallest core that will just provide the required inductance with an acceptable temperature rise and minimum voltage drop.

The stored energy J is $\frac{1}{2} L \cdot I^2$ joules. To maximize J with a defined maximum load current, L must be maximum. Further, for a defined core size and material, $L \propto N^2$, and hence the number of turns must be maximum.

However, the larger the number of turns, the greater the copper loss and temperature rise. Hence, to obtain the optimum inductance on a particular core size, a compromise must be made between temperature rise and inductance.

3.B.2. PACKING FACTOR K_u

The window area of the core cannot be fully utilized for winding copper because of the need for a bobbin and insulation. Further, the wire shape and wastage mean that only part of the available space is occupied by copper.

In chokes, a single winding is normally used, and the need for insulation is minimal. The nomogram assumes that round wires will be used, and the ratio of a round conductor to the occupied rectangle is $\pi(r)^2/(2r)^2 = \pi/4 = 0.785$. In the middle of the wire range (#20 AWG), if heavy-grade insulation is used, the ratio of the copper area to the overall wire area is 0.83 (the remainder being taken up by the wire insulation). This reduces K_u to 0.65, and to allow a margin for end wastage and insulation, a packing factor K_u of 0.6 will be used. If the wire is to be wound on a bobbin, this packing factor must be applied to the usable window area of the bobbin a_{wb} rather than to the window area of the core.

3.B.3. RESISTANCE FACTOR R_x

The resistance factor R_x is defined as the resistance of a single turn with a cross-sectional area that will just fill the available window area at a packing factor of 0.6. To calculate the resistance of this single turn, it is necessary to obtain its effective mean length l_m and its area. For this, the mean diameter and window area of the particular bobbin must be used. Hence, the resistance factor R_x has a specific value for each core size and allows the resistance of a fully wound bobbin R_w to be quickly established for any number of turns, as follows:

$$R_w = N^2 \cdot R_x \qquad (3.B.1)$$

where R_w = the resistance of the winding, assuming a fully wound bobbin, $\mu\Omega$
N = total turns
R_x = resistance factor (defined above), and shown in Table 3.3B.1

Note: The resistance increases as N^2, not N, because the bobbin is always assumed to be fully wound. Hence if the turns double, the copper area must halve, and the total length of the winding doubles.
The volume resistivity ρ of copper at 0°C is 1.588 $\mu\Omega/cm^3$.

Note: The volume resistivity is defined as the resistance across opposite faces of a 1-cm cube of copper at 0°C (+273 K), in microhms. Hence the resistance of a length of copper wire at 0°C, R_{Cu}, is given by

$$R_{Cu} = \rho \cdot \frac{l}{A} \qquad \mu\Omega \qquad (3.B.2)$$

The temperature coefficient of copper referred to 0°C, R_T, is between 0.00427 and 0.00393 $\Omega/(\Omega \cdot °C)$, depending on how soft it is. For this example a nominal value of 0.004 $\Omega/(\Omega \cdot °C)$ will be used. Hence the resistivity of the copper wire at 100°C, ρ_{tc}, may be calculated as follows:

$$\rho_{tc} = \rho(1 + R_T \cdot T) \qquad \mu\Omega/cm^3$$

where ρ_{tc} = resistivity of copper at $T°C$, $\mu\Omega/cm^3$
ρ = resistivity of copper at 0°C, $\mu\Omega/cm^3$
R_T = temperature coefficient of copper at 0°C
T = working temperature, °C

Hence at 100°C, ρ_{tc} is

$$1.588(1 + 0 \cdot 004 \cdot 100) = 2.22 \ \mu\Omega/cm^3 \qquad (3.B.3)$$

Using this value of resistivity, the published window area of the EC range of bobbins A_{wb}, and a packing factor K_u of 0.6 for round wire, the resistance factor R_x for the EC range of bobbins was calculated, and is shown in Table 3.3B.1.

$$R_x = \frac{\rho_{tc} \times l_m}{A_{wb} \times K_u} \qquad \mu\Omega \qquad (3.B.4)$$

TABLE 3.3B.1   Resistance Factor R_x and Effective Area Product EAP for the EC Range of Cores, Using Round Magnet Wire at a Working Temperature of 100°C

	Core type					
	EF 20	EF 25	EC 35	EC 41	EC 52	EC 71
A_{wb}	0.20	0.34	0.606	0.834	1.29	2.86
l_m	4.12	5.2	5.0	6.0	7.3	9.5
R_x	46	34	18.72	16.33	12.8	7.54
EAP			0.715	1.46	2.96	9.91

where R_x = resistance factor
  ρ_{tc} = copper resistivity at 100°C, $\mu\Omega/cm^3$
  l_m = mean length of turn, cm
  A_{wb} = bobbin window area (one side), cm^2
  K_u = packing factor (0.6)

Note: R_x is in microhms.

3.B.4. NUMBER OF TURNS IN A FULLY WOUND BOBBIN

The number of turns that will fit into a fully wound bobbin is simply determined by dividing the useful window area of the bobbin, $A_{wb} \times K_u$, by the area of the copper for the selected wire gauge, A_x. Hence

$$\text{Turns} = \frac{A_{wb} \times K_u}{A_x} \qquad (3.B.5)$$

3.B.5. RESISTANCE OF A FULLY WOUND BOBBIN

From Eq. (3.B.1) and the listed resistance factor, the resistance of a fully wound EC bobbin can be quickly established as follows:

$$R_w = N^2 \times R_x \qquad \mu\Omega \qquad (3.B.6)$$

APPENDIX 3.C
DERIVATION OF NOMOGRAM 3.3.1

The nomogram in Fig. 3.3.1 is developed for the Micrometals T range of iron powder cores, with the initial magnetization condition $H = 50$ Oe. From Fig. 3.2.1 it can be seen that this gives an initial permeability of 50% for the #26 iron powder material. This provides the 2:1 inductance swing for the "swinging choke" requirement.

From the magnetizing equation,

$$H = \frac{0.4\pi N \cdot I}{l_e} \qquad (3.C.1)$$

where H = magnetizing force, Oe
  N = turns
  I = peak current, A
  l_e = effective magnetic path length, cm

From Eq. (3.C.1),

$$N = \left(\frac{H \times l_e}{0.4\pi}\right) \times \frac{1}{I} \qquad (3.C.2)$$

Figure 3.3.1 shows a plot of N vs. I for each core at a constant magnetizing force H of 50 Oe.

The inductance for the specified number of turns may be obtained from the published A_L value or from Eq. (3.A.17). The inductance has also been plotted on Fig. 3.3.1.

Note: The discontinuity between cores T38-50 and T184-225 is caused by a change in the geometry of the cores between these sizes.

CHAPTER 4
SWITCHMODE TRANSFORMER DESIGN (GENERAL PRINCIPLES)

4.1 INTRODUCTION

The design of the switchmode transformer, more than any other single factor, will determine the overall efficiency and cost of a modern switchmode supply. Unfortunately, the transformer design process also tends to be the most poorly understood area of switchmode design.

The intention of this section is to throw some light on the major parameters which control high-frequency transformer design. The subject is vast, and many volumes have already been devoted specifically to it. It is beyond the scope of this book to deal with more than some general aspects, but these are covered in sufficient detail to allow the production of working prototypes which will not be too far from an optimum design.

Expert transformer engineers will already be familiar with the many volumes available on the subject, and will be using one of the more sophisticated core geometry, area product, or computer-aided design techniques. This section is not intended for them, and they will hopefully forgive the nonrigorous approach used in the interest of simplicity. The subject is dealt with here in a relatively general way, as specific design examples have already been given in the appropriate converter sections.

This chapter is broken down into five parts as follows:

1. Sections 4.2, 4.3, and 4.4, transformer size
2. Section 4.5, flux-density swing
3. Section 4.6, turns
4. Section 4.7, core loss
5. Section 4.8, copper loss
6. Section 4.9, skin effects

The design procedures apply in the frequency range 20 kHz to 100 kHz and assume the use of low-loss transformer ferrites and bobbin windings. It is also assumed that commercial-grade, convection-cooled transformers are to be produced. Typically these transformers will use AIEE class A or class B insulation (105 or 130°C maximum operating temperature); also, they will meet UL, CSA, and in some cases IEC and VDE transformer safety specifications.

4.2 TRANSFORMER SIZE (GENERAL CONSIDERATIONS)

Although it is generally understood that the transformer size is related to the power output, it is not always fully appreciated that there are no fundamental electrical formulas linking a transformer's size solely to its electrical parameters. It is the second-order effects of core loss, copper loss, skin and proximity effects, cooling efficiency (air flow), insulation, core geometry, finish, and location within the equipment which define the temperature rise and hence size of the transformer. The need to limit the temperature rise to an acceptable value under real working conditions defines the required surface area and hence the size.

If losses are very small, and an efficient means of extracting the heat is provided, the transformer can be very small and still meet the electrical requirements. For example, by using low-resistance magnet wire (superconductors), low-loss core material (amorphous alloy), and efficient vapor-phase liquid cooling techniques (heat pipes), miniature high-efficiency transformers with superior electrical parameters can be, and are, produced. This point is made to illustrate that transformer size is very much a function of the many practical factors that control the temperature rise, as well as of the electrical parameters.

In practical switchmode applications, copper magnet wire will be used, together with class A or B insulating materials and ferrite cores. In many applications free air convected cooling is used. In such applications, the transformer size will be defined by the internal power loss of the transformer and the surface area required to meet the temperature rise limitations.

Many manufacturers provide nomograms for their cores. These indicate the typical output power for each core type, under push-pull and single-ended high-frequency operation. (A typical example of a power nomogram is shown in Fig. 3.4.1.) These nomograms, or the area product factors, provide a good starting point for selecting the transformer size.[1,2]

4.3 OPTIMUM EFFICIENCY

Optimum efficiency will occur when core and copper losses are minimized. In general terms, for a given power throughput, the core losses increase as the core size and flux density swing increase, and the copper loss increases as the core size and flux density swing decrease. Hence, to obtain maximum efficiency, an optimum balance between core loss and copper loss must be maintained.

Clearly, these two requirements are in conflict, and a compromise selection must be made. It is normally assumed that optimum efficiency will occur when core and copper losses are equal. In practice, the precise loss apportionment for maximum efficiency depends on the core material, core geometry, and operating frequency.

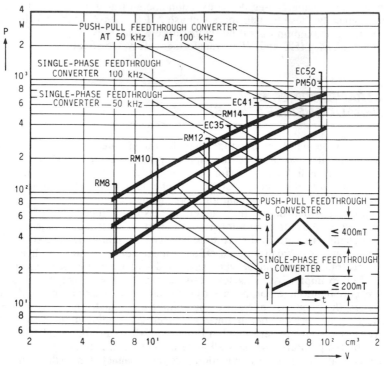

FIG. 3.4.1 Nomogram giving power ratings of ferrite cores as a function of core volume, with core size and frequency as parameters. (*Siemens AG*)

Figure 3.4.2 shows the copper loss, core loss, and total loss of a N27 ferrite EC41 core at 20 kHz and 50 kHz, as a function of flux density swing.

At 50 kHz the maximum efficiency (lowest loss) occurs when the core loss is 44% and the copper loss 56% of the total loss. At 20 kHz the optimum efficiency occurs near where core and copper losses are equal. However, the minimum loss curve is quite broad-based, and near-optimum designs for both frequencies will result if copper and core losses are kept nearly equal.

Further factors affecting core size are operating frequency, core material, number of windings, number and location of screens, and any special insulation requirements.

The manufacturer's power nomograms normally assume single primaries and secondaries, with minimum insulation requirements. Furthermore, the nomogram is developed for a defined operating frequency and temperature rise. However, in real applications, to obtain an optimum design, the designer must make allowances for all the major design factors when selecting core size.

In general, transformers requiring a large number of secondary windings and/or special insulation requirements (for example, transformers that are required to meet the 6-mm creepage distance for UL, IEC, and VDE safety specifications) will require larger cores. Transformers to be operated at higher frequencies and those with simple windings and insulation requirements will require smaller cores.

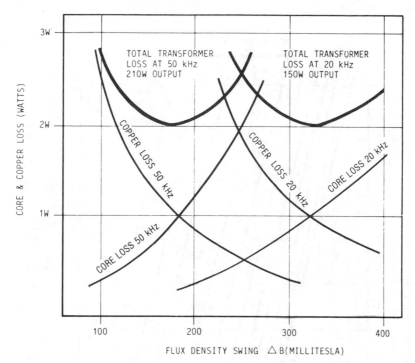

FIG. 3.4.2 Copper and core losses as a function of flux density swing, with frequency as a parameter, showing the conditions for minimum total loss in SMPS transformers, using an EC41 ferrite core.

Finally, the choice of converter and rectifier topology also affects the copper utilization factor and hence transformer size.

4.4 OPTIMUM CORE SIZE AND FLUX DENSITY SWING

With so many interdependent variables, it is difficult for the designer to make an optimum core size and flux density choice. The nomogram shown in Fig. 3.4.3 has been developed by the author to provide a more comprehensive approach to core size and flux density selection for half- and full-bridge topologies. It has been developed from the information provided for Siemens N27 ferrite (a typical power ferrite) and gives the optimum flux density swing and core size (in terms of area product) for a known power requirement and operating frequency. The temperature rise using this approach should be near 30°C in free air.

The nomogram may also be used as a general guide for core size in single-ended topologies, although the power rating should be reduced by approximately 35%. (The actual reduction depends on the duty ratio.) In single-ended designs, the optimum flux density swing shown at the base of the nomogram would apply only to high-frequency core-loss-limited designs.

Although the nomogram provides a good starting point, some iteration may

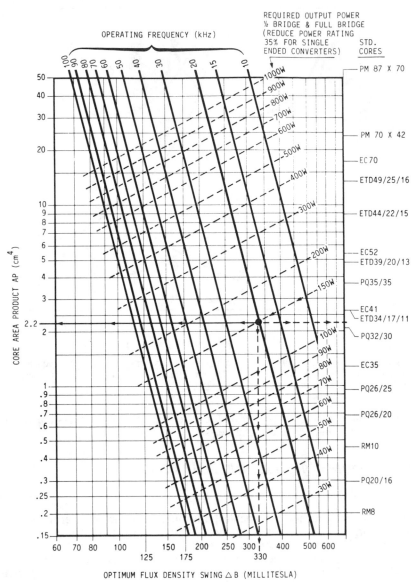

OPTIMUM FLUX DENSITY SWING △B (MILLITESLA)

FIG. 3.4.3 Nomogram for ferrite cores, giving area product (and hence core size) and optimum flux density swing as a function of output power and frequency.

still be required for optimum design. (The use of this nomogram is demonstrated in Chap. 5.)

The area product calculations used for the development of Fig. 3.4.3 are shown in Appendix 4.A and allow for a sandwiched construction using split primaries (see Fig. 3.4.8*a*, *b*, and *c*), interwinding safety or Faraday screens, and

6-mm creepage distances as shown in Fig. 3.4.9. Hence, the power ratings for the cores are somewhat lower than the manufacturers' nomograms would indicate. The reduced power ratings are in accordance with the author's findings in practical applications.

Some of the more sophisticated core geometry design methods or computer-aided design programs are intended to provide the optimum size without the need for multiple iterations. However, even with the aid of such programs, in most practical designs some iteration will still be required.

For the power supply designer who does not have computer-aided design programs at hand, the nomogram shown in Fig. 3.4.3, the manufacturers' nomograms, or the area product approach shown in Sec. 4.5 will provide a good design starting point and will reduce the amount of iteration required to optimize the design.

4.5 CALCULATING CORE SIZE IN TERMS OF AREA PRODUCT

The area product AP is the product of the winding window area and the cross-sectional area of the core. It can be shown (Appendix 4.A) that the area product is related to the power rating of the core. By limiting the number of variables, an equation linking core area products to power outputs can be developed. This allows the initial core size to be calculated in terms of its area product.

Equation (4.A.14) shows how, once the required temperature rise and wire current density have been defined, the area product of a core is related to its power-handling capacity. Since most manufacturers now give the area product values for their cores, a core size can be quickly selected once the required area product has been calculated. If the area product is not given, it can be established from the core dimensions ($AP = A_w \times A_e$).

The following equation shows the area product in terms of output power, flux density swing, operating frequency, and an overall copper utilization factor K. [See Eq. (4.A.14) and Table 3.4.1.]

$$ AP = \left(\frac{11.1 \cdot P_{\text{in}}}{K' \cdot \Delta B \cdot f} \right)^{1.143} \quad \text{cm}^4 $$

where AP = area product, cm^4
P_{in} = input power, W
K' = overall copper utilization factor
ΔB = flux density swing, T
f = frequency, Hz

In the development of this equation, the following assumptions have been made. (Values are summarized in Table 3.4.1.)

1. *Temperature rise.* The temperature rise for natural free air convection cooling is to be approximately 30°C.
2. *Wire current density I_a.* The copper magnet wire is operated at a current density commensurate with a temperature rise of 30°C. (This is based on the empirically developed value of 450 A/cm^2 for an area product of unity.) Larger

TABLE 3.4.1   Overall Copper Utilization Factors K' for Standard Converter Types

Converter type	Primary form	Secondary form	K_p (A_p/A_{wb})	K_u	K_t (I_{dc}/I_p)	K' $(K_p \cdot K_u \cdot K_t)$
Forward	SE	CT	0.32	0.4	0.71	0.091
	SE	SE	0.4	0.4	0.71	0.114
Full and half	SE	CT	0.41	0.4	1.0	0.164
bridge	SE	SE	0.5	0.4	1.0	0.2
CT push-pull	CT	CT	0.25	0.4	1.41	0.141
	CT	SE	0.295	0.5	1.41	0.208

transformers must have a lower current density, as the ratio of cooling surface area to heat-generating volume is lower. [See Eq. (4.A.12) and Refs. 1 and 2.]

3. *Efficiency.*   The input/output efficiency is assumed to exceed 98%. (A well-designed transformer should achieve this value.)

4. *Overall copper utilization factor K'.*   The overall copper utilization factor $K' = K_u \times K_p \times K_t$. The nomogram shown in Fig. 3.4.3 has been developed for the bridge and half-bridge topologies, and a conservative overall copper utilization factor K' of 0.12 has been used. (See Table 3.4.1.) More correctly, the copper utilization factor K' depends upon the converter type, rectification methods, insulation, wire form, and window utilization factor. It is developed from a number of subfactors described in Secs. 4.6, 4.7, and 4.8 as follows.

4.6   PRIMARY AREA FACTOR K_p

This factor is the ratio A_p/A_{wb} and relates the effective primary copper area A_p to the available winding window area A_{wb}. The effective copper area A_p depends on the converter topology. For example, a full-wave center-tapped winding is only 50% utilized, as the current only flows in one half of the winding at each half cycle. Hence, to keep the copper loss power density the same in primary and secondary, 59% of the window area would be used for the center-tapped primary winding. The K_p factors for other topologies are given in Table 3.4.1.

4.7   WINDING PACKING FACTOR K_u

In practice, only about 40% of the available core window area A_w is filled by copper. Although this may sound very low, it should be remembered that round wires can occupy only 78% of the available window area because of the inevitable gaps between wires, and of that, only 80% of the wire is copper because of the insulation layer on the wire. A further part of the window is to be used for interwinding insulation material, creepage distance requirements, and RFI and safety screens. Hence K_u is dependent on a number of practical factors related to winding and insulation needs.

Note: The area product is normally quoted for the core window area A_w. When bobbins are used, a further 20% to 35% of the core window area is taken up by the bobbin. This effect has been allowed for in the development of Fig. 3.4.3, reducing K_u to 0.4. That means that only 40% of the core window area A_w can be used for conducting copper wire.

4.8 RMS CURRENT FACTOR K_t

This factor, I_{dc}/I_p, relates the effective DC input current to the maximum rms primary current. It depends on the converter topology and operating mode. Typical values of K_t for the common converter topologies are shown in Table 3.4.1.

4.9 THE EFFECT OF FREQUENCY ON TRANSFORMER SIZE

The classic transformer equation is usually given as

$$N = \frac{10^6 \, V}{4.44 f \, \hat{B} A_e} \quad \text{(for sine-wave operation)}$$

where N = minimum turns
V = winding voltage, V rms
f = minimum frequency, Hz
$\hat{B}$ = peak flux density, T (*Note:* 10,000 G = 1 T)
A_e = effective core area, mm^2

This formula indicates that turns N and frequency f are inversely proportional, and on this basis it could be assumed that doubling the frequency would halve the number of turns, leading to a much smaller transformer. However, in practice this does not occur, as most core materials show a rapid increase in core loss as the frequency increases. Consequently, to maintain core and copper losses nearly equal (for maximum efficiency), a much smaller flux density excursion is used at the higher frequency, and B will be reduced.

Since the number of turns cannot be reduced as much as may have been expected, the anticipated reduction in core size will not be fully realized, unless special low-loss high-frequency ferrites are used in the higher-frequency application. Manufacturers are constantly improving the performance of their ferrite materials, and the designer would do well to investigate the latest materials and optimum flux density swing where high-frequency operation is required.

4.10   FLUX DENSITY SWING ΔB

For maximum efficiency, the flux density swing ΔB should be selected such that the core loss is equal to the copper loss. Unfortunately, this optimum selection is not always possible, as core saturation may limit the flux density swing to a lower value, preventing an optimum selection.

Figure 2.13.6a shows the normal saturation characteristic of a typical ferrite material at 25°C and 100°C. It should be noted that to allow a working margin, the peak flux density should not exceed 250 mT. (In push-pull this would be a maximum flux density swing ΔB of 500 mT peak-to-peak.)

Figure 3.4.2 shows a plot of copper loss, core loss, and total loss for an EC41 core and bobbin, when operated at 20 kHz and 50 kHz with output powers of 150 and 210 W, as a function of flux density swing. (This design is covered in Chap. 5.) At 20 kHz the maximum efficiency occurs near where the core and copper losses are equal, giving a total loss of 2 W and a temperature rise of 30°C for this size core. The flux density swing for optimum efficiency at 20 kHz is 320 mT peak-to-peak.

For 50-kHz operation, to obtain the same core loss (and hence temperature rise), the flux density must be reduced to 180 mT. However, the frequency has increased by more than the reduction in flux density, so the turns will be fewer. The reduction in number of turns and increase in wire size reduce the copper winding resistance, allowing a larger winding current for the same copper loss. The transformed power may be increased to 210 W to give the same copper loss and temperature rise as in the previous 20-kHz example. Hence, increasing the frequency results in a net increase in the transformer power rating, but a reduction in optimum flux density swing.

It is clear from Fig. 3.4.2 that for optimum efficiency the flux density swing must be quite carefully selected to suit the operating frequency, power output, and permitted temperature rise. The total transformer loss will be equivalent to twice the core or copper loss when the transformer is designed for optimum efficiency, since core and copper loss are nearly equal at this optimum condition.

Figure 3.4.4 shows the flux density swing ΔB as a function of core loss for N27 ferrite, with frequency as a parameter, in the range 5 to 200 kHz. The recommendations shown in Fig. 3.4.3 provide a good starting point for the selection of flux density swing ΔB. Remember, in push-pull applications the peak flux density $\hat{B}$ will be only half the total swing provided that the sweep is centered. (See the discussions of staircase saturation and flux doubling effects in Chaps. 6 and 7.)

In single-ended forward converters, only the first quadrant of the B/H loop is utilized. With ferrite materials at low frequencies (below 40 kHz and 100 W), even if the total available flux excursion is fully utilized, it is unlikely that the core loss will be equal to the copper loss for normal core geometry. A design of this type is said to be saturation-limited. Further, unless current-mode control (or one of the special input voltage compensated control chips) is used, the working flux density in the forward converter may need to be further reduced to prevent saturation during start-up or transient operation. (During start-up, maximum volt-second conditions could well be applied to the core.)

In push-pull applications (full-bridge and half-bridge) at low frequencies, the full B/H characteristic range can in theory be utilized. However, once again, this flux density swing may need to be reduced to prevent saturation during start-up and transient operating conditions. Current-mode control overcomes these start-up and transient limitations, allowing a larger flux density swing to be utilized. Various methods of controlling transient and start-up conditions are discussed in

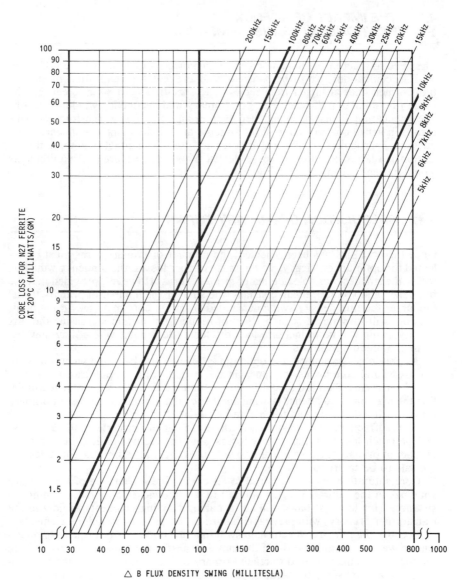

FIG. 3.4.4 Core loss for N27 ferrite material as a function of flux density swing and frequency. (*Siemens AG.*)

the appropriate converter sections. In high-frequency applications, the flux density swing may be very much limited by the optimum efficiency requirements, and special soft-start circuits may not be required. The converter topology, method of operation, power output, and frequency must be considered before selecting the operating flux density swing.

4.11 THE IMPACT OF AGENCY SPECIFICATIONS ON TRANSFORMER SIZE

The need to meet insulation and creepage distance requirements, where UL and VDE specifications are to be satisfied, can prevent the realization of smaller transformer sizes at higher frequencies. The specified 4- to 8-mm creepage distance (the minimum distance between primary and secondary windings for off-line applications) must be maintained even in high-frequency transformers. This results in very poor utilization of the window area and an increase in leakage inductance, particularly when smaller cores are used. The effect is to force the selection of a larger core than the pure electrical and temperature rise requirements would normally demand. (See Fig. 3.4.9.)

4.12 CALCULATION OF PRIMARY TURNS

Once the core size has been selected, the number of primary turns must be selected for optimum efficiency. To minimize copper losses, the tendency would be to use the smallest possible number of primary turns. However, provided that the frequency and voltage remain constant, the smaller the number of primary turns, the larger the flux density swing demanded from the core material. In the limit, the core will saturate. A second effect of reducing turns and increasing the flux density swing will be to increase the core losses to a point where they may become the predominant loss.

As previously explained, the optimum efficiency will be found where the copper losses and core losses are approximately equal. In push-pull transformers at high frequencies, the need to satisfy this optimum efficiency requirement will define the maximum flux density swing and hence the minimum number of primary turns. Such a design is said to be core-loss-limited.

At low frequencies, particularly with single-ended converters, the core loss will be much less than the copper losses, and the factor limiting the minimum number of primary turns will be the need to prevent core saturation. Such designs are said to be saturation-limited.

Core saturation must be avoided at all costs. The impedance of the primary windings in the saturated region will fall to a value close to the DC winding resistance. This low resistance will allow damagingly high currents to flow in the transformer primary, with inevitable failure of the primary switching elements.

Because the primary waveform in switchmode converters is a square or quasi-square wave, a modified form of the classic transformer equation (derived from Faraday's law) may be used to relate primary or secondary turns to the core parameters and transformer operating parameters. In this equation the turns are related to the applied volt-seconds as follows:

$$N = \frac{V \times t}{\Delta B \times A_c}$$

where N = primary turns
V = DC voltage applied to winding when switching device is "on"
t = "on" time of half period, μs
ΔB = maximum flux density swing, T
A_c = core cross-sectional area, mm^2

Note: In saturation-limited designs, the minimum core area A_c should be used to prevent saturation of any part of the core. In core-loss-limited designs, the effective core area A_e should be used to more correctly reflect the bulk core loss.

Under steady-state conditions, each cycle is identical, and a single period is sufficient to define the operating parameters. From the preceding equation, it will be noted that the primary turns N are directly proportional to the primary voltage V and the time that the voltage is applied to the primary windings, time t. They are inversely proportional to the flux density swing ΔB and the core cross-sectional area A_c.

It would now seem to be a simple matter to establish the primary turns by inserting the appropriate constants into this equation. However, a further complication now arises in the selection of the constants.

In some voltage-controlled converter circuits, it is possible, under start-up or transient conditions, for the maximum primary voltage and maximum "on" period to coincide. If this type of converter topology is used, to prevent saturation of the core, the maximum primary voltage and maximum "on" period must be used in the equation to calculate the primary turns.

If current-mode control is used, the onset of core saturation is controlled, and the maximum "on" time will only coincide with the minimum primary voltage; thus these values will be used in the equation to calculate the primary turns. This will result in a smaller number of turns in a saturation-limited design.

Some duty-ratio-controlled systems apply primary input voltage feedforward compensation, fast primary current limiting, or slew rate control. In such cases, the same conditions as for current-mode control will apply to the transformer design.

In push-pull applications, a saturation problem can arise on initial start-up. The flux excursion for the first half cycle will be in the first or third quadrant of the B/H loop only. (The core will have restored to the remanent flux density B_r near zero when the supply was previously turned off.) Unless precautions are taken to limit the flux excursion for the first few cycles of operation (soft start) or current-mode control is employed, the push-pull transformer can saturate on the first half cycle (the so-called "flux doubling effect"). If soft start or current-mode control is not used, the transformer must be designed for a smaller flux swing, resulting in an increased number of primary turns. Hence, to realize the improved efficiency that is normally possible with push-pull transformers because of their larger flux density swing, appropriate soft-start methods, slew-rate control, or current-mode control techniques must be used to prevent core saturation during start-up.

Remember, under steady-state conditions, in the push-pull transformer, it should be possible to swing the flux density from the positive first quadrant through to the negative third quadrant, doubling the possible flux excursion compared with the single-ended transformer. In the ideal case, this would halve the number of primary turns and improve the transformer efficiency. In practice, it is usually not possible to utilize this full flux density swing, as some margin must be provided for start-up and transient operation, and at high frequencies the flux excursion may be limited by core loss considerations.

For core-loss-limited applications using N27 or similar transformer ferrite materials, select a value of flux density swing, as recommended in Fig. 3.4.3, for initial design purposes. For other materials, calculate the loss permitted for the temperature rise required [Eq. (4.A.16) or (4.A.18)]. Select a flux density swing to give a core loss of half this value (in an optimum design, the other half will be used for copper loss). The manufacturer's core material loss curves will provide core loss and flux density swing information, and the optimum flux density swing can be established.

After all these aspects are considered, the appropriate constants are entered in the equation and the number of primary turns calculated.

4.13 CALCULATING SECONDARY TURNS

When the number of primary turns has been calculated, the number of secondary turns may be established from the primary-to-secondary voltage ratio. In buck-derived converters, the secondary voltage will exceed the output voltage as defined by the duty ratio. A further allowance must be made for diode drop and choke voltage drop. These calculations are usually made for minimum input voltage and maximum pulse width. Some adjustment of primary turns may be required to eliminate partial secondary turns; in the case of saturation-limited designs, the turns adjustment must be to the next higher integer.

In closed-loop converter topologies that employ current-mode control, or in duty ratio systems with primary voltage pulse-width compensation, the primary voltage–"on" time product, $V_{dc} \times t_{on}$, remains constant as a result of the control circuit action. It is often more expedient in this case to calculate the secondary turns first so as to avoid partial turns. Since the output voltage is maintained constant by the control loop, it has a defined (known) value. This output voltage and the maximum "on" period will occur at minimum input voltage, and these values would be used in the equation to calculate the minimum number of secondary turns. It should be remembered that in saturation-limited designs, the number of turns established from the equation is the minimum number of turns that may be used, and any rounding process to eliminate partial turns must result in an increase in turns rather than a decrease. In core-loss-limited designs, the rounding process may be in either direction, increasing or decreasing the core loss as desired.

In multiple-output applications, the number of secondary turns for the minimum output voltage is normally calculated first. The normal requirement is that partial turns must be avoided, and so this winding is rounded to the nearest integer. (In saturation-limited designs this would be the next higher integer.) The primary turns and the remaining secondary turns are then scaled accordingly.

4.14 HALF TURNS

Where E cores are used, special techniques can be employed for half-turn requirements on major outputs. (See Sec. 4.23.) For low-power auxiliary outputs, half turns are sometimes used on the center pole or core legs, and in this event a small gap must be introduced in each leg of the transformer, to ensure good coupling to the half turn and reduce flux imbalance in the outer legs. (Typically a gap of 0.1 mm would be sufficient.) Alternatively, external inductors may be used for the adjustment of auxiliary voltages. (See Part 1, Chap. 22.)

4.15 WIRE SIZES

The selection of primary and secondary wire gauges and the overall winding topology is a most important and often the most difficult part of the transformer design. A large number of practical and electrical parameters control the choice of winding topology and wire size.

The initial selection of the core size was based upon a wire current density selected to give a 30°C temperature rise. One approach in selecting wire size is to calculate the current density required to meet this criterion [from Eq. (4.A.12)] and hence obtain the required cross-sectional area of the wire. However, at this stage, the "die has already been cast," because the core size has been selected. Hence the winding window area is already defined by the bobbin.

To maintain the primary and secondary losses equal, the winding area occupied by a primary should be the same as that occupied by the secondary. It was initially assumed that 50% of the available bobbin window area would be occupied by copper (25% each for primary and secondary), the remainder of the window space being occupied by space between windings (because round wire is used) and the insulation and screening between primaries and secondaries. Since the numbers of primary and secondary turns are known, it is better at this stage to select a wire size that will make the best use of the available window area. However, before this can be done, it is necessary to make an assessment of the winding topology, and to consider the implications of skin and proximity effects. At low frequency the complete window will be used, but at high frequencies with push-pull operation, lower losses may result from using thinner wire and fewer layers because of the improved F_r ratio. In this case not all the window area will be used. (The F_r ratio is the ratio of the effective ac resistance of the wire to its DC resistance. See Appendix 4.B.)

4.16 SKIN EFFECTS AND OPTIMUM WIRE THICKNESS

Before the final wire gauge can be selected, skin and proximity effects must be considered. (See Appendix 4.B.) In simple terms, at high frequencies, the combined effects of the internal field within the wire and the proximity of fields from adjacent turns is to force the current to flow in a thin cusp at the surface and to the upper or lower edge of the conductor. In a simple open wire, this thin surface conduction layer is annular, with a thickness called the "penetration depth" Δ. The penetration depth is frequency-dependent, and the current density will have fallen to approximately 37% at a depth defined by

$$\Delta = \frac{66}{\sqrt{f}}$$

where Δ = penetration depth, mm
$\quad f$ = frequency, Hz

Hence in the simple (open wire) case, if the radius of the wire exceeds the penetration depth, there will be a poor copper utilization factor, giving excessive copper losses.

In the transformer, the situation is more complex because of the fields from adjacent windings; hence, the winding topology plays an important role in the selection of wire size. The proximity of adjacent wires and layers forces the current into an even smaller cusp at the inner or outer edge of the wire, so that the ef-

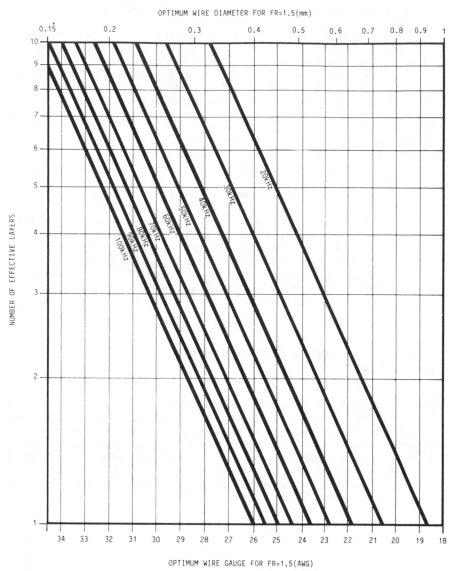

OPTIMUM WIRE DIAMETER FOR FR=1.5(mm)

NUMBER OF EFFECTIVE LAYERS

OPTIMUM WIRE GAUGE FOR FR=1.5(AWG)

FIG. 3.4.5 Optimum wire gauge (AWG) and diameter (mm) as a function of the number of effective layers in the winding, with frequency as a parameter.

fective conduction area is further reduced. Therefore, when more than one layer is used, the wire size should be further reduced.

In practice, the minimum F_r ratio (the ratio between the DC resistance and the effective ac resistance) will approach a minimum of 1.5 in a well-designed winding. To achieve this, the wire diameter or strip thickness must be optimized for the operating frequency and number of layers. Figures 3.4.5 and 3.4.6 indicate

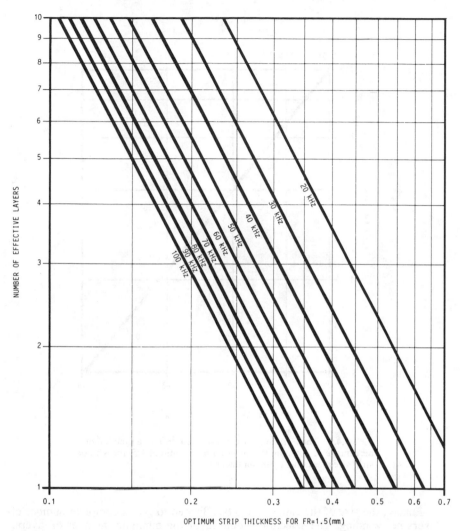

FIG. 3.4.6 Optimum copper strip thickness (mm) as a function of the number of effective full-width layers in the winding, with frequency as a parameter.

the maximum wire diameter (as a single strand) or maximum strip thickness that should be used in the transformer winding, as a function of the number of effective layers, with frequency as a parameter, to give an F_r ratio of 1.5.

If the cross-sectional area of the wire that would just fill the available window space, or the size indicated by the current density, exceeds the size indicated by Fig. 3.4.5 or 3.4.6, then two or more wires should be used to make up a cable of the appropriate cross-sectional area. These multifilament windings may be applied as a single multifilar layer or as a twisted cable. For very high frequency applications, the specially woven Litz wire should be considered.

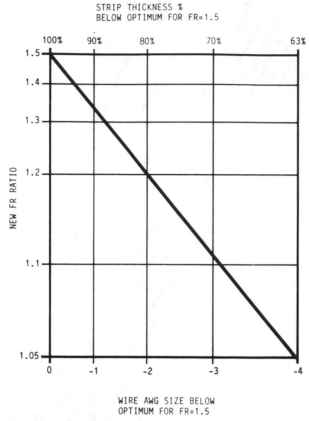

STRIP THICKNESS %
BELOW OPTIMUM FOR FR=1.5

WIRE AWG SIZE BELOW
OPTIMUM FOR FR=1.5

FIG. 3.4.7 The F_r ratio (ac/DC resistance ratio), as a function of a percentage of optimum thickness for an F_r ratio of 1.5, for wire or strip sizes less than optimum thickness.

Finally, the size of the cable should be adjusted to give a complete number of layers per winding. The selection should favor the minimum number of layers, since a reduction in number of layers improves the F_r ratio (somewhat compensating for the increased resistance of the smaller cable). Also, fewer layers will reduce the leakage inductance.

The effective F_r ratios for wires which are less than the diameter used for an F_r ratio of 1.5 are shown in Fig. 3.4.7.

4.17 WINDING TOPOLOGY

The winding topology has a considerable influence on the performance and reliability of the final transformer.

To reduce leakage inductance and proximity and skin effects to acceptable limits, the use of sandwich winding construction is almost inevitable in high-frequency transformers.

Figure 3.4.8*a* shows the distribution of magnetomotive force (mmf) in a simple wound transformer, and Fig. 3.4.8*b*, that in a sandwiched wound transformer. In the simple winding, the primary magnetizing mmf builds up with increasing ampere-turns to a maximum at the primary-to-secondary interface. The secondary ampere-turns are nearly equal and opposite, reducing the mmf to zero at the outer limits of the winding. Large values of mmf increase proximity effects and leakage inductance. In the sandwiched winding, the maximum mmf is reduced; it is zero in the center of the secondary winding.

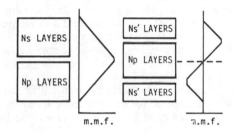

SIMPLE AND SANDWICHED TRANSFORMER WINDING
ARRANGEMENTS AND m.m.f. DIAGRAMS

(a)   (b)

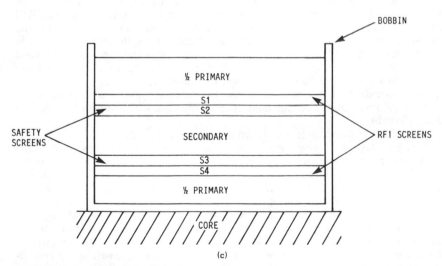

(c)

FIG. 3.4.8 (*a*) and (*b*) Distribution of magnetization force in simple and sandwiched transformer winding topologies. (*c*) Makeup of a sandwiched (split) winding topology, showing the position of windings, safety screens, and RFI screens.

4.17.1 Effective Layers

The effective layers referred to in Figs. 3.4.5 and 3.4.6 are the number of layers between a plane of zero mmf and a plane of maximum mmf. Hence, in the case of the secondary winding in the sandwiched construction of Fig. 3.4.8b, the effective layers are only half the total number of secondary layers, because a zero mmf occurs in the center of the winding.

Since the maximum mmf and proximity effects are lower in the sandwiched construction, the copper losses and leakage inductance are reduced. If the secondary is a single layer, giving an effective half layer, the wire or strip may be twice as thick as the optimum given for a single layer in Fig. 3.4.5 or 3.4.6.

In sandwiched construction, the normal design approach is to split the primary into two halves with the secondary windings sandwiched between them, as shown in Fig. 3.4.8c. In some multiple-output applications, particularly those in which the secondary windings have low-voltage, high-current, relatively constant loads, there is some advantage in splitting the secondary windings into two sections, with the primary in the middle. In this second arrangement, the winding carrying the highest secondary current would be placed close to the core, as the mean turn length will be smaller, resulting in lower copper losses. A second advantage of this topology is that windings which are close to the core will now have lower ac voltages, reducing the RFI coupling from primary to core, secondary, and case. However, this approach should be used with caution, because the secondary windings must be selected for equal ampere-turns loading above and below the primary and the loads must be constant; otherwise large leakage inductances can occur.

In this example the primary has been split into two equal parts, one above and one below the secondary windings. Four screens have been incorporated between the primary and secondary windings. The screens adjacent to the primary, S1 and S4, are Faraday screens, fitted to reduce RFI coupling between the high-voltage primary windings and the safety screens S2 and S3. The Faraday screens are connected to the primary common line, to return capacitively coupled RFI currents to the primary circuit. The safety screens S2 and S3 are connected to the chassis or ground line, to isolate secondary outputs from the primary circuit in the event of an insulation failure. These screens, although necessary to meet safety and emission requirements, occupy considerable space and increase the primary-to-secondary leakage inductance. (See Sec. 4.22.)

After deciding on the winding topology, calculate the space occupied by screens and insulation. The remaining space is then available for primary and secondary windings.

Some further constraints are placed upon the winding design.

It is preferable that windings occupy a discrete number of layers. In the case of the split primary winding, the layers should be an even number to allow equal splitting of the half primaries. Partly wound layers both are inefficient in the use of the winding space and promote insulation breakdown where the terminating wire is brought across the top of the underlying layers. Since the volts per turn can be quite large in high-frequency transformers, a terminating wire which spans several turns will be subjected to a higher breakdown voltage stress. Further, the terminating wire is subject to considerable mechanical stress, because it forms a discontinuity or "bump" in the winding, and the remaining layers apply considerable pressure to this bump. Most insulation breakdown failures in switching transformers can be traced to this type of winding discontinuity or to bad termination practices in which wires are crossed over each other in the terminating process. Good winding practice dictates that all layers should be complete, that terminating wires be brought out with additional insulation, and that terminating wires not cross over other windings or terminating wires wherever practical.

To meet the creepage distance and spacing requirements demanded by VDE, UL, IEC, and CSA specifications, it is necessary to leave a creepage distance of up to 8 mm between the primary and secondary windings. (See Fig. 3.4.9a, b, and c.)

To meet the creepage distance requirement more easily, it is good practice to terminate primary windings on one side of the bobbin and secondary windings on the other. This also has the advantage that primary and secondary terminating

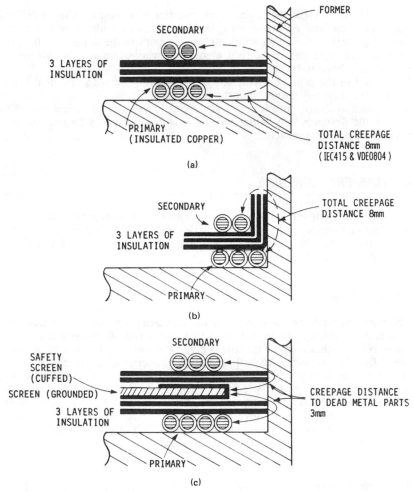

FIG. 3.4.9 (a), (b), and (c) Some methods of insulation and winding arrangements, meeting safety creepage distance requirements, in agency approved types of transformer makeup.

wires are well separated. At high frequencies, where window space is limited, the technique shown in Fig. 3.4.9b may help. If a grounded safety screen is fitted, the creepage distance may be reduced to 3 mm (see Fig. 3.4.9c). The screen must be connected to ground (chassis) and rated to carry the maximum fault current to clear fuses or other protective devices.

Although the design intention is to occupy the space equally with primary and secondary windings, some deviation from this equality is acceptable, as an increase in loss in the primary will be partially compensated by a decrease of loss in the secondary, or vice versa. The overall efficiency of the transformer is not greatly compromised by small deviations from the ideal, although the hot spot within the greatest loss winding will be somewhat higher in a nonbalanced situation.

It has been shown that the selection of wire size is a complex choice defined by many practical and electrical considerations. It is incumbent on the transformer design engineer to either wind or closely supervise the winding of the prototype transformer, to ensure that it is a practical proposition and that his design intentions have been fully satisfied.

The layout of the printed circuit board should conform with the ideal transformer pin terminations, rather than the reverse.

In the final analysis, core losses and copper losses should be calculated to check that the design is fully optimized. Several iterations may be necessary to achieve optimum performance.

4.18   TEMPERATURE RISE

Under convection-cooled conditions, the temperature rise depends on the total internal loss and surface area. The original design aim was a temperature rise of 30°C, and this should be checked by calculating the losses, and hence the temperature rise, in the final design.

4.18.1   Core Loss

The core loss depends on the core material, flux density swing, frequency, and core size. Figure 3.4.4 shows the core loss for a typical transformer ferrite (Siemens N27). In this example the loss is given in terms of milliwatts per gram to a base of flux density swing ΔB with operating frequency as a parameter.

Most loss diagrams assume push-pull operation and are plotted for peak flux density. They assume a symmetrical flux swing about the origin, and the losses are plotted for a flux density swing of twice the peak value $\hat{B}$. When calculating core loss for a single-ended converter using such diagrams, divide the transformer flux density swing by 2 and enter the graph with this as the peak value to obtain the core loss for single-ended applications.

4.18.2   Copper Loss

The copper loss depends on the winding resistance, the F_r ratio (ac to DC resistance ratio), and the rms current.

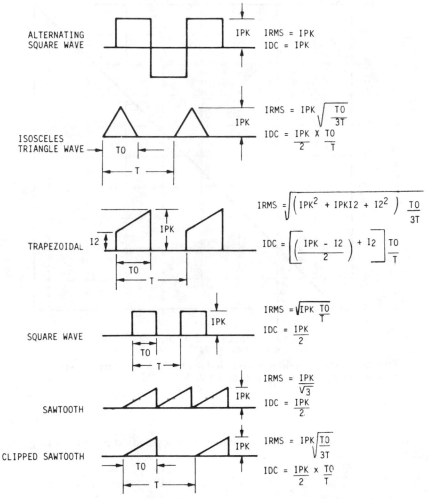

FIG. 3.4.10 Common switchmode power converter waveforms, showing effective rms and DC values.

Figure 3.4.10 gives the rms conversions for the more common converter waveforms; Table 3.1.1, the standard winding data; and Fig. 3.4.11, the temperature-resistance correction factors.

The DC winding resistance is calculated for the total winding length, using the mean turn length for each section of the winding, turns, and wire gauge. This DC resistance is then multiplied by the F_r ratio, obtained from Fig. 3.4.7 or 3.4B.1, and the copper temperature resistance factor, from Fig. 3.4.11, to obtain the ef-

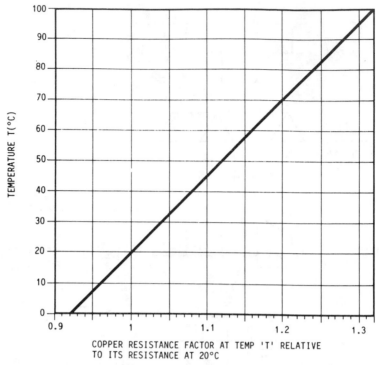

FIG. 3.4.11 Copper resistance factor as a function of temperature, showing ratio of resistance at temperature T compared with that at 20°C.

fective ac resistance R_e at the operating frequency and estimated working temperature.

$$P_w = I_{rms}^2 \times R_e \qquad W$$

where P_w = winding copper loss, W
$\quad I_{rms}$ = rms winding current, A
$\quad R_e$ = effective ac resistance of winding, Ω

The copper losses should be calculated for each winding to check for reasonable loss distribution. The sum of copper and core losses gives the total transformer loss.

4.18.3 Temperature Rise Nomogram

The temperature rise may be checked using the area product–surface area–temperature nomogram shown in Fig. 3.1.7.

Enter the top of the nomogram with the area product AP. The nearest horizontal to the intersect with the dashed "AP line" gives the surface area to the left. Enter the lower scale with the total loss, and the nearest diagonal to the intersect with the surface area line gives the temperature rise.

Alternatively, calculate the temperature rise from Eq. (4.A.16) or (4.A.18):

$$\Delta T = \frac{800 \cdot P_t}{A_s} \quad °C$$

or

$$\Delta T = \frac{23.5 \cdot P_t}{\sqrt{AP}} \quad °C$$

where ΔT = temperature rise, °C
P_t = total internal power loss, W
A_s = surface area, cm^2
AP = area product, cm^4

This approximate formula holds well for temperatures in the range 20 to 50°C.

4.19 EFFICIENCY

The efficiency η may be established in the normal way from the calculated losses and transferred power.

$$\eta = \frac{\text{output power} \times 100}{\text{output power} + \text{losses}} \quad \%$$

4.20 HIGHER TEMPERATURE RISE DESIGNS

Where a temperature rise in excess of 30°C is permitted, the design process used in Fig. 3.1.7 may be reversed, giving a smaller transformer with a higher temperature rise.

In this case, enter Fig. 3.1.7 with the required area product and establish the horizontal surface area line. The intercept with the required (higher) temperature rise gives the permitted total dissipation. For an optimum design, the core loss will be half this value, and Fig. 3.4.4 gives the optimum flux density swing for this core loss.

The transformer design example shown in Sec. 3.5 illustrates the design procedure.

4.21 ELIMINATING BREAKDOWN STRESS IN BIFILAR WINDINGS

In a bifilar winding, two or more insulated magnet wires are wound together, in parallel, to give independent windings. These windings, although isolated, are closely magnetically coupled.

Bifilar windings on high-voltage converters are a source of potential break-down, so this technique should be limited to low-voltage applications. No problem exists where wires are eventually to be connected in parallel to form a multifilar winding, or where the stress voltage between isolated windings is low.

A typical bifilar winding application would be the energy recovery winding in a single-ended forward converter. In some configurations an energy recovery winding will be required to restore the transformer during the flyback period and recover the flyback energy. A bifilar winding will normally be preferred, since any leakage inductance between the main winding and the flyback winding will result in an excessive voltage overshoot on the collector of the switching device.

In off-line applications, it is common practice to provide a bifilar-wound energy recovery winding on the primary of the transformer. However, this winding is a possible cause of failure, since a high stress voltage exists between adjacent turns of the two windings. If there is a weak point anywhere in the insulation covering, this is a potential cause of failure. Although the insulation on the magnet wire may be rated at several thousand volts, a single flaw along the length of the winding can result in breakdown, since the voltage stress, at all points, is considerable. Moreover, careless winding techniques which allow two bifilar wires to cross will result in high mechanical stress at the crossover point, which may cause failure under high-temperature conditions.

Hence, if bifilar windings are considered essential, only the best-quality high-temperature insulating materials should be used, and considerable care must be taken in winding, insulation, and material handling. The operatives should understand the problem.

In the above example, separating the windings into two isolated and insulated layers would normally be unacceptable because of the increase in leakage inductance. However, it is possible, by suitable circuit techniques, to obtain good performance without the need to bifilar-wind the energy recovery winding. One such technique (discussed in Part 2, Sec. 8.5) will be used as an example here.

As shown in Fig. 2.8.1, a separate energy recovery winding (not bifilar-wound) may be used without the leakage inductance becoming a problem if the energy recovery diode D3 is placed in the top end of the energy recovery winding. A capacitor C_c links the junction of this diode and winding to the collector of the switching transistor Q1. The two equal turns windings on the transformer are on separate layers, isolated from each other, so inevitably there will be some leakage inductance between them.

During a cycle of operations, if no leakage inductance were present, the starts of the two windings would exactly track each other (with DC offset of V_{cc}), so that the voltage across the capacitor C_c would remain constant at the supply voltage. However, the leakage inductance will tend to produce voltage spikes, but any tendency for the collector voltage to overshoot is now very effectively clamped by the path provided through C_c and D3.

The value of C_c is chosen to be large compared with the transferred energy so that the voltage change across C_c during a clamping period will be insignificant.

In this topology, although the leakage inductance has not been eliminated, it is no longer a problem. The energy stored in the leakage inductance is returned to the supply line through C_c and D3 during the turn-off transient. (This is more fully explained in Sec. 8.5 of Part 2.)

Using this circuit topology, the transformer reliability problems that would have been inherent with a high-voltage bifilar winding have been eliminated. This example demonstrates the importance of integrating the circuit and transformer design processes if optimum designs are to be produced.

4.22   RFI SCREENS AND SAFETY SCREENS

To prevent RF currents flowing from primary to secondary or ground through the interwinding capacitance, it is necessary to fit a screen against the primary windings. This should be connected to the common input point to return capacitively coupled currents to the primary. This common point will usually be either the positive or the negative high-voltage input line.

Since this is not a safety screen, very thin copper screen material can and should be used. Thick copper is not the best selection, as its low resistance gives a large eddy-current loss. For this application, a higher-resistance nonmagnetic material, such as phosphor bronze or manganin, should be considered. Also, minimum-thickness insulation should be used. Excessive buildup in the screens and insulation should be avoided to minimize leakage inductance.

In high-voltage or "off-line" applications, a further safety screen should be provided between the RFI screen and the secondary. This screen, together with its terminating wire, must be of sufficient gauge to carry the fusing current of the supply (a safety regulation requirement). Insulation type and thickness must be selected to meet specified safety requirements. Where both screens are fitted, some reduction in the overall insulation requirements may be obtained by isolating the RFI screen from the input supply common point with an approved high-voltage series capacitor. A value of 0.01 μF is adequate, and the voltage rating should be as required by the specified safety regulations. The total insulation thickness for all screens then becomes accountable for safety requirements.

Where high-voltage isolated secondary outputs are required, it may be necessary to provide a third screen to return the capacitively coupled output winding to screen currents, back to the appropriate output winding. Fortunately, in most cases this third screen is not essential, because in most applications the output windings common either is connected directly to the safety ground (chassis) or may be returned to same through a low-impedance capacitor.

Note: Such connections or capacitors must be fitted as close as possible between the safety screen and the offending output winding to reduce the length of the circulating current path.

The ends of the screens must be suitably insulated to prevent a shorted turn, and the minimum overlap should be used to minimize the overlap capacitance. (Note that at high frequencies a large overlap capacitance will make the screen look like a shorted turn.)

The screen terminating wire should be taken from the center of the screen to minimize the inductive coupling of the screen return currents. (Capacitively induced but inductively coupled screen currents then cancel as they flow in opposite directions in each half of the screen winding.) These second-order effects are more important at higher frequencies.

4.23   TRANSFORMER HALF-TURN TECHNIQUES

In some applications, it would be an advantage to be able to adjust the transformer windings to the nearest half turn.

However, the simple expedient of winding half a turn on the center pole of an E core at best results in poor coupling to the half turn and bad transformer regulation and under some conditions can cause saturation of one leg of the transformer core.

An analysis of this simple half-turn arrangement (see Fig. 3.4.12a and b) shows that it is similar to placing a turn on one leg of the transformer core, since the external current loop must be closed back to the start of the main winding eventually (even though it may be via the external circuit). In Fig. 3.4.12, a turn and a half is shown, but the effect is the same. Since under balanced conditions, half the flux in the center pole can normally be expected to flow in each leg, the leg half turn would appear on first inspection to give the desired half flux linkage and hence half voltage coupling. However, as soon as any load is applied to this winding, a back mmf is generated, increasing the reluctance of the leg to which the half turn couples. (This is shown in Fig. 3.4.12b, leg C).

Since a low-reluctance alternative path for the flux is presented in the other leg of the E core (path A), the magnetic flux from the center pole will be redirected into this side and the apparent effective half-turn coupling will be lost; consequently, the voltage generated by the half turn will very quickly decay as the winding is loaded, giving poor regulation.

The situation can be improved by gapping the core (including the outer legs), so that the difference in reluctance caused by the unbalanced mmf is swamped by the large reluctance introduced by the air gap. In flyback converters, this air gap may already be the normal situation. Provided that the mmf generated by the half turn is small compared with the reluctance introduced by the air gap, then reasonable coupling to the half turn will be maintained, and for low-power auxiliary outputs, this simple half-turn winding with an air gap compensation may be quite satisfactory.

However, for higher-power applications, the technique shown in Fig. 3.4.12c and d should be adopted. An analysis of this arrangement shows that two half turns are now effectively applied to each leg of the transformer. These half turns operate in parallel and are phased such that the back mmf will equal and oppose the prime flux in each leg, so that flux balancing is maintained irrespective of the loading. Regulation is now very good, and an air gap is not essential.

For current transformers, this special half-turn arrangement can be particularly useful in high-current applications, since it will halve the secondary turns required. In this case, of course, the half turn would be the primary current winding, and the full turn shown on the center leg would not be required.

By a similar process, X cores and some pot cores with four ports provide for balanced quarter turns.

4.24 TRANSFORMER FINISHING AND VACUUM IMPREGNATION

Wound components are resin-impregnated for three major reasons:

1. To exclude moisture from the windings and prevent degradation of the insulation by fungicidal attack.
2. To stabilize the position of the windings and prevent mechanical creepage and noise.
3. To exclude air voids and provide a homogeneous mass for best thermal properties, to eliminate hot spots, and to prevent discontinuity in interwinding capacitance, eliminating corona-induced failure.

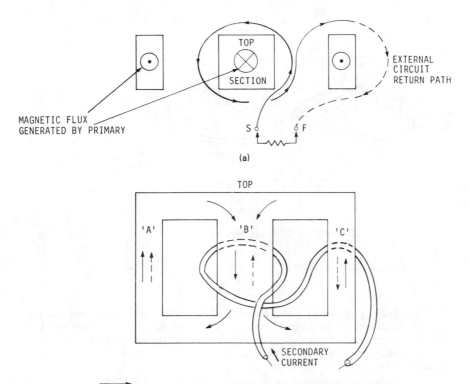

MAIN FLUX PATH (GENERATED BY PRIMARY CURRENT)

BACK mmf AS A RESULT OF SECONDARY CURRENT

UNBALANCED 1½ TURN SECONDARY

(b)

(c)

FIG. 3.4.12 (*a*) and (*b*) Diagram showing the loss of flux linkage, and flux imbalance, caused by one method (often used) of winding half turns on an E core transformer. (*c*) and (*d*) A special E core half-turn winding arrangement, giving good flux linkage with balanced core flux density operation.

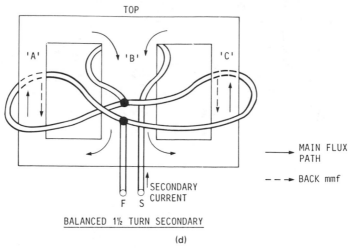

FIG. 3.4.12 (*Continued*)

A number of proprietary varnish and resin materials are available for this purpose, and care should be taken to select an approved material in accordance with the final specifications and operating temperature of the transformer. To ensure thorough impregnation of small wound components, a resin with a low viscosity should be chosen where possible.

The wound bobbin should be thoroughly dry before encapsulation; an oven drying process at 110°C for at least 4 h is recommended. This also tends to stabilize and anneal the windings and remove stresses which would otherwise be locked in by the encapsulation process.

4.24.1 Impregnation Problems with Ferrite Materials

With ferrite materials and some grain-oriented iron materials, it is particularly important that the core not be included in the vacuum impregnation process. With some materials, there can be considerable degradation of the magnetic properties as a result of the mechanical stresses set up by the hardening resin. Some cores must be free to magnetostrict, and this is particularly important with square-loop ferrite toroids and HCR tape wound cores. For this type of component, the resin finishing should be limited to the winding only.

For square-loop toroids, it may be possible to eliminate the need for impregnation by using multifilar, thin-walled insulated wire in place of magnet wire. A high-temperature insulation is preferred, such as PTFE or irradiated PVC.

Toroidal cores that have been Parylene coated are less susceptible to change when impregnated, as the Parylene coating, which is stress-free, prevents ingress of resin into the normally porous ferrite.

Cores with bobbins can suffer additional problems because the varnish fills up the space provided for bobbin expansion. This causes further mechanical stress and possible breakage of the core at high temperatures because of the different rates of expansion of the various materials.

In general, ferrites are not improved by varnish or resin, and vacuum impregnation should be limited to the winding and bobbin only, where possible. Expect considerable changes in the properties of the core material when impregnation of the complete transformer is unavoidable.

High-voltage stress testing should not be carried out until the winding has been dried out and the varnish or resin has been completely cured.

4.25 PROBLEMS

1. Why is it so important for the switchmode power supply designer to have a very good working grasp of the design of switchmode transformers?

2. What is the difference between an AIEE class A and an AIEE class B transformer?

3. Why is it difficult to relate transformer size directly to power throughput?

4. What are the loss conditions normally assumed to give optimum transformer efficiency?

5. Why does the choice of converter topology and rectifier arrangements affect transformer core size?

6. What property of the core material normally defines the minimum number of primary turns on a medium- or low-frequency transformer?

7. What property of the core material normally controls the number of primary turns in a high-frequency transformer design?

8. Why is it often difficult to reduce the transformer size in direct-off-line converter applications, even when the operating frequency is very high?

9. What are the basic steps that should be taken to reduce primary-to-secondary leakage inductance?

10. What is the core area product AP?

11. In what way is the area product dimension useful in transformer design?

12. What would be the typical efficiency of an optimum-design switchmode transformer?

13. Why are skin and proximity effects so important in the selection of magnet wire sizes for high-frequency switchmode transformers?

14. Why is the transformer DC winding resistance not meaningful in terms of calculating the copper losses in a high-frequency transformer?

15. Why are half turns a problem in conventional E-core transformer designs?

16. What is the advantage of a bifilar winding, and how is it constructed?

17. What are the disadvantages of a bifilar winding?

18. What is the difference between a transformer RFI screen and a safety screen, and how are these constructed?

19. Is it possible to fit half turns to a conventional E transformer without causing flux imbalance?

APPENDIX 4.A

DERIVATION OF AREA PRODUCT EQUATIONS FOR TRANSFORMER DESIGN

4.A.1. DERIVATION OF TRANSFORMER AREA PRODUCT (AP)

The transformer input power P_{in} depends on the output power P_{out} and efficiency η. Hence

$$P_{in} = \frac{P_{out}}{\eta} \qquad (4.A.1)$$

The average DC input current to the converter transformer I_{dc} depends on the input power P_{in} and the DC input voltage V_{in}. Hence

$$I_{dc} = \frac{P_{in}}{V_{in}} \qquad (4.A.2)$$

The maximum rms primary current I_{pm} occurs when the input voltage is minimum $V_{in(min)}$ and the pulse width is maximum. Factor K_t relates the DC input to the rms primary current dependent on the converter topology. $K_t = I_{dc}/I_{pm}$. Hence

$$I_{pm} = \frac{I_{dc}}{K_t} \qquad (4.A.3)$$

Substituting Eq. (4.A.2) for I_{dc} at minimum input voltage,

$$I_{pm} = \frac{P_{in}}{V_{in(min)} \times K_t} \qquad (4.A.4)$$

The usable window area A_p (available for the primary winding) depends on the total window A_w, the window area reserved for the primary, given by the "primary area factor" K_p, and the primary area "utilization factor" K_u. Hence

$$A_p = A_w \times K_p \times K_u \tag{4.A.5}$$

The number of primary turns N_p that will just fill the primary window space A_p at a wire current density of J depends on the primary current; hence

$$N_p = \frac{A_p \times J}{I_{pm}} \tag{4.A.6}$$

Substituting Eq. (4.A.5) for A_p and Eq. (4.A.4) for I_{pm},

$$N_p = \frac{A_w \cdot K_p \cdot K_u \cdot J \cdot V_{\text{in(min)}} \cdot K_t}{P_{\text{in}}} \tag{4.A.7}$$

or ·

$$A_w = \frac{N_p \cdot P_{\text{in}}}{K_p \cdot K_u \cdot J \cdot V_{\text{in(min)}} \cdot K_t}$$

From Faraday's law, $$E_{dt} = Nd\,\Phi \tag{4.A.8}$$

Hence

$$V_{\text{in(min)}} \cdot t_{\text{on(max)}} = N_p \cdot \Delta B \cdot A_e$$

or

$$A_e = \frac{V_{\text{in(min)}} \cdot t_{\text{on(max)}}}{N_p \cdot \Delta B}$$

where t_{on} = "on" period
ΔB = flux density change during "on" period
A_e = effective core area

The maximum "on" time is one half period at the operating frequency f; hence

$$t_{\text{on(max)}} = \frac{1}{2f} \tag{4.A.9}$$

Substituting Eq. (4.A.9) into Eq. (4.A.8),

$$A_e = \frac{V_{\text{in(min)}}}{N_p \cdot \Delta B \cdot 2f} \tag{4.A.10}$$

Now $$AP = A_w \times A_e$$

Combining Eqs. (4.A.7) and (4.A.10),

$$AP = \frac{P_{\text{in}}}{K_t \cdot K_u \cdot K_p \cdot J \cdot \Delta B \cdot 2f} \qquad \text{m}^4 \tag{4.A.11}$$

If the transformer is limited to a 30°C temperature rise under convection-cooled conditions, the wire current density J is given by the empirical relation[1,2]

$$J = 450 \times 10^4 \times AP^{-0.125} \qquad A/m^2 \qquad (4.A.12)$$

(For a constant temperature rise, the current density must fall as the transformer size increases, as the ratio of volume to surface area falls with increasing size.)
Substituting Eq. (4.A.12) into Eq. (4.A.11) and converting AP to centimeters,

$$AP = \frac{P_{in} \times 10^8}{K_t \cdot K_u \cdot K_p \times 450 \times 10^4 \times AP^{0.125} \times \Delta B \times 2f} \qquad cm^4$$

$$AP^{(1 - 0.125)} = \frac{P_{in} \times 10^4}{K_t \cdot K_u \cdot K_p \times 450 \times \Delta B \times 2f} \qquad cm^4$$

Therefore

$$AP = \left(\frac{P_{in} \times 10^4}{K_t \cdot K_u \cdot K_p \times 450 \times \Delta B \times 2f} \right)^{1.143} \qquad cm^4 \qquad (4.A.13)$$

Since $K' = K_t \cdot K_u \cdot K_p$ (see Table 3.4.1), Substituting K' in Eq. (4.A.13) and simplifying gives

$$AP = \left(\frac{11.1 \, P_{in}}{K' \cdot \Delta B \cdot f} \right)^{1.143} \qquad cm^4 \qquad (4.A.14)$$

Hence the size of the transformer, in terms of area product AP, can be found knowing the input power P_{in}, flux density swing ΔB, frequency f, and a constant of topology K' for a free air temperature rise of 30°C.

4.A.2. TOPOLOGY FACTORS K'

The topology factor K' depends on the type of converter, the type of secondary winding and rectification, insulation and screening requirements, and current waveforms.
   K' is made up of three subfactors as follows:

$$K' = K_p \cdot K_u \cdot K_t \qquad (4.A.15)$$

4.A.3. PRIMARY AREA FACTOR K_p

This is the ratio of the winding area provided for the primary to the total window area (A_p/A_w). Although the window area is normally split equally between primary and secondary, the primary area is not always fully used for the main primary winding. For example, in the forward converter, an energy recovery winding is usually bifilar-wound, with the main primary taking up part of the winding area. Further, in the center-tapped push-pull topology only half the primary is active at any time, reducing the effective primary to 25% of the total window area. In the same way, a center-tapped secondary winding has the same 25% utility factor.

4.A.4. WINDOW UTILIZATION FACTOR K_u

This is the ratio of the area of window occupied by copper to the total available window area. With round wires and normal insulation, this factor is typically 0.4 (40%). When bobbin windings are used, this may be as low as 30%.

4.A.5. CURRENT FACTOR K_t

This is the ratio of the DC input current to the maximum primary current (I_{dc}/I_p). It depends on the topology of the converter and the shape of the primary current waveform. For simplicity, rectangular waveforms are assumed; this introduces little error in practice.

The winding form of the primary is defined by the type of converter. However, a choice exists for the secondary winding form depending on the rectification circuit. Bridge rectifiers require a single winding and biphase rectifiers a center-tapped winding with a lower copper utilization factor.

4.A.6. TEMPERATURE RISE

The temperature rise of the transformer under free air convection-cooled conditions depends on the total internal losses (core loss plus copper losses) and the effective surface area. Figure 3.1.7 is developed from measured results and information published in References 1, 2, and 15. It shows the relationship of surface area to area product for typical switchmode ferrite cores. It also predicts the temperature rise above ambient as a function of total internal dissipation with area product or surface area as a parameter. The temperature rise predictions assume free air cooling and an ambient air temperature of 25°C. The predicted temperature rise (in the range 20 to 70°C) may be obtained directly from the nomogram by entering with the surface area of the transformer and internal dissipation. If the area product is known, an indication of the surface area may be obtained from the same nomogram using the "AP line" area product intersect.

Alternatively, for a small temperature rise in the range 20 to 50°C, the following approximate formula developed from Fig. 3.1.7 may be used:

$$\Delta T = \frac{800 \cdot P_t}{A_s} \quad °C \tag{4.A.16}$$

where ΔT = temperature rise, °C
P_t = total internal loss, W
A_s = surface area of transformer, cm^2

The surface area A_s is related to the area product AP as follows:

$$A_s = 34 \times AP^{0.51} \quad \text{cm}^2 \tag{4.A.17}$$

Substituting for A_s in Eq. (4.A.16),

$$\Delta T = \frac{23.5 \times P_t}{\sqrt{AP}} \quad °\text{C}$$

from which the thermal resistance R_t (normalized for a 30°C temperature rise) will be

$$R_t = 23.5AP^{0.5} \quad °\text{C/W} \tag{4.A.18}$$

APPENDIX 4.B

SKIN AND PROXIMITY EFFECTS IN HIGH-FREQUENCY TRANSFORMER WINDINGS

4.B.1. INTRODUCTION

The information presented here provides some explanation and justification for the design methods used in Chap. 4. For a more complete background, see References 1, 2, 8, 15, 31, 58, 59, 60, 65, 66, and 67.

To optimize the efficiency of high-frequency switchmode transformers, suitable wire gauges, strip sizes, and winding geometries must be used. Filling up the available window area with a gauge of wire that will fit simply will not do if optimum efficiency is to be obtained. The simple design rules used for line-frequency transformers are inadequate for optimum design of high-frequency transformers.

Figures 3.4B.1, 3.4B.5, and 3.4B.6 show how the effective ac resistance of a winding is related to frequency, wire size, and number of layers. Hence, at high frequencies, when two or three layers of wire are used in a winding, the F_r ratio (the ratio of the effective ac resistance of the winding to its DC resistance) could quite easily be a factor of 10 or more in a poor design. That is, the effective resistance of the winding at the working frequency could be 10 times greater than its DC resistance. This would give excessive power loss and temperature rise.

The intuitive temptation to use as large a wire as possible often leads to the wrong result in a high-frequency application. Using too large a wire results in excessive loss as a result of skin and proximity effects. Hence too large a wire gauge, giving many layers and excessive buildup, is just as inefficient as having too small a gauge. It will be shown that because of skin and proximity effects, an ideal wire size or strip thickness exists, and this must be used if optimum efficiency is to be obtained.

A brief examination of skin and proximity effects would perhaps be helpful at this stage.

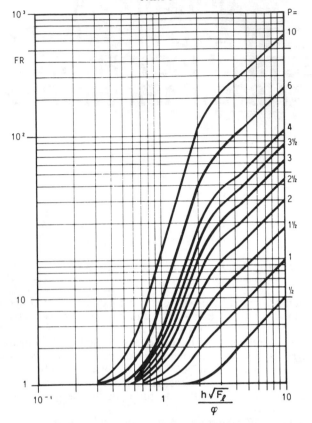

FIG. 3.4B.1 F_r ratio (ratio of ac/DC resistance as a result of skin effect) as a function of the effective conductor thickness, with number of layers P as a parameter (*After Dowell*[31].)

4.B.2. SKIN EFFECT

Figure 3.4B.2 shows how an isolated conductor carrying a current will generate a concentric magnetic field. With alternating currents a magnetomotive force (mmf) exists, generating eddy currents in the conductor. The direction of these eddy currents is such as to add to the current at the surface of the wire and subtract from the current in the center. The effect is to encourage the current to flow near the surface of the conductor (the well-known skin effect). The majority of the current will flow in an equivalent surface skin thickness or penetration depth Δ, defined by the formula

$$\Delta = \frac{K_m}{\sqrt{f}} \tag{4.B.1}$$

where Δ = penetration depth, mm
f = frequency, Hz
K_m = material constant (K_m ranges from 75 for copper at 100°C to = 65.5 for copper at 20°C)

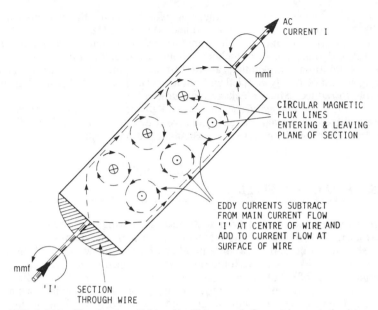

FIG. 3.4B.2 Showing how "skin effect" is caused. Current is constrained to flow in the surface layer of the conductor as a result of concentric magnetic fields in the body of the conductor caused by the current flow.

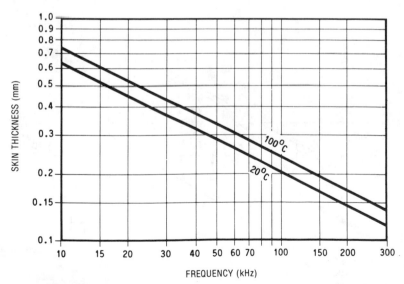

FIG. 3.4B.3 Effective skin thickness as a function of frequency, with temperature as a parameter.

Figure 3.4B.3 shows the skin thickness or penetration depth for copper at 20°C and 100°C plotted over the frequency range 10 to 300 kHz.

For connections to and from the transformer, individual wire diameters in excess of 2 or 3 times the skin thickness should be avoided. For high-current applications, multifilament windings are preferred. Go and return wires to the same winding should be tightly coupled and run as parallel or twisted pairs. This is also desirable to reduce external leakage inductance.

Remember, the copper losses increase in proportion to the current density squared; therefore, a small increase in current density at the surface will have a significant effect on the effective ac resistance ratio (F_r ratio).

4.B.3. PROXIMITY EFFECTS

In a transformer, the simple current distribution resulting from the skin effect will be further modified by proximity effects from adjacent conductors.

As shown in Fig. 3.4B.4, when a number of turns are wound to form one or more layers, a magnetomotive force (mmf) is developed in line with the plane of the winding. The effect of this mmf is to develop eddy currents whose direction is such as to add to the current flow toward the primary-to-secondary winding interface and reduce the current on the side of the winding away from the interface. As a result of these proximity effects, the useful area of a conductor is further reduced.

The proximity effect is most pronounced where the mmf is maximum, that is, at the primary-to-secondary interface. Figure 3.4.8*a* and *b* (p. 3.81) shows the dis-

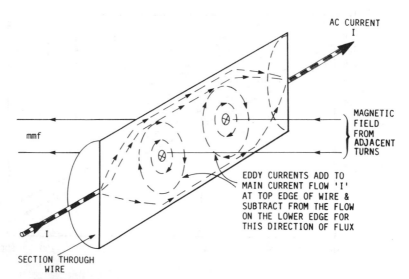

FIG. 3.4B.4 Showing how "proximity effects" are caused. Current is constrained to flow toward the interface of the windings as a result of incident magnetic fields from nearby turns.

tribution of mmf in a simple winding configuration and that in a sandwiched construction. In the sandwiched form, the maximum mmf is halved, and the center of the middle winding has an mmf of zero. As a result, the proximity effects in the center of the winding are also zero. Hence, for the determination of F_r, only half the layers and turns of the center winding need be considered.

4.B.4. DETERMINATION OF OPTIMUM WIRE DIAMETER OR STRIP THICKNESS

Figure 3.4B.1 shows the F_r ratio (ac resistance/DC resistance) plotted against the equivalent conductor height φ, with number of layers as a parameter.

In general,

$$\varphi = \frac{h}{\Delta \sqrt{F_l}} \qquad (4.B.2)$$

where φ = effective conductor height, mm

h = thickness of strip (or effective diameter of round wire)

F_l = copper layer factor

Note: To simplify the mathematical treatment, a round conductor of diameter d is replaced by a square one of the same area with an effective thickness h; e.g.,

$$\text{Area of round wire} = \pi r^2 \qquad \text{or} \qquad \pi(d/2)$$

$$\text{Area of square wire} = h^2$$

Hence

$$\pi \left(\frac{d}{2} \right)^2 = h^2$$

Therefore,

$$h = d \sqrt{\frac{\pi}{4}} \qquad (4.B.3)$$

The copper layer factor F_l is a function of the effective wire diameter, spacing between wires, turns, and useful winding width:

$$F_l = \frac{N \cdot h}{b_w}$$

where N = turns per layer

b_w = useful winding width, mm

4.B.5. OPTIMUM STRIP THICKNESS

In the simple case of a full-width copper strip winding, at a fixed frequency, φ goes to h/Δ, since $F_l = 1$; also, the number of layers is equal to the number of turns.

The ideal strip thickness can now be established from Fig. 3.4B.1 as follows:

$$R_{ac} = F_r \cdot R_{dc}$$

but (for a defined bobbin size and number of turns)

$$R_{dc} = \frac{N \cdot \rho \cdot l}{A} \propto \frac{\rho}{b_w \cdot h}$$

where ρ = resistivity of copper, Ω/cm
$\quad A$ = area of wire, cm^2
$\quad l$ = length of wire, cm
$\quad N$ = turns

Hence

$$R_{ac} = F_r \cdot R_{dc}$$

Therefore

$$R_{ac} \propto F_r \cdot \frac{\rho}{b_w \cdot h} \propto \frac{\rho}{b_w \cdot \Delta} \times \frac{F_r}{\varphi} \qquad \text{(since } \varphi = h/\Delta\text{)}$$

and

$$\frac{R_{ac} \cdot b_w \cdot \Delta}{\rho} \propto \frac{F_r}{\varphi} \qquad\qquad\qquad (4.B.4)$$

By plotting F_r/φ against φ with turns (or layers with a strip winding) as a parameter (Fig. 3.4B.5), the minimum ac resistance point for each number of turns can be seen. The minimums fall close to a line (dashed in the figure) where $F_r = 4/3$. For a given number of turns (layers), the optimum strip thickness h is obtained from the lower scale as a multiple of the skin thickness at the operating frequency. For example, with two turns, the minimum ac resistance is found where $H/\Delta = 1$ and the optimum strip thickness is the same as the skin thickness.

It has been shown that for a simple strip winding, the optimum strip thickness is a function of the frequency and number of turns (layers), and will be found near an F_r ratio of 1.33.

4.B.6. OPTIMUM WIRE DIAMETER

With round wire windings, the determination of optimum wire diameter is more complex than that for the strip winding, shown above; however, by a similar process, it has been shown[58,59,60,66] that the optimum wire diameter will be found near $F_r = 1.5$.

Figures 3.4B.6 and 3.4B.7 show the optimum copper strip thickness or wire diameter to use to obtain the ideal F_r ratio. They are plotted against the number of "effective layers," with frequency as a parameter.

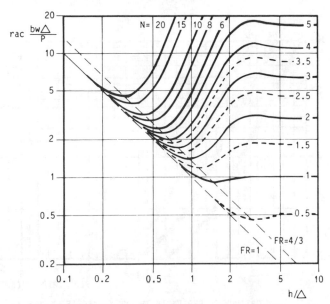

FIG. 3.4B.5 Plot of $R_{ac\ -\ h}/\Delta$ with number of layers as a parameter, showing the development of the conditions for optimum F_r ratio and optimum conductor thickness. (*J. Jongsma 1882 Mullard Ltd. Ref. 58.*)

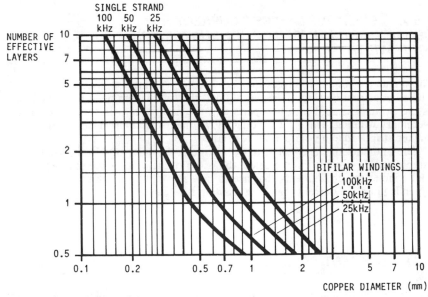

OPTIMUM COPPER WIRE DIAMETER (FOR F_R=1.5)

FIG. 3.4B.6 Optimum wire diameter for an F_r ratio of 1.5, as a function of the number of effective layers in the winding, with frequency as a parameter (*Mullard Ltd.*)

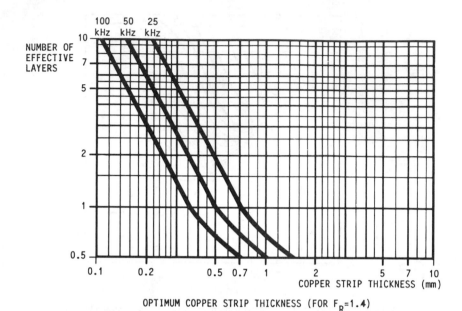

OPTIMUM COPPER STRIP THICKNESS (FOR F_R=1.4)

FIG. 3.4B.7 Optimum strip thickness for an F_r ratio of 1.4, as a function of the number of effective layers, with frequency as a parameter. (*Mullard Ltd.*)

4.B.7. "EFFECTIVE LAYERS," CURRENT DENSITY, AND NUMBER OF CONDUCTORS OR STRIP WIDTH

The number of effective layers depends on the winding topology. In the sandwiched construction (Fig. 3.4.8c), only half the total layers need be considered, as the primary is split into two parts and the mmf is zero in the center of the secondary winding. This split winding allows the use of a larger-diameter wire or thicker strip, reducing the overall resistance.

The optimum thickness of strip or maximum diameter of wire has now been established. It remains only to select the width of the strip or the number of parallel conductors or filaments that are to be used for the windings. This choice depends upon the current that each winding is to carry.

4.B.8. CURRENT DENSITY I_a

As a first approximation, the current ratings for the wire shown in Table 3.1.1 may be used. These are based on a current density of 450 A/cm^2, which is the optimum density for a core with an area product of 1 cm^4 and a 30°C temperature rise.

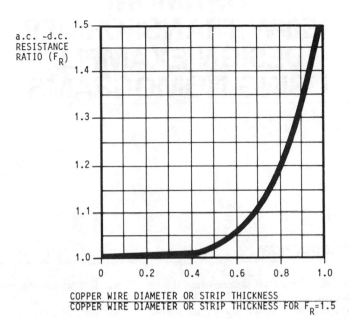

FIG. 3.4B.8 F_r ratio for wires below optimum thickness. (*Mullard Ltd.*)

More correctly, with larger cores the current density I_a should be reduced, as the heat-dissipating surface area increases less rapidly than the heat-generating volume. In general:

$$I_a = 450 \cdot AP^{-0.125} \qquad \text{A/cm}^2 \qquad (4.\text{B}.5)$$

Practical limitations may not allow the use of the optimum wire size. For nonoptimum conditions, the effective ac resistance may be established from Fig. 3.4B.1 for a thickness greater than optimum, and from Fig. 3.4B.8 for a thickness smaller than optimum.

CHAPTER 5
OPTIMUM
150-W TRANSFORMER
DESIGN EXAMPLE
USING NOMOGRAMS

5.1 INTRODUCTION

The following example will demonstrate the rapid optimization possible using the various nomograms shown in Chap. 4.

The transformer is a self-oscillating half-bridge square-wave DC-to-DC converter operating at 20 kHz and producing 150 W. To minimize leakage inductance and skin effects, a sandwiched construction as shown in Fig. 3.4.8c is to be used. The transformer is required to meet UL and VDE safety requirements, and will have two grounded safety screens.

The input voltage is 200 V, and the output is to be 25 V at 6 A, using bridge rectification. The temperature rise is to be 30°C in free air cooling conditions.

5.2 CORE SIZE AND OPTIMUM FLUX DENSITY SWING

For an output power of 150 W at 20 kHz with a 30°C temperature rise, for half-bridge or push-pull applications, the nomogram shown in Fig. 3.4.3 indicates an EC41 core with a flux density swing of 330 mT for optimum efficiency. (The example is shown on the nomogram.)

5.2.1 Use of Nomogram 3.4.3

Step 1. Enter the nomogram on the right-hand side with the required power (150 W), and at the top with the required operating frequency (20 kHz), as shown in the example drawn on the nomogram.

Step 2. The horizontal line from the intersection of the frequency and power lines gives the area product *AP* (left) and some examples of standard switchmode cores (right). In this example, $AP = 2.2$, and suitable cores would be EC41 or ETD34/17/11. Choose one of the recommended cores, or select a different core type using the area product value. In this example, the EC41 (FX 3730) core is chosen.

Step 3.   The vertical projection from the intersection of the power line with the frequency line to the lower flux density swing scale indicates a ΔB value of 330 mT. Hence, for this example, an EC41 core with a flux density swing of 330 mT is chosen.

5.3 CORE AND BOBBIN PARAMETERS

Core type = EC41

Area product of EC41 (core) = 2.6 cm^4

Window area of bobbin = 134 mm^2

of core = 215 mm^2

Width of bobbin = 24 mm

Effective core area = 121 mm^2

Topology factor K' (from App. 4.A) = 0.164

Total weight (cores) = 52 g

Optimum flux density swing ΔB from
Fig. 3.4.3 = 330 mT

Frequency = 20 kHz

Half period = 25 μs

5.4 CALCULATE PRIMARY TURNS

The converter is in square-wave full-conduction-angle operation, so the maximum "on" period for each drive device is a half cycle, or 25 μs. (This type of converter is sometimes referred to as a DC transformer; see Part 2, Chap. 17.) A single half-period square pulse gives the maximum primary volt-seconds stress; hence the turns can be established from a single half cycle using the volt-seconds approach.

From Faraday's law of induction,

$$N_p = \frac{V \cdot t}{\Delta B \times A_e}$$

where N_p = primary turns
 V = primary voltage (200 V)
 t = "on" period (25 μs)
 ΔB = flux density swing (0.33 T)
 A_e = effective core area (121 mm^2)

Hence

$$N_p = \frac{200 \times 25}{0.33 \times 121} = 125 \text{ turns}$$

5.5 CALCULATE PRIMARY WIRE SIZE

The bobbin window area A_w for the EC41 is 134 mm².

K_{ub}, the bobbin window copper utilization factor, is 50% (Appendix 4.A). For equal primary and secondary loss, 25% will be used for each winding. Hence 25% of A_w is used for the primary winding window A_{wp}. The rest is used for insulation and screens.

Hence primary window area A_{wp} is

$$A_{wp} = A_w \times 25\% = 134 \times 0.25 = 33.5 \text{ mm}^2$$

The area available for each turn is

$$\frac{A_{wp}}{\text{Turns}} = \frac{33.5}{125} = 0.268 \text{ mm}^2$$

Although round wire, which would normally utilize only 78% of the window area, is to be used in this example, the K_u factor already corrects for this, and the calculated area is the area of the round wire. Hence, from Table 3.1.1, the nearest insulated wire size is #24 AWG, with a diameter of 0.57 mm (including the insulation thickness).

The bobbin width is 24 mm. Allowing a creepage distance of 3 mm at each side, the usable width is 18 mm.

The maximum number of turns per layer is

$$\frac{18}{0.57} = 31.5 \text{ turns}$$

With a split primary, an even number of primary layers must be used. In this example, four layers of 31 turns will be used, giving a total of 124 turns for the primary.

5.6 PRIMARY SKIN EFFECTS

From Fig. 3.4.5, the maximum wire diameter for an F_r ratio of 1.5 can be found. (F_r is the ratio of the effective ac resistance to the DC wire resistance.) At 20 kHz, using two layers (the first-half primary section), the maximum wire diameter would be 0.7 mm for an F_r ratio of 1.5.

From Table 3.1.1, the selected #24 AWG wire diameter (copper only) is 0.51 mm, or only 73% of the value for an F_r of 1.5.

From Fig. 3.4.7, the F_r value will be 1.12 (negligible skin effect), and this F_r value will be used in calculating copper loss for the primary.

5.7 SECONDARY TURNS

A bridge rectifier will be used on the secondary so that a single winding may be used.

The DC voltage is to be 25 V. The secondary voltage must allow for two diode drops and winding resistance.

Allowing for two diode voltage drops of 0.8 V, the approximate secondary voltage will be $25 + 1.6 = 26.6$ V. The primary volts per turn are

$$\frac{P_v}{N} = \frac{V_P}{N_P} = \frac{200}{124} = 1.613 \text{ V/turn}$$

Hence the secondary turns are

$$\frac{V_s}{P_v/N} = \frac{26.6}{1.613} = 16.5 \text{ turns}$$

with a half-turn allowance for winding resistance; therefore 17 turns will be used.

5.8 SECONDARY WIRE SIZE

With 17 turns and a window area of 33.5 mm, the area available for each turn is 1.97 mm^2.

From Table 3.1.1, the nearest standard wire is #15 AWG, with a copper diameter of 1.45 mm.

5.9 SECONDARY SKIN EFFECTS

With a split winding, only half the secondary layers need be considered; assume 1 layer at this stage. From Fig. 3.4.5, the maximum wire diameter for an F_r ratio of 1.5 is 1 mm, and the single #15 AWG wire at 1.56 mm diameter is too large for effective use.

It is interesting to note that even at this low 20-kHz frequency, the skin effects are beginning to control the selection of wire size.

Dropping down three wire gauges will halve the area of the wire, and a bifilar winding of #18 AWG will have the same total area as the #15 AWG original selection. The copper diameter of the #18 AWG wire is 1.02 mm, giving the required F_r ratio of 1.5.

The usable bobbin width is 18 mm, so that a bifilar winding of 9 turns/layer may be used, giving nearly two layers for the secondary winding.

5.10 DESIGN NOTES

At this stage the design is essentially complete for most practical purposes. A half turn was added to round up the secondary, but no allowance has yet been made for voltage drop due to the winding resistance. The voltage drops will be small (usually less than 2%), and the primary turns may be adjusted to compensate.

For the most accurate results, the winding resistances must be calculated and an accurate adjustment of the primary turns can then be made. In the final analyses the performance and temperature rise must be measured in the finished product to include all the intangible thermal effects resulting from location and nearby components.

In this example, the losses will be calculated to prove the design.

5.11 DESIGN CONFIRMATION

In this example, the core and copper losses will be calculated to prove the design. From the calculated figures, the efficiency and the temperature rise will be predicted, and in the process, the core and copper losses will be obtained. From this the optimum design efficiency will be checked. Finally, with the winding resistances established, any need for a turns correction can be accurately determined.

5.12 PRIMARY COPPER LOSS

Calculate the mean turn length M_{lt} for the inner primary section:

$$M_{lt} = \pi \times d = 3.14 \times 1.5 = 4.71 \text{ cm/turn}$$

Total length of half primary $l_p/2 = N_P/2 \times M_{lt}$.

$$\frac{l_p}{2} = \frac{125}{2} \times 4.71 = 294 \text{ cm}$$

From Table 3.1.1 and Fig. 3.4.11, the resistance of #24 AWG wire at 50°C is 0.00095 Ω/cm, giving a half primary resistance of 0.279 Ω at DC.

At the operating frequency, the working resistance R_{ac} will be greater because of skin and proximity effects. From Fig. 3.4.5, the primary F_r is 1.12; hence

$$R_{ac} = R_{dc} \times F_r = 0.297 \times 1.12 = 0.333 \ \Omega$$

Because the converter has full duty cycle square-wave conduction, the primary rms current is the same as the DC input; hence

$$I_{p\,(rms)}^2 = \frac{P_{in}}{V_{dc}} = \frac{150}{200} = 0.75 \text{ A}$$

Copper loss in the inner half primary is $I_p^2 R_p = (0.75)^2 \times 0.333 = 0.187$ W.

In a similar way, the loss in the larger-diameter outer half primary is 0.22 W, giving a total primary loss of 0.407 W.

5.13   SECONDARY COPPER LOSS

The mean diameter of the secondary winding is 1.8 cm. The length of the secondary winding is

$$\pi \times d \times N_s = 3.14 \times 1.8 \times 17 = 96 \text{ cm}$$

The DC resistance of #18 AWG wire at 50°C is 0.00023 Ω, giving a resistance of 0.022 Ω for each wire (0.011 Ω for the bifilar winding).

The F_r ratio is 1.5 for the secondary, giving an ac resistance of

$$1.5 \times 0.011 = 0.0165 \ \Omega$$

The secondary copper loss $I_s^2 R_s$ is $6^2 \times 0.0165 = 0.594$ W.

The total copper loss is $0.407 + 0.594 = 1.001$ W.

5.14   CORE LOSS

Figure 3.4.4 shows the core loss as a function of flux density swing for N27 ferrite, with frequency as a parameter. Enter the graph at the bottom with the flux density swing (330 mT from Fig. 3.4.3). The horizontal projection from the flux density–frequency intersection gives the core loss in milliwatts per gram (in this case, 21 mW/g).

The total core weight is 52 g, giving a total core loss of

$$52 \times 21 = 1.09 \text{ W}$$

Total Loss

The total transformer loss is W_{copper} plus W_{core}:

$$1.001 + 1.09 = 2.091 \text{ W}$$

5.15   TEMPERATURE RISE

The temperature rise may be established from the nomogram shown in Fig. 3.1.7 or from Eq. (4.A.18).

Using Fig. 3.1.7, proceed as follows: Enter the nomogram at the top with the core area product (2.6 for the EC41). The horizontal projection from the intersection of the AP value with the dashed AP line gives the surface area, 54 cm^2 (left scale). Enter the lower scale with the total dissipation, 2.09 W. The intersection of the dissipation line with the surface area line gives the temperature rise on the diagonal (30°C in this example).

From Fig. 3.1.7, the temperature rise is predicted to be 30°C. From Eq. (4.A.18), the predicted rise is 30.46°C.

5.16 EFFICIENCY

$$\text{Efficiency } \eta = \frac{\text{output power}}{\text{output power + losses}}$$

$$\eta = \frac{150}{150 + 2.09} = 98.6\%$$

Since the copper losses and core losses are nearly equal (Sec. 5.14), this is clearly an optimum design.

CHAPTER 6
TRANSFORMER STAIRCASE SATURATION

6.1 INTRODUCTION

The term "staircase saturation" describes a dynamic transformer saturating effect that is particularly prevalent in push-pull converters. In such converters, the transformer primary is actively driven in both directions, and both quadrants of the B/H characteristic are utilized.

For maximum-efficiency operation, full use must be made of the transformer core, and at lower frequencies large symmetrical flux excursions would normally be required.

As a result of variations in saturation voltage, switching times, rectifier voltage drops, and transformer winding resistance, the transformer flux excursion in one direction is not always exactly balanced by the flux excursion in the reverse direction. Therefore, the mean working point for the core may move slightly away from the center point (zero flux) at the end of a cycle of operations. In push-pull forward converters, because of output diode flywheel action, the winding voltage is clamped to zero during the "off" period; hence there is no DC restoration of the core between cycles. As a result, any flux density offset progressively increases with each cycle, and the mean flux level will "staircase" toward saturation.

In effect, the imbalanced transformer polarization results in a net DC current in the transformer windings, and even a slight imbalance will quickly drive a high-permeability core to saturation.

Fortunately, the effect tends to self-limit, because the increasing magnetizing current which occurs during the saturating half cycle tends to reduce the width of the power pulse during that half cycle. (The storage time of the transistor is reduced.) However, the mean working point for the core will have been offset from zero, and if no corrective action is taken, the available flux excursion in one direction is reduced. This limits the ability of the supply to respond to transient changes, reducing the utility of the core.

In duty-ratio-controlled push-pull applications, some imbalance in alternate half cycles, and hence flux excursions, is almost inevitable. In practice there will always be some small drive or output diode imbalance. Thus, unless steps are taken to ensure flux balancing, there will inevitably be partial saturation of the core in one direction or the other.

6.2 METHODS OF REDUCING STAIRCASE SATURATION EFFECTS

By careful selection of components, the saturating effects of pulse asymmetry can be reduced to barely acceptable limits. Matching of transistors and diodes offers a partial solution to the problem. So, too, may adjustment of the differential drive conditions and the introduction of a small air gap in the core. (With an air gap, some small DC current offset can be tolerated without severe saturation.) However, these measures can give only limited success.

The same asymmetry problem can occur in the half-bridge circuit, although some improvement may be obtained by breaking any DC path in the primary circuit with a small series capacitor. It is quite common to find such a capacitor in half-bridge converters.

Unfortunately, this primary "blocking" capacitor does not completely eliminate the staircase saturation problem. A second DC path usually exists in the secondary winding and output circuit. This DC path is more difficult to break. For example, in a buck-derived push-pull output stage, the output choke forces secondary current conduction, even during the "off" period of the primary power switches. During this period, current continues to flow in both output rectifier diodes and secondary windings. Since both diodes conduct, if there is a difference in diode forward voltages, an effective net DC voltage will be applied to the secondary winding during the "off" period. Once again, this will result in a net DC polarizing current in the secondary, and this secondary current can also cause core saturation.

Under the conditions where the secondary is unbalanced, the primary blocking capacitor will introduce a second undesirable effect. It takes up a net DC charge, so that alternate half cycles will now be of different voltage amplitudes. This introduces subharmonic ripple into the output filter, increasing output ripple.

Hence, even though these commonly used techniques can reduce the staircase saturation effects to workable limits, severe restrictions are placed on transient performance as a result of the limited flux density swing in one direction. Immediate saturation of the core can occur if the "on" period, or applied voltage, is suddenly increased during the critical, near-saturated half cycle.

To overcome these problems, current-mode control or other active flux balancing techniques should be used, particularly in high-power push-pull converters. In duty ratio control systems, a forced flux balancing technique can be used, as described in the following section.

6.3 FORCED FLUX BALANCING IN DUTY-RATIO-CONTROLLED PUSH-PULL CONVERTERS

When a core is operating with nonsymmetrical flux excursions, two effects will be noticed in the primary. First, alternate primary current pulses will have a DC offset. Second, in more severe cases, there will be a sudden increase in the primary current at the end of the "on" period during one half cycle. This increase occurs for the partially saturated direction of drive, because of the rapid curva-

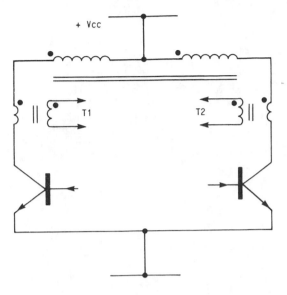

PUSH-PULL

FIG. 3.6.1 Basic push-pull power circuit, showing current transformer locations (T1 and T2) for forced flux density balancing in push-pull topologies.

ture of the B/H characteristic toward core saturation. Hence, peak and mean currents will be different for alternate half cycles.

By differentially comparing the currents flowing in the two half cycles, small differences may be detected and used to differentially control the pulse width of the drive to the switching devices and thus minimize the offset.

Figures 3.6.1 and 3.6.2 show the basic elements of a forced flux balancing circuit.

To provide the required information on first- and third-quadrant flux excursion, two separate current transformers T1 and T2 are required. These are fitted in series with the two switching devices, so as to measure the forward and reverse transformer primary currents. The actual position of the current transformers depends on the converter topology; Fig. 3.6.1 shows a typical push-pull example.

Figure 3.6.2 shows the drive output circuit of a typical pulse-width-modulated (duty ratio control) system. The normal oscillator A6, voltage control A5, pulse-width modulators A3 and A4, and output gates U2 and U3 are shown.

To this duty ratio control system has been added a forced flux balancing circuit, T1, T2, A1, and A2. This operates as follows.

The outputs of the two current transformers T1 and T2 are rectified by D1 and D2 and compared via R1 and R2 to develop a voltage on C1 which is proportional in direction and amplitude to the mean difference between forward- and reverse-going current pulses in the two power switches.

The voltage amplifier A1 gives an output proportional to the mean DC voltage on C1. Heavy integration of the current pulses is provided by feedback components R1 and C1. This makes the output of amplifier A1 essentially DC, propor-

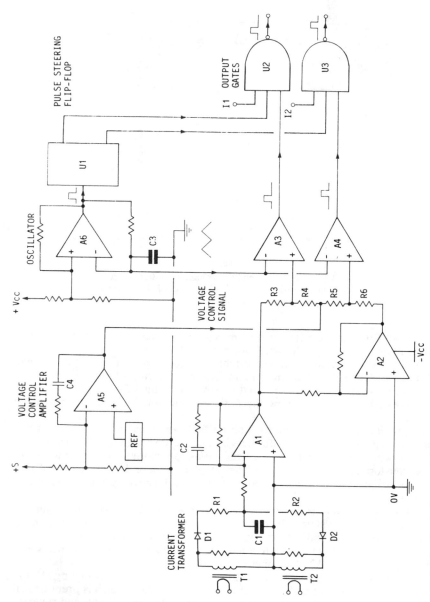

FIG. 3.6.2 Circuit diagram of a duty cycle, voltage-controlled, push-pull drive section with forced flux balancing.

tional in amplitude and direction to any imbalance detected between the collector currents.

The output of A1 is fed directly to the top of the resistor chain R3, R4, R5, R6 and also to amplifier A2, where it is inverted and fed as an equal but opposite polarity to the bottom of the resistor chain.

Note that under this condition, the control input at the center of the resistor chain (junction R4, R5) is not changed in response to a current imbalance. However, the inputs to the ramp comparator amplifiers A3 and A4 are differentially adjusted, with alternate pulse widths changed to restore current asymmetry.

Hence, the current balancing circuit is able to increase the pulse width for one side and decrease the other, depending upon the direction and amplitude of any imbalanced condition, without changing the mean output voltage.

The voltage control amplifier A5 will continue to adjust the common pulse width (the output of both ramp comparators) so as to control the output voltage in the normal way.

Note: Changing the voltage at the center of the divider chain R3, R4, R5, R6 (voltage control input) changes both output pulse widths equally, resulting in a change of output voltage without changing the differential current control. Hence, the two control loops operate independently of each other, an important stability criterion.

This type of circuit is able to introduce a net primary DC current to compensate for any secondary asymmetry, provided that a DC current path is provided in the primary circuit. Hence a DC blocking capacitor *must not* be fitted in the primary when this circuit is used. With half-bridge circuits, special techniques will be required to provide a DC current path in the primary. (See Sec. 10.10 in this Part.)

The same current transformers can be used for current limiting purposes by using the mean current analogue voltage information available at the outputs of D1 or D2. (The extra inputs to gates U2 and U3, I1 and I2, are additional inhibit inputs. These may be used to prevent cross conduction; see Part 1, Chap. 19.)

6.4 STAIRCASE SATURATION PROBLEMS IN CURRENT-MODE CONTROL SYSTEMS

Although current-mode control can provide automatic balancing of the flux density in push-pull systems, it also requires a DC current path in the primary. This is hardly surprising, because with current-mode control, the peak current remains constant, and any imbalance in the primary or secondary volt-seconds results in a differential pulse-width change to correct the asymmetry. Hence, the original volt-second asymmetry must be compensated for with an ampere-second asymmetry. This requires a compensating DC current flow in the primary circuit, and a DC blocking capacitor must *not* be used in the primary circuit. Once again, special techniques will be required in the half-bridge circuit to meet this need. (See Sec. 10.10.)

6.5 PROBLEMS

1. What is meant by the term "staircase saturation"?
2. What methods may be used to reduce staircase-saturation effects?
3. What method of switchmode control is recommended to eliminate the problems of staircase saturation?
4. A DC blocking capacitor must not be used in the primary when current-mode control is used. Why is this?

CHAPTER 7
FLUX DOUBLING

The term "flux doubling" refers to a possible saturation hazard that exists in push-pull systems.

Under steady-state conditions, a balanced push-pull transformer should be able to sustain a maximum flux excursion of nearly twice the peak flux density (from $-\hat{B}$ to $+\hat{B}$). In many low-frequency designs, maximum advantage is taken of this large potential flux swing, so that the number of primary turns may be reduced and improved efficiency obtained.

Under steady-state operation, the starting position for the flux at the beginning of each half cycle will be either $+\hat{B}$ or $-\hat{B}$, the core having been clamped at this value during the "off" period by flywheel action of the output choke and rectifier diodes. Hence the maximum flux density swing during steady-state half cycles will be $2\hat{B}$. However, in a system running at this large level of flux density swing, a possible hazard exists on initial switch-on and under transient conditions.

The starting point for the initial flux excursion (when the system is first switched on or following a condition of very light loading when the pulse width has been very narrow), will be near zero. (See Fig. 2.9.2.) From this starting point, a sudden flux excursion of $2\hat{B}$ (the steady-state excursion) would result in the core saturating for the first half cycle, often with damaging results.

To prevent this so-called "flux doubling" effect, either the initial choice of working flux density swing must be lower than $\hat{B}$, reducing the utility of the core, or the control circuit must recognize the potential hazard and reduce the pulse width until the correct working conditions are established. (See "Soft Start," Part 1, Chap. 9, and "Current-Mode Control," Part 3, Chap. 10.)

CHAPTER 8
STABILITY AND CONTROL-LOOP COMPENSATION IN SMPS

8.1 INTRODUCTION

Any closed-loop control system that has the potential for a loop gain of unity and a frequency-dependent internal phase shift of 360° has a potential for instability. In switchmode supplies, because a low-pass filter will be used in the power stages to eliminate noise and provide smooth DC outputs, an inevitable phase shift is introduced which reduces the phase margins and can lead to instability.

Almost all switchmode power supplies will have a closed-loop negative feedback control system, to provide good performance. In the negative feedback system, the control amplifier will be connected in such a way that an intentional phase shift of 180° is introduced. Hence, any perturbation internally generated within the feedback loop will normally result in antiphase feedback to eliminate the undesirable changes.

If the phase of the feedback were to remain at 180°, then the control loop would always be stable and the design engineer's life would be very much simplified. Of course, this is not the case in the real world, where various switching delays and reactances introduce additional phase shifts, which may lead to instability if the correct loop compensation is not used.

8.2 SOME CAUSES OF INSTABILITY IN SWITCHMODE SUPPLIES

Consider Fig. 3.8.1. This shows a typical closed-loop negative feedback switching supply with the major elements separated into three parts.

Block 1 is the converter section, which will have a transfer function that depends on the converter topology and output filter.

Block 2 is the pulse-width modulator, which will provide most of the voltage gain to be found external to the control amplifier.

Block 3 is the control comparator amplifier, the reference voltage, and the loop compensation networks.

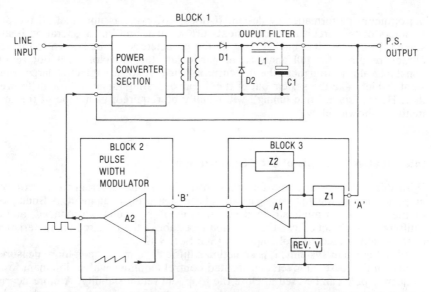

FIG. 3.8.1 Block schematic diagram of the control loop, for a forward (buck-derived) switchmode power converter. Showing power section block 1, pulse-width modulator block 2, and control circuit block 3.

Assume that a square-wave perturbation is introduced at node A. This square wave will contain energy distributed as an infinite series of odd harmonic components. If the response of a real system is examined for progressively increasing harmonics, it will be seen that the gain and phase shift change as the frequency increases. If the gain is equal to unity at a frequency at which total additional phase shift is 180° (this additional 180°, together with the original fixed 180° starting value, gives 360° total phase shift), then sufficient energy will arrive back at the input of the system, in phase, to sustain the original perturbation, and the system will oscillate at that frequency.

However, normally the control amplifier will have feedback compensation components Z2 arranged to reduce the gain at higher frequencies such that stability is maintained at all frequencies.

8.3 METHODS OF STABILIZING THE LOOP

Several quite different but equally valid and successful methods can be used to stabilize the control loop.

By Circuit and Mathematical Analysis

For design engineers who are comfortable with the theoretical and mathematical analysis of closed-loop systems, this is probably the most suitable approach, giving optimum performance by design. It does, however, assume that all the parameters of the circuit and components are known, and linear operation is assumed. In practice it is unlikely that all the parameters will be known to the accuracy required for full analysis; in particular, the inductors will not retain constant values throughout the full current range. Also, the effect of large transients which take the system out of the linear operating range is difficult to predict. Hence, some "fine tuning" will usually be required, using one of the test methods shown below.

Interrogative Methods of Loop Stabilization

With this approach, the transfer function of the circuit, external to the control amplifier, is investigated using gain and phase measuring equipment. A Bode plot of the pulse-width modulator and power converter circuits is produced, and a "difference technique" can then be used to establish the required characteristics of the compensated control amplifier. (See Sec. 8.12.)

With high-gain systems, it is sometimes difficult to make open-loop measurements. In this case, an overcompensated control amplifier with a dominant low-frequency pole can be used to close the loop and retain stability. A more nearly optimum compensation network can then be established by measuring the voltage at the input to the pulse-width modulator, node B, and comparing this with the output voltage and phase at node A. The optimum compensation network can then be obtained by "mirror image" techniques.[15]

Empirical Methods of Obtaining Loop Stability

In this approach, the control loop is closed using an overcompensated control amplifier with a dominant low-frequency pole to obtain initial stability. The compensation network is then dynamically optimized by using transient pulse-loading techniques. This method is very fast and effective, but the correct general form of the compensation network required to suit the converter topology must be known. (See Figs. 3.10.8 and 3.10.9.) One advantage of this approach is that large-signal conditions can be dynamically investigated. For this reason, transient load testing is recommended as a final proving test, whatever technique is used to establish the compensation network.

The major disadvantage of the empirical approach is that it will not be known whether optimum performance has been obtained. However, if the performance is well within the required specification requirements, it probably is not important that it always be optimum. Several units should be examined to obtain spread margins. This method of testing will not always expose loops which are only conditionally stable.

Combination of Design and Measurement

This approach uses a combination of the above methods, depending on the designer's skill and experience, the type of product, the equipment available, and the designer's preference.

8.4 STABILITY TESTING METHODS

Transient Load Testing

The theoretical analysis of loop stability criteria in switchmode power supplies is complicated by the fact that the transfer function changes under different loading conditions. The effective inductance of the various wound power components will often change significantly with load current. Further, when large-signal transient conditions are considered, it may be found that the control circuits are displaced to nonlinear operation (hit the end stops), and linear analysis will not give the complete picture.

Transient load testing provides a very quick and powerful tool for the examination of the overall loop response at various loads and under large-signal dynamic conditions.

8.5 TEST PROCEDURE

Figure 3.8.2 shows a typical "transient load test" setup. Typically, the power supply under test is connected directly to a fixed load R1, and via a fast semiconductor switch to load R2. Both loads are resistive and adjustable. It is important that the switched component of the load be noninductive. (Noninductive carbon pile adjustable resistors may be used in this position.)

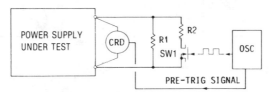

FIG. 3.8.2 A pulse loading test circuit, used for transient load testing of power supplies.

The fixed load is adjusted to provide at least the minimum current rating for the power supply, typically 10 or 20%.

The switched component of load may then be adjusted from zero to 80 or 90%, and the waveform under transient conditions examined on the oscilloscope.

8.6 TRANSIENT TESTING ANALYSIS

Typical transient current and voltage waveforms are shown in Fig. 3.8.3. Waveform (*a*) shows an underdamped response, with the control loop ringing after the transient edge. A power supply showing this type of response will have poor gain and phase margins, and may be only conditionally stable. With a

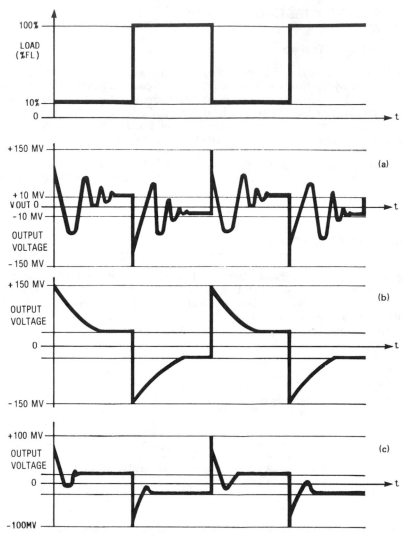

FIG. 3.8.3 Typical output current and voltage waveforms, for switchmode converters under pulse loading conditions: (*a*) underdamped performance, (*b*) overdamped performance, (*c*) optimum performance.

performance of this type, oscillation may occur under some loading conditions, or as a result of additional phase change or increase in gain at higher temperatures. Hence, this response is not recommended, and the compensation network should be adjusted to roll off at a lower frequency.

Waveform (*b*) shows an overdamped response, which, although very stable, does not give the best transient recovery performance. The roll-off frequency should be increased.

Waveform (*c*) is closer to an optimum condition, giving good performance and stable transient response, with adequate gain and phase margins for most applications. However, conditionally stable systems can give good transient performance, and a Bode plot is also advisable. A good symmetrical waveform for both positive- and negative-going current edges is also desirable, as this indicates that the control and power sections are well centered on the control range and that the slew rate is the same for both increasing and decreasing load conditions. It may not be possible to get symmetrical slew rates with buck regulators. The header voltage from the transformer will often be low, and the rate of change of current in the output inductor for increasing currents will be less than that for reducing currents. (The pulse width cannot increase by the same ratio as it can decrease.)

The transient load switch will normally be adjusted for unity mark space ratio. However, with high-current power supplies, it is sometimes an advantage to make the "on" period shorter than the "off" period, thereby reducing the power requirements for the switched load. The repetition rate should be adjustable so that initial measurements can be started at a low frequency and increased as the loop is optimized. The rate of change of current on the load switch dI/dt should be defined. Industry standards include rates of 5 A/μs and 2 A/μs. Transient response figures that do not state the magnitude and rate of change of load current are meaningless.

Once the compensation components have been optimized for full load, the load and input voltage should be reduced to obtain results for the full specified operating range.

Several units should be examined to give an indication of spreads in performance. Finally, the loop should be examined, using a gain and phase measuring technique (Bode plot) to obtain the gain and phase margins and to ensure that the system is not conditionally stable.

8.7 BODE PLOTS

Bode plots of the completed power supply are an excellent method of indicating the overall dynamic response of the system. A Bode plot is a graphical representation of the gain in decibels and phase angle in degrees for any part of, or the total, open or closed loop, plotted to a base of log frequency. A plot of the overall system shows the phase margins when the gain is unity (0 dB) and the gain margins when the phase shift is 360° total (an extra 180° phase shift). A Bode plot will also show any tendency for unwanted deviations in response, which can be caused by input and output parasitic filter resonances. For optimum performance and stability margins, it is recommended that the phase margin be at least 45° at unity gain.

8.8 MEASUREMENT PROCEDURES FOR BODE PLOTS OF CLOSED-LOOP POWER SUPPLY SYSTEMS

Most stabilized power supplies will have very large DC gains, to give well-defined output voltages. This high gain makes it difficult to make open-loop measurements. In open-loop operation a very small DC or ac input voltage change tends to take the output to the limit of the range and into nonlinear operation, and the correct DC operating conditions are not maintained. Hence, it is more normal to carry out the Bode plot with the unit in closed-loop operation.

For this reason, the measurement procedures to be considered here will allow the phase and gain to be plotted without breaking the feedback loop.

Figure 3.8.4 shows one basic method for making noninvasive measurements of gain and phase for a closed-loop Bode plot.

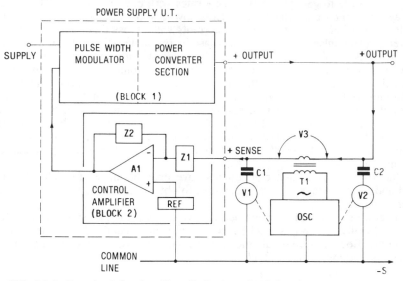

FIG. 3.8.4 Test circuit for closed-loop Bode plots of switchmode converters.

Block 1 is the converter and pulse-width modulator, and block 2 is the compensated control amplifier within the power supply. The measurement equipment, consisting of transformer T1, voltmeters V1 and V2, and oscillator V3, is connected between the external output and the remote sensing terminals "+ OUT" and "+ S." Transformer T1 is designed to have a very low output impedance and wide bandwidth. It also provides a DC link so that the control loop is not broken for DC conditions.

The oscillator introduces a very small series-mode voltage V3 into the loop via transformer T1. The effective ac input to the control amplifier is measured by voltmeter V1. The ac output voltage of the supply is measured by voltmeter V2. (Capacitors C1 and C2 provide DC isolation.) The ratio V2/V1 (in decibels) is the voltage gain of the system. The difference in phase (in degrees) is the phase shift around the loop (after making allowance for the fixed 180° negative feedback phase reversal).

The injected signal level must be sufficiently small that no part of the control loop is taken outside its normal linear behavior range. Typical injection levels for V1 may be 10 mV or less at the input to the amplifier "+ Sense." The frequency is then changed in small increments. At each measurement frequency, the input signal voltage V1 is measured, and this is compared with the signal arriving at the output terminals of the supply V2. Both the amplitude ratio V2/V1 in decibels and the phase difference in degrees are plotted.

The plot of amplitude and phase is made against a log base of frequency, after making due allowance for the 180° phase shift built into the comparator connections. A typical Bode plot is shown in Fig. 3.8.5.

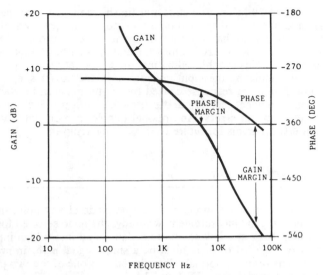

FIG. 3.8.5 Example of a Bode plot for a switchmode power converter, showing good phase and gain margins.

8.9 TEST EQUIPMENT FOR BODE PLOT MEASUREMENT

The basic equipment required for this type of plot is

1. A variable-frequency oscillator covering the range from low frequencies—10 Hz or less—to 50 kHz or more
2. Dual, narrowband selective peak or rms reading voltmeters covering the same band as the oscillator
3. Specialized gain and phase meters

Because of the large system gain, the input voltage V1 will be very small, and the voltmeters will use narrowband tuned amplifiers to extract the required signal with a good signal-to-noise ratio. These voltmeters must be adjusted to the oscillator frequency for each point on the frequency spectrum to be examined. This is a time-consuming exercise. Several specialized measuring instruments are available which have sweep oscillators and tracking voltmeters, and many of these will display and print out the Bode plot automatically.

Alternatively, one of the many spectrum analyzers may be used. These will have a built-in sweep generator and comparative voltmeters and phase detectors; also, they will give an automatic sweep over the required band. This second method is preferred, not only because it is much quicker, but because very small increments of frequency are easily achieved, producing a nearly continuous plot.

Suitable instruments include the Hewlett-Packard 3594A/3591A oscillator and selective voltmeter and the Bafco Model 916H Frequency Response Analyzer.

Among the particularly suitable spectrum analyzer instruments is the Anritsu Network/Spectrum Analyzer Type MS420A. This instrument has a frequency range from 3 Hz to 30 MHz. The start and finish frequencies can be specified, and the gain and phase are displayed on a linear or log frequency scale. Internal calibration is provided so that any gain and phase errors introduced by the test method, transformer T1, or terminations C1 and C2 can be eliminated from the final measurements. This is achieved by making a calibration sweep into a resistive termination simulating the input impedance of the power supply amplifier. Any deviation from a flat response caused by the termination reactance is then retained in the equipment memory and subtracted from the final measurement, eliminating measurement errors. An optional bubble memory interface is available to retain test programs for future use, reducing setup time.

8.10 TEST TECHNIQUES

In theory, the Bode plot measurements may be made at any point in the loop. However, to obtain good measurement accuracy, the node selected for signal injection must look back at a low source impedance and into a high input impedance in the next stage. Also, there must be a single signal path. In practice, the measurement transformer is normally introduced into one of the two positions in the control loop shown in Figs. 3.8.4 and 3.8.6.

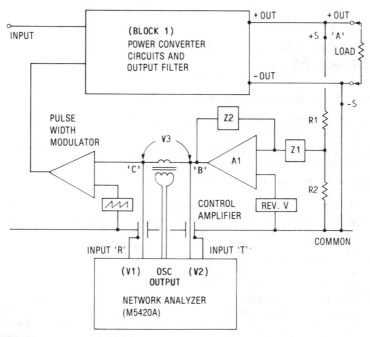

FIG. 3.8.6 A closed-loop Bode plot, showing an alternative injection point and using a network analyzer method.

The position for T1 shown in Fig. 3.8.4 meets the above criteria. The source impedance (in the direction of the injected signal) is the low output impedance of the power section, and the input impedance of the next stage is the high input impedance of the control amplifier A1. A second position in the control section meeting this criterion is shown in Fig. 3.8.6, between the low-impedance output of the amplifier A1 and the high input impedance of the pulse-width modulator.

To provide the best signal-to-noise ratio for good measurement accuracy, the signal level should be as large as possible. It must, however, be small enough to ensure that the loop remains in linear operation and does not "bottom out" at any part of the circuit at any frequency. This can be checked by observing the waveform at the power supply output terminals, where any distortion during the frequency sweep indicates an overload condition.

The phase margin is measured where the gain is 0 dB (unity gain), and the gain margin is measured where the phase change is 360° total (180° measured). The gain at this frequency should be less than unity.

8.11 MEASUREMENT PROCEDURES FOR BODE PLOTS OF OPEN-LOOP POWER SUPPLY SYSTEMS

As previously mentioned, the open-loop low-frequency and DC gains of power supplies are usually very large, making it difficult to make true open-loop measurements. If the interrogative method (Sec. 8.3) is to be used, then the open-loop Bode plot of the power section and pulse-width modulator is required, so that the optimum compensation network can be determined by the "difference technique." (See Sec. 8.12.)

One method that can usually be used to obtain the effective open-loop Bode plot above a few hertz is shown in Fig. 3.8.7.

To define the DC conditions, the loop is closed at DC, using an overcompensated control amplifier with a dominant very low frequency pole. This will ensure that the loop remains stable for measurement purposes and allows the required load currents and output voltages to be set up. The transient performance would, of course, be very poor, but it is not important at this stage.

An interrogating signal V1 is now injected at the input to the pulse-width modulator, node B. The output of the loop V2 is measured at the output terminals of the supply, node A.

Note: The input to the control amplifier, node C, should not be used for input or output measurements, as the voltage at this node will always be constant at V_{ref}. Hence, in a high-gain closed-loop amplifier, there will not be any measurable ac voltage at node C.

A Bode plot of the pulse-width modulator input V1 to power supply output V2 gives the open-loop transfer function of the power-loop and pulse-width modulator stages. The control amplifier and feedback components have been excluded from this measurement. From this information, the optimum compensation network for A1 can be calculated.

In lower-gain systems, it is possible to operate in true open loop by replacing the amplifier A1 by an adjustable DC polarizing supply. If this approach is used, the DC supply should have a large decoupling capacitor C1, to ensure a low ac source impedance. The voltages would again be injected and measured as shown in Fig. 3.8.7, except that the output DC value would be set by adjusting the po-

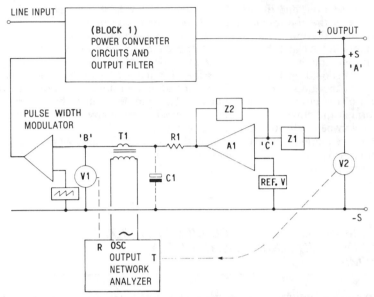

FIG. 3.8.7 A quasi open-loop Bode plot, showing the transfer function of the power and modulator sections of a switchmode converter and using a network analyzer method.

larizing supply voltage. Some test equipment (for example, the Solartron FRA) has a DC polarizing output for this purpose.

8.12 ESTABLISHING OPTIMUM COMPENSATION CHARACTERISTIC BY THE "DIFFERENCE METHOD"

Once the transfer characteristic (open-loop Bode plot) of the pulse-width modulator and power converter sections (that is, the complete loop less the control amplifier) has been measured, the requirements of the compensated control amplifier can be established.

On the same Bode plot, roughly define the required optimum transfer characteristic. The difference between the optimum and measured open-loop characteristics is the desired response of the compensated control amplifier.

Of course, it may not be possible to completely satisfy the optimum characteristic with a real amplifier, and the aim should be to get as close as possible. Proceed as follows:

1. Put zeros in the compensation network near the frequencies where excess poles occur in the open-loop plot, so that the phase shift is less than 315° up to the crossover frequency (a phase margin of at least 45°).

2. Put poles in the compensation network near the frequencies where ESR zeros

occur in the open-loop plot. (Otherwise these zeros will flatten the gain characteristic and prevent it from falling off as desired.)

3. If the low-frequency gain is too low to give the desired DC regulation (because of the zeros added in step 1), add a pole-zero pair to boost the gain at low frequencies.

In most cases, some "fine tuning" will be required, and this is best carried out using transient load testing (Sec. 8.6).

8.13 SOME CAUSES OF STUBBORN INSTABILITY

8.13.1 Gain-Forced Instability

If the control amplifier is connected so that the inverting input is connected to the reference voltage (a common mode of operation when using optocouplers; see Fig. 3.8.8a), then *the gain of the amplifier cannot be less than unity even with a 100% feedback factor.* (The amplifier tends to a voltage follower as the negative feedback increases. See Fig. 3.8.8b.)

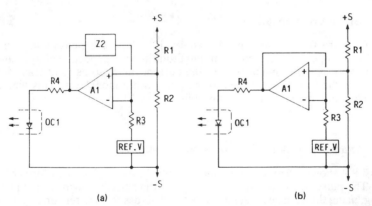

FIG. 3.8.8 A diagram of an often-used control amplifier configuration, showing how the loop gain in this arrangement is limited to a minimum of unity.

If the gain of the rest of the circuit is greater than unity at high frequency, then it will be impossible to stabilize the loop using conventional roll-off components around the amplifier section only.

The correct approach here is to reduce the gain of the pulse-width modulator and power converter section well below unity. Alternatively, the configuration of the amplifier can be changed to the more normal inverting input connection using a further unity gain inverting amplifier to restore the required phase to the optoisolator input. (See Part 3, Chap. 11, for suitable optocoupler circuits.)

8.13.2 Subharmonic Instability in Current-Mode Control Systems

In the examples dealt with so far, the assumption has been made that voltage-mode control has been used, and in these systems the subharmonic instability problem will not exist. However, this will not be the case with current-mode control.

When current-mode control is used with continuous inductor current topologies (for example, the flyback, boost, and buck converters or regulators), a form of subharmonic instability will occur if the duty cycle exceeds 50%, that is, if the maximum "on" period exceeds 50% of the total period. In this instability mode, alternate pulses will be wide and then narrow; although not damaging, this is undesirable, as it will increase the output ripple and may result in transformer saturation during the wide pulse.

In some applications, it is necessary to provide full output for a very wide input voltage range. Limiting the maximum pulse width to less than 50% for this requirement results in excessively narrow "on" periods for high input voltages. This results in a loss of efficiency as a result of high peak primary currents. Therefore, it would be useful, where the input voltage range is large, to allow the drive pulse width to exceed 50% without developing subharmonic instability.

There are a number of ways of achieving this. Clearly, reverting to voltage-mode control would eliminate the difficulty, but this is not always desirable, since the advantages of the current control method would then be entirely lost.

8.13.3 Slope Compensation

The usual cure for subharmonic instability is "slope compensation." In this method of compensation, a constant-amplitude time-dependent voltage ramp is summed with the voltage analogue of the current ramp, so as to at least double the slope of the waveform applied to the pulse-width modulator. This will completely eliminate the subharmonic instability. (See Chap. 10.)

8.13.4 Example of Slope Compensation Methods

Figure 3.8.9 shows current-mode control applied to a single-ended forward converter. The pulse-width modulator responds to the current flowing in the primary winding, using the voltage analogue developed across R3. Q1 remains "on" until the current reaches the limiting value, at which point voltage applied to the inverting input of the PWM exceeds the control voltage, and Q1 will turn off. The current analogue is summed with a compensation voltage ramp from R1 to eliminate subharmonic instability.

As an alternative, in forward converters, a gap can be introduced in the transformer core, increasing the primary magnetization current. This magnetization current is proportional to the input voltage, and hence can provide the required ramp compensation. The input voltage and time-dependent magnetizing current ramp will be added to the secondary inductor current ramp in the transformer. This approach has the advantage of improving the input ripple rejection and stability, but reduces the efficiency as a result of losses in the energy recovery circuit and the increased switching losses.

Some converter topologies, namely Ćuk, flyback, and boost converters when operated in the incomplete energy transfer mode (continuous inductor current mode), will display a tendency to instability that will not respond to normal compensation techniques. This instability is often caused by the "right-half-plane zero" inherent in the topology. (See Chap. 9.)

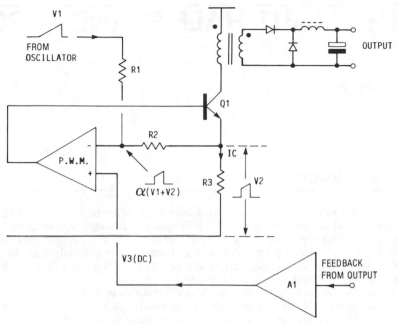

FIG. 3.8.9 A current-mode control section used in forward converters, showing the ramp compensation input derived from the oscillator section.

8.14 PROBLEMS

1. Describe the basic stability factors required in any closed-loop control system for an unconditionally stable loop.

2. What are the criteria for a conditionally stable loop?

3. What are the dangers of a conditionally stable loop?

4. Why are dynamic methods of loop stability analysis, such as Bode plots and transient analysis, recommended in addition to normal mathematical analysis?

5. Explain why it is so difficult, in practice, to obtain open-loop analysis of most switchmode or linear power supply control loops.

6. Explain the basic criteria controlling the point in the loop at which the interrogating signal is injected in interrogative methods of loop stabilization measurements. Give an example for a switchmode supply.

7. In interrogative methods of loop stability measurement, what defines the amplitude of the injection signal, and how would the operator know whether a suitable signal is being used?

8. Why is it usually considered essential to use a tracking oscillator and detector for loop stability measurement in switchmode power supplies?

9. Why is transient load testing considered useful for interrogating performance of a closed-loop power supply system?

10. Give a possible cause of stubborn instability in a control loop which will not respond to the normal loop compensation techniques (an effect particularly prevalent in boost and continuous-mode flyback converters).

CHAPTER 9
THE RIGHT-HALF-PLANE ZERO

9.1 INTRODUCTION

For many years power supply engineers have been aware of the difficulty of obtaining a good stability margin and high-frequency transient performance from the continuous-inductor-mode (incomplete energy transfer) flyback and boost converters. For stable operation of such converters, it is generally necessary to roll off the gain of the control circuits at a much lower frequency than would be the case with the buck regulator topologies.

It has been demonstrated mathematically[15] that this problem is the result of a negative zero in the small-signal duty cycle control to output voltage transfer function. The negative sign locates this zero in the right half of the complex frequency plane. Although a rigorous mathematical analysis is essential for a full understanding of the problem, for many, the mathematical approach alone will not provide a good grasp of the dynamics of the effect, and the following explanation by Lloyd H. Dixon Jr. will be found most helpful.

9.2 EXPLANATION OF THE DYNAMICS OF THE RIGHT-HALF-PLANE ZERO

A Bode plot of the right-half-plane zero has the characteristic of a rising 20 dB/decade gain with a 90° phase lag instead of the more usual phase lead. It is considered impossible to compensate this effect by normal loop compensation methods, and the designer is obliged to roll off the gain at a lower frequency, giving poor transient response.

In simple terms, the right-half-plane zero is best explained by considering the transient action of a continuous-mode flyback converter. In this type of circuit, the output current from the transformer secondary is not continuous; it flows only during the flyback period, when the primary power switching device is "off."

When a transient load is applied to the output, the first action of the control circuit will be to *increase* the "on" period of the power switch (so as to increase the input current in the primary inductance in the longer term). However, the large primary inductance will prevent any rapid increase in primary current, and several cycles will be required to establish the final value.

However, with a fixed-frequency converter, the first, and immediate, effect of *increasing* the "on" period is to *reduce* the flyback period. Since the primary current, and hence the flyback current, will not have changed much in the first few cycles, the mean output current will now immediately *decrease* (rather than *increasing,* as was required). This reverses the normal control action during the transient, giving an additional 180° of phase shift, and this is the cause of the right-half-plane zero.

It would seem that the only cure for this effect is to change the pulse width slowly over a large number of cycles so that the inductor current can follow the change. Under these conditions, the dynamic output reversal will not occur; however, the transient response will be rather poor.

The following discussion by Lloyd H. Dixon, Jr., provides a more complete explanation. (Adapted from the "Unitrode Power Supply Design Seminar Manual," Reference 15. Reprinted with permission of Unitrode Corporation.)

9.3 THE RIGHT-HALF-PLANE ZERO—A SIMPLIFIED EXPLANATION

In small-signal loop analysis, poles and zeros are normally located in the left half of the complex *s*-plane. The Bode plot of a conventional or left-half-plane zero has the gain magnitude rising at 20 dB/decade above the zero frequency with an associated phase lead of 90°. This is the exact opposite of a conventional pole, whose gain magnitude *decreases* with frequency and whose phase *lags* by 90°. Zeros are often introduced in loop compensation networks to cancel an existing pole at the same frequency; likewise, poles are introduced to cancel existing zeros in order to maintain total phase lag around the loop less than 180° with adequate phase margin.

The right-half-plane (RHP) zero has the same 20 dB/decade rising gain magnitude as a conventional zero, but with 90° phase *lag* instead of lead. This characteristic is difficult if not impossible to compensate. The designer is usually forced to roll off the loop gain at a relatively low frequency. The crossover frequency may be a decade or more below what it otherwise could be, resulting in severe impairment of dynamic response.

The RHP zero never occurs in circuits of the buck family. It is encountered only in flyback, boost, and Ćuk circuits, and then only when these circuits are operated in the continuous-inductor-current mode.

Figure 3.9.1 shows the basic flyback circuit operating in the continuous mode

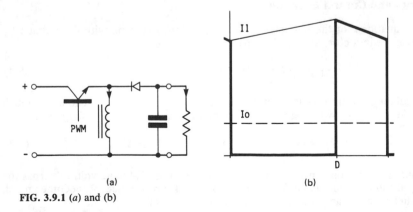

(a)   (b)

FIG. 3.9.1 (*a*) and (b)

with its current waveforms. In flyback as well as boost circuits, the diode is the output element. All current to the output filter capacitor and load must flow through the diode, so the steady-state DC load current must equal the average diode current. As shown in Fig 3.9.1b, the inductor current equals the peak diode current, and it flows through the diode only during the "off" or free-wheeling portion of each cycle. The average diode current (and load current) therefore equals the average inductor current I_L times $(1 - D)$, where D is the duty ratio (often called duty cycle).

If D is modulated by a small ac signal d whose frequency is much smaller than the switching frequency, this will cause small changes in D from one switching cycle to the next. Figure 3.9.2 shows the effects of a small increase in duty ratio (during the positive half cycle of the applied signal).

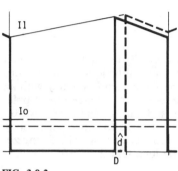

The first effect is that the temporarily larger duty ratio causes the peak inductor current to increase each switching cycle, with an accompanying increase in the average inductor current. If the signal frequency is quite low, the positive deviation in duty ratio will be present for many switching cycles. This results in a large cumulative increase in inductor current, whose phase lags d by 90°. This change in inductor current flows through the diode during the "off" time, causing a proportional change in output current, in phase with the inductor current.

FIG. 3.9.2

The second effect is more startling: The temporary increase in duty ratio during the positive half cycle of the signal causes the diode conduction time to correspondingly *decrease*. This means that if the inductor current stays relatively constant, the average diode current (which drives the output) actually decreases when the duty ratio increases. This can be clearly seen in Fig. 3.9.2. In other words, the output current is 180° out of phase with d. This is the circuit effect which is mathematically the right-half-plane zero. It dominates when the signal frequency is relatively high so that the inductor current cannot change significantly.

Duty Ratio Control Equations

The equations for the flyback circuit are developed starting with the voltage V_L across the inductor, averaged over the switching period:

$$V_L = V_i D - V_o(1 - D) = (V_i + V_o)D - V_o \tag{9.1}$$

Modulating the duty ratio D by a small AC signal d whose frequency is much smaller than the switching frequency generates an ac inductor voltage $\hat{v}_L$:

$$\hat{v}_L = (V_i + V_o)\, d \; - \hat{v}_o(1-D) \; \overline{\wedge} \; (V_i + V_o)\, d \tag{9.2}$$

Assuming V_i is constant, $\hat{v}_L$ is a function of d and of $\hat{v}_o$, the ac voltage across the output filter capacitor. At frequencies above filter resonance, $\hat{v}_o$ becomes much smaller than $\hat{v}_L$, and the second term may be omitted.

The ac inductor current $\hat{\imath}_L$ varies inversely with frequency and lags $\hat{v}_L$ by 90°. Substituting for $\hat{v}_L$ in Eq. (9.2) gives $\hat{\imath}_L$ in terms of $\hat{d}$:

$$\hat{\imath}_L = \frac{\hat{v}_L}{j\omega L} = -j\frac{V_i + V_o}{\omega L}\hat{d} \tag{9.3}$$

Referring to Fig. 3.9.1, the inductor provides current to the output through the diode only during the "off" portion of each cycle:

$$I_o = I_L(1 - D) \tag{9.4}$$

Differentiating Eq. (9.4), the ac output current $\hat{\imath}_o$ has two components (see Fig. 3.9.1)—one component in phase with $\hat{\imath}_L$ and the other 180° out of phase with $\hat{d}$:

$$\hat{\imath}_o = \hat{\imath}_L(1 - D) - I_L\hat{d} \tag{9.5}$$

Substituting for $\hat{\imath}_L$ in Eq. (9.3) gives $\hat{\imath}_o$ in terms of the control variable $\hat{d}$. In a continuous-mode flyback circuit, $(1 - D) = V_i/(V_i + V_o)$:

$$\hat{\imath}_o = -j\frac{(V_i + V_o)(1 - D)}{\omega L}\hat{d} - I_L\hat{d} = -j\frac{V_i}{\omega L}\hat{d} - I_L\hat{d} \tag{9.6}$$

The first term is the inductor pole, which dominates at low frequency. Its magnitude decreases with frequency, and the phase lag is 90°. At a certain frequency the magnitudes of the two terms are equal. Above this frequency, the second term dominates. Its magnitude is constant, and the phase lag is 180°. This is the RHP zero, occurring at frequency ω_Z where the magnitudes are equal.

Figure 3.9.3 is a Bode plot of this equation (arbitrary scale values). Above f_Z, the rising gain characteristic of the RHP zero cancels the falling gain of the inductor pole, but the 90° lag of the RHP zero *adds* to the inductor pole lag, for a total lag of 180°. The Bode plot of the entire power circuit would also include the output filter capacitor pole, which combines with the inductor pole, resulting in a second-order resonant characteristic at a frequency well below the RHP zero. The ESR of the filter capacitor also results in an additional conventional zero.

The RHP zero frequency is calculated by equating the magnitudes of the two terms in Eq. (9.6) and solving for ω_Z:

$$\omega_Z = \frac{V_i}{LI_L} \tag{9.7}$$

Substitute Eq. (9.4) for I_L and V_o/R_o for I_o. In a flyback circuit, $V_i/V_o = (1 - D)/D$; $(1 - D) = V_i/(V_i + V_o)$:

$$\omega_Z = \frac{R_o V_i(1 - D)}{LV_o} = \frac{R_o(1 - D)^2}{LD} = \frac{R_o V_i^2}{LV_o(V_i + V_o)} \tag{9.8}$$

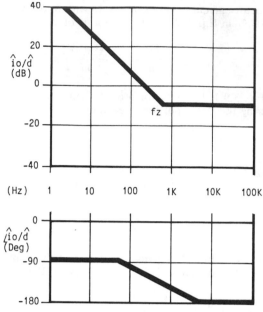

FIG. 3.9.3

Current-Mode Control Equations

Equations (9.1), (9.2), (9.4), and (9.5) pertain to the flyback continuous-mode power circuit and are valid for any control method, including current-mode control. Equation (9.3) is valid for current-mode control, but it applies to the inner, current control loop. Solve Eq. (9.3) for $\hat{d}$ in terms of $\hat{i}_L$ and substitute for $\hat{d}$ in Eq. (9.5):

$$\hat{i}_o = \hat{i}_L(1 - D) - j\frac{\omega L I_L}{(V_i + V_o)}\hat{i}_L = \frac{V_i}{(V_i + V_o)}\hat{i}_L - j\frac{\omega L I_L}{(V_i + V_o)}\hat{i}_L \qquad (9.9)$$

Equations (9.6) and (9.9) are the same, except that in Eq. (9.6) the control variable is $\hat{d}$ for duty ratio control, whereas in Eq. (9.9) the control variable is $\hat{i}_L$, established by the inner loop and consistent with current-mode control.

Unlike in Eq. (9.6) for duty ratio control, the first term in Eq. (9.9) is constant with frequency and has no phase shift. This term dominates at low frequency. It represents the small-signal inductor current, which is maintained constant by the inner current control loop, thus eliminating the inductor pole. The second term *increases* with frequency, yet the phase *lags* by 90°, characteristic of the RHP zero. It dominates at frequencies above ω_z where the magnitudes of the two terms are equal. The RHP zero frequency ω_z may be calculated by equating the two terms of Eq. (9.9). The result is the same as Eq. (9.7) for duty ratio control.

Figure 3.9.4 is the Bode plot of Eq. (9.9). The output filter capacitor will of course add a single pole and an ESR zero. Because the inductor pole is eliminated by the inner loop, the outer voltage control loop does *not* have a two-pole

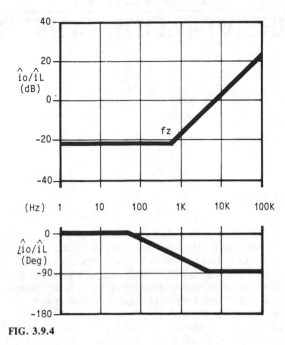

FIG. 3.9.4

resonant (second-order) characteristic. However, the RHP zero is clearly still present with current-mode control.

9.4 PROBLEMS

1. Explain the cause of the right-half-plane zero.
2. Which power supply topologies display a right-half-plane zero in their duty ratio to output transfer functions?
3. In simple terms, explain the dynamics of the right-half-plane zero as applied to a fixed-frequency duty-ratio-controlled boost converter.
4. What methods are normally used to prevent instability in systems which have a right-half-plane zero in the transfer function?

CHAPTER 10
CURRENT-MODE CONTROL

10.1 INTRODUCTION

Although current-mode control had been in use in various forms for many years (the original invention being attributed to Thomas Froeschle of Bose Corp. in 1967), it was not generally recognized as a fundamentally different control mode until 1977, when A. Weinburg and D. O'Sullivan published a paper in which the fundamental differences were highlighted. Since then, many aspects of this control technique have been more fully investigated, and it is becoming the control mode of choice in many new designs.

Previously, in constant-frequency switching regulators or switching-mode power converters, output regulation would normally be provided by duty ratio control (that is, by adjusting the ratio of the "on" period to the "off" period for the power switching devices, in response to input or output voltage changes). In this respect, conventional duty ratio control and current-mode control are similar; both control methods adjust the duty cycle to achieve output regulation. However, whereas duty ratio control adjusts the ratio only in response to output voltage changes, current-mode control initially adjusts the ratio in response to the main (power) inductor current changes.

This apparently simple change in the initial control parameter has very far-reaching effects on the behavior of the overall closed-loop system.

10.2 THE PRINCIPLES OF CURRENT-MODE CONTROL

To more easily explain the operating principle of current-mode control in its simplest terms, a complete energy transfer mode (discontinuous-mode), open-loop flyback converter will be considered.

A further advantage of choosing the discontinuous-mode flyback converter for this example is that it does not have a right-half-plane zero and will not display the subharmonic loop instability problems inherent in the large duty ratio continuous-inductor-current topologies. (See Sec. 10.7.)

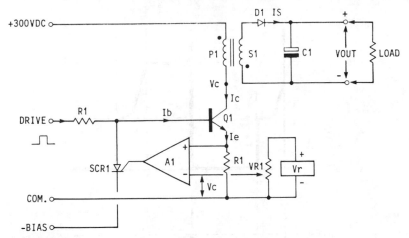

FIG. 3.10.1 Open-loop flyback converter, showing the principles of current-mode control.

Figure 3.10.1 shows the major elements of a simple open-loop flyback converter, which operates as follows.

Transformer (inductor) P1, S1, transistor Q1, and components C1 and D1 form the power sections, the transistor being driven by a constant-frequency square-wave drive via resistor R1.

When Q1 is turned on, the transformer primary current will increase linearly from zero as shown in Fig. 3.10.2. Assuming that I_b is negligible compared with I_C, a voltage analogue of the primary current will be developed across the emitter resistor R1.

When the current has increased to a value at which the voltage across R1 (as applied to comparator A1) exceeds the control voltage V_c, the output of A1 will go high, turning on SCR1 and removing the drive to Q1, terminating the "on" pulse. (SCR1 will be reset during the following "off" period of the drive.)

Hence, the control loop is open as far as output voltage control is concerned, but closed to maintain a fixed peak current in P1 as far as current-mode control is concerned. The fixed value of voltage V_c defines the peak current I_{cp}, and the first advantage of current-mode control becomes apparent.

Because I_{cp} is defined, the input energy to the flyback transformer is also defined. (This energy $= \frac{1}{2} \cdot L_p \cdot I_{cp}^2$, as functionally P1 is just an inductor during this phase.) Since this is a fixed-frequency complete energy transfer converter, the output power is also defined. Hence, provided that the load resistance remains fixed, the output voltage and current will also remain constant, without the need to close the loop for voltage control.

Thus the open-loop current-mode control holds the output constant, even if the supply voltage changes, because I_{cp} still remains constant. This is more clearly demonstrated in Fig. 3.10.3a and b, where it can be seen that an increase in V_c results in an increase in the slope of I_e (which is the same as I_c), with a corresponding reduction in pulse width. However, $I_{c(peak)}$ remains constant for each cycle, and since the frequency is constant, the output power will also remain constant.

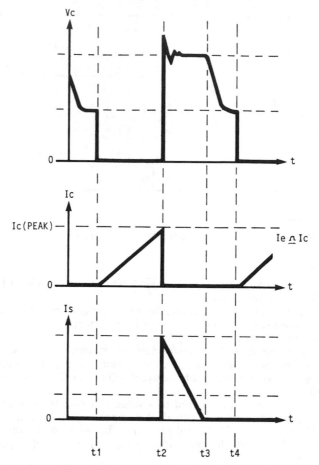

FIG. 3.10.2 Voltage and current waveforms of a discontinuous-mode flyback converter.

Because the primary current-mode control action operates on a pulse-by-pulse basis, a second advantage now becomes apparent. If the maximum value of the control voltage is limited (V_c cannot exceed V_r in this example), then the peak primary current (and hence, for this type of converter, the maximum throughput power) is also limited. Thus the converter has an inherent, fast-acting, pulse-by-pulse overload protection.

It has been shown that primary current-mode control, as applied to the discontinuous flyback converter, intrinsically provides a very fast acting, constant power control, which initially maintains the output voltage constant; it has good input ripple rejection without closed-loop voltage control. Further, it is able to respond to current programming (changes in V_c) very rapidly, and has the ability to switch the transferred power pulse between zero and maximum in a single cycle of operation. By closing the voltage control loop to adjust V_c, rapid voltage control can be applied, as discussed in the following section.

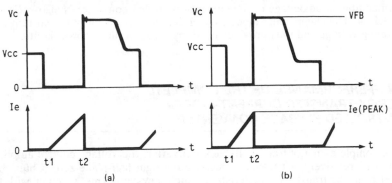

FIG. 3.10.3 Voltage and current waveforms of a discontinuous flyback converter when current-mode control is applied, showing pulse width and peak current with (a) low and (b) high input voltages.

10.3 CONVERTING CURRENT-MODE CONTROL TO VOLTAGE CONTROL

In current-mode-controlled supplies, the output voltage can be controlled by the addition of a second outer voltage control loop which adjusts the current programming voltage V_c. This is provided (as shown in Fig. 3.10.4) by amplifier A2. This amplifier compares the output voltage with a reference, and adjusts the current programming voltage V_c for the inner current control loop so as to maintain the output voltage constant. This outer voltage control loop is relatively slow compared with the inner current control loop. Hence the response to load variations (which depends on the voltage control loop) will not be as fast as the inherent input voltage transient and ripple rejection performance (which depend on the faster inner current control loop).

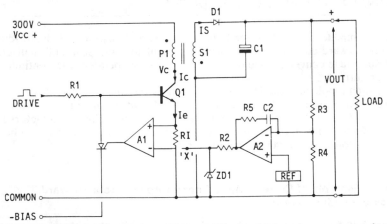

FIG. 3.10.4 Discontinuous flyback converter with closed-voltage-loop current-mode control, showing a method of input power limiting by clamp zener diode (ZD1).

An additional advantage of the current-controlled converter is also shown in Fig. 3.10.4. The clamp zener diode ZD1 limits the upper limit of the current programming voltage and thus provides pulse-by-pulse primary power limiting.

10.4 PERFORMANCE OF THE COMPLETE ENERGY TRANSFER CURRENT-MODE-CONTROLLED FLYBACK CONVERTER

In the complete energy transfer flyback converter, the voltage control amplifier can have a relatively fast response because the right-half-plane zero (Chap. 9) is absent from the transfer characteristic of such a system. This, together with the elimination of the inductor from the small-signal model as a result of current-mode control, permits a good high-frequency roll-off in the loop compensation network R5, C2 with relatively small values of capacitance C2. This will give good load and line transient performance. Although the discontinuous topology described above has been useful to demonstrate the current-mode control action, this topology does not gain as much from the technique as does the continuous-mode forward or buck regulator. In fact, the same performance may be obtained from the discontinuous-mode flyback converter simply by using duty ratio control with input voltage feedforward to the ramp comparator.

Note: In the duty-ratio-controlled case, the feedforward will provide immediate compensation for input voltage changes, outside of the slower voltage control loop. This gives advantages similar to those of the current-mode control technique. Several switchmode control ICs provide for this type of compensation; the Unitrode UC 3840 is a typical example.

10.5 THE ADVANTAGES OF CURRENT-MODE CONTROL IN CONTINUOUS-INDUCTOR-CURRENT CONVERTER TOPOLOGIES

The incomplete energy transfer flyback converter (continuous-inductor-current mode), forward converter, buck switching regulator, and push-pull, half-bridge, and full-bridge converter topologies are all normally operated with continuous inductor current.

In such topologies, the DC output current is the time average of the inductor current. Also, to obtain the maximum range of control, the duty ratio will often exceed 50% (the "on" pulse exceeds the "off" pulse). These two factors introduce effects which must be considered when current-mode control is to be used for these continuous inductor conduction topologies.

10.5.1 Example of Current-Mode Control Applied to a Forward (Buck-derived) Converter

Figure 3.10.5 shows a "forward" (buck-derived) converter with primary current-mode control. This circuit operates as follows.

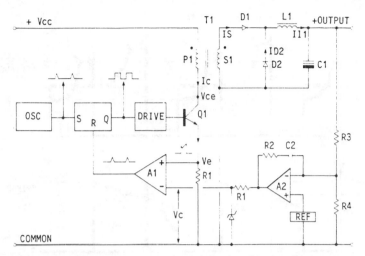

FIG. 3.10.5 Forward converter (buck-derived), with closed-voltage-loop current-mode control.

When Q1 turns on, the starts of all windings on T1 go positive, and diode D1 will conduct. The current flowing in D1 and L1 is transformed to the primary of T1 to flow as I_c in the collector of Q1. Assuming that the drive current is negligible compared to the collector current, a voltage analogue of the transformed output inductor current will be developed across the emitter resistor R1. (The primary and secondary waveforms are shown in Fig. 3.10.6.)

When the current in R1 has increased to a value such that the voltage across R1 exceeds the control voltage V_c, the drive to Q1 will be turned off. Hence the *peak* current flowing in L1 will be controlled. However, the effective DC output load current will be the *average* of the inductor current, and the average-to-peak ratio changes with input voltage (the ripple current changes). The error introduced by the difference between the peak and average currents causes two major problems, discussed in the next section.

10.5.2 Subharmonic Instability and Input Ripple Rejection in Continuous-Inductor-Current Topologies with Current-Mode Control

When current-mode control is used in a forward converter (or other topologies in which the inductor current is continuous), if the input voltage is increased while the current programming voltage V_c is maintained constant, the *peak* inductor current will be maintained constant. However, average (DC) output current will decrease because the peak-to-peak ripple current is larger. (This is shown in Fig. 3.10.6, where $I_{c \text{ (ave)}}$ #2 is less than $I_{c \text{ (ave)}}$ #1 and I_{load} #2 is less than I_{load} #1.)

Hence, without compensation, the intrinsic open-loop input ripple rejection of the forward current-mode-controlled converter will not be very good. Further, a more subtle and much more problematical instability effect will occur if the duty cycle exceeds 50%.

For stable steady-state operation, the current in L1 must start and finish at the

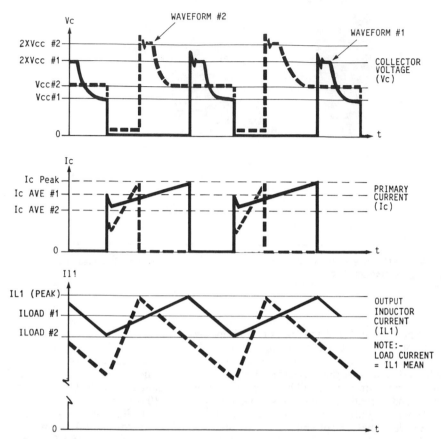

FIG. 3.10.6 Voltage and current waveforms for continuous-inductor-current, current-mode-controlled forward converters.

same amplitude over a cycle of operation. Figure 3.10.7*a* shows how, when the duty cycle exceeds 50%, a small perturbation introduced to the current waveform will grow to become larger (but reversed in direction) at the end of the cycle; hence the original small perturbation will continue to grow. This results in subharmonic instability. Fortunately both of the above problems can be entirely eliminated by the correct amount of slope compensation.

10.6 SLOPE COMPENSATION

In continuous-inductor-current, current-mode-controlled converters, subharmonic instability and poor input ripple rejection can be simultaneously eliminated by introducing "slope compensation" to the control voltage V_c (or the ramp comparator input).

If, instead of V_c being maintained constant during a cycle, it is made to ramp

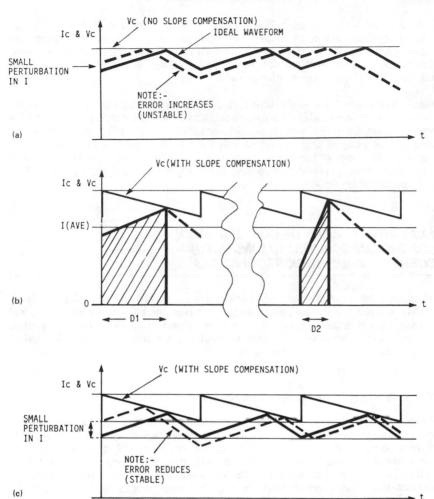

FIG. 3.10.7 Current waveforms for continuous-inductor-current buck-derived converters, showing (a) subharmonic instability at duty ratios exceeding 50%. (b), (c) Also shown is the corrective effect of slope compensation in current-mode control.

downward with a slope which is exactly half of the output inductor current slope (or, more correctly, half of the analogue voltage developed across R1 by the transformed inductor current), then the average (DC) output current will no longer change as the pulse width is changed. (The compensation makes the peak current change with pulse width so as to maintain the average current constant.) This will restore good open-loop input ripple rejection because V_c no longer requires adjustment by the control circuit to compensate for the former peak-to-average error. Further, at duty cycles exceeding 50%, a perturbation introduced into the current waveform will now decrease over a cycle to fade away and give stable operation.

This effect is more clearly shown in Fig. 3.10.7b. Here, two different duty ratios D1 and D2 are used to demonstrate that the average primary current during a conduction period remains constant for the two different pulse widths (provided that the average value of current programming voltage V_c also remains constant and that 50% slope compensation is used).

Note: The dashed line is the transferred secondary decay current in L1 as referred to the primary. This decay slope remains constant for different duty cycles provided that the output voltage is constant ($dI/dt = V_{out}/L$). Therefore the slope compensation can be exactly correct only for a defined and fixed output voltage. However, the slope of the increasing current, when Q1 is "on," will change even if the output voltage is constant, because the supply voltage V_{cc} can change. This will not affect the slope compensation.

10.7   ADVANTAGES OF CURRENT-MODE CONTROL IN CONTINUOUS-INDUCTOR-CURRENT-MODE BUCK REGULATORS

It has been shown above that when current-mode control (with 50% slope compensation) is applied to the continuous-inductor-current buck-derived converter, it will provide current limiting, good line regulation, and good input ripple rejection. However, a more important advantage is the effective elimination of the filter inductor from the small-signal voltage control loop.

10.7.1   Voltage Control Loop Compensation

When current-mode control is used, in effect, the fast-acting inner current control loop turns the converter into a fast-response, voltage-programmed, constant-current source. Since the filter inductor L1 is effectively in series with the constant-current source and inside the current control loop, it is effectively "taken out" and is eliminated from the small-signal voltage transfer function. As a result, the voltage control loop has only the single pole of the output filter capacitor and load resistance to compensate. Figure 3.10.8a shows the Bode plot of such a system.

Because the 90° phase lag of a single pole is inherently stable, it is now very easy to get high loop gain and excellent small-signal dynamic performance, using only a single small capacitor to "take out" the output capacitor's ESR zero. This simple compensation network is shown in Fig. 3.10.8b.

10.7.2   Small-Signal Dynamic Performance

When properly compensated, the small-signal dynamic performance of the buck regulator with conventional duty ratio control may be nearly as good as that of the same regulator with current-mode duty cycle control.

For high-frequency converters, small-signal response times of the order of 100 μs or so can be achieved with either control method. However, with conventional duty ratio control, the output *LC* power filter has a two-pole second-order

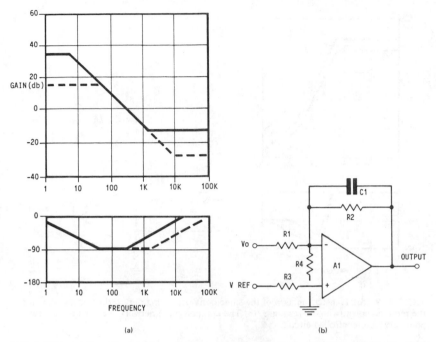

FIG. 3.10.8 (*a*) Transfer function of a current-mode-controlled converter, showing the single pole response. (*b*) Simple single-pole compensation network used in the current-mode control circuit.

characteristic, as shown in Fig. 3.10.9*a*. There is an abrupt 180° phase lag at filter resonance, which will cause ringing and instability if not effectively compensated. Also, with duty ratio control, a larger gain-bandwidth product is required from the error amplifier, together with large compensation capacitors. The compensation network will have time constants of the order of milliseconds, and these long time constants become a problem when large-signal conditions are considered. The compensation network for the duty-ratio-controlled case is shown in Fig. 3.10.9*b*.

10.7.3 Large-Signal Dynamic Performance

Unlike the small-signal behavior, when large-signal dynamic conditions are considered, there is a dramatic difference in the performance of the duty ratio and current-mode control circuits.

The large filter inductance used for effective filtering and wide dynamic current range in continuous-mode circuits limits the output current slew rate so that rapid changes in load cannot be accommodated. For large changes, this limit applies regardless of the control method used. As a result, when large dynamic load changes are made, the output voltage must change significantly, driving the error amplifier beyond its linear range (to the end stops). This temporarily opens the control loop and charges the compensation capacitors C1, C2, and C3 to voltages which are totally unrelated to normal operation.

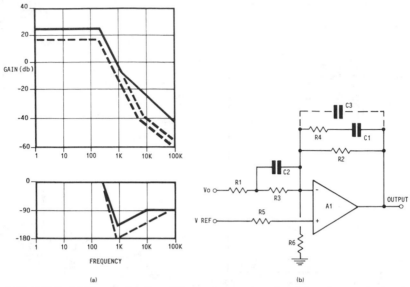

(a)   (b)

FIG. 3.10.9 (*a*) Transfer function of the same converter using duty ratio control, showing the more complex two-pole response. (*b*) The compensation network required for the two-pole duty-ratio-controlled circuit.

When the inductor current reaches the new value of load current, the output voltage will be restored, and the control amplifier will start to regulate again. However, the output voltage will now be at a new and incorrect value because of the offset voltage across the compensation capacitors. With conventional duty ratio control, the time required to restore the correct voltage on the relatively large compensation capacitors to bring the output voltage back to normal can be several milliseconds. Hence, the compensation capacitors necessary for good small-signal performance with duty ratio control will cause poor large-signal performance.

In contrast, current-mode control achieves excellent small- and large-signal performance because compensation capacitors are not required (other than one small capacitor required to cancel the output capacitor's ESR zero). Further, the control circuit recovers accurate regulation much more rapidly after a large load transient, because the single small capacitor is less than 10% of the value required for duty ratio control.

The simplicity of the single-capacitor compensation circuit required for current-mode control is evident from a comparison of Fig. 3.10.8*b* with the more complex circuit required for duty ratio control, Fig. 3.10.9*b*. (At very low temperatures, both compensation networks may require additional high-frequency compensation to offset the effects of increased output capacitor ESR.)

10.7.4 Stability Limitations with Conditionally Stable Systems

During large-signal transient conditions (conditions which drive the control amplifier out of linear operation), the time-averaged loop gain is reduced because the amplifier is not working for part of the time. As a result, the crossover frequency is lowered. This may cause serious instability problems if the loop is under-compensated and only conditionally stable.

If the loop is only conditionally stable (phase shift exceeding 180° at some frequency below crossover), the reduction in crossover frequency for large transients may initiate large-signal oscillation which will be maintained. For this reason, conditionally stable loops should be avoided in switching power supplies.

Current-mode control, with its reduced phase shift, is more easily made unconditionally stable.

10.7.5 Advantages of Current-Mode Control in Parallel System Operation

With current-mode control, it is relatively easy to parallel several supplies. (This may be required for high-reliability parallel-redundant power supply systems.)

If all the supplies to be paralleled have identical current sense resistors and identical current control loops, a single control voltage (common to all supplies) will cause them to deliver identical output currents. If a single reference and error amplifier is used to provide the control voltage V_c to all units, then the outputs may be paralleled to a common load and will share this load equally.

10.8 DISADVANTAGES INTRINSIC TO CURRENT-MODE CONTROL

Very few disadvantages apply to current-mode control in discontinuous-mode applications. However, current-mode control also has little to offer in this operating mode. For continuous-mode operation, there are few disadvantages that are intrinsic to current-mode control as such, but some of the longstanding problems common to continuous-conduction boost-derived converters will still exist in current-mode-controlled versions. These are more fully described below.

10.8.1 Poor Noise Immunity

In continuous-inductor-current-mode circuits, current-mode control suffers from poor noise immunity because the slope of the voltage analogue of the current ramp is quite small and never far away from the control voltage V_c. Hence small noise voltages can cause spurious operation of the ramp comparator.

As shown in Fig. 3.10.6, the inductor ramp sits on top of a large-amplitude square wave (the reflected load current), and the slope is quite shallow, especially when the load is large and the input voltage is low. Under these conditions, a small noise spike can cause premature termination of the conduction period.

Great care must be taken in the design of the pcb layout to reduce the noise level injected into the ramp comparator. Differential comparators should be used, and inputs should be taken directly to the current sense resistor terminations. A small RC filter will normally be required to further eliminate noise and remove the inevitable "spike" on the leading edge of the current pulse. (These spikes will be caused by snubber components, diode reverse recovery currents, and distributed capacitance. All these effects should be reduced to the minimum for good performance.)

Note: If these spikes are to be removed by a low-pass *RC* filter, the time constant of the filter should be as small as possible to prevent loss of control at light loads when the pulse is very narrow. The components should be mounted close to the comparator inputs. Very often common-mode noise problems are best eliminated by using a small current transformer to feed the ramp comparator, in place of the current sensing resistor.

Hence, to obtain the maximum ramp slope for best noise immunity and improved transient response, the filter inductor should be as small as is consistent with maintaining the continuous mode at minimum load current. To obtain lower output ripple voltages, it is better to use larger low-ESR output capacitors, rather than large inductance values.

10.8.2 Transfer Function Irregularities Caused by Current-Mode Control in Multiple-Output Applications

Figure 3.10.5 shows a current-mode-controlled circuit of the buck family with a single output. It is clear in this example that the current in the output filter inductor is directly controlled by the primary current comparator.

When multiple outputs are required, it is normal practice with duty ratio control to provide additional secondary windings on the transformer, with rectifiers and *LC* output filters to provide the extra outputs.

However, when current-mode control is used, the transformer tends to look like a constant-current source driving all the outputs in parallel. This high-impedance drive is not a problem at DC or low frequency, where all the output voltages will be defined by the duty cycle and transformer turns in the same way as in duty-ratio-controlled converters. However, at frequencies above the lowest filter resonance, the story is quite different, and stability problems can occur.

Normally only one output is sensed and fed back to become part of the voltage control loop. The input of the *LC* filter of this controlled output is driven from the high-impedance primary current source. However, the *LC* filters of the other outputs are also effectively attached in parallel to this same driving point (the transformer secondaries). At the series resonant frequency of each filter, the driving point is shunted by the low impedance of the particular resonant output. Hence the source is no longer a constant-current source at this resonant frequency.

Under this condition, the inductor in the closed-loop voltage-controlled output is no longer eliminated from the small-signal model of the outer voltage control loop, and additional phase shift is introduced, which may lead to instability. This problem is particularly severe if the resonant shunt filter is only lightly loaded, making its *Q* high.

The ideal solution is to couple the filter inductors by winding them on a common core. The output filters are now no longer independent and do not have separate resonances. This integrated output inductor (more correctly choke, as it has a dc component) also dramatically improves the dynamic cross regulation. Hence, integrated output chokes are preferred for multiple-output forward converters, particularly when current-mode control is used.

10.8.3 The Right-Half-Plane Zero in Current-Mode-Controlled Converters

In the boost-derived family of continuous-inductor-current-mode regulators (incomplete energy transfer mode), the DC output current is a function of both the

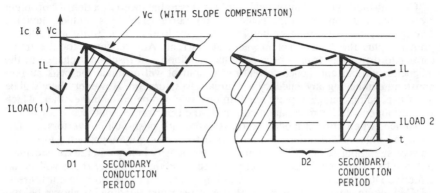

FIG. 3.10.10 Primary and secondary current waveforms of a boost converter, showing that the immediate effect of increasing the "on" pulse width from D1 to D2 is a reduction in the secondary conduction period, and hence the transferred energy (the cause of the right-half-plane zero).

average inductor current and the duration of the "off" (secondary conduction) period. (This is due to the discontinuous nature of the output current, which flows in the output rectifier diodes only when the power switches are "off," as shown in Fig. 3.10.10.)

However, for any continuous-mode regulator, the duty ratio, and hence the "off" period, is a direct function of V_{in}. Hence, if the input voltage changes, the average inductor current must also be changed to maintain the output current constant. Therefore, unlike in the buck regulator, the open-loop line regulation of boost and flyback continuous-mode regulators using current-mode control is very poor, even if slope compensation is applied.

When the input voltage increases, the duty ratio will decrease to maintain a constant output voltage in the longer term. Unfortunately, the inductor current cannot change rapidly, and the immediate effect of an *increase* in input voltage is to reduce the duty ratio, which results in an *increase* in the "off" period and the output diode conduction period. Since the inductor current will not have changed very much in this period, the immediate effect is to *increase* (rather than decrease) the output voltage. This dynamic reversal in the required effect will continue until the inductor has time to adjust to a lower current.

This reversal in the dynamic response is the cause of the right-half-plane zero and is intrinsic to the boost-derived topologies when operating in continuous-inductor-current mode. Unfortunately, current-mode control (even with slope compensation) does *not* eliminate the right-half-plane zero in the boost and flyback continuous-mode converter topologies. (See Chap. 9.)

10.9 FLUX BALANCING IN PUSH-PULL TOPOLOGIES WHEN USING CURRENT-MODE CONTROL

In any transformer-coupled push-pull circuit using voltage control, "staircase saturation" of the main switching transformer is a well-known and often severe problem. (See Chap. 6.)

Current-mode control intrinsically solves this unbalanced flux density problem by sensing and controlling the peak primary currents in the switching devices. The primary currents are made up of the inductor load current, transformed to the primary, plus the transformer magnetizing current. Any tendency for the transformer to drift away from a balanced flux condition will result in a change in the magnetization current; hence current-mode control will maintain the peak current constant, eliminating any tendency to staircase to saturation. However, there will be a corresponding change in pulse width in one side compared with the other, which will introduce a DC compensation current in the transformer winding.

Hence, when current-mode control is used with push-pull converters, an unbalanced pulse *width* differential is created to correct any asymmetry in the diodes or switching devices. This maintains flux density balance in the transformer, but results in an unbalanced ampere-second condition. Hence, there will be an effective DC current in the primary winding. This can cause problems if there is a DC blocking capacitor in series with the transformer winding, as shown below.

Note: A DC blocking capacitor is often fitted in duty-ratio-controlled push-pull converters to prevent transformer saturation. The blocking capacitor is in fact intrinsic to the topology of the half bridge.

10.10 ASYMMETRY CAUSED BY CHARGE IMBALANCE IN CURRENT-MODE-CONTROLLED HALF-BRIDGE CONVERTERS AND OTHER TOPOLOGIES USING DC BLOCKING CAPACITORS

As mentioned above, when current-mode control is applied to any push-pull circuit, the peak current will be identical for each side. In order to correct any volt-second asymmetry that results from diode or switching device imbalance, a small differential offset in pulse width will be created. However, any difference in pulse width now results in a small difference in the ampere-seconds, or charge, drawn alternately through the primary switching devices.

In half-bridge circuits (or any push-pull application in which DC blocking capacitors are fitted in series with the transformer winding), the unbalanced charge in each half cycle will cause a voltage to build up across any series capacitors. Unfortunately, the direction of the voltage buildup is such that it tends to reinforce the original volt-second asymmetry, and a runaway situation quickly develops. The series capacitors will charge toward one of the supply voltages, and alternate half cycles will have unequal voltage amplitudes. Hence there must *not* be a DC blocking capacitor in series with the transformer winding if current-mode control is to be used.

10.10.1 DC Restoration Techniques

In the half-bridge circuit, the capacitors are intrinsic to the topology, and steps must be taken to provide a DC path to the primary winding by some other means. A suitable method is shown in Fig. 3.10.11a, where a separate winding P1 on the main switching transformer with two catching diodes D1 and D2 restores the cen-

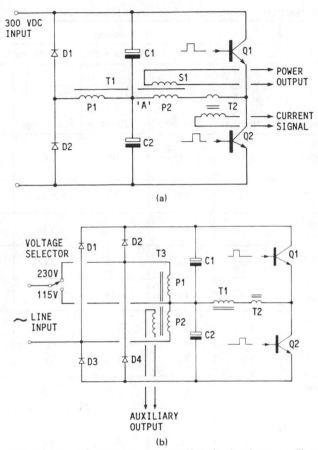

(a)

(b)

FIG. 3.10.11 (*a*) A DC charge restoration circuit using an ancillary winding on the main switching transformer. DC restoration is required for current-mode-controlled half-bridge converters. (*b*) A DC charge restoration circuit using a center-tapped winding on the 60-Hz auxiliary transformer to restore the C1, C2 center point voltage.

ter point voltage on C1 and C2, compensating for any unequal ampere-seconds in the switching devices. The wire gauge in this winding and the diodes can be quite small, as they carry only small restoration currents. The number of turns on the separate winding should be the same as the primary turns. If they are bifilar-wound, they have the additional advantage of providing a leakage inductance energy recovery action.

A different method of providing DC restoration of the center point voltage is shown in Fig. 3.10.11*b*. In this circuit a small 60-Hz auxiliary transformer is used to provide the auxiliary supply to the control circuits. The dual-voltage primary winding on this auxiliary transformer also provides the DC restoration of C1 and C2 when they are connected in series for 230-V operation. (DC restoration is provided by the supply when they are linked for 115-V voltage doubler operation.)

TABLE 3.10.1 Summary of Performance for Current-Mode Control Topologies

Advantages and limitations	Flyback complete energy transfer	Topology			
		Continuous-inductor-current modes			
		Flyback	Boost	Half-bridge	Forward full-bridge full-wave p—p
Pulse-by-pulse automatic current limiting	Yes	Yes	Yes	Yes	Yes
Good open-loop line regulation	Yes	No	No	Medium	Medium
Poor open-loop load regulation	Yes	Yes	Yes	Yes	Yes
Slope compensation required	No	Yes	Yes	Yes	Yes
$(1 - D)$ compensation required	No	Yes	Yes	No	No
Simplified loop compensation	Yes	Yes	Yes	Yes	Yes
Poor noise immunity	No	Yes	Yes	Yes	Yes
Right-half-plane zero	No	Yes	Yes	No	No
Voltage offset transformer DC restoration required	No	No	No	Yes	No, provided DC blocking capacitor not fitted
Loop irregularities with multiple outputs	No	No	No	Yes	Yes
Integrated output inductor required for multiple outputs	No	No	No	Yes	Yes
Automatic symmetry correction in push-pull circuits	—	—	—	Yes	Yes
Current sharing for paralleled supplies	Yes	Yes	Yes	Yes	Yes

10.11 SUMMARY

Provided that the correct amount of slope compensation is used, current-mode control is superior in many ways to the conventional voltage programmed duty ratio control.

In all topologies it provides inherent fast-acting pulse-by-pulse current or power limiting, a major contribution to reliable performance. Further, changing the low-impedance source of duty ratio control to the high-impedance constant-current source of current-mode control eliminates the inductor from the small-signal model, allowing the outer voltage control loop to be fast and stable. This, together with the provision of slope compensation, provides good immunity to input voltage variations and good open-loop load transient response.

It is important to recognize when and what type of compensation is required for each type of topology. Table 3.10.1 gives a summary of the noncompensated performance of the major topologies. In most cases where a poor performance is indicated, the application of the correct compensation will very much improve the performance. For example, the poor open-loop load regulation indicated for the forward converter would be greatly improved by closing the voltage control loop and using slope compensation. In the case of the continuous-mode flyback converter, input voltage feedforward compensation would also be required. However, this converter has a right-half-plane zero, which limits the transient performance, as the loop gain must roll off at a low frequency, irrespective of the type of control mode used.

The design of multiple-output supplies using current-mode control is complicated by the need for integrated output inductors to eliminate loop irregularities (see Sec. 10.8.2). However, the design time is well justified, as there is a bonus in improved cross regulation.

Acknowledgment

Some parts of this section are adapted from "Current-Mode Control of Switching Power Supplies" by Lloyd H. Dixon Jr. Reprinted with the permission of the author and Unitrode Corporation.

10.12 PROBLEMS

1. What are the basic elements of a current-mode control system?
2. How is current-mode control used to control the output voltage?
3. Give three major advantages of current-mode control.
4. What is the purpose of slope compensation in current-mode control?
5. What percentage of slope compensation will guarantee stability under any conditions?
6. Why is current-mode control more suitable for parallel operation?
7. Why is a DC current path required in the primary circuit of a current-mode-controlled half-bridge converter?

8. Why is current-mode control not recommended for multiple-output applications when separate output filter inductors are used?

9. Describe a method of eliminating loop gain irregularities in a multiple-output current-mode-controlled supply.

10. Does current-mode control eliminate the right-half-plane zero problem?

CHAPTER 11
OPTOCOUPLERS

11.1 INTRODUCTION

In switchmode power supplies, optocouplers—or, more precisely, "opto-electronic coupling and isolating elements"—are often used to convey information from the secondary output circuits back to the input primary control circuits without compromising the galvanic isolation between the two.

However, optocouplers have a number of parameter variations and limitations, which must be considered at the design stage if problems are to be avoided. Of particular interest are variations in transfer ratio with device type, operating temperature, stability problems caused by the nonlinear current transfer ratio, and the considerable variations between devices of the same family. Further, a particular optocoupler may change its parameters considerably throughout its working life (aging). Finally, the interelectrode capacitance, although small, can cause noise problems because of the high gain in the optotransistor.

As a result of these limitations, the optical coupler should *not* be used in an open-loop mode, where such changes would have a direct effect on the performance. (It was probably the use of these devices in an open-loop mode that gave optocouplers a bad name in early applications.) Optocouplers are now more often used inside a closed control loop with considerable negative feedback; hence the variations in device parameters will not significantly alter the transfer function of the loop.

In many direct-off-line switchmode power supplies, a control loop with negative feedback from the DC output back to the primary pulse-width modulator is provided to maintain the output voltage constant. It is usually necessary to close the control loop to the output, while at the same time providing galvanic isolation between the primary and secondary circuits to meet safety and application needs.

Optocouplers will very conveniently provide this isolated information link, although only a few types meet all the safety agency requirements.

To ensure that full control will be maintained for all conditions, it is essential that the drive circuitry to the optocoupler diode have sufficient drive current margin to take up the many variations that will occur between different devices and the reduction in transfer ratio that can occur with age.

11.2 OPTOCOUPLER INTERFACE CIRCUIT

Figure 3.11.1 shows a typical optical coupler drive circuit. In this example a 5-V secondary output, on the right-hand side, is to be controlled by a pulse-width modulator in the primary circuit (on the left).

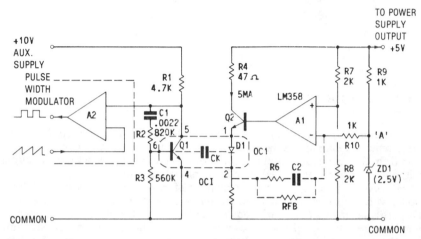

FIG. 3.11.1   Optically coupled voltage control loop, using a voltage comparator amplifier A1 and voltage reference on the secondary, optically coupled to the primary pulse-width modulator A2.

The amplifier A1 compares the reference voltage on ZD1 (node A) with the output voltage via the divider network R7, R8. The conduction state of Q2 is thus controlled to define the current in the optodiode D1 and via optical coupling the collector current in optotransistor Q1. Q1 then defines the pulse width and output voltage, compensating for any tendency for the output voltage to change.

To prevent any loss of control as the optocoupler ages and gain (transfer ratio) falls, it is essential to provide an adequate drive current margin in Q2.

Consider the Motorola MOC1006 optocoupler/isolator. (The typical transfer characteristics are shown in Fig. 3.11.2.) This device has a specified minimum current transfer ratio of 10%, and a maximum diode current rating of 80 mA continuous.

The LM 358 amplifier A1 is rated for a maximum output current of only 10 mA. Consequently, to get the maximum range of control for the optocoupler (0 to 80 mA), a buffer transistor Q1 is connected to the output of A1. A limiting resistor R1 ensures that the maximum current rating of OC1 cannot be exceeded during current limit or transient conditions.

To eliminate the effects of variations and nonlinearity in the forward voltage drop of the optodiode D1, negative feedback to amplifier A1 is taken across resistor R5.

In this example, the chosen operating current for the optodiode is 5 mA. The large current accommodation range provided by Q2 should be adequate to take up any variations in devices or reduction in transfer ratio resulting from aging of the optocoupler. Even so, optocouplers with well-defined transfer ratios should

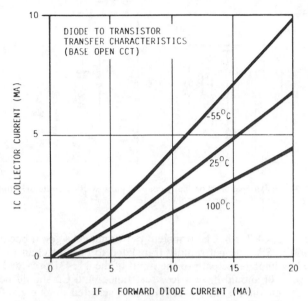

FIG. 3.11.2 Typical current transfer function of an optical coupler (collector current, as a function of diode current, in common-emitter mode), showing how the transfer ratio is temperature-dependent.

be used, as unspecified devices can have extremely wide tolerances. Large changes in the open-loop gain make it difficult to define the transfer function of the overall control loop.

The total system control loop is closed by the remaining power and control circuitry. Consequently, the optocoupler is inside the feedback loop, and the effect of variations in the transfer ratio is reduced by the negative feedback such that the closed-loop gain remains nearly constant.

Note: The voltage gain for a series voltage negative feedback amplifier is

$$A' = \frac{A}{(1 + \beta) - A\beta}$$

where A = loop gain without feedback
 A' = gain with feedback
 β = feedback factor

This tends to $-1/\beta$ when $A \geq A'$.

Hence, large variations in the optocoupler tolerances and aging will not cause significant degradation of the overall performance provided that the open-loop gain is large and the optocoupler changes are within the range of the control circuit.

Figure 3.11.3 shows an alternative optocoupler drive circuit using the TL 431 shunt regulator IC. This regulator contains the reference voltage and can deliver up to 100 mA of drive into the optodiode D1, eliminating the need for the buffer transistor. In this circuit the TL 431 operates as a transconductance amplifier, the current in the optodiode D1 being proportional to the input voltage as a result of

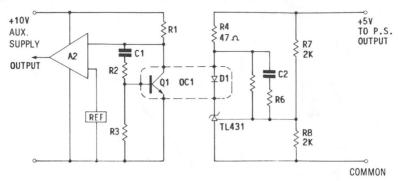

FIG. 3.11.3   An example of an optically coupled pulse-width modulator, using the TL 431 shunt regulator IC, as the control element.

the divider network R7, R8. The transient performance is good, because changes in voltage on the 5-V line get rapidly translated to current changes in D1, as a result of the low-impedance path to the input of the TL 431 provided through R4 and D1. Hence, R4 should have a low resistance, and C2 should be small. (R4 also defines the maximum current in D1 and is selected to give a safe limit.)

11.3 STABILITY AND NOISE SENSITIVITY

Consider Fig. 3.11.1 once again. Negative feedback via R5, R6, and C2 reduces the high-frequency gain of A1, and this, together with the additional roll-off provided by C1 and R2 provides the gain and phase margins required to maintain overall loop stability. (See Chap. 8). It should be noted that the local feedback around A1 is proportional to the current in R5 and hence to the optocoupler diode current, making the A1-Q2 combination a transconductance amplifier. Further, in this type of control circuit, the output voltage divider network R7, R8 is applied to the noninverting amplifier input. This has a considerable effect on the loop performance because *the gain of the A1-Q2 combination cannot be less than unity, even with 100% negative feedback.*

As a result, to maintain stability, the loop gain of the remainder of the control and power circuits must be less than unity at the crossover frequency. However, if full range control is to be maintained, the DC voltage gain of the remaining circuit must be greater than unity, and additional roll-off will be required elsewhere in the loop. (In this example it is provided by C1 and R2 on the base of Q1.)

The local negative voltage feedback provided by the base collector components C1 and R2 reduces the variations in optocoupler ac gain. The DC gain is normalized by R3, to give more consistent results with different optocouplers.

Although the interelectrode capacitance C_k in the optocoupler is very small, it is a common cause of output noise in optocoupled units. There can be considerable noise and ripple voltages between the primary and secondary output circuits. Hence, the optocoupler interelectrode capacitance C_k, although small, can be a major problem. The optotransistor Q1 has a very high gain; hence very small currents injected into the base of Q1 via the interelectrode capacitance will cause noise modulation of the Q1 collector current and hence the output voltage. This

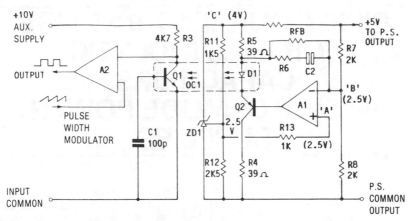

FIG. 3.11.4 Optically coupled pulse-width modulator, with the control amplifier configured for loop gains of less than unity.

effect will be reduced by fitting a base-emitter capacitor on Q1, or by the "Miller" feedback provided by C1 and R2.

The value of R1 in the collector of Q1, together with the amplitude of the ramp comparator ramp voltage, controls the gain of the ramp comparator circuit. The value of R1 is often a compromise selection. It should be high enough to reduce the range of collector current required in Q1 to maintain control, but not so high that excessive loop gain leads to instability. Further, a large value of R1 will also increase the noise sensitivity.

Figure 3.11.4 shows a drive circuit configuration in which the gain of the amplifier A1 can be less than unity. In this circuit, the voltage feedback is applied to the inverting input of A1. Phase reversal is provided by connecting the optodiode in the top end of Q2, which is now a PNP transistor. In this arrangement it is essential to stabilize the supply to D1 and Q2 to remove the common-mode input voltage variations (otherwise the minimum gain will again be unity). The TL 431 is suitable for this application. It provides a shunt regulator action to node C, maintaining this at 4 V while still providing a 2.5-V reference to amplifier A1. This circuit will also benefit from a capacitor C1 on the optotransistor base, to reduce noise sensitivity.

11.4 PROBLEMS

1. Why are optocouplers often used in switchmode power supply control systems?

2. What are the major disadvantages of optocouplers?

3. What precautions should be taken in the design of the drive circuit when optocouplers are to be used?

CHAPTER 12

RIPPLE CURRENT RATINGS FOR ELECTROLYTIC CAPACITORS IN SWITCHMODE POWER SUPPLIES

12.1 INTRODUCTION

It is generally well known that for reliable performance and long life, electrolytic capacitors must be selected with generous voltage and temperature margins. Less well known is the need for adequate ripple current ratings. To better understand this requirement, the basic construction of an electrolytic capacitor should be considered.

In a typical electrolytic capacitor, two strips of aluminum foil will be spirally wound between layers of absorbent material saturated with an electrolyte. The capacitance is formed at the interface of the aluminum and the conducting electrolyte by means of a very thin film of insulating dielectric which is formed and maintained by the polarizing voltage in the presence of the fluid electrolyte. If the electrolyte begins to dry out, the resistance of the absorbent separator increases, the dielectric begins to break down, and the capacitor will rapidly degrade.

To prevent the loss of electrolyte, the capacitor will have hermetically sealed end caps, lead-out wires, and connections. The integrity of these seals is put under great strain at high temperatures, when the electrolyte tends to vaporize and pressurize the case. Further, the losses in the capacitor increase at high temperatures, leading to a runaway effect. Therefore, the temperature of the capacitor becomes a major concern for longer-term reliability. Three major factors combine to define the internal temperature of the capacitor. These are

1. Ambient operating temperature
2. Thermal design and environment (air flow)
3. Internal dissipation

The ambient temperature is a matter of application and specification, generally outside the designer's control.

The thermal design is a major factor which is more often under the control of the designer. The location of the higher-temperature components, layout, heat sink design, size, and methods of cooling (forced air or convected cooling) often have a much greater effect on the temperature rise of the electrolytic than the internal dissipation. The designer must keep the need for minimum thermal stress in the electrolytic capacitors in mind at all times if good MTBF figures are to be maintained.

The internal dissipation within the electrolytic capacitor is generally quite low and will be controlled by the voltage stress, temperature, and, in particular, the ripple current.

To assist the designer, the manufacturer specifies, as a general guide, a maximum rms ripple current rating, usually at a frequency of 120 Hz and in a free air temperature of 85°C or 105°C. A typical example is shown in Table 3.12.1.

TABLE 3.12.1 Typical Electrolytic Capacitor Ripple Current Ratings

Capaci-tance μF	RV							
	10	16	25	35	50	63	80	100
470							950	1030
1,000					1300	1420	1650	1770
2,200			1780	1940	2120	2270	2480	2770
3,300		2120	2370	2600	2780	3030	3390	
4,700	1970	2310	2770	2970	3240	3620		
6,800	2570	3050	3670	3900	4350			
10,000	3200	3660	4820	5280				
15,000	4570	5470	6470					
22,000	5540	6790						

The manufacturer establishes these ripple current ratings by operating the capacitors with a DC polarizing voltage and sine-wave ripple current stress at the test frequency (usually 120 Hz). The figures quoted are therefore based on a low-harmonic-content (sine-wave) rms ripple current that will cause a defined maximum internal dissipation and temperature rise within the capacitor. The permitted temperature rise depends upon the design of the capacitor and usually will be of the order of 8°C maximum. The actual permitted temperature rise due to internal dissipation is not usually quoted but can be obtained from the manufacturer. It is important not to exceed the internal dissipation limit irrespective of the operating temperature, as there can be a thermal runaway as a result of an increase in internal loss at high ripple currents.

As the internal losses are lower at lower temperatures, higher ripple currents are permitted. This results in a temperature-dependent correction factor for the ripple current. A typical example is shown in Fig. 3.12.1.

The internal dissipation of the electrolytic capacitors also tends to fall at higher frequencies, and a further increase in ripple current is permitted for high-frequency operation. A typical frequency correction factor is shown in Fig. 3.12.2.

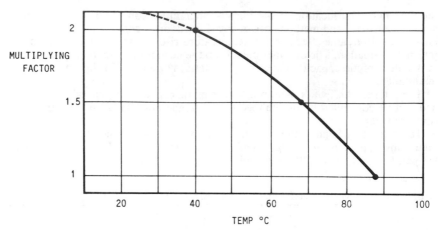

FIG. 3.12.1 Typical ripple current multiplying factor for electrolytic capacitors as a function of ambient temperature.

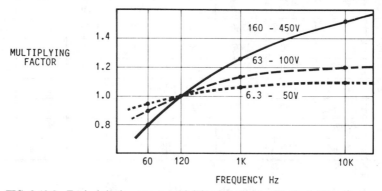

FIG. 3.12.2 Typical ripple current multiplying factors, as a function of frequency, with voltage rating as a parameter, for electrolytic capacitors.

Figures 3.12.1 and 3.12.2 show typical correction factors which can be applied to one range of commercial-grade capacitors based upon a permitted internal temperature rise (above ambient) of 8°C. The ripple ratings assume rms sine-wave conditions.

12.2 ESTABLISHING CAPACITOR RMS RIPPLE CURRENT RATINGS FROM PUBLISHED DATA

This may appear simple. For example, assume a requirement for a 2200-μF 25-V capacitor to operate in a switchmode application at an ambient of 40°C and a frequency of 10 kHz.

From Table 3.12.1, the basic ripple current rating is 1780 mA at 85°C.

From Fig. 3.12.1, at 40°C the ripple multiplying factor is × 2.

From Fig. 3.12.2, an additional multiplying factor of × 1.1 applies at 10 kHz for 25-V capacitors.

Hence the total rms ripple current rating would appear to be

$$1780 \times 2 \times 1.1 = 3920 \text{ mA}$$

This would indeed be the ripple rating for 10-kHz sine-wave operation (remember, the ripple current indicated by data is an rms value assuming sine-wave operation). However, in switchmode applications the waveform is far from being a sine wave, and although the calculations give a good starting point, the ripple will be far from sinusoidal and, because of the harmonic components, the rms value will be higher. This must be considered in the selection of filter capacitors for switchmode applications, as discussed in the next section.

12.3 ESTABLISHING THE EFFECTIVE RMS RIPPLE CURRENT IN SWITCHMODE OUTPUT-FILTER CAPACITOR APPLICATIONS

An approximate rms ripple current can be calculated by assuming ideal secondary current waveforms. Typical examples of output waveforms with their conversion factors are shown in Fig. 3.4.10. However, because of the considerable deviation of the actual waveforms from the ideal shapes, the results are not very accurate, particularly for flyback converters at low voltages and high currents. In most cases the real rms ripple will be lower than the calculated value because of the tendency for peak current limiting in the real case. This can become important in high-power systems.

Often, the effects of the layout, leakage inductance, output capacitor ESR, and circuit losses are unknown. As a result, it may be more expedient to measure the ripple current in the prototype unit and establish, or (if it has previously been calculated) confirm the final rms values.

Note: In general ripple currents are best measured using current transformers and true rms current meters. (See Chaps. 14 and 15 for suitable current transformer designs.) Be sure to use a current transformer designed for unidirectional current pulses if there is a rectifier in the line, or ac current transformers for capacitor ripple currents.

The current transformer should be connected to a true rms milliammeter. (RF thermocouple instruments are ideal for this application.) If an rms voltmeter is to be used, then the current transformer must be terminated in the correct burden resistor, and the rms voltmeter must have a good crest factor rating and must respond to the DC component for unidirectional measurements. (Take care. Many of the modern digital, so-called true rms instruments do not respond to a DC component.) However for capacitor ripple current measurements, the current is always ac under steady-state conditions.

12.4 RECOMMENDED TEST PROCEDURES

Irrespective of the method used to establish the ripple rating and hence capacitor size, it is recommended that the temperature be measured in the final application because the final temperature rise is a function of the internal dissipation resulting from the ripple current, the proximity effects of surrounding components, and the thermal design. Radiated and convected heat from nearby components will often cause a greater temperature rise in the capacitor than the internal dissipation.

The maximum temperature rise permitted in the capacitor, as a result of ripple current and peak working temperature, varies with different capacitor types and manufacturers. In the examples used here, the maximum rise allowed for ripple current is only 8°C in free air. (It is this limit that the manufacturer uses to qualify the ripple current rating.) The rating applies to an ambient free air temperature of 85°C, giving a case temperature of 93°C. This sets the absolute limit of operation irrespective of the cause of the temperature rise. The life of the capacitor at this temperature is not good, and lower operating temperatures are recommended.

Very often, the effective rms current will not be known, and although this can be calculated or measured at the operating frequency, it does not always help much in establishing the final temperature rise of the capacitor. The harmonic content is often very high with switchmode applications, and the capacitor losses vary with the frequency and amplitude of each harmonic (the ESR changes in a nonlinear way with frequency); hence the loss component is frequency-dependent, usually in an unknown way. Hence, the final test procedure described in the following section is recommended.

12.4.1 Confirm the Selection, by Measuring the Temperature Rise in the Final Application

1. Measure the temperature rise of the capacitor under normal operating conditions away from the influence of other heating effects. (If necessary, connect the capacitor to a short length of twisted cable away from the heating effects of other components, or insert a thermal barrier between the hot components and the capacitor.) Measure the temperature rise of the capacitor resulting from the ripple current alone, and compare this temperature rise with the manufacturer's limiting values. (This figure is not always given on the data sheets but can be obtained from the manufacturer. The maximum temperature rise allowed for internal dissipation is typically between 5 and 10°C.)

2. If the temperature rise resulting from ripple current is acceptable, mount the capacitor in its normal position and subject the power supply to its highest temperature stress and load conditions. Measure the surface temperature of the capacitor and ensure that it is within the manufacturer's maximum temperature limit. Several samples should be tested.

The most important parameter for long-term reliability of electrolytic capacitors is undoubtedly the temperature rise of the capacitors in the working environment. Electrolytic capacitors quickly become more lossy at high temperatures, increasing the internal dissipation and leading to thermal runaway. There is no substitute for measuring the performance and temperature rise in the finished product.

12.5 PROBLEMS

1. Why are ripple current ratings so important in the selection of electrolytic capacitors?
2. What effect does frequency have on the ripple rating of the electrolytic capacitor?
3. What effect does ambient temperature have on the ripple rating of electrolytic capacitors?
4. At what temperature is the ripple rating normally specified?
5. Why is it important to measure the temperature rise of the capacitor in the final application?

CHAPTER 13
NONINDUCTIVE CURRENT SHUNTS

13.1 INTRODUCTION

Normal current transformers are not particularly suitable for measuring ac currents at very low frequencies, or even high-frequency ripple currents where there is a large DC component. (Special DC current probes can be used at low frequencies; see Chap. 14.)

The reason is that in order to achieve a good response at low frequencies, the inductance of the secondary winding should be very high. This requires cores with high permeability and a large number of secondary turns. (If the inductance is too low, there will be a large magnetizing current in the primary, and this will be subtracted from the measured secondary current, giving considerable error.)

However, in applications in which there is a large DC component, a high-permeability core will easily saturate, even in high-frequency current transformers; also, the DC component will not be transformed to the secondary and is lost to the output measurement. Consequently, for applications with high DC current components, resistive current shunts are normally preferred, although special DC current transformers can be used if necessary. (See Sec. 14.9.)

13.2 CURRENT SHUNTS

A current shunt is a four-terminal low-value resistive element which is inserted in series with the line in which the current is to be measured. A voltage analogue of the current is developed across the resistance. Hence, for minimum insertion loss or interference with the action of the circuit, this shunt should have a very low resistance and ideally zero self-inductance. However, all conductors, even a straight wire, will display some inductive effect, and this must be considered when measuring currents in high-frequency circuits, as the following example will show.

13.3 RESISTANCE/INDUCTANCE RATIO OF A SIMPLE SHUNT

Consider a current shunt made from a straight piece of #22 AWG manganin resistance wire, 1 in long. The resistance of the wire will be 37 mΩ at 20°C. Hence, this shunt would be expected to develop a voltage of 370 mV from end to end in a 10-A circuit.

However, at high frequency, it is the total impedance of the shunt that controls the developed voltage, and this includes some inductance effects even in the simple straight wire.

The following formula gives the approximate inductance of a straight wire where the length is much greater than the diameter:

$$L = 2 \cdot l_g \left(\log_e \frac{2 \cdot l_g}{r} - 0.75 \right) \cdot 10^{-7}$$

where l_g = wire length, m
r = radius of wire, m

Hence the inductance of the 1-in #22 AWG wire is 19.7 nH. The reactance at, say, 60 kHz will be 7 mΩ.

Hence, the ratio of inductance to resistance in this example becomes significant even for the fundamental frequency component at typical switchmode frequencies. Since many waveforms have square transitions, the harmonics extend to much higher frequencies, and in low-resistance, high-current shunts a significant error will occur. The transients will be distorted, as the voltage generated across the shunt will have a significant $L \cdot di/dt$ component superimposed on the required resistive component.

13.4 MEASUREMENT ERROR

Consider a typical flyback power supply delivering an output of 5 V at 10 A. The rectified diode current is to be investigated.

If a 50% mark space ratio is used, the flyback current will be approximately 4 times the DC output, in this case 40 A. The output will increase to this value very rapidly, limited only by the leakage inductance. The period will be typically less than 1 μs.

Figure 3.13.1a shows the typical secondary current waveform without distortion. If a straight resistance wire (as described above) is used in a shunt to measure this current, the waveform shown in Fig. 3.13.1b will be obtained. In this example the reactance error is so large that the information is of little value.

13.5 CONSTRUCTION OF LOW-INDUCTANCE CURRENT SHUNTS

The inductive component of the resistive element can be considerably reduced by one of the following methods.

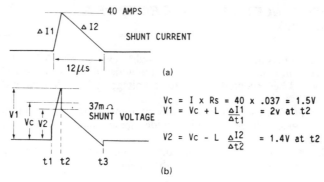

FIG. 3.13.1 Applied current waveform and measurement distortion caused by inductance in a resistive current measurement shunt when operated at high frequency. (*a*) Applied current waveform; (*b*) distorted voltage waveform.

Double Helix Construction

Figure 3.13.2*a* shows a suitable method for low-current shunts—up to, say, 10 A. The resistive element, which should be low-temperature-coefficient insulated resistance wire, is twisted into a tight double helix, so that the magnetic fields will cancel. For convenience, this helix may be further spiralized.

Sandwich Sheet Resistance Construction

For higher-current applications, a flat sheet of resistance metal is folded as shown in Fig. 3.13.2*b*. An insulator is fitted between the two sides. For long lengths the "sandwich" may be spiralized to form a double spiral.

Both these arrangements will give much better results at high frequencies than the more commonly used straight current shunts.

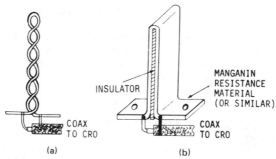

FIG. 3.13.2 Two fabrication methods used for high-frequency, low-inductance current shunts. (*a*) Double helix; (*b*) folded strip.

13.6 PROBLEMS

1. Why is the inductance of a current shunt often very important in switchmode applications?
2. Describe two methods of reducing the inductance of a resistive current shunt.

CHAPTER 14
CURRENT TRANSFORMERS

14.1 INTRODUCTION

When the switchmode power supply designer thinks in terms of current transformers, it is more often in the context of the less precise power supply protection and control applications, rather than the highly accurate instrument-type applications.

In most switchmode power supplies, the current transformers will be indicating trends, changes, or peak values, rather than absolute quantities. Consequently, a high degree of accuracy is not essential, and very simple design and winding techniques can be used.

The advantages of current transformers for control and limiting applications should not be overlooked. They give good signal-to-noise ratio, provide isolation between the control circuit and the line being monitored, provide good common-mode rejection, and do not introduce excessive power loss in high-current applications.

As previously shown, current shunts are not very satisfactory for high-frequency current limit applications, because the resistance of the shunt must be very low to reduce the insertion loss, giving a poor signal-to-noise ratio. Further, the resistance-to-inductance ratio is very poor, and the voltage analogue signal generated across the shunt is dependent on the rate of change of the current pulse, rather than its amplitude. Consequently, the information available for current-limiting purposes from resistive current shunts is often distorted and will change with the conditions of operation. Also, isolation is not provided, and the power loss is much greater. (However, even with these limitations, current shunts are sometimes used in current-limiting applications.)

Current transformers are particularly useful when independent output current limiting is required in multiple-output power supplies. It is also useful in those applications in which galvanic isolation is required between the measured quantity and the control circuit, for example, between the primary and secondary circuits.

14.2 TYPES OF CURRENT TRANSFORMERS

Depending on the application, to obtain the best performance, one of four basic types of current transformer will normally be used. Two types are in common use and will be considered first, the third type is used where current pulses are very short, and the last is a special DC current transformer for high-current applications. The four types are described below.

Type 1, Unidirectional Current Transformers

This first type measures unidirectional current pulses, such as output rectifier diode currents, switching regulator currents, and the currents that would flow in the primary or secondary of a forward converter transformer.

Type 2, AC Current Transformers

In the second type, the transformer is used to indicate ac currents where the measured current flows equally in both directions and there is no net DC component. A typical application would be to measure the current in series with the primary winding of a half-bridge push-pull converter.

Type 3, Flyback-Type Current Transformers

This type of current transformer is used in the flyback mode, and is particularly useful for applications in which the current pulse can be very narrow.

Type 4, DC Current Transformers

This very useful, less well known DC current transformer can be used to measure current in high-current DC output lines, with very low loss.

14.3 CORE SIZE AND MAGNETIZING CURRENT (ALL TYPES)

Selecting the core size is probably the most difficult part of the design exercise. A compromise must be struck between the ideal performance and practical considerations of size, cost, and number of turns.

In general, for current transformers, the larger the inductance, the smaller the magnetizing current and the more accurate the measurement. The magnetizing current component increases during the pulse duration and will be subtracted from the quantity to be measured. Consequently, at the end of the conduction pulse, the magnetizing current should be small compared with the measured quantity. For current limit applications, a magnetizing current of 10% is a typical design limit. This magnetization effect is most easily shown in the unidirectional current transformer.

Figure 3.14.1a shows a typical unidirectional current transformer and secondary circuit. This would be used for current limiting in, say, a forward converter. The primary has a single turn on a toroidal core. (This primary turn is often a

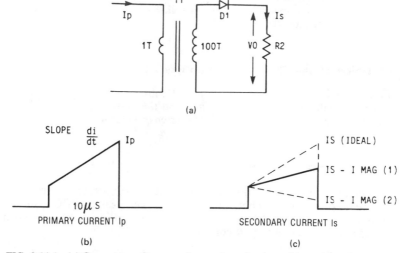

FIG. 3.14.1 (*a*) Current transformer and secondary circuit used for unidirectional current pulse measurement. (*b*) Applied unidirectional primary current waveform, and (*c*) developed secondary current waveforms on R2, showing the effect of current transformer magnetization current.

connecting lead or output bus which is simply passed through the center of the toroid.) The primary turn is thus in series with the line to be monitored.

The secondary has a large number of turns, which are terminated in a ballast resistor R2 via a diode D1. The intention is that a true voltage analogue of the primary forward current pulse be developed across R2. (D1 blocks the reverse recovery voltage.) However, it will be seen from Fig. 3.14.1*b* and *c* that the secondary waveform is distorted as a result of the magnetizing current component.

Figure 3.14.1*b* shows the applied unidirectional (all positive) primary current pulse. Figure 3.14.1*c* shows the corresponding secondary voltage pulse developed across R2. The effect of two values of secondary magnetizing current, a small value I_{mag1} and a large value I_{mag2}, shows how the magnetizing current is effectively subtracted from the ideal transformed current analogue $I_{s(ideal)}$. It will be clear from this diagram that if the peak value of the current at the end of the conduction pulse, I_p', is to be useful for current-limiting purposes, then the secondary inductance of the current transformer must be high enough to ensure that at least a positive slope remains on the net secondary waveform. This means that a large secondary inductance is needed, and so a large number of secondary turns, a large core, and high-permeability core material are required. In general, the largest-permeability core material will be used, leaving a trade-off between turns and core size.

Note: Losses in amplitude resulting from winding resistance, diode loss, and magnetizing current amplitude can be corrected to some extent by adjusting R2, provided that the slope remains positive.

A second major factor that influences the current transformer magnetizing current will be the magnitude of the secondary voltage. This voltage is the sum of the selected signal voltage V_0 (say 200 mV in this example) and the D1 rectifier diode forward voltage drop (say 0.6 V). The secondary voltage should be as small as possible (consistent with a good signal-to-noise ratio), since large values of V_0 will result in large magnetizing currents. To this end, small schottky diodes should be considered for D1.

If a very small toroid is chosen for the current transformer, then to get the required inductance, a large number of secondary turns will be required. If the number of secondary turns is too large (say in excess of 200), then there will be significant interwinding capacitance, and the high-frequency response (response to narrow current pulses) will be degraded.

Hence, the core size is a compromise between cost and performance. A good compromise is to select a core that will require approximately 100 turns on the secondary (in a single layer) and will give the required minimum inductance.

14.4 CURRENT TRANSFORMER DESIGN PROCEDURE

General Requirements

A unidirectional current transformer would be used for monitoring the discontinuous current pulses in, say, the output rectifiers of a high-current forward converter or the primary current of a single-ended forward converter. A typical application is shown in Fig. 3.14.2a, where the current transformer T1 is in series with the primary of a single-ended forward converter. The design steps are as follows:

Step 1

Calculate (or observe), the peak primary current to be measured and the slope di/dt on the top of the current waveform. This will be used to calculate the minimum current transformer inductance.

Step 2

Select the current transformer secondary voltage at the limiting current value. (This should be kept as low as possible, typically < 1 V including the diode drop.)

Step 3

Select a high-permeability core material and initial size.

Note: For unidirectional current pulse applications, the core material should have two properties:

1. High permeability, so that a large inductance will be obtained for the minimum number of turns.

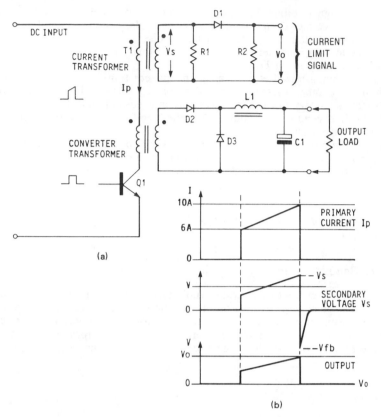

(a)

(b)

FIG. 3.14.2 (*a*) Basic circuit of a single-ended forward converter, showing a unidirectional current transformer in the primary of the main power transformer. (*b*) Primary current and secondary voltage waveforms of current transformer, together with the current analogue signal voltage developed across R2. This may be used for current-mode control and current limiting.

2. A low remanence B_r, so that the core will restore to a low flux level when the current pulse drops to zero. This will ensure that the core will not saturate after a few cycles of operation.

Unfortunately, these two requirements tend to be mutually exclusive, so a compromise choice must be made. A material such as H5B2 (Fig. 2.15.4*b*) would be a good compromise. The core should be insulated to reduce capacitance between the winding and the core and thereby reduce interwinding capacitance.

At larger currents, the physical requirements for the prmary wire diameter may define the minimum core size. If a large output voltage is required, then it is recommended that a larger core be used, allowing a greater number of secondary turns. For practical reasons, the primary will be maintained at one turn (that is, the primary wire is passed straight through the center hole of the toroid).

14.5 UNIDIRECTIONAL CURRENT TRANSFORMER DESIGN EXAMPLE

Figure 3.14.2b shows the waveforms that would be expected from a current monitoring transformer of type 1 (unidirectional pulse type) in the forward converter of Fig. 3.14.2a.

In this example, when the primary power transistor is "on," the forward current in T1 takes the starts of all windings positive, and secondary diode D1 conducts. The current in R2 will be a transform of the primary current, and an analogue voltage of the primary current will be developed across R2.

When Q1 turns off at the end of the forward current pulse, rapid recovery of the current transformer core occurs because D1 blocks and the secondary flyback load resistance R1 is high. As a result, the flyback voltage is large, and this gives a rapid core restoration between forward pulses. That is, the flux density returns to the residual value B_r during the "off" period, ready for the next forward pulse.

Note: This rapid core restoration allows closely spaced forward current pulses to be monitored accurately, without saturation of the core. Clearly, the value of R1 will be chosen to get the required minimum recovery time, and the voltage rating of D1 must be selected to block the reverse flyback voltage across R1.

Step 1, Calculate the Primary Ampere-Turns

In this example, the primary current is 10 A in a single turn. This gives a primary magnetizing force of 10 ampere-turns due to the forward forcing current that flows in the primary of the forward converter transformer shown in Fig. 3.14.2.

Step 2, Define Secondary Turns and Calculate Secondary Current

This design will use the preferred winding not exceeding 100 turns of #34 AWG in a single layer (to give a small core size and good high-frequency performance). Therefore, to provide an equal depolarizing magnetic force of 10 ampere-turns, the secondary current will be 100 mA.

Step 3, Define the Required Secondary Voltage

The secondary voltage will be 0.8 V, made up of the 0.6-V diode drop and the 0.2-V current analogue signal developed across R2. Consequently, with 100 turns on the secondary and a single current-linked turn on the primary, the voltage drop on the primary winding will only be 8 mV. The primary insertion loss is thus very small.

For forward converter applications, the peak value of the primary current at the end of an "on" period is the required current-limiting value (this indicates the peak current flowing in the output inductor). Hence the secondary current must reflect a rising current during the "on" period, and this sets a limit on the magnetizing current. A pulse-by-pulse voltage analogue of the current is given by the voltage across R2. Hence, a fast-acting current limit may be introduced at this point in the control circuit, using the peak analogue voltage information, or the current signal may be used in a current-mode control circuit.

Step 4, Check Magnetizing Current

The current to be measured is 10 A, and the core chosen as the initial selection is the TDK #T6-12-3 (Table 2.15.1) in a H5B2 material. This material has a permeability of 7500 and a low remanence value B_r of 40 mT. The next step is to check that the magnetization current is acceptably small.

In this example, a single-layer secondary winding of 100 turns of #34 AWG is to be used on the secondary.

To calculate the magnetization current, the secondary inductance is required. If the A_L value of the core is known, then the inductance will be given by

$$L = N^2 A_L$$

In this example the A_L value is not known, and so the inductance will be calculated from the basic formula:

$$L = \mu_r \cdot \mu_0 \cdot N^2 \frac{A_e}{l_e}$$

where $\mu_r = 4\pi \times 10^{-7}$
  μ_0 = permeability of core
  N = number of secondary turns
  A_e/l_e = core factor (ratio of effective core area to effective magnetic path length, m)

For the H5B2 material, the permeability is 7500 and the core factor is $(1/30.2) \times 10^{-2}$, and so the inductance can be calculated:

$$L = 4\pi \cdot 10^{-7} \cdot 7500 \cdot N^2 \cdot \frac{10^{-2}}{30.2} = 31 \text{ mH}$$

The slope of the magnetization current, dI/dt, can be calculated from

$$e = -L\frac{dI}{dt}$$

Therefore

$$\left|\frac{di}{dt}\right| = \frac{e}{L}$$

where

$$e = 0.8 \text{ V}$$

Hence

$$\left|\frac{dI}{dt}\right| = 25.8 \text{ A/s}$$

At the end of the 10-μs pulse, $I_{mag} = 0.258$ mA. This magnetization current transforms to the primary as 25.8 mA, a negligible error in 10 A.

It has been shown that the secondary current will be a true transform of the primary current with a loss of only ¼ mA in 100 mA. Although this indicates that a smaller core could be used, it can be difficult to wind the secondary on a core much smaller than this.

The value of the burden resistor R2 required to develop the chosen signal voltage of 0.2 V can be calculated:

$$R2 = \frac{V_0}{I_s}$$

where V_0 = signal voltage
I_s = secondary current

In this example, $I_s = (I_p/N) - I_{mag}$, or approximately 100 mA, as referred to the secondary. Hence

$$R2 = \frac{0.8}{0.1} = 8 \ \Omega$$

Flux Density

Although the primary current pulses are unidirectional with a large mean DC value, the core will not saturate, as the flux density falls to the remanence value B_r between each current pulse. To permit this reset action, the voltage on both primary and secondary must reverse (fly back) during the "off" period. This requires that the reverse currents be blocked except for that flowing in R1. Clearly diode D1 will block the reverse current in the secondary during the flyback (off) period, but to block the primary current the core must be placed in a suitable position in the primary circuit. In the example shown in Fig. 3.14.2, reverse current is blocked by Q1, which is "off" for the reset period.

The reset voltage and time are determined by R1 and the reset time must be less than the shortest "off" time.

Note: Increasing R1 reduces reset time but increases the flyback voltage.

Each primary pulse can now be considered a single event, and the flux density excursion during a pulse can be calculated from the mean applied volt-seconds as follows:

$$B = \frac{V_m \cdot t}{N \cdot A_e}$$

where B = flux density, T
V_m = mean secondary voltage
A_e = effective core area, mm^2
t = pulse duration, μs

In this example,

$$V_m = 0.16 + 0.6 = 0.76 \ \text{V}$$

(the mean signal voltage plus the diode drop). Hence

$$B = \frac{0.76 \times 10}{100 \times 8.65} = 8.78 \ \text{mT}$$

This is a very small flux density swing and is typical of current transformers in switchmode supplies. Usually a smaller core cannot be used for practical winding reasons. The small flux level also results in a very small core loss.

14.6 TYPE 2, CURRENT TRANSFORMERS (FOR ALTERNATING CURRENT) PUSH-PULL APPLICATIONS)

Figure 3.14.3 shows the typical arrangement of an ac current transformer in se-ries with the primary of a half-bridge converter transformer. In this position, it is able to recognize current pulses in both directions by using a bridge rectifier in the current transformer secondary. The same current flows for forward and re-turn pulses, as the series capacitor C3 ensures that there will not be a DC com-ponent, and hence the core will not saturate. The design approach is very similar to that used in Sec. 14.4.

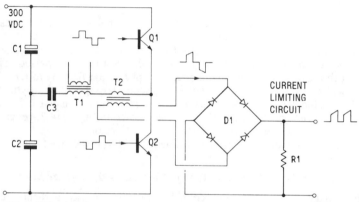

FIG. 3.14.3   A full-wave current transformer used in push-pull and half-bridge circuits.

The core material need not have a low remanence for this application, as there is a forced reversal of flux each half cycle.

If there is any possibility of a DC component, then a core with a lower per-meability will be chosen so that it can withstand the expected DC component without saturation. Normally one would avoid this problem in the current trans-former and the main transformer by either introducing a series DC blocking ca-pacitor or using a forced flux balancing system in the main converter.

14.7 TYPE 3, FLYBACK-TYPE CURRENT TRANSFORMERS

It should be understood that the term "flyback," as applied to current transform-ers in the following section, refers to the mode of operation of the current trans-former, not to its application. In fact, the type of flyback current transformer de-scribed here is not very suitable for flyback converters.

The flyback current transformer is of particular value where the current pulse is very narrow. Typical applications would be current-limiting circuits in fixed-frequency forward converter and buck regulators. In these types of converters, to prevent excessive output current when the output is short-circuited, the current pulse must be very narrow.

Figure 3.14.4 shows a typical buck regulator power output stage. There are two possible positions for the current transformer. In position T1(a), the current in the series switch Q1 is directly monitored. When the output is short-circuited, the pulse width must be reduced to a very narrow pulse to maintain control.

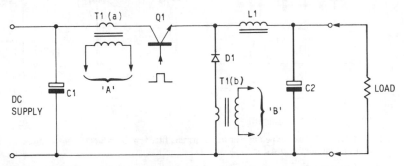

FIG. 3.14.4 Two possible (optional) positions, A and B, for a current transformer in a buck regulator circuit.

With the current transformer in position T1(b), the current pulse under short-circuit conditions is wide, but the direct control of the peak current in transistor Q1 has been lost (Q1 must turn off before the current is commutated to D1 and T1 can "see" the current). Hence, current limit action will be lost if Q1 locks "on." Also, in position T1(b), the current pulse width will be very narrow when the input voltage is low and the load is high, so that control tends to be lost at high loads, where it is most needed.

In conventional current transformers, the response to narrow pulses is often very poor, because of the limited frequency response in the current transformer and the use of low-pass noise filters in the current limit circuit. This loss of response at narrow pulse conditions can result in an increase in output current (loss of current limit control) for short-circuit conditions. This problem can be overcome by using a flyback current transformer.

The flyback type of current transformer is an interesting departure from normal current transformer design. It uses the flyback action of the current transformer to provide the required current-limiting information.

Figure 3.14.5 shows a typical application for the flyback transformer where it is applied to position T1(a) in the buck regulator power circuit.

14.7.1 Operating Principles (Flyback Current Transformers)

The transformer is wound on a very low permeability core (e.g., $\mu60$ MPP or iron powder), and the windings are phased in such a way that diode D2 will be blocked when Q1 is conducting. The core size and permeability are chosen so that the core will not saturate with a single primary turn at the maximum current limit value.

A high-voltage diode must be used for D1, as the voltage across the secondary winding will be large during the turn-on edge of Q1 ($n \times V_{in}$). This high voltage is desirable to provide rapid setting of the core. The leakage inductance and distributed capacitance will normally prevent excessive secondary voltages being de-

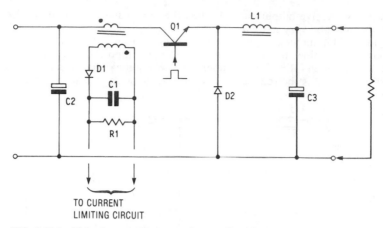

TO CURRENT
LIMITING CIRCUIT

FIG. 3.14.5 Flyback-type current transformer, fitted in the collector of the buck regulator switching transistor.

veloped, but if required, an additional loading resistance may be added across the secondary winding. The secondary voltage should not be clamped to less than 200 V.

Because of the very low inductance of T1 and the large volts/turn applied to the primary when Q1 turns on, there will be only a small delay on the leading edge of the current power pulse during the turn-on edge.

At the end of a conducting period, energy of $\frac{1}{2} LI_p^2$ will have been stored in the current transformer. This energy will be transferred by flyback action into capacitor C1 via D1 when Q1 turns off. Hence, at a fixed frequency, the power transferred to the output will be

$$P = \frac{1}{2}LI_p^2 \times f = \frac{V_c^2}{R1}$$

Therefore
$$I_p \propto V_c$$

This power must be dissipated in the loading resistor R1, which makes the output voltage on C1 proportional to I_p (where I_p is the peak current flowing in the primary of the current transformer immediately before Q1 turns off).

The response of this flyback-current-limiting transformer to narrow pulses is very good, as the energy transferred each cycle is proportional to the peak current, not the duration of the pulse.

Low-permeability ($\mu_r < 60$) iron dust cores or Molypermalloy cores will be found suitable for this application. Ferrite toroids are generally unsuitable, since they will not store sufficient energy and would saturate, except for very low current application. However, gapped ferrite cores may be used.

14.8 TYPE 4, DC CURRENT TRANSFORMERS (DCCT)

It is not generally well known that it is possible to measure DC output currents using special DC current transformers.

The general arrangement of this interesting class of DC current transformers and the required polarizing circuits are shown in Fig. 3.14.6.

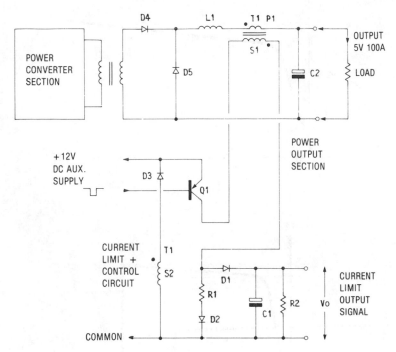

FIG. 3.14.6 A DC current transformer and polarizing circuit, fitted in the secondary of a forward converter.

The term "DC transformer" may sound like a contradiction in terms; however, it will be shown that it is possible to use transformers for measuring DC if suitable polarizing and control circuits are added.

DC current transformers can be used to control or limit the DC current in the outputs of any type of switchmode supply or other DC lines. They are particularly valuable for high-current applications (100 A or more) or where isolation between the output line and the control circuit is required.

In the example shown in Fig. 3.14.6, the DC transformer T1 is in series with L1 in the DC output line of a 100-A buck regulator.

14.8.1 Operating Principle (DC Current Transformer)

Assume that the primary winding T1 P1 has been conducting the output DC current for a sufficient time for a stable saturation flux density to have been established in the core. (This is shown as point P1 on the B/H curve in Fig. 3.14.7a.) The corresponding primary magnetizing force H_1 is also shown.

Transistor switch Q1 is now turned on, applying the low-resistance auxiliary 12-V supply to the secondary winding. Provided that Q1 is maintained "on" for a defined period and the voltage across the secondary remains at 12 V, the flux

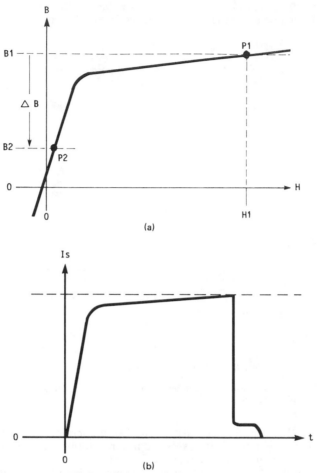

FIG. 3.14.7 (*a*) Magnetization characteristic (*B/H* curve) of a DC current transformer in the forward (core set) direction. (*b*) Primary (core reset) current waveform.

density must be incremented by a defined amount ΔB from B_1 to, say, B_2, and the core will be reset from P1 to P2 on the *B/H* characteristic.

During this reset period, Q1 is "on," and a constant-voltage low-impedance 12-V supply is applied to the secondary winding; hence the rate of change of flux density dB/dt must be constant. To maintain this change, the magnetic field strength H must move from H_1 to near zero. However, the DC primary current has remained constant during this reset period, and to offset this, there will be a rapid increase in the reset current as the flux density falls, so that when $B = 0$, the secondary ampere-turns are equal and opposite to the primary ampere-turns. At this point the core has been fully reset. The reset current is shown plotted against time in Fig. 3.14.7*b*. (The reset current is a mirror image of the original primary magnetizing current.)

The peak value of the reset current is thus proportional to the DC current in the primary. A voltage analogue of the reset current is developed across R1. (D2 provides voltage and temperature compensation for the forward drop of diode D1.) This signal voltage is peak-rectified by D1 and stored on capacitor C1. Constant updating of this current limit signal voltage is provided by interrogating the core with a series of pulses at the converter switching frequency.

Provided that the core has a very square B/H loop and the hysteresis loss is small, the absolute value of ΔB is not critical so long as the core is reset to some point on the vertical (high-permeability) part of the loop. The corresponding change in H remains nearly the same, and the peak value of the reset current is proportional to the DC current.

It is relatively easy to apply this technique to switchmode systems, as suitable square-wave drive signals for Q1 can usually be obtained from the supply drive circuitry. The drive to Q1 should be well defined, with a constant pulse width. The auxiliary voltage should also be reasonably constant.

The secondary turns and voltage must be selected to ensure that the core will always be fully restored during the reset period. The primary will normally be a single turn. (Typically the output bus will be passed through the center of the toroid to form a single turn.) When Q1 turns off, the core will be set back to its original working point P1 under the forcing action of the DC current in the primary. To prevent excessive voltages from being generated on the secondary during this set action, a clamp winding S2 is provided. The energy taken from the auxiliary supply during the resetting of the core is now returned to the auxiliary line during the set action. Hence, the total energy demanded from the auxiliary line will be quite small. To ensure that the core is fully set during the "off" period of Q1, either the "off" period must exceed the "on" period or the clamp winding S2 must have fewer turns than the secondary S1.

14.8.2 Selection of Core Type and Material

For this application, a square-loop core with high permeability and stable parameters should be chosen. Since the core will be operated over its full hysteresis range, it should have low loss and minimum weight and size. Small toroidal ferrite cores will be found ideal for this application.

In this example, a TDK T6-14-3 toroid in H5B2 material will be considered. Many similar materials are available from the various manufacturers of ferrite materials. Because a very square loop material was chosen, the reset point P2 is not critical, as the change in H over the nonsaturated range of B (contribution from magnetizing current) will be small. Hence, the shape of the top of the reset current waveform will be nearly square.

14.8.3 Primary Current Range and Turns

In high-current applications, the primary will usually be a single turn (the output wire or bus being taken straight through the toroid center hole). On lower-current applications, it may be necessary to use more primary turns to ensure a good working point for P1; this should be well into saturation. This also ensures that the primary magnetizing current is small compared with the measured value.

The saturating magnetic field strength H is about 2 Oe (159 At/m) for the H5B2 material. A working value for H of approximately 5 times this (say 800 At/m) is recommended to bring the working point to P1 at the full output current limit.

For the T6-12-3 toroid, the primary current required to give a magnetization H of 800 At/m in a single turn can be calculated:

$$H = N \cdot \frac{I}{l_e}$$

where H = magnetic field strength, At/m
$\quad N$ = primary turns
$\quad I$ = primary current, A
$\quad l_e$ = effective length of magnetic path around toroid, m

In this example,

$\quad H = 800$ At/m
$\quad N = 1$ turn
$\quad l_e = 0.026$ m

Hence

$$I = H \cdot l_e = 800 \times 0.026 = 20.8 \text{ A}$$

Hence, this core would be suitable for a current limit in the 10- to 30-A range (the upper limit being set by the primary wire size and the lower by the need for a good load current to magnetizing current ratio).

14.8.4 Secondary Turns

The secondary turns are best calculated using a volt-seconds (Faraday's law) approach for a defined value of ΔB. The value of ΔB should be chosen such that full recovery of the core takes place at the lowest operating temperature and negative saturation will not occur at the highest temperature.

It is important that negative saturation is not entered under any conditions; otherwise there will be a large increase in current at the end of the reset pulse, giving false information. The saturating value of $\hat{B}$ is temperature-dependent for the H5B2 material, and a safe value of 200 mT is chosen for ΔB. The reset period and auxiliary voltage must be defined and constant. In this example,

$\quad t = 16$ μs
$\quad V = 12$ V $\quad$ (auxiliary voltage)
$\quad A_e = 8.6$ mm^2

The secondary turns may now be calculated from Faraday's law:

$$N = \frac{V \cdot t}{B \cdot A_e} = \frac{12 \times 16}{0.2 \times 8.6} = 111 \text{ turns}$$

(t in microseconds and A_e in millimeters)

The position of the primary in the DC output circuit is important. During the set and reset action of the core, voltage transients will be generated in the primary winding which will be in series with the DC output. Consequently, a position must be chosen at which these transients will be effectively filtered. Further, transformer action will take place during the reset of the core, and this could result in excessive loading of the drive circuit and a measurement current error, unless the transformer primary is in a high ac impedance path. In the buck regulator, a suitable position meeting both of these requirements will be in series with L1. Alternatively, an extra inductor may be fitted in series with the primary of the current transformer.

Figure 3.14.8 shows the transfer characteristic of the DCCT. The graph is offset as a result of the forward voltage drop in D1. An additional diode D2 will correct this offset and provide temperature compensation if required.

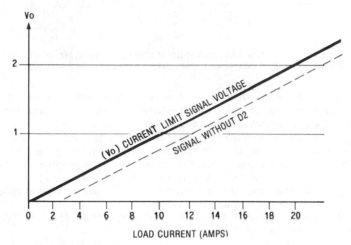

FIG. 3.14.8 Transfer characteristic of DC current transformer. Output signal voltage is a function of DC converter output (load) current.

14.9 USING CURRENT TRANSFORMERS IN FLYBACK CONVERTERS

Current limiting can be more difficult in flyback converters than in the forward case. The output current is the mean value of the primary or secondary currents, which have very triangular waveforms. Limiting the current to a peak value gives constant power limiting and results in a large output current at low output voltages—i.e., when the output is a short circuit.

Figure 3.14.9 shows a type 1 current transformer fitted in the secondary of a flyback transformer circuit.

It is the average value of the current flowing in the secondary that is of interest, and this is obtained by fitting a large capacitor C1 across the burden resistor R2 in the current sensing circuit.

The average voltage on C1 will be an analogue of the average output current. The delay introduced by this capacitor means that rapid changes of output current will not be recognized immediately. The important overpower protection

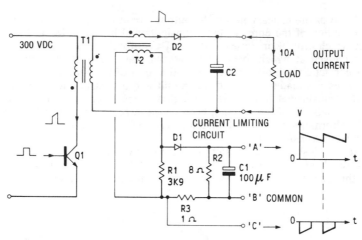

FIG. 3.14.9 A unidirectional current transformer in the secondary of a flyback converter, providing average and peak current signals.

provided by the pulse-by-pulse current limiting will be lost. However, if an extra resistor R3 is introduced in series with the burden resistor R2, information on both the peak and mean currents becomes available to the control circuit.

The peak primary or secondary current should not be used for output constant current limiting in discontinuous flyback converters. The peak current will give only constant power limiting, since the output power is proportional to $\frac{1}{2} LI^2$ Constant power limiting gives very large currents on short circuit.

The type 1 current transformer design for the flyback converter follows the example given in Sec. 14.4 in all other respects.

The circuit and current waveforms for the current transformer in flyback applications are shown in Fig. 3.14.9.

CHAPTER 15
CURRENT PROBES FOR MEASUREMENT PURPOSES

15.1 INTRODUCTION

In the development of switchmode power supplies, it is often found necessary to measure the current flowing in various components. In fact, measurement may be the only way to obtain the required information with the degree of confidence necessary for full qualification approval.

Currents ranging from a few milliampere to many hundreds of amperes DC or ac with frequencies from a few hertz to hundreds of kilohertz may be found in switchmode supplies. As a further complication, combinations of DC and ac with waveforms ranging from square to sine waves, both unidirectional and balanced, will be found. It can be very difficult to measure these currents by conventional means.

Chapter 13 has already described some of the problems encountered when conventional current shunts are used for high-frequency measurement, and methods of minimizing the errors introduced by the inductance of the shunt. In practice, no single device is ideal for the very wide range of measurements described above, although "Hall effect" probes perhaps come close. However, these tend to be very expensive.

The engineer can obtain very good results with inexpensive equipment by choosing the correct methods for the range of measurements to be made. Clearly, for DC current measurements, a standard four-terminal shunt can be used. Current transformers can be used for ac measurement, but several types of transformers will be required to cover the wide frequency range found in typical applications. When a combination of DC and ac is expected, noninductive current shunts or special low-permeability current transformers can be used. If the current pulse chain is unidirectional (for example, diode currents), a different type of current transformer will be required. The following section describes the design and construction of suitable current transformers for the above applications.

For true ac currents up to 10 A in the medium- to high-frequency range, the Tektronix current probe #P6021 or a similar probe may be used. This probe has the advantage that it is fitted with a sliding gate on the current transformer, which allows it to be easily slipped onto the wire in which the current is to be measured. This probe is also useful for the calibration of any "in-house" manufactured special-purpose current probes, as described below.

15.2 SPECIAL-PURPOSE CURRENT PROBES

Since the accuracy required for switchmode power supply ac current measurements is generally not particularly critical (5% would normally be adequate), current transformer probes for oscilloscope display are quite easily fabricated.

The general principles of current transformer design used here are similar to those shown for the current-limiting transformers described in Chap. 14. However, in order that these current probes may interface directly with the oscilloscope input and give true waveform reproduction, the terminating networks are more critical.

For easy fabrication, and to better maintain the calibration, the current transformers discussed here will be of toroidal form, and the wire that is to be measured must be disconnected and passed through the toroid for measurement purposes. Although this is somewhat inconvenient, it has the advantage that there is no gap in the core (to perhaps vary), so that the measurements and calibration accuracy are more stable; also, the probe is more easily fabricated.

Note: Those of a more ingenious mechanical bent may be able to devise methods of clipping current transformer probes onto the wire to be measured. The various cores that are now manufactured with built-in permanent magnetic bias have possibilities for this application, in that the pole pieces will stick together by magnetic attraction and the built-in magnetic bias makes the core more suitable for unidirectional pulse measurement.

However, the following discussion assumes that nongapped toroidal HCR, ferrite, and MPP cores will be used. For low-frequency ac applications (say 20 Hz to 10 kHz), HCR cores will be used. These cores should have low loss, high permeability, and high saturation flux density. For medium frequencies (10 to 100 kHz), high-permeability tape-wound or ferrite cores will be used. For unidirectional high-frequency pulse measurement, the residual flux level should be low to ensure that the core will restore during the quiescent period, and so ferrite cores will be used. (The TDK H5B2 material, or similar, is very suitable.) For high-frequency ac measurements, low-loss ferrite cores will be used. For ac measurements on lines carrying both ac and DC, a lower-permeability material should be chosen so that the DC current component can be tolerated without causing core saturation. For this application, Permalloy or low-permeability ferrite, such as Siemens #N30 or similar, is more suitable. The design and construction of these current transformers and probes will now be considered in more detail.

15.3 THE DESIGN OF CURRENT PROBES FOR UNIDIRECTIONAL (DISCONTINUOUS) CURRENT PULSE MEASUREMENTS

Very often there will be a diode in series with the line to be measured (for example, the output winding of a flyback converter). The current pulses in such a line are unidirectional and will have a mean DC component. If normal ac current transformers are used in this application, the core will often saturate, and the transformer will give misleading output information. Moreover, even if the transformer does not saturate, the restoration of the core during the quiescent period will give misleading results (an apparent reversal of current during the quiescent period).

Note: This effect occurs when the Tektronix current probe #P6021 is used, since it is designed for ac conditions only.

Hence, for unidirectional current pulse measurement, a special current probe is required that will respond correctly to the unidirectional pulse and will not saturate. Figure 3.15.1a shows a suitable circuit for a unidirectional current probe. This probe will faithfully reproduce a unidirectional current pulse, as shown in Fig. 3.15.1b. It operates in the following way.

Unidirectional pulses I_P will pass through the toroid in the direction shown in the diagram. By current transformer action, a secondary current I_s of magnitude I_P/N_s will flow through diode D1 into burden resistor R1. A voltage analogue of the primary forward current will be developed across R1. The value of R1 is selected for the required oscilloscope scaling factor.

Note: The voltage across the series network D1, R1 should be small, and the turns ratio large, so that the magnetizing current and insertion loss will be small. (The voltage across the single primary turn will then be very small.)

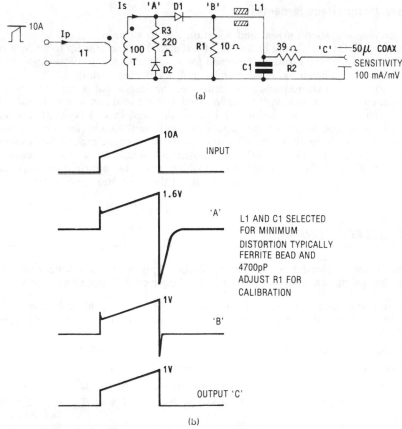

(a)

(b)

FIG. 3.15.1 (a) Low-cost oscilloscope current probe for high-frequency unidirectional current pulse measurement. (b) Input current waveforms and output voltage analogues of oscilloscope current probes: Top, primary input current; A, voltage waveform, secondary; B, voltage waveform after rectification; C, voltage waveform after shaping.

During the quiescent ("off") period, rapid restoration of the core takes place by flyback action, because diode D1 will block and a large flyback voltage will be developed across R3 and D2 (the resistance of R3 being much higher than that of R1). The value of R3 is chosen to just prevent the flyback voltage from exceeding the voltage rating of D1. For fast recovery in high-current transformers, a voltage-dependent resistor or zener diode has some advantages in this position. Diode D2 will assist the recovery process at low currents. The recovery waveform is shown in Fig. 3.15.1(b) as waveform A.

Diode D1 blocks the majority of the reverse recovery voltage from the oscilloscope input, as shown in waveform B of Fig. 3.15.1(b). Padding and matching components R2, C1, and L1 will match the low-impedance output from R1 to the coaxial input impedance. The padding components will also reduce the spike introduced by the recovered charge in D1 during the switching edges, so that a true voltage analog of the primary forward current pulse is indicated on the CRO, as shown in waveform C.

General Design Requirements

To keep magnetization current and insertion loss to a minimum, the secondary voltage across D1, R1 should be as low as possible. For currents up to 20 A, it is recommended that the probe be designed for a scaling factor of 10 mA/mV. For higher-current applications, a value of 100 mA/mV is more suitable.

For extended high-frequency performance, the interwinding capacitance of the secondary should be low. It is recommended that the secondary turns do not exceed 100 and that this winding be in a single, well-spaced layer. The core should be insulated to reduce the capacitance between winding and core, since this will also act to increase the effective interwinding capacitance. Fast diodes should be chosen for D1 and D2, and resistor R1 should be a low-inductance type. A small ferrite bead is fitted in L1 position, and C1 is adjusted to minimize edge distortion; these should be selected during the calibration procedure.

15.4 SELECT CORE SIZE

The core size is selected to suit the application. Larger currents, longer pulses, and higher probe sensitivity will all require a larger core. Proceed as follows:

1. Assume the number of secondary turns will be 100 turns. (This choice is made to keep interwinding capacitance to a low value by allowing a single-layer winding.)
2. Choose the required maximum current range—say, 0 to 100 A.
3. Choose the required sensitivity—say, 10 mV/A (1 V/100 A).
4. Choose the maximum range of pulse widths—say, 0 to 30 μs.
5. Select a working value for the flux density B—say, 10% of B_{sat} at 20°C = 40 mT.

   Note: A very low flux density is used (10% or less of B_{sat}) to give low loss and low magnetization currents.

6. Calculate the maximum secondary voltage required from the transformer as follows:

$$V_s = V_{out} + \text{diode D1 voltage drop}$$

In this example the maximum output voltage at 100 A is 1 V, and the diode drop is 0.6 V. Therefore $V_s = 1.6$ V.

15.5 CALCULATE REQUIRED CORE AREA

From Faraday's law,

$$A_e = \frac{V_s \cdot t}{B \cdot N} = \frac{1.6 \times 30}{0.04 \times 100} = 12 \text{ mm}^2$$

where A_e = effective core area, mm^2
V_s = secondary voltage
t = pulse duration (max), μs
B = change in flux density, T
N = secondary turns

From Table 2.15.1, the nearest core with an area of 11.8 mm^2 is the TDK T7-14-3.5, and so this is chosen.

The maximum flux density was chosen at only 10% of the saturating value. This should result in a low magnetization current and also yield a convenient core size for the chosen current. (A 100-A primary wire would not pass through a much smaller toroid.)

The actual value of the magnetization current will depend on the permeability of the core material and the change in flux density. To maintain the probe accuracy within the 5% tolerance, the magnetization current should be less than 5% of the measured current; therefore, it is necessary to check the magnetization current for the selected core.

The magnetization current may be obtained from the B/H characteristic or calculated from the primary inductance value. In this example, the core A_L factor is available, and the inductance approach provides a quick solution.

15.6 CHECK MAGNETIZATION CURRENT ERROR

The chosen core is the TDK T7-14-3.5. A low-loss, high-permeability material is required, and H5B2 or a similar material would be suitable. (See Part 2, Chap. 15.) From the manufacturer's data (Table 2.15.1), the cross-sectional area of the toroid is 11.8 mm^2, and the core A_L factor using H5B2 material is 3.5 μH. The primary is only one turn; hence the primary inductance will be 3.5 μH.

The slope of the magnetization current during the pulse, dI/dt, can be calculated:

$$V_p = -\frac{L \, dI}{dt}$$

where V_p = primary voltage drop (1.6/100)

L = primary inductance (3 μH)

Hence

$$\frac{dI}{dt} = \frac{0.016}{3 \times 10^{-6}} = 5.33 \times 10^3 \text{ A/s}$$

At the end of the 30-μs period, the primary magnetizing current is 160 mA, a negligible error at currents in excess of 10 A.

Calibration

Assuming that 100 turns will be used on the transformer secondary, the value of R1 may be calculated as follows:

$$R1 = S_f \cdot N_s$$

where S_f = scaling factor

N_s = secondary turns

Note: The scaling factor depends on the oscilloscope sensitivity required—say, 10 mA/mV = 0.1 V/A = S_f of 0.1 or 100 mA/mV = 0.01 V/A = S_f of 0.01.

Hence, for current probes up to 20 A, R1 will be 10 Ω, and for higher currents, R1 will be 1 Ω. The final values will be adjusted on calibration.

For calibration purposes, the probe should be placed in series with the Tektronix probe #P6021 or a similar probe in an ac pulse generating circuit. To prevent saturation, the pulse current should be kept below 10 A and the period 20 μs or less. The waveforms from the two probes should be displayed on a double-beam scope; small adjustments can be made to R1 to adjust the measured amplitude as required. The padding components L1 (which is usually one or two 30-nH ferrite beads) and C1 are adjusted for best transient performance. For a final check, the two waveforms should be superimposed and the inputs to the oscilloscope amplifiers changed over, to ensure that there are no channel difference errors in the CRO.

This type of current probe is suitable for measuring currents which are unidirectional *and discontinuous*. A nonconducting period of sufficient duration (10%) is essential to allow the core to be reset to its remanence value B_r before the next forward pulse. Otherwise core saturation can occur.

15.7 CURRENT PROBES IN APPLICATIONS WITH DC AND AC CURRENTS

To observe currents in applications in which the current is unidirectional and continuous (always flowing in the forward direction)—for example, the ripple current in a buck regulator inductor—a low-inductance resistive shunt or Hall effect probe should be used. If it is only necessary to view the ac component of current, then a special current transformer can be used.

For this application, a different type of current transformer which can tolerate the DC current component is required. To prevent core saturation in this application, a low-permeability material such as MPP must be used. Alternatively, an air gap could be introduced into the magnetic path.

The main disadvantage of using a low-permeability core is that a large number of turns will be required on the secondary if a reasonable ratio of measured current to magnetization current is to be obtained. A large number of secondary turns will have considerable interwinding capacitance, and the high-frequency cutoff point will be lower than in the previous example.

Note: For this discussion, the cutoff frequency is that frequency at which the transfer ratio of the probe has fallen by 5% of the midband figure.

Hence these DC current tolerant types of current probes are usually designed for a specific narrow band of frequencies, and will cover a limited DC polarization current component. The design process is similar to that above, except that the lower permeability value would be used. Also, the DC saturation current must be established as shown in Sec. 3.2.1.

15.8 HIGH-FREQUENCY AC CURRENT PROBES

Figure 3.15.2a shows a typical circuit for an ac current probe. The design is very similar to that of the unidirectional pulse probe, except that the rectifier diode is omitted. The choice of transformer material is somewhat different, since it is better that the probe does not saturate if by chance a small DC current component is present in the primary. Consequently, a lower-permeability material is usually chosen. In other respects, the design approach is very similar to that used in the example in Sec. 15.3, and a single-layer winding of approximately 100 turns will be found satisfactory. The lower permeability of the core material results in smaller inductance, and consequently the low-frequency cutoff point will be somewhat higher than in the previous example. The high-frequency cutoff point will be unchanged, and the probe will have a narrower bandwidth.

Current probes should be used with care and some knowledge of the conditions. If a large continuous DC current component is present, then the current transformer may saturate, and output information will be lost.

15.9 LOW-FREQUENCY AC CURRENT PROBES

Low-frequency probes are required for line inrush and reservoir capacitor ripple current measurements. The frequency here would be in the range 50 to 120 Hz.

For this application a much larger iron core would be used, and the secondary would have a large number of turns, say 1000 turns or more. The requirement is to make the inductance large enough to minimize the magnetization current and reduce the measurement error. The larger interwinding capacitance will not be a problem because the frequency is so low.

The probe circuit would be the same as that shown in Fig. 3.15.2.

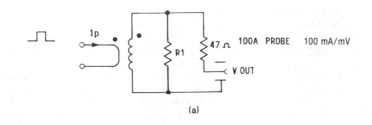

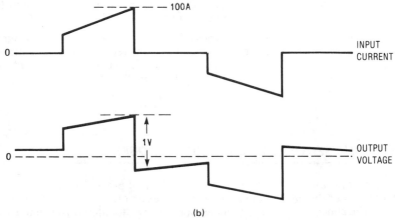

FIG. 3.15.2 (*a*) High-current ac input current probe circuit. (*b*) Input current waveform and output voltage waveform of ac current probe, showing the effect of current transformer magnetization current on low-frequency performance.

15.10 PROBLEMS

1. What is the basic function of the current transformer?
2. Explain how a current transformer is essentially different from a voltage transformer.
3. What is the function of a unidirectional current transformer?
4. Under what conditions can an ac current transformer be used?
5. What is the basic function of a flyback-type current transformer?
6. Explain the action of a DC current transformer.
7. Why are high-permeability core materials normally used for current transformers?
8. Why are relatively simple current transformers used in switchmode applications?
9. Where would flyback current transformers normally be used?

CHAPTER 16
THERMAL MANAGEMENT
IN SWITCHMODE POWER SUPPLIES

Note: The imperial system of units is used in this section because most thermal information is still presented in this form. Temperatures are in degrees Celsius, except for the radiant heat calculations, where the absolute Kelvin temperature scale (starting at −273°C) is used. Dimensions are in inches (1 in = 25.4 mm).

16.1 INTRODUCTION

It is well known that to obtain maximum reliability in any electronic equipment, and in particular in switchmode supplies, good management of the cooling requirements is essential. However, although this need is well known, the thermal design still tends to get less attention from both user and designer than the electrical design. This is unfortunate, since high component temperatures are undoubtedly one of the major causes of premature failure in switchmode systems.

The prudent engineer will keep the need for effective and efficient cooling very much in mind throughout the design process. The layout, size, shape, selection of components, and mechanical and enclosure design, together with the complex thermal interaction of the various parts of the circuit and other nearby equipment, are intimately linked to the thermal performance and long-term reliability.

Some of the more basic thermal design decisions must be made before the electrical design is started, because the whole design approach may depend on the type of cooling to be employed. For example, is the system to be contact cooled (requiring a machined flat thermal interface with thermal shunts)? Or is it forced-air cooled (requiring large-surface-area heat exchangers)? Is free air convection cooling required? What is the temperature range? If it is air cooled, to what altitude? The answer to these basic thermal questions, and many that are more complex, will clearly control the general mechanical and electrical design of the system.

A brief examination of the effect of temperature on the predicted failure rate for semiconductors will indicate the general effect of higher temperatures on long-term reliability.

16.2 THE EFFECT OF HIGH TEMPERATURES ON SEMICONDUCTOR LIFE AND POWER SUPPLY FAILURE RATES

Semiconductor failure rates have been very well established by many years of testing and are found to be temperature-dependent. Figure 3.16.1 is a reproduction of a graph from MIL-HDBK 217 A showing the predicted failure rate for silicon NPN transistors as the temperature is increased above 25°C, compared with the failure rate expected at 25°C.

Although the graph is a statistical prediction that is specific to silicon NPN

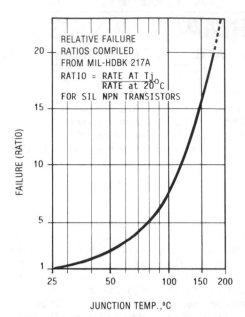

FIG. 3.16.1 Relative failure ratios of NPN silicon semiconductors as a function of temperature.
(From MIL-HDBK 217A.)

transistors, it indicates the general trend for most electronic components in that
the failure rate increases rapidly with increasing temperature. The effect of high
temperatures is dramatic.

For example, a transistor operated at a junction temperature of 180°C has only
$\frac{1}{20}$ the life of a transistor operated at 25°C, or is twenty times more likely to fail in
a given time.

Clearly many different types of components are involved in a complete power
supply system, and all play a part in the overall reliability. A complete power
supply will have a much higher failure rate than a single transistor because of the
large number of components involved and because some components (for example, electrolytic capacitors) are more sensitive to high temperatures. In most
cases, the electrolytic capacitors, or even the wound components, will limit the
maximum permitted operating temperature to values well below the 200°C shown
for the transistors.

The intrinsic failure prediction depends on the number of components, stress
factors, and temperature. However, in general, the failure rate for a typical
SMPS doubles for each 10 to 15°C temperature rise above 25°C. Hence a typical
unit operating at 70°C will have a predicted MTBF (mean time between failures)
of approximately 10% of its MTBF at component temperatures of 25°C.

From the above, it will be clear that good thermal design and low operating
temperatures are crucial to long-term reliability. Although the absolute maximum
temperature rating for silicon semiconductors may be near 200°C, much better
long-term reliability will be obtained if the junctions are operated at much lower
temperatures. The same general rules apply to other components.

16.3 THE INFINITE HEAT SINK, HEAT EXCHANGERS, THERMAL SHUNTS, AND THEIR ELECTRICAL ANALOGUES

16.3.1 Infinite Heat Sink

Although the term "heat sink" is often applied to the familiar finned aluminum extrusions, this term more correctly applies to the heat sinking medium, often free air, to which the heat is to be finally transferred. For all practical purposes, it is assumed that this near infinite medium will not change temperature (from its ambient value), regardless of how much heat energy is transferred to it. Since the medium has nearly infinite thermal capacity, it becomes an infinite "heat sink" (it can drain away as much heat energy as we require without changing temperature).

The electrical analogue would be ground (its potential remains constant).

16.3.2 Heat Exchangers

As the name implies, heat exchangers pass heat from one medium to another—for example, from metal to air. The familiar finned metal extrusions are one example of a heat exchanger. Other examples include vehicle "radiators," heat pipes, and liquid-cooled "heat sinks." As you can see, the popular terms are often misleading.

In general, the quality of a heat exchanger should be gauged by the efficiency of its heat exchange process rather than by its intrinsic thermal resistance. Hence fin design, surface area, and finish all play a part in the overall performance. The thermal efficiency of the interface is defined in units of thermal resistance.

The electrical analogue is resistance.

16.3.3 Thermal Shunts

Thermal shunts, or heat conductors, provide a low-thermal-resistance heat transfer path between the "hot spot" (often the junction in a semiconductor) and the heat exchanger. Thermal shunts include all the things between the hot spot and the interface of the heat exchanger. This includes the mounting of the semiconductor, any insulation material, and the body of the heat exchanger. The heat exchanger, of which the finned aluminum extrusion is an example, is part of this heat transfer path. However, to be effective, the heat exchanger requires the final interface to the "infinite free air heat sink." It is this last requirement that is sometimes overlooked, with disastrous results.

The electrical analogue of the thermal shunt (conductor) is electrical conductor. (The resistance is in electrical ohms.)

Note: A common misconception: It should be clearly understood that making the chassis, heat sink, or heat exchanger thicker or larger or of better material (such as copper) will not help if the heat transfer to the final infinite (free air) heat sink is not improved.

Larger heat exchangers just take longer to reach their final, but same, temperature if the heat energy is not carried away. Hence a good air flow (water flow, heat pipe, heat exchanger surface, etc.) is required if the thermal path is to be complete. This may be better understood by considering the electrical analogue of the thermal circuit.

16.4 THE THERMAL CIRCUIT AND EQUIVALENT ELECTRICAL ANALOGUE

Figure 3.16.2a shows a typical cooling problem: a rectifier diode mounted on a heat exchanger with an insulator between the case of the diode and the heat exchanger. In this example, the temperature of the junction of the diode, when it is mounted on the heat exchanger in a free air cooling environment, is to be calculated,

Figure 3.16.2b shows the thermal circuit, and Fig. 3.16.2c and d shows its electrical analogue. The circuit design engineer will probably prefer to use the electrical analogue, but before this can be done the analogue conversions must be considered.

16.4.1 Thermal Units and Equivalent Electrical Analogue

Thermal Unit Parameter	Units	Electrical Analogue	Units
Time t	s	Time t	s
Temperature difference T_d	°C	Potential difference P_d	V
Thermal resistance* R_θ	°C/W	Resistance R	Ω
Thermal conductivity* K	W/°C	Electrical Conductivity	S
Heat energy P_q	J	Electric energy P	J
Heat flow Q	J/s (W)	Current I	A
Heat capacity* C_h	J/°C	Capacitance C	F

* Of item or interface

16.4.2 Heat Generator (Analogous to Constant-Current Generator)

Consider Fig. 3.16.2b. At the left-hand side, heat is being generated in the junction of the diode at a constant rate of 10 J/s (10 W). Under steady-state conditions (when thermal equilibrium has been established), the temperature of the junction is constant, and the heat flowing away from the junction must equal the heat being generated at the junction. The temperature will continue to rise until this state of equilibrium has been established. Hence, the constant heat generator is analogous to the constant-current generator shown in Fig. 3.16.2c, an important analogy.

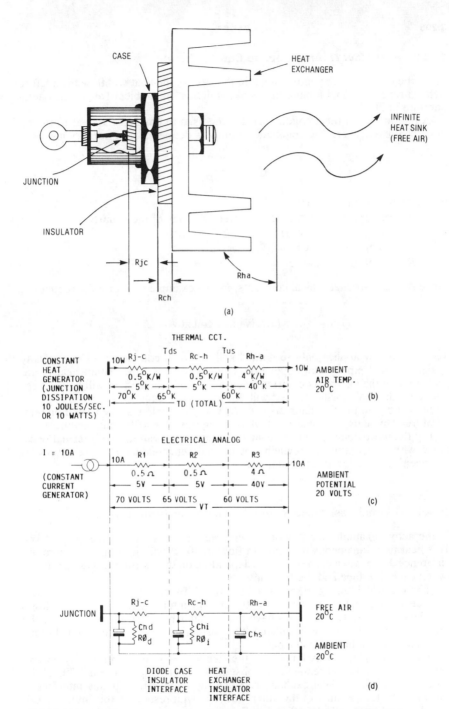

FIG. 3.16.2 (*a*) Thermal resistance example: stud-mounted D04 diode on finned heat exchanger. (*b*) Example of thermal resistance circuit. (*c*) Example of electrical analogue of thermal resistance circuit. (*d*) Thermal resistance model, showing the effect of thermal capacity (specific heat of heat shunts and heat sinks) and local heat loss.

16.4.3   Heat Flow Q (Analogous to Current Flow)

Since the junction is the hottest point in the circuit, the heat will flow from left to right to reach the final infinite heat sink, which is the ambient free air at a temperature of 20°C.

A thermal shunt (heat conductor) is being used to conduct the heat to the remote heat exchanger. The conduction rate Q is defined from Fourier's law as

$$Q = \frac{A \times T_d}{L \times R_\theta}$$

where Q = heat flow, J/s (W)

$\quad T_d$ = temperature difference between the ends of the shunt, °C

$\quad A$ = cross-sectional area

$\quad L$ = length of conductor

$\quad R_\theta$ = thermal resistance

Since A and L are mechanical constants, in this example this formulia reduces to

$$Q \propto \frac{T_d}{R_\theta} \qquad \text{which is analogous to} \qquad I = \frac{V}{R}$$

Note: This law applies only to normal solid thermal conductors. The specially designed "heat pipes" that depend on a change of state (i.e., latent heat of vaporization of the internal coolant) for their heat conduction effect will have a very nonlinear thermal resistance, and will not follow this equation.

In heat pipes, the thermal resistance R_θ will go very low at the transition temperature. This must be considered when using this type of thermal shunt.

For the more commonly used heat sink metals, the variation of thermal resistance with temperature is negligible at normal semiconductor temperatures. It has been neglected in these examples.

16.4.4   Thermal Resistance R_θ (Analogous to Resistance R)

In the above example, the junction is dissipating 10 J/s (and hence Q = 10 W). This heat flow (analogous to a current flow of 10 A) will develop a temperature difference T_d between each interface, depending on the thermal resistance R_e between each interface and the heat flow.

(The electrical analogue shows a potential difference V between each interface, depending on the resistance R between each interface and the current flow.)

When steady-state conditions have been established, the temperatures at the various interfaces may be calculated by considering the heat flow and thermal resistances in the heat transfer path.

In this example, it is assumed that the free air, by virtue of its nearly infinite bulk and free flow, will remain at a constant ambient temperature of 20°C at the surface of the finned heat exchanger. Since the temperature at this interface is constant, the temperature of the other junctions with respect to this interface can be calculated from right to left in Fig. 3.16.2b.

(The electrical analogue of a free air temperature of 20°C is a ground potential of 20 V, in this example.)

There are three thermal resistors to be considered in Fig. 3.16.2b. First (and usually the most important, because it is the largest), there is the thermal resistance of the free air interface itself, that is, from the finned heat exchanger surface to the surrounding free air. (This is designated R_{h-a}.)

Second, there is the thermal resistance from the finned heat exchanger surface, through the mica insulator, to the case of the diode. (This is designated R_{c-h}.)

Finally, there is the thermal resistance from the case of the diode to the internal junction. (This is designated R_{j-c}.)

(The electrical analogue shows resistors R3, R2, and R1 in the same positions.)

For convenience, the thermal resistance of each section will be considered separately, starting with the heat exchanger interface.

From the manufacturer's data, the finned heat exchanger has a thermal resistance R_{h-a} of 4°C/W in free air.

The diode is mounted on an insulator to provide electrical isolation. This mica insulator also has a defined thermal resistance between the diode case and the mounting surface of the heat sink R_{c-h} of 0.5°C/W.

Finally, from the diode mounting surface to the internal junction (where the heat is being generated), the resistance R_{j-c} is given in the manufacturer's data as 0.5°C/W for this diode.

(The electrical analogues are 4 Ω, 0.5 Ω, and 0.5 Ω.)

Hence, the total thermal resistance R_θ from junction to free air is the sum of these three, or $4 + 0.5 + 0.5 = 5$°C/W (or 5 Ω), and this total resistance (R_θ) is used to calculate the total temperature difference T between the junction and free air.

From the previous equation,

$$T = Q \times R_\theta = 10 \times 5 = 50°C$$

where T = temperature rise (above ambient), °C

Q = dissipation in junction, W

R_θ = total thermal resistance, junction to free air, °C/W

Electrical analogue:

$$V = I \times R = 10 \times 5 = 50 \text{ V}$$

Since T is the temperature rise above ambient, the junction temperature will be 70°C, and the analogous voltage would be 70 V.

Clearly the electrical analogue is hardly necessary in this simple example; however, it serves to demonstrate the principle and will be found very useful in more complex applications. The engineer will make few errors in thermal design if this simple model is kept in focus.

16.5 HEAT CAPACITY C_h (ANALOGOUS TO CAPACITANCE C

The concept of heat capacity tends to get little attention in thermal design, although it is significant in magnitude. It is the confusion between thermal capacity (specific heat) and true thermal resistance that leads to a common error. It is often assumed that a copper heat exchanger will perform better than, say, an aluminum heat exchanger with the same surface area. This error stems from the fact that the copper does not appear to get hot as quickly as the aluminum. In fact, what is being observed here is the effect of the increased thermal capacity of the copper. The copper heat exchanger will eventually end up at the same temperature. (Although copper is a better heat conductor, it is the surface area which predominates and defines the thermal resistance.)

This effect will become clear from the more complete model shown in Fig. 3.16.2d. The various thermal capacitors, C_{hd}, C_{hi}, and C_{hs}, have been included, together with the previously neglected direct heat losses from the surface of the various bodies, $R_{\theta d}$ and $R_{\theta i}$.

The heat losses from the surface of the components, $R_{\theta d}$ and $R_{\theta i}$, are normally neglected, as the direct heat loss is negligible because of the small exposed area of the diode and insulator. However, this is not the case with the thermal capacitors C_{hd}, C_{hi}, and C_{hs}. In the example shown, the electrical analogue capacity will effectively be hundreds of farads. (Even at 10 W input it can take several minutes for the heat exchanger to reach final thermal stability.)

From Table 3.16.1, it will be noted that the heat capacity of common heat conductors can be very large (for example, 57.5 J/°C for a 1-in copper cube). Hence, for the example shown in Fig. 3.16.2, if 10 in³ of copper were used in the construction of the heat exchanger (quite realistic), then with a heat input of 10 W (10 J/s), it would take 57 s for the temperature to increase by only one degree. Thus, it would take several minutes to reach the final temperature. The heat sink's thermal mass (thermal capacitance) will *not* affect the value of the steady-state temperature, only the time taken to reach thermal stability.

However, if the heat input is of a transient nature, with a small duty ratio (al-

TABLE 3.16.1 Heat Storage Capacity and Thermal Resistance of Common Heat Exchanger Metals

Common heat sink materials	Heat storage capacity, J/in³/°C	Thermal resistance (block 1″ × 1″), R_θ,°C/W
Aluminum (6061)	40.5	0.23
Copper 110	57.5	0.10
Steel C1040	63	0.84
Brass 360	50	0.34

lowing plenty of cooling time), then the larger thermal capacity (or greater specific heat) will be effective in reducing the maximum variation in temperature during a thermal load transient. Since thermal capacity will not affect the final steady-state temperature, it is not considered further in this example.

16.6 CALCULATING JUNCTION TEMPERATURE

In the previous example, the diode junction temperature was easily established because the dissipation was known. However, in practice the dissipation in switchmode applications can often be very difficult to establish, as some factors, like diode reverse recovery losses, can be difficult to establish with any real degree of confidence. Under these conditions, any known thermal resistance in the heat conduction path can be used to establish the heat flow (and hence the junction dissipation) by measuring the temperature differential across the interface of the known thermal resistance.

Consider again the electrical analogue shown in Fig. 3.16.2c. In the same way that the voltage difference between two parts of a circuit is given by $I \times R$, the temperature difference is given by the product of heat flow in joules per second (watts) and thermal resistance. For the example shown in Fig. 3.16.2b and c, the heat flow is known, and the temperature difference for each element of the thermal shunt can be calculated as follows:

$$\Delta T = W_j \times R_\theta$$

where ΔT = temperature difference (across element)

$\quad W_j$ = heat flow (power dissipated at the junction)

$\quad R_\theta$ = thermal resistance (of element)

The temperatures at the various interfaces may be calculated as follows:
Temperature of heat sink surface T_h:

$$T_h = (W_j \times R_{h-a}) + T_{amb} = (10 \times 4) + 20 = 60°C$$

where R_{h-a} = heat exchanger to ambient thermal resistance

$\quad T_{amb}$ = ambient air temperature, °C

Temperature of diode surface T_{ds}:

$$T_{ds} = [W_j \times (R_{h-a} + R_{c-h})] + T_{amb} = [10 \times (4 + 0.5)] + 20 = 65°C$$

where R_{c-h} = thermal resistance from device case to heat exchanger surface

The junction temperature T_j will be the total temperature difference T across all the various series heat shunt elements, plus the ambient temperature; hence

$$T_j = [W_j \times (R_{h-a} + R_{c-h} + R_{j-c})] + T_{amb} = [10 \times (4 + 0.5 + 0.5)] + 20$$

$$= 70°C$$

It has been shown that if the power dissipated in the junction and the thermal resistance to the heat shunts or heat exchanger are known, then the junction and interface temperatures can be calculated. Clearly, if the temperature of the heat exchanger is measured and the thermal resistance is known, then the heat flow and junction dissipation can be calculated.

16.7 CALCULATING THE HEAT SINK SIZE

In many practical cases the power dissipated in the junction will be known, and the thermal resistance of the heat exchanger will need to be calculated for a defined junction temperature rise. The design procedure would be as follows.

Assume that a finned heat exchanger as shown in Fig. 3.16.3 is to be used to free-air-cool a T0-3 transistor dissipating 20 W, and that the junction temperature is not to exceed 136°C when the ambient air temperature is 50°C.

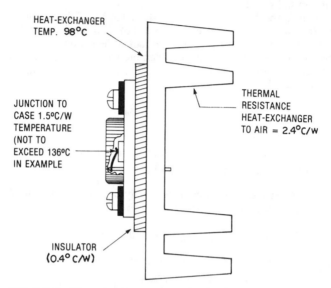

FIG. 3.16.3 Thermal resistance example, showing a T03 transistor on a finned heat exchanger.

From the manufacturer's data, the thermal resistance between junction and case of the T0-3 transistor, R_{j-c}, is 1.5°C/W. An insulating mica washer is also to be used, and this has a thermal resistance of, say, 0.4°C/W. (The thermal resistance of the insulator may be established from the basic material properties in Table 3.16.2 or from Table 3.16.3.)

The maximum temperature permitted at the interface of the insulator and the heat sink when the junction temperature is 136°C can be calculated as follows:

The thermal resistance R_{j-h} from junction to insulator–heat exchanger interface is

$$R_{j-h} = R_{j-c} + R_{c-h} = 1.5 + 0.4 = 1.9°C/W$$

The temperature difference ΔT between the junction and the heat exchanger interface is the product of the thermal resistance and heat flow Q:

$$\Delta T = R_{j-h} \times Q = 1.9 \times 20 = 38°C$$

The temperature at the heat exchanger interface T_h will be $T_{j(max)}$, less the difference ΔT from junction to heat exchanger:

$$T_h = T_{max} - \Delta T = 136 - 38 = 98°C$$

TABLE 3.16.2 Thermal Resistance, Maximum Operation Temperatures, and Dielectric Constant of Common Insulating Materials

Common insulating materials	Thermal resistance (block 1″ × 1″), °C/W	Maximum temperature, °C	Dielectric constant, 25°C
Mica	62–91	550	6.5–8.7
Aluminum oxide	1.43	1700	8.9
Beryllium oxide*	0.15–0.27	2149	6.5
Polyimide plastic	270	400	3.5
Silicone rubber	151	180	1.6
Thermal epoxy	25–50	90	6
Still air	1430		1

*Warning: Beryllium oxide is highly toxic if fragmented into small particles.

TABLE 3.16.3 Typical Thermal Resistance of Case to Mounting Surface of T0-3 and T0-220 Transistors When Using Standard Insulator Kits and Materials with Thermal Mounting Compound

Standard insulator kits	Device type, case	Insulator thickness, in	Typical thermal resistance R_{c-h}, °C/W	Maximum working temperature, °C
Mica	T0-3	0.006	0.4	> 200
	T0-220	0.006	1.8	> 200
Aluminium oxide	T0-3	0.062	0.34	> 200
	T0-220	0.062	1.53	> 200
Beryllium oxide	T0-3	0.062	0.2	> 200
	T0-220	0.062	1.0	> 200
Polyimide plastic (Thermofilm)	T0-3	0.002	0.55	> 200
	T0-220	0.002	2.3	> 200
Silicone rubber	T0-3	0.008	1.0	180
	T0-220	0.008	4.5	180

The maximum permitted temperature difference ΔT_h from the heat exchanger surface to free air at 50°C is

$$\Delta T_h = T_h - T_{amb} = 98 - 50 = 48°C$$

The thermal resistance of the heat exchanger, R_{ha}, is the temperature difference divided by the heat flow:

$$R_{ha} = \frac{\Delta T_h}{Q} = \frac{48}{20} = 2.4°C/W$$

Hence a heat exchanger extrusion of 2.4°C/W will be chosen. The manufacturer's data provide information on the heat exchanger thermal resistance for various extrusions or heat exchanger designs, and a suitable size can be calculated.

16.8 METHODS OF OPTIMIZING THERMAL CONDUCTIVITY PATHS, AND WHERE TO USE "THERMAL CONDUCTIVE JOINT COMPOUND"

In the example shown in Fig. 3.16.3, the largest thermal resistance is from the heat sink to free air, R_{h-a}. (This will often be the case with convected cooling.) Since the total thermal resistance of the heat shunt from the junction to free air is the sum of the various elements, this final large thermal resistance swamps the effects of all the others. For example, a 50% increase or decrease in the resistance of the mounting arrangements would affect the temperature at the junction by only 2.5°C. Hence, in this example, there would be little advantage to using thermal compound to reduce the thermal resistance of the mounting arrangements—the effect would be negligible.

It is interesting to note from the above example that the messy (and expensive) practice of using thermal mounting compound on small air-cooled heat exchangers is probably not very effective in most cases.

The designer should locate the interface with maximum thermal resistance and reduce this to a value compatible with the other elements in the path. In the above example, a large improvement would come from an increase in the heat exchanger surface area or an increase in cooling air flow, but not very much from a reduced mounting interface resistance.

In the second example, Fig. 3.16.4a, a highly dissipating transistor (for example, an active load) is to be mounted on an efficient water-cooled heat exchanger. This heat exchanger can be considered an infinite heat sink. (For practical purposes, it may be assumed that the surface temperature of the heat exchanger will not exceed 20°C regardless of how much heat is conducted to it.)

Assume that the transistor dissipation is 100 W. The equivalent thermal resistance diagram is shown in Fig. 3.16.4b. In this example, the junction-to-case thermal resistance is 0.5°C/W, and the case-to-heat-sink thermal resistance (because an insulator is used) is higher, 1°C/W. (Since the heat exchanger in this example is the infinite heat sink, its thermal resistance is zero.)

With 100 W dissipated, the temperature drop across the insulator will be 100°C, giving a case temperature of 120°C. The temperature rise within the transistor from case to junction will be 50°C, giving a junction temperature of 170°C.

In this example, there would indeed be a considerable reduction in tempera-

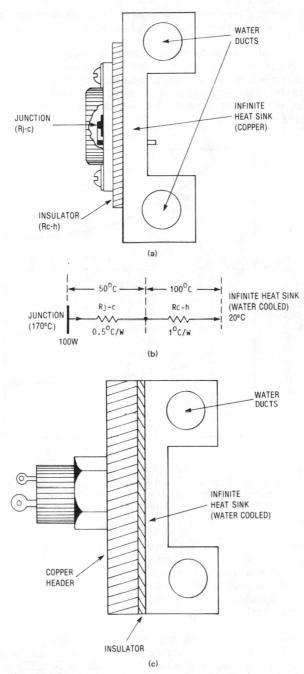

FIG. 3.16.4 (*a*) Thermal resistance example, showing a T0-3 transistor mounted on a water-cooled (near-infinite) heat sink. (*b*) Equivalent thermal resistance circuit model for T0-3 transistor on ''near infinite'' heat sink. (*c*) T0-111 stud transistor on ''near infinite'' heat sink, with a copper header.

ture at the junction if the thermal properties of the mounting arrangement were improved. (The insulator now has the highest thermal resistance in the series chain.)

Figure 3.16.4c shows a suitable modification to reduce the mounting thermal resistance while retaining galvanic isolation. The transistor (now in a T0-59 case) is screwed directly into a copper block, and the copper block is then insulated from the heat sink. Thermal compound would be used on all interfaces to exclude any air voids, and the mounting screws should be tightened to the recommended torque (see Fig. 3.16.5).

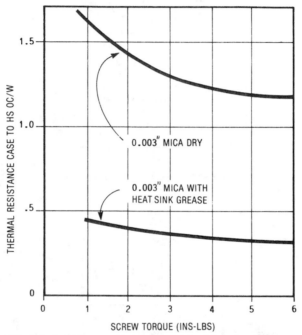

FIG. 3.16.5 Effective thermal resistance of interface between T0-3 transistor and heat sink when using a standard mica insulator, as a function of screw torque, with and without heat sink compound.

The area of the unavoidably high thermal resistance insulation interface is now 5 times greater than it was in the previous example, and the effective thermal resistance of the insulator interface is now only 0.2°C/W. Hence, for the same dissipation conditions and insulating material thickness, the junction temperature will now be 90°C, a considerable improvement. Alternatively, an insulating material of lower thermal resistance, such as beryllium oxide, may be used (see Table 3.16.2).

This example demonstrates the importance of identifying the point of maximum thermal resistance. This resistance is the one which should be reduced if effective improvement is to be obtained. (In any series circuit, it is the highest resistance that predominates.)

16.9 CONVECTION, RADIATION, OR CONDUCTION?

Of the three major heat exchanger mechanisms, convection and conduction are of major interest to the power supply designer. Radiation within the supply is generally a nuisance, as heat radiated away by one component is generally absorbed by adjacent components. Very often the power supply will be mounted inside the user's enclosure, together with the dissipating loads, where it will receive as much (or even more) radiant energy as it gives off. Consequently, the radiation properties of the power supply are often of little value.

16.9.1 Convection Cooling

If a free air flow is available, then convection, or forced air (fan), cooling is by far the most cost-effective way of removing unwanted heat. Figure 3.16.6 shows the dramatic improvement that will be obtained when forced air cooling is used. Heat exchangers with good forced air cooling properties will be chosen, and these usually have a large effective surface area as a result of having many cooling fins.

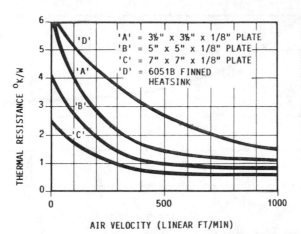

FIG. 3.16.6 Thermal resistance as a function of air velocity for various heat exchanger sizes.

Convection cooling becomes less effective at high altitude because of the reduction in the air density. Figure 3.16.7 shows this altitude effect. It should be noted that there is a 20% reduction in cooling efficiency at 10,000 ft.

Further, in nonforced convection cooling, the thermal resistance of the heat exchanger is not linearly proportional to the size. (Larger surfaces will not be as effectively cooled, as the air will be heated as it passes over the surface of the exchanger.) Figure 3.16.8 shows how the thermal resistance of a vertical finned extrusion varies with the length. Very little improvement is obtained beyond a length of 12 in.

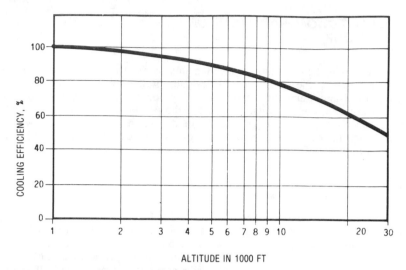

FIG. 3.16.7   Free air cooling efficiency as a function of altitude.

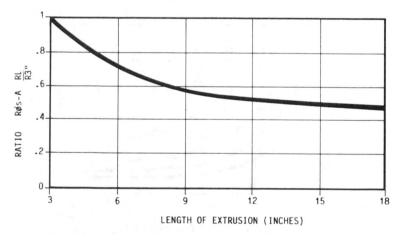

FIG. 3.16.8   Ratio of the thermal resistance of a 3-in length of finned heat sink extrusion to the resistance of longer lengths as a function of heat sink length, with the device mounted in the center.

16.9.2 Conduction Cooling

Where little air flow is available, conduction cooling is a viable alternative. For conduction cooling, the finned heat sinks will be replaced by thermal shunts (bridges) between the heat generating components and the chassis. The thermal conduction properties of the material chosen for the heat shunts will be important. (See Table 3.16.1.) The chassis, in turn, must have a good thermal contact to some external heat exchanger—for example, the case of the equipment.

For conduction cooling, it should be remembered that aluminum is only half as good a thermal conductor as copper, and steel is only 25% as good as aluminum.

Tables 3.16.1, 3.16.2, and 3.16.3 give the thermal properties of commonly used heat exchanger and insulating materials. The typical thermal resistance of T0-3 and T0-220 packages using standard mounting insulators and procedures is also shown.

16.9.3 Radiation Cooling

As previously explained, radiation is not usually a very effective method of cooling in switchmode supplies. Radiant heat is an electromagnetic wave phenomenon, and as such travels in straight lines, and good "line of sight" free radiant paths are not often provided in switchmode applications. Radiant energy from hot spots falling on the case or other components either gets reflected back or simply raises the temperature of the other components and the environment. However, when a good radiant path can be established, this mode of cooling should be considered.

For a perfect blackbody radiator, the Stefan-Boltzmann law states that the rate of energy radiated is proportional to the fourth power of the absolute temperature differential (in Kelvins) between the hot and cold bodies. The cold body in this case is the environment.

Hence, *where a good radiant path can be established,* radiation can provide a considerable proportion of the total heat loss, particularly for high-temperature components where the air flow is restricted. Under these conditions, the radiation properties (emissivity) of the heat exchanger surface become important.

From the Stefan-Boltzmann constant, the watt loss per second per square inch of radiating surface Q is

$$Q = e \times 36.77 \times 10^{-12} \times T^4$$

where Q = watt loss per second per square inch
e = emissivity of surface
T = temperature difference, K

The emissivity of a surface is the ratio of the surface's radiation properties to those of a true blackbody radiator. Table 3.16.4 shows the emissivity of some commonly used heat exchanger materials. It should be noted that the emissivity is dependent on the finish as well as on the type of material. Glossy surfaces will not be as good as matt surfaces. Further, since the radiation is in the infrared

TABLE 3.16.4 Typical Emissivity of Common Metals as a Function of Surface Finish and Color

Material	Surface finish and color	Typical emissivity e
True blackbody	True black	1.0
Aluminum	Polished	0.04
Aluminum	Painted (any color)	0.9
Aluminum	Rough	0.06
Aluminum	Matt anodized (any color)	0.8
Copper	Rolled bright	0.03
Steel	Plain	0.5
Steel	Painted (any color)	0.8

range, the color in the visual range is not important. For example, matt anodized aluminum has the same 0.8 emissivity value *in any color*.

If oil paints are to be used, the thickness, finish, and thermal properties of the paint should be considered. It may not always be possible to realize the full potential of the emissivity figures if the paint is thick or glossy.

Figure 3.16.9 shows the heat loss due to radiation alone from both sides of a 1-in-square plate in a free radiation environment of 293 K (20°C) compared with a true blackbody radiator. Notice that there is a considerable difference between the bright glossy aluminum and the matt painted surface.

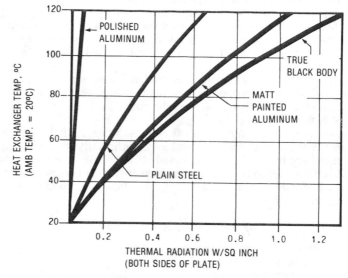

FIG. 3.16.9   Thermal radiation properties as a function of heat sink temperature differential, with surface finish as a parameter.

Poor radiators are also good reflectors, and this can be used to advantage in protecting heat-sensitive components from nearby hot components. For example, polished aluminum foil can be placed between a hot resistor and an electrolytic capacitor to reduce the radiant heating of the capacitor. Since the reverse is also true, good radiant heat dissipators are good heat absorbers, and power supplies designed for radiant cooling should be kept out of direct sunlight.

16.10   HEAT EXCHANGER EFFICIENCY

The heat exchanger designs available to the power supply designer are legion. Manufacturers claim various advantages for their designs, and there is a tendency to assume that a heavy sink with many fins or fingers must be more efficient.

When free air convected cooling is to be used, the difference between heat exchangers that *occupy the same effective overall volume* is in fact quite small, probably less than 10% in most cases. The reason for this is that the radiation from one fin or finger is picked up by the facing fin or finger, so that the effective

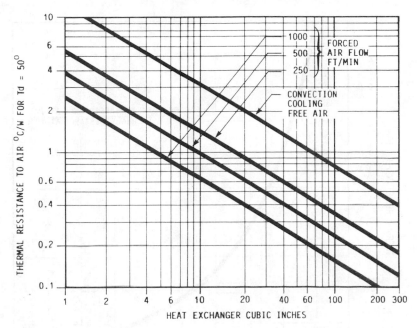

FIG. 3.16.10 Thermal resistance of finned heat exchangers as a function of enclosed heat exchanger volume, with air flow as a parameter.

radiation surface is given by the surface of the silhouette. Hence, the surface of a box of a volume that will just contain the heat sink will have similar heat exchanger and radiation properties in free air. For natural convection cooling, the air flow over the fins is such that the increased surface area is not very effectively utilized.

Figure 3.16.10 shows a log/log graph of the thermal resistance of various commercial heat exchangers plotted against the effective volume (developed from the published values) for a temperature rise of 50°C. Very few of the conventional heat exchanger designs deviate far from this result for natural convection cooling. (Some of the more expensive "pin fin" exchangers claim up to 30% improvement.) However, when forced-air cooling is used, the effect of the fins or fingers can be dramatic. Graph D in Figure 3.16.6 shows the improvement that can be obtained when a finned heat exchanger is used in a forced-air condition—the thermal resistance falls from more than 6°C/W to less than 1.5°C/W with an air flow of 1000 ft/min

16.11 THE EFFECT OF INPUT POWER ON THERMAL RESISTANCE

The effective thermal resistance from a heat exchanger to the ambient air is not constant, but will fall as the input power and hence the thermal differential increases. This is due both to the increase in radiated heat as the temperature rises (Stefan-Boltzmann law) and the increase in convection turbulence at the higher

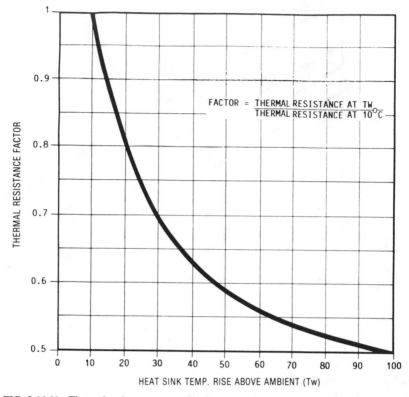

FIG. 3.16.11   Thermal resistance correction factor as a function of heat exchanger temperature differential (heat exchanger to free air).

temperatures. The quoted thermal resistance should be adjusted in accordance with the graph shown in Fig. 3.16.11 to get the effective thermal resistance at the working temperature.

16.12   THERMAL RESISTANCE AND HEAT EXCHANGER AREA

The thermal resistance does not fall in direct proportion to the surface area, as one might expect. This is due to the need to conduct the heat to the remote regions of the heat exchanger, which gives a temperature differential. Also, the air will be progressively heated as it passes over the surface of the heat exchanger. The graph shown in Fig. 3.16.12 gives a general indication for flat square metal plates with the heat input in the center and various surface finishes. As might be expected, the thermal resistance also changes with the attitude of the plate, and for this reason a finned extrusion should be used in the vertical plane for maximum efficiency in free air convected cooling conditions. However, the difference

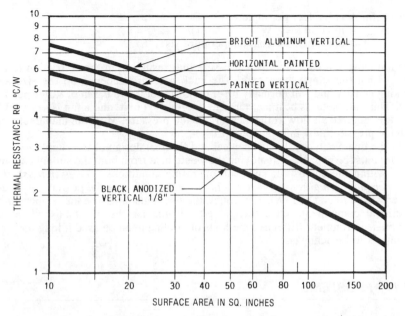

FIG. 3.16.12 Thermal resistance as a function of surface area (both sides of a ⅛-in-thick flat metal plate), with a dissipation of 20 W, in convection-cooled conditions, with surface finish and mounting plane (vertical or horizontal) as parameters.

in thermal resistance between horizontal and vertical fins is approximately 10%, not as large a difference as might be expected.

16.13 FORCED-AIR COOLING

In forced-air cooling, a fan or duct forces the air to flow through the supply enclosure at a much increased rate. There are many advantages to this method of cooling. Apart from the obvious improvement provided by the more rapid exchange of air, the flow direction can be used to advantage by placing the hot components in the exhaust port so that heat is carried away from the rest of the unit. Further, the air flow can be directed to prevent any static air buildup, an effect that is difficult to avoid in a free-flow convection-cooled supply.

Power supplies with outputs of 500 W or more will normally have some form of forced-air cooling.

The amount of air required for cooling depends on the air density, the power to be dissipated, and the permitted temperature rise. At sea level the following equation applies:

$$\text{Air flow (cfm)} = \frac{1.76 \times W_{\text{loss}}}{\Delta T}$$

where cfm = cubic feet per minute
$\quad W_{loss}$ = internal loss, W
$\quad\quad \Delta T$ = permitted internal temperature rise, °C

The fan must overcome the back pressure that results from the restriction to the flow within the supply enclosure. The back pressure depends on the size and packing density and will normally be of the order of 0.1 to 0.3 in of water per 100 cfm. This parameter is best measured in the finished unit, and a fan is selected to give the required air flow at the measured back pressure. Most fan manufacturers provide pressure/flow information.

In the final analysis, the mechanical and thermal design of the supply *and enclosure* must be capable of removing the waste heat from both the supply and any loads within the enclosure, under all operating conditions. *The user must also consider the thermal requirements;* otherwise the efforts of the power supply designer will be to no avail. When in doubt, the user must measure the temperatures of critical components once thermal equilibrium has been established in the working environment. Effective methods of cooling must be used if long-term reliability is to be achieved.

16.14 PROBLEMS

1. Why is thermal management so important in switchmode power supply design?
2. How is the MTBF related to operating and component temperatures?
3. When using an electrical model for thermal design, what would be the electrical analogue of a heat-generating transistor junction?
4. What is the electrical analogue of thermal resistance?
5. What is the electrical analogue of specific heat?
6. What is the electrical analogue of temperature differential?
7. What is the essential difference between convection cooling and radiation cooling?
8. Why is the common practice of using thermal compound on small printed circuit board mounted heat sinks generally of little value?
9. Is the thermal resistance a fixed parameter?
10. Is the thermal resistance of a heat exchanger directly proportional to its size?
11. Does the color of the heat exchanger affect its thermal properties?
12. In free air conditions, does the orientation of a finned heat exchanger have much bearing on its performance?
13. What is the effect of increasing altitude on heat exchanger performance?
14. Under convected-cooling conditions, is a heat exchanger with many closely spaced fins better than a heat exchanger of equivalent size with fewer fins?
15. Why is a small amount of forced air cooling so valuable for improving the life of switchmode power supplies?

P · A · R · T · 4

SUPPLEMENTARY

CHAPTER 1
ACTIVE POWER FACTOR CORRECTION

1.1   INTRODUCTION

The electrical utility companies produce electric power from various prime energy sources, typically coal, oil, gas, hydro, and nuclear power. Clearly, it is in everyone's interest that the conversion efficiency be as high as possible, to minimize thermal and other pollution problems.

Generating and distributing equipment uses rotating machinery, 60/50-Hz transformers, and power distribution lines, all of which are more efficient when the loads are purely resistive, taking sine wave currents in phase with the applied voltage and free of harmonics.

Reactive/resistive load combinations, although taking a sinusoidal current, introduce a phase shift between voltage and current, which reduces the efficiency of the distributing equipment. One has only to consider a purely capacitive load; this will take current from the line, which will cause copper loss in the distributing equipment, but produces no useful power at the load. A pure inductor would do the same thing. Inductive loads are common; a typical example would be the older type of magnetic ballast, where an inductor is used to control the current in fluorescent lighting applications.

Fortunately, the phase error caused by reactive loading can be corrected quite simply, using passive power factor correction. (In the case of an inductive load, a shunt capacitor of suitable value will correct the phase error and the combination will simulate a purely resistive load.)

More insidious problems are caused by nonlinear loads, which distort the current waveform and introduce harmonic currents in the supply and distribution system.

The harmonics (distortion) caused by nonlinear loads cannot be eliminated by the traditional passive power factor correction methods. If the harmonic currents are allowed to flow in the supply system, they will cause additional transformer and distribution loss, and the odd harmonic components result in a compensating current flow in the neutral line of a three-phase, four-wire distribution system. This latter effect is often the most problematical, as the neutral line is not designed to carry such currents.

Clearly, it is incumbent on the responsible engineer to minimize power line pollution. Moreover, many countries have long been proposing to enforce regulations to limit such pollution; typical requirements are defined under the IEC-555-2 harmonic limits, or more recently IEC 1000-3-2, which in 1998 was the generally accepted standard. Legislation to enforce such standards has been delayed many times, perhaps as a result of the reluctance or inability of industry to meet the correction needs.

Although passive low-pass filters in the power line can be used to reduce the harmonic content to acceptable limits, a more viable solution at high power levels can be found, using one of the various so-called active power factor correction methods.

Sections 1.2 and 1.3 of this chapter provide some background based on well-known passive power correction methods; this material will be of more value to those readers who are not very familiar with the general principles. Section 1.4

describes the principles of active power factor correction, and Sec. 1.10 gives an applied design example, as applied to a 2.2-kW switching power supply for use in industrial lighting applications.

1.2 POWER FACTOR CORRECTION BASICS, MYTHS, AND FACTS

1.2.1 Power Factor Basic Definition

Power factor (PF) is defined as the ratio of the true power dissipated in the load to the apparent power taken by the load, irrespective of the waveform, as shown in Eq. (4.1.1):

$$PF = \frac{\text{true power}}{\text{apparent power}} \tag{4.1.1}$$

The true power is the time-averaged power, which results in heating or mechanical work being done, the rate of work being measured in true watts.

Apparent power is the time-averaged product of the rms voltage and rms current, measured at the input to the load, without adjustment for phase shift or distortion. It is known as the input voltamperes (VA), and the rate of apparent work has units of apparent watts, as shown in Eq. (4.1.2):

$$\text{Apparent power (VA)} = V_{\text{rms}} \times I_{\text{rms}} \tag{4.1.2}$$

In the special case of a purely resistive load, the voltage and current will both be sinusoidal and in phase, and the true power is equal to the apparent power, giving the ideal power factor ratio of unity. Although power factor ratios correctly range from 0:1 to 1:1, it is quite common to quote a percentage, where 100% is equal to a ratio of 1:1.

1.2.2 Power Factor in Sine Wave Applications

In the simple case of nondistorting reactive plus resistive load combinations, the current and voltage will both remain sinusoidal, but a phase shift is introduced by the reactive component (with the current lagging the voltage for inductive loads and leading it for capacitive loads).

Typical nondistorting reactive loads are shown in Fig. 4.1.1. Figure 4.1.1a shows a series inductive and resistive load combination, with a shunt capacitor C1 to provide power factor correction. Figure 4.1.1b shows a shunt capacitive and resistive load combination, with a series inductor L2 to provide power factor correction. Other series or shunt configurations are possible, depending on the application.

Figure 4.1.2 shows the voltage and current waveforms for a typical inductive load. The current lags the applied voltage by an amount depending on the ratio of resistance to inductance, giving a lagging phase angle of between 0 and 90°.

Figure 4.1.3 shows a vector diagram of the real and apparent power developed in the load. It demonstrates how the real power will be less than the apparent power if there is a reactive component.

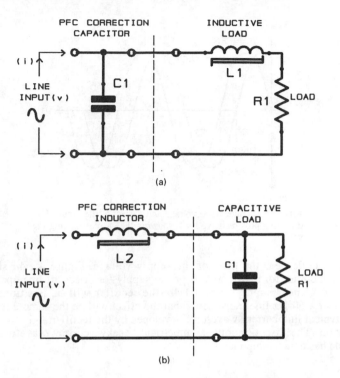

FIG. 4.1.1 *(a)* Passive power factor correction for a linear inductive plus resistive load, using a power factor correction shunt capacitor. *(b)* Passive power factor correction for a capacitive plus resistive load, using a series power factor correction inductor.

By inspection of Fig. 4.1.3, it is clear that the ratio of the true power to the apparent power is given by cos Ø, where Ø is the phase angle between the voltage and the current. This gives rise to the well-known power equation for true sine waves shown in Eq. (4.1.3):

$$\text{True power} = V_{\text{rms}} \times I_{\text{rms}} \times \cos \varnothing \tag{4.1.3}$$

1.2.3 Distortion Effects

When the current (or voltage) is not a true sine wave, it is said to be distorted. Fourier analysis will show the waveform to be made up of a number of harmonically related sine waves of differing amplitude and phase. In the case of symmetrical distortion, only the odd harmonics will be found.

It was shown above that the power factor with true sine waves can be equal to or less than unity, depending on the phase angle; however, when distortion is present, the power factor will always be less than unity.

Unfortunately, any load which has rectifiers and capacitors in the power path (this includes linear and switching power converters) will tend to draw current from the supply in a distorted form; typically, the current will flow in a narrow

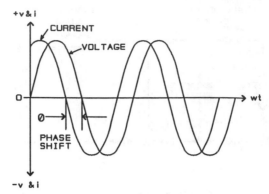

FIG. 4.1.2   Sine wave voltage and current at the input of an inductive plus resistive load, showing the current lagging the voltage.

conduction angle near the peak of the supply voltage. Figure 4.1.4*a* shows the typical input circuit for a switchmode power supply for direct-off-line operation. In the case of a linear supply, Fig. 4.1.4*b*, the rectifier will be positioned in the secondary of a 60-Hz line transformer, but the effect will be the same. Figure 4.1.5 shows a typical line-current waveform developed by the rectifier and capacitor load combination. The discontinuous, symmetrical "peaky" current waveform is very rich in odd harmonic components.

1.2.4 Myth (a Common Error)

Upon inspection of Fig. 4.1.5, the rectifier current waveform appears to be generally sinusoidal and in phase with the voltage. This leads to a rather common error. There is a tendency to assume that with this type of waveform the power will be given by the simple VA product, $V_{rms} \times I_{rms}$. However, this is far from being true.

As mentioned above, distortion introduces harmonics, and in this example the power factor is likely to be in the range 0.5 to 0.8 (a potential 50% error on the calculated VA value). The discontinuous-current waveform, although it may appear sinusoidal, is rich in odd harmonics, and power transfer takes place only at the fundamental frequency. However, the rms current measurement includes the harmonics. For such waveforms, a true wattmeter instrument is essential to measure true input power. In calculating power, the phase and amplitude of all harmonics must be included.

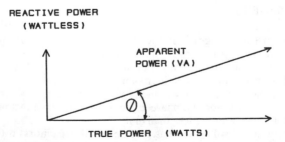

FIG. 4.1.3   Vector diagram showing how the apparent power will exceed the real power in a resistive plus reactive load.

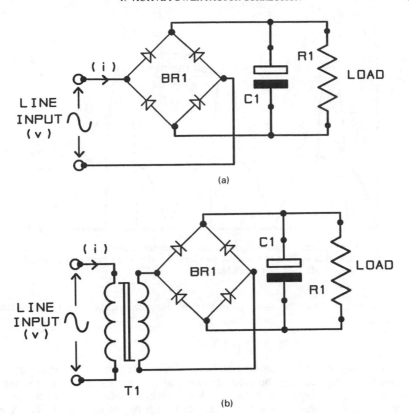

FIG. 4.1.4 *(a)* A typical rectifier capacitor input stage for a direct-off-line switchmode power supply. *(b)* A typical 60-Hz transformer rectifier capacitor input stage for an isolated linear power supply.

1.2.5 True Power Measurement

True wattmeter instruments are used to measure the true power of phase-shifted and/or distorted waveforms. The older type, analog moving-coil instruments (dynamometer wattmeters), relies on the magnetic interaction Φ between two air-cored coils, one that is fixed and carries the load current I_L (current coil) and a second (voltage coil) that is free to rotate against a return spring. The second coil carries a much smaller current proportional to the applied voltage. This current is supplied via a limiting resistor from the voltage supply terminals I_v.

The dynamics of such an arrangement are very simple but effective. The instantaneous magnetic moment applied between the coils is proportional to the instantaneous product of the in-phase components of I_L and I_v. This moment is time-averaged by the inertia of the relatively heavy moving parts, and the mean moment works against the return spring on the moving coil to provide a deflection (and hence a reading) proportional to the mean moment. This arrangement satisfies the basic requirements for power measurement, as shown in Eqs. (4.1.4) and (4.1.5).

$$\text{Deflection} = \frac{1}{T} \int \Phi I(t) \times \Phi V(t) \ dt \qquad (4.1.4)$$

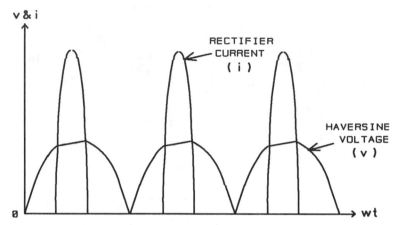

FIG. 4.1.5 Typical rectifier output voltage and current waveforms when a large capacitive load is applied to the input rectifier, showing large peak currents and discontinuous conduction, rich in odd harmonics.

By inspection, this has the same form as the electrical equation for power for a single-phase system:

$$\text{True power} = \frac{1}{T} \int VI(t) \times \Phi I(t) \ dt \tag{4.1.5}$$

The old dynamometer-type moving-coil instruments are becoming rather rare, as they are being replaced by the more modern, lower-cost electronic instruments. However, the author has found that for development applications (where power measurements are often made at the breadboard stage before input RFI filters have been fitted), some digital instruments may be upset by the large RF noise levels found in switching applications and can give large errors. Hence it is prudent to at least compare the test results with a dynamometer-type instrument, which just cannot respond to the high-frequency noise. Therefore, if you can find a good dynamometer instrument, cherish it!

Unless it is known for sure that the applied voltage is a pure sine wave and the load is purely resistive, true power measurements require the use of a wattmeter, or one of the digital power analyzers, that will correctly compensate for the distortion and/or phase shift effects.

1.2.6 Power Factor (Fact!)

With distorted waveforms, it is possible to satisfy the power factor requirements with various power factor correction methods without meeting the harmonic limitations given in IEC-555-2 or IEC 1000-3-2.

This is possible because the IEC specifications limit the amplitude of each harmonic, and it is possible to have one or more harmonics out of limit and still meet the overall power factor requirement.

Hence, it is necessary to measure both power factor and harmonic distribution and amplitude to be sure of meeting all IEC requirements. Remember, the voltage waveform is used as a reference for the current waveform in most power factor correction (PFC) systems, and hence any distortion of the applied voltage should be considered in the measurement process.

1.2.7 Efficiency (Myth!)

It is often claimed that power factor correction improves efficiency. In general, this is not true. In most topologies, the correction circuit dissipates additional power and does little to improve the efficiency of the following switching power circuits.

Very often, two power conversions are carried out in the active PFC system, and hence the total power loss normally exceeds that of a noncorrected system with only a single power conversion stage.

Therefore, the efficiency of a corrected unit is generally lower than that of a noncorrected unit, and for the same output power and size, the working temperature of the PFC unit will often be higher.

1.2.8 Line Utilization (Fact!)

In general, it is possible to obtain more power from a limited source (say, a standard wall outlet) when using a power factor corrected unit. This is possible because the rms input current of the corrected unit remains less than that of the noncorrected unit, in spite of the lower overall efficiency of the corrected unit.

For example, the standard 120-V, 15-A wall socket is limited by UL ratings to run continuously at 12 A. With a typical noncorrected rectified power unit, the power factor will be of the order of 0.65 and the efficiency, say, 80%. At 12 A, this will provide a maximum output power of 750 W.

With a power factor corrected unit, the power factor can exceed 98%, although the efficiency will generally be lower, say, 73%. Hence, at 12 A, the output power will now be of the order of 1000 W for the corrected unit (250 W more than for the noncorrected unit).

I believe this effect is the cause of the confusion over efficiency. It is important to notice that the power dissipated within the noncorrected unit at 750 W output at 80% efficiency will be 187 W, while the corrected unit at 73% efficiency and the same 750-W power output would dissipate 275 W, an extra 88 W. At 1000 W output, the loss in the corrected unit will be as high as 370 W. Hence improved cooling may be required in the PFC unit.

1.3 PASSIVE POWER FACTOR CORRECTION

Passive power factor correction involves the use of linear inductors and capacitors to improve the power factor and minimize harmonic components. This method works very well with simple nondistorted reactive loads, where the undesirable reactive component (or phase shift) can be taken out with the addition of simple equal but opposite reactive components.

However, the harmonics (or distortion) produced by nonlinear loads do not respond very well to this approach. To eliminate the higher-frequency harmonics in distorted waveforms, we require a low-pass power filter in series with the power line.

At higher power levels, large amounts of power must be stored and manipulated by this filter, and hence it requires large inductors and capacitors, which are not very cost-effective. However, at low power levels, such filters are used to good effect in passive power factor correction systems.

Typical applications include ballasts for fluorescent lamps, where the large input

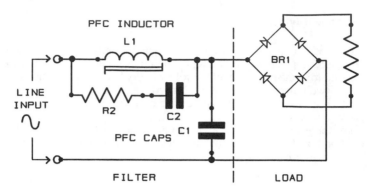

FIG. 4.1.6 Passive *LCR* input filter for harmonic reduction, typically used
in passive power factor corrected electronic ballasts.

filter inductor is sometimes integrated into the lamp transformer design. Power
factors exceeding 0.9 (90%) can be obtained with such methods, but they are
normally limited to power levels of 100 W or less. Figure 4.1.6 shows a typical filter
used at the input of a 70-W fluorescent ballast. It will be noticed that the design is a
compromise. Practical limitations of size and cost normally limit the inductor L1 to
about 250 mH and C1 to between 2 and 4 µF, giving a cutoff frequency of about
200 Hz. C2 is added to form a notch for the higher-frequency harmonics. Although
this is not ideal, the improvement can be dramatic, moving the power factor from
less than 0.6 to above 0.9. The harmonic content can remain quite large, but for low-
power applications the agency limitations on harmonic content are more tolerant.

A side effect to bear in mind when using this type of passive power factor
correction circuit is that L1 and C1 tend to form a series resonant circuit, which
together with the wider conduction angle of the rectifier results in a higher DC
output voltage from the bridge rectifier.

Further, because of this resonant effect, it has been found that the voltage across
L1 can exceed the supply voltage by up to 25%, and this should be considered when
designing L1. Equation (4.1.6) may be used to calculate the minimum number of
turns required on L1, which would normally be a laminated iron core with an air
gap, to make the inductance more linear.

$$N_{min} = \frac{10^6 \, V_e}{4.44 \times f \times B \times A_e} \qquad (4.1.6)$$

where for laminated iron cores $V_e = V_{in} + 25\% \, V_{in}$, V rms
f = line frequency, Hz (typically 50 or 60 Hz)
B = maximum flux density, T (typically 1.3 T)
A_e = effective core area, mm² (typically 0.6 × pole
area)

1.3.1 "Valley-Fill" Power Factor Correction Methods

The so-called valley-fill method relies on the use of extra diodes and capacitors to
change the effective circuit at various stages of the charge and discharge cycle of the
storage capacitors so as to improve the power factor. It is not truly passive (there is
no *L/C* filter), and is active only inasmuch as there is a switching action of the
diodes at various parts of the cycle.

The method was described by Spangler[86] in 1988, and more recently, computer modeling of a voltage-doubled version of the Spangler circuit by KitSum[85] indicates that power factors of 98% should be possible.

It has potential as a low-cost method in lower-power applications such as fluorescent lamps, where the original Spangler methods have been used for some years. It is a good, practical, low-cost, efficient method which should not be neglected.

Figure 4.1.7 shows the original Spangler circuit, and Fig. 4.1.8 shows the computer-modeled current waveform expected at the input to this circuit. Figure 4.1.9 shows the new voltage-doubler version of the Spangler circuit, and Fig. 4.1.10 shows the computer-modeled current waveform expected at the input of the voltage-doubled version.

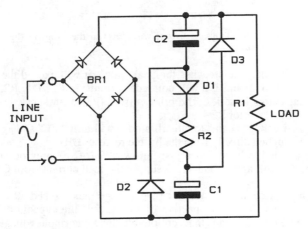

FIG. 4.1.7 "Valley-fill" power factor correction circuit used in low-power applications. *(Spangler.[86])*

1.3.2 Function

In simple terms, the valley fill circuit of Fig. 4.1.7 functions as follows:

Consider a condition where the input sine wave is just passing through zero. Let the output voltage applied to the load R1 be approximately 1/3 of the peak applied voltage; C1 is providing current to the load via D3, and at the same time, C2 is supplying current to the load via D2. Hence C1 and C2 are effectively in parallel, supplying the load current between them. Diode D1 is reverse-biased and is not conducting.

Since the output voltage of the main bridge rectifier BR1 exceeds the supply voltage, the bridge diodes are reverse-biased and the input current will be zero, corresponding to the start of the waveform shown in Fig. 4.1.8.

When the input voltage exceeds the output, BR1 will conduct to increase the output voltage. At this time, diodes D2 and D3 will block and capacitors C1 and C2 will stop providing current to the load. Hence, the load current will now be provided by the bridge rectifier directly from the supply. D1 will not be conducting at this time, as the applied voltage is less than the sum of the two voltages on C1 and C2.

Until the applied voltage reaches the sum of the voltages on C1 and C2, the only load applied to the bridge rectifier output will be the linear load resistance of the output load, and the input current will now follow the input-voltage sine-wave shape.

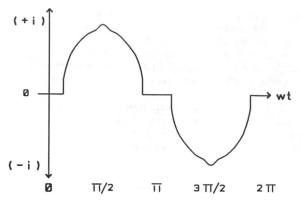

FIG. 4.1.8   Typical current waveform at the input to the Spangler circuit.

As the applied voltage approaches the peak value, it will exceed the sum of the voltages on C1 and C2, and D1 will conduct current via C2, D1, R2, and C1 to recharge the capacitors in series. The brief current flow at the peak of the applied voltage is limited by resistor R2.

As the applied voltage begins to fall, all diodes will turn "off," and again the load current will be supplied directly from the bridge rectifier BR1.

When the applied voltage drops just below 50% of the previous peak value, diodes D3 and D2 will again conduct, to supply the load current from C1 and C2 in parallel.

With this type of circuit, the output ripple voltage must exceed 50% of the peak rectified haversine voltage, irrespective of the size of the capacitors. Hence the method is suitable only for loads that can tolerate this large ripple voltage.

With the enhanced circuit shown in Fig. 4.1.9, the voltage-doubling effect provided by small capacitors C3 and C4 causes conduction at a lower applied voltage, to partly fill in the initial gap in the current waveform, as shown in Fig. 4.1.10, giving a slight reduction in distortion. C3 and C4 will be much smaller than C1 and C2.

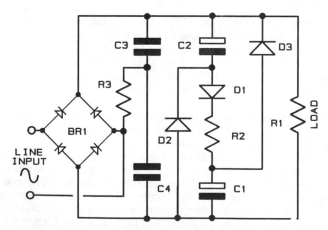

FIG. 4.1.9   An improved "valley-fill" circuit.   (*Spangler and KitSum.*[85,20])

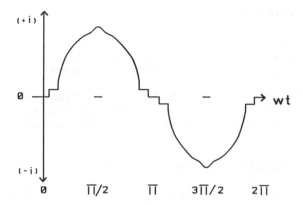

FIG. 4.1.10 Current waveform at the input of the improved Spangler circuit.

1.4 ACTIVE POWER FACTOR CORRECTION

Up to this point, we have considered only the manipulation of various passive linear components to give the required phase correction, or filtering, so as to obtain near-sine-wave input currents and reasonable power factor correction. We have seen that these methods are more suitable for lower-power applications. In the following sections, we consider some of the basic concepts of active power factor correction, which not only provide better performance but may be used for much higher-power applications.

In general terms, it is possible to force the line current to follow the applied sine-wave voltage by using various so-called active current control power factor correction methods. For good efficiency, switchmode power control systems are normally used for this application.

In principle, if the current waveform into any type of load (for example, a switchmode power converter) can be constrained to follow the shape and phase of the applied voltage, then this load will be made to simulate a purely resistive load, and the conditions for unity power factor will have been satisfied.

Further, provided that the applied voltage is a pure sine wave and the current is constrained to follow it, the conditions for zero harmonic content will also have been satisfied.

The shape of the applied voltage waveform is very important, because most of the active PFC methods in common use assume that the applied voltage will be a pure sine wave. The voltage waveform is used as a reference for the current waveform. In fact, the assumption of a pure sine-wave supply voltage is rarely true, as other nonlinear loads on the same distribution system will often introduce distortion to the voltage waveform (via the common line impedance). This effect must be taken into consideration when measuring the performance of an active power factor correction system. If the line voltage is distorted, then the line current will have at least the same amount of distortion. (Clearly, this would be true even with a real pure resistive load.)

1.4.1 Basic Concepts

Because most power factor converter topologies require a unidirectional input, it is normal to full-wave rectify the ac input (without a storage capacitor), as shown in Fig. 4.1.11*a*. This will produce a unidirectional haversine waveform, as shown in Fig. 4.1.11*b*.

The unidirectional haversine is then applied to the power conditioning components. At the same time, the voltage is processed within the control circuit to produce a haversine reference signal for the current control circuits.

1.4.2 Boost Power Factor Correction Topology

The basic boost circuit shown in Fig. 4.1.12 will be used to explain the principles of an active power factor correction system. In general, the same principles will apply to all the topologies shown in this chapter.

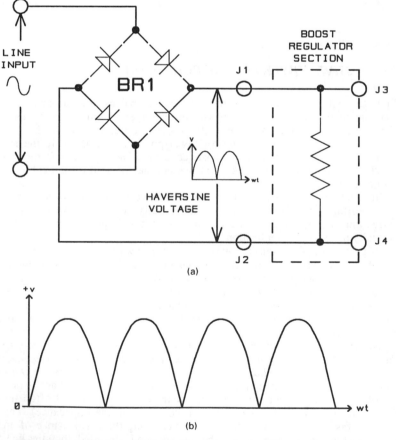

FIG. 4.1.11 (*a*) Bridge rectifier (without capacitor) used at the line input to produce a haversine voltage waveform at the input to an active power factor correction system. (*b*) A typical haversine voltage waveform.

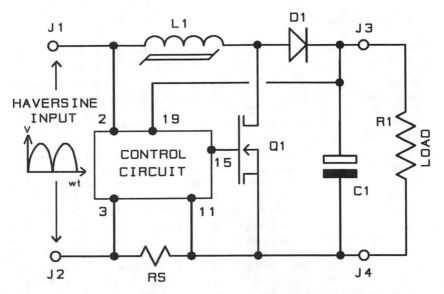

FIG. 4.1.12 A basic boost regulator, showing the essential control elements. The negative line is common to input and load (except for a low-resistance current shunt R_s).

Figure 4.1.11 shows that the line input is rectified by BR1 to produce a unidirectional haversine, as shown in Fig. 4.1.11b. This haversine is applied to the input of the high-frequency boost regulator shown in Fig. 4.1.12. The same haversine voltage is also applied to the control circuit (pin 2) and processed in the control circuit, to be used as a reference signal to define the required phase and shape for the input current.

The input current flows into L1 and returns via the current-sensing resistor R_s, which provides a sample of the actual current waveform to the control circuit on pins 3 and 11.

To get the required mean 120-Hz haversine current waveform in L1, the power switch Q1 will be turned "on" and "off" (or duty cycle modulated) at high frequency (typically 50 to 200 kHz), as the instantaneous input voltage changes relatively slowly during the haversine period.

If the high-frequency switching process is correctly controlled in the short term (say over one cycle), the mean current in L1 (and R_s) will be made to track the haversine voltage signal in relative shape and phase during this cycle. In this way, L1 and following circuits are made to simulate a purely resistive load during the cycle.

The amplitude of the mean haversine current in L1 is also adjusted in the longer term to compensate for any long-term input voltage changes and to compensate for any load changes so as to meet the power requirements of the load.

To accomplish this, the control circuit slowly adjusts the mean haversine current amplitude over a few cycles to satisfy the average load power, while maintaining a near haversine shape during each cycle. This longer-term adjustment will compensate for any long-term input voltage or load changes, and is best done by maintaining the output voltage on C1 constant.

1.4.3 Input Variables

We can now see that there are four basic input variables which control the shape, phase, and amplitude of the input haversine current. These are as follows:

1. *Reference haversine voltage (pin 2)*. This is a reference haversine waveform that is in phase with and proportional to the applied haversine input voltage. This defines the potential phase and haversine shape of the current. (This reference is normally derived directly from the input voltage haversine on pin 2 by a resistive divider and is a real-time signal.)

2. *Power demand feedback (pin 19)*. This is a much slower input variable that defines the longer-term amplitude of the current haversine so as to satisfy the output power requirements. (This is normally defined by a voltage feedback from the boost section output capacitor C1, and adjusts the mean haversine current so as to maintain the output voltage constant as the output load changes.)

3. *Input voltage compensation (pin 2)*. This is an input variable which will adjust the longer-term mean haversine current amplitude as the longer-term mean input voltage changes. It will be shown later that this requirement is best satisfied by changing the gain of the control modulator while maintaining the output voltage constant. This gain change is necessary to compensate for the intrinsic change in gain of the power section as the input-to-output voltage differential changes. (We will see later that this gain change is inherent to the boost-type power section.)

4. *Current feedback signal (pin 3)*. A current feedback signal is required to close the current control loop. Ideally, we require a signal that is proportional to the actual instantaneous input current in L1. For convenience, this is more often obtained from a resistive shunt R_S in the common return line. Current transformers and Hall effect current transducers can be used in higher-current applications.

1.4.4 Input Current Waveform

By the steering action of the diodes in the input bridge BR1, the 120-Hz unidirectional haversine current pulses at the input to L1 are commutated back to the input of the bridge rectifier to form an alternating 60-Hz sine wave current, in phase with and of the same polarity as the applied sine wave input voltage. Two unidirectional current haversines are required for one cycle of the input. Hence the input to the rectifier BR1 simulates a resistive load.

1.4.5 Switching Control Modes

There are many ways in which the mean haversine current waveforms can be achieved, and also many different types of control ICs are available to provide the control functions.

Referring again to Fig. 4.1.12, pin 15 provides the drive to Q1. Although Q1 can be controlled in various ways to get the required mean haversine current, it is important to remember that for efficient operation, Q1 can only be either fully "on" or fully "off," and it is only the timing and duration of the "on-off" function that is adjustable.

Hence, the many and various ICs and more or less complex control systems all result in a simple "on-off" switching action of Q1. If you were to use a microcontroller for this application, this switching action is the parameter you would control.

However, the way this apparently trivial switching action is controlled has far-reaching effects on the overall performance of the system. We will look at some of the more common modes used to control Q1.

1.4.6 Complete Energy Transfer mode

This is probably the most straightforward mode. Figure 4.1.13 shows the current waveform.

For the complete energy transfer mode (discontinuous mode), the inductance of L1 will be relatively small, and Q1 will turn "on" when the current in L1 reaches zero and "off" when the current in L1 reaches twice the mean required current, as shown in Fig. 4.1.13.

This leads to a variable conduction period and hence a frequency which will change during the haversine. It has the advantage of using only a small inductor, and since the inductor current is a control variable, the inductor is taken out of the small-signal transfer function, leading to a more easily stabilized system.

Low-cost 8-pin ICs are available for this application. The method has the disadvantage of larger peak currents in Q1, D1, and C1 and larger input ripple currents into L1 at the switching frequency. These high-frequency currents must be filtered out by the line input filter. For this reason, this method is normally reserved for lower-power applications, such as power factor corrected electronic ballasts for fluorescent lighting and other PFC applications up to 100 W or so.

1.4.7 Incomplete Energy Transfer Mode

For the incomplete energy transfer mode (continuous mode), the inductance will be relatively large, so that the inductor current changes only slightly during a switching period. Figure 4.1.14 shows a typical current waveform.

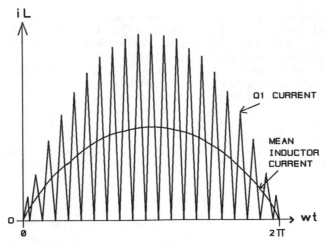

FIG. 4.1.13 Typical switching frequency ripple current waveform at the input of a discontinuous-mode power factor correction boost stage for a single haversine period. The mean inductor current is shown.

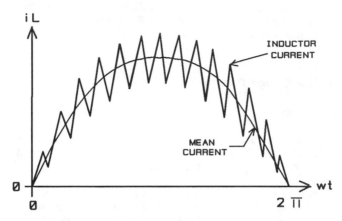

FIG. 4.1.14   Typical switching frequency ripple current waveform at the input of a continuous-mode power factor correction boost stage for a single haversine period. The mean inductor current is shown.

Consider a condition where the drive to Q1 is at a fixed high frequency with duty ratio control. Q1 turns "on" or "off" at the beginning of a period, depending on leading-edge or trailing-edge modulation. The duty ratio will be adjusted to make the average inductor current (as sensed across R_s) track the reference waveform, developed from the haversine input voltage via the modulator. This is perhaps the most popular control method at the time of printing.

Because the frequency is fixed and the inductance is finite, a discontinuous mode of operation will occur at low currents at each end of the haversine waveform.

1.4.8   Other Control Methods

Other, more complex control methods can be used. Remember, all methods can control only the timing of the switching of Q1. However, the different timing methods can introduce quite different transfer functions, which in turn will require different control functions in the closed-loop current control circuits. A typical example is the so-called hysteretic control.[87] In this method, both the peak and minimum inductor currents are made to track near the ideal reference current, both the maximum and minimum current deviation being defined.

This, once again, leads to variable periods, and the frequency will change during the haversine. Compensation can be applied to reduce the range of frequency variation. Since the inductor current is now included as a control variable, the inductance is taken out of the small-signal transfer function, and in a current-mode-controlled system, the control loop is inherently stable. A typical current waveform is shown in Fig. 4.1.15.

1.4.9   Methods of Pulse-Width Modulation

1. *Trailing-edge pulse-width modulation.* For trailing-edge pulse-width modulation, Q1 is turned "on" at the beginning of a period and will stay "on" until the current

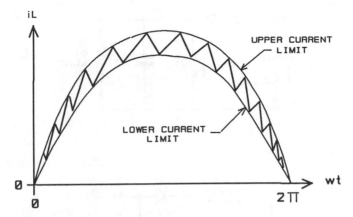

FIG. 4.1.15 Typical switching frequency ripple current waveform at the input of a hysteretic-mode power factor correction boost stage for a single haversine period. The upper and lower current limits are shown.

sensed by R_s increases to the reference current signal supplied by the control circuit, at which point Q1 will turn "off." (Remember, the reference signal is developed from the four input variables supplied to the modulator, as described in Part 4, Sec. 1.4.3.)

2. *Leading-edge pulse-width modulation.* In this case, Q1 is turned "off" at the beginning of a period and turns "on" when the current sensed by R_s falls to the reference current value.

In either of the above cases, when Q1 turns "off," the inductor current continues to flow via D1 into the output capacitor. Since the output capacitor voltage always exceeds the input voltage (a basic requirement for the boost topology), the current in L1 will decay during the "off" period of Q1.

A combination of the above modulation methods can be used to reduce ripple currents and cross modulation in combinations of converters. (This method is used later in the combination converter described in Part 4, Sec. 1.10, to reduce the ripple current in the buck output capacitor.)

1.5   MORE REGULATOR TOPOLOGIES

Figure 4.1.16 shows some of the boost regulator topologies that may be used for power factor correction. All except the flyback share a common input-to-output power line; hence the output is not isolated and is electrically linked to the ac line supply.

The simple boost regulators shown in Fig. 4.1.16a to d cannot provide overcurrent protection, as the power switching device is shunt-connected and there is a direct path from the supply through L1 and D1 to the output.

Since in many applications overcurrent limiting is required and isolation and/or additional voltage regulation is also required, it is normal to follow the boost PFC regulator with one of the many transformer-isolated switching converters or with a nonisolated buck regulator stage.

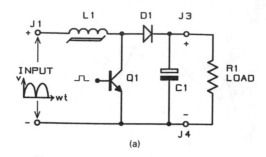

(a)

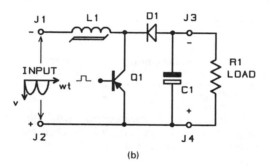

(b)

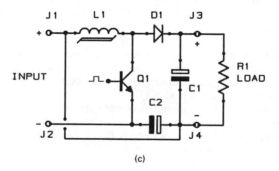

(c)

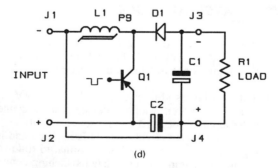

(d)

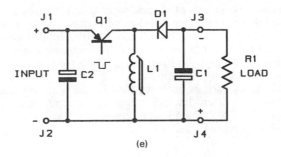

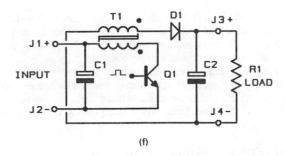

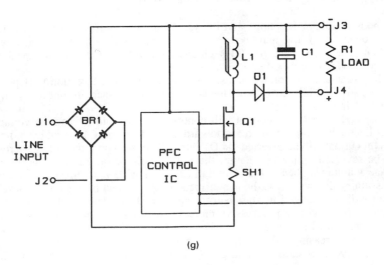

FIG. 4.1.16 Basic power topologies. *(a)* Nonisolated, three-terminal positive boost regulator. *(b)* Nonisolated, three-terminal negative boost regulator. *(c)* Positive nonisolated, three-terminal bootstrap regulator. *(d)* Nonisolated, three-terminal negative buck-boost regulator. *(e)* Nonisolated, three-terminal negative buck-boost regulator. *(f)* Isolated, four-terminal boost regulator, often called flyback convertor. *(g)* Nonisolated, four-terminal power factor correction bootstrap boost circuit, showing essential control elements.

The basic transfer properties of several topologies are given in this section. It should be noticed that only the boost and boost-derived family is able to maintain control of the input current over the full range of the applied haversine voltage. Also, only the boost family has the potential for continuous input current, which reduces input ripple and supply RFI problems. Hence the boost family is preferred for power factor correction applications.

Figure 4.1.17 shows various buck regulators, which are often used as output voltage regulators and current-limiting stages in converter combinations. Figure 4.1.18 shows some useful combinations of converters. This section (1.5) describes the basic parameters of boost regulators.

1.5.1 Positive Boost Regulator

Figure 4.1.16*a* shows the basic power elements of a positive boost regulator circuit. This is probably the preferred preregulator for power factor correction, and we will recognize it as being the power section used in Fig. 4.1.12. Many ICs are dedicated to this topology, and excellent results can be obtained.

Typically, for power factor correction applications, the input to this circuit would be a positive haversine waveform from a line-connected bridge rectifier.

It should be noticed that the input and common line is negative, and the output is positive with respect to this line. Hence the name positive boost regulator.

1.5.2 Principle of Operation (Boost Regulator)

The boost regulator functions as follows: Assume an initial condition where the input voltage is positive and above zero and current is flowing in L1 and Q1, with Q1 "on." The output capacitor C1 is charged to a voltage exceeding the peak input voltage, and D1 is reverse-biased.

The instantaneous voltage across L1 is in the forward direction (left positive, right negative). While Q1 is "on," the current in L1 will increase linearly with time (flowing left to right).

When Q1 turns "off," current will continue to flow into L1 and the voltage on Q1 collector will rapidly increase toward the voltage on C1, at which point the inductor current will be diverted via D1 into C1. At this point, the voltage across L1 will be reversed (since by design, the output voltage on C1 exceeds the input voltage at all times); hence, the current in L1 will now decrease linearly with time.

Assuming that we are on the increasing side of the input haversine, the next "on" period will be arranged to start before the current has decreased to its previous starting value, giving a progressive increase in current for each cycle so that the mean current in L1 tracks the rising haversine voltage shape.

On the decreasing side of the haversine, the current will be allowed to drop below the previous starting point, giving a progressive reduction in current. In this way, the required mean haversine current will be traced out.

The input current is continuous for most of the haversine, provided that L1 is large and the switching frequency is high compared with the line frequency.

1.5.3 Output Capacitor C1

The output current from D1 into C1 is discontinuous at the switching frequency, requiring a low-impedance output capacitor (or a combination of capacitors) at the

C1 position. Further, for power factor correction applications, up to 50% of the 120-Hz haversine input current flows in this capacitor (the rest flows in Q1), so a large capacitance is also required for C1.

In the 2.2-kW PFC example used in this chapter, C1 is made up of electrolytic capacitors shunted by low-ESR film capacitors to better handle the high- and low-frequency components of this ripple current.

Since for steady-state conditions, the long-term volt-seconds applied to L1 in the forward direction must equal the reverse volt-seconds, the output voltage must exceed the input voltage at all times, and a margin of at least 10 V is preferred. Hence for PFC applications at the industry universal range of 80 to 264 V rms input, the DC output should be at least 380 V.

1.5.4 Transfer Function (Stability)

As with all boost topologies, energy is transferred to the output only when Q1 is "off." Hence, when running L1 in the continuous-conduction mode, a transient increase in load requires an increase in L1 current. Hence, Q1 must increase its "on" period to increase the forward current in L1. The immediate effect of this increase in the "on" period is to reduce the energy-transferring "off" period, and since the current in L1 cannot change rapidly, the output current is reduced before it recovers to the required value. This transient phase reversal translates to a noncompensatable right-half-plane-zero in the transfer function (see Part 3, Chap. 9).

For stability reasons, the gain of the voltage control loop must fall below unity well before the RHP-zero frequency. For power factor correction, this low-frequency rolloff is not normally a problem, because the outer voltage control-loop crossover frequency is below 30 Hz for other reasons to be discussed later.

1.5.5 Negative Boost Regulator

Figure 4.1.16b shows the negative boost regulator. This has exactly the same properties as its positive counterpart, except that the common line is the positive input line. As before, the output voltage must exceed the input voltage, but in this case it is negative. This regulator is used in some converter combinations.

1.5.6 Positive Bootstrap Boost Regulator

Figure 4.1.16c shows this regulator. The topology is sometimes referred to as a buck-boost, although by inspection it is clearly still a positive boost topology, with the DC output C1 stacked on top of the input voltage C2.

The current waveforms on all components are exactly the same as those of the conventional positive boost regulator. The difference is that the DC output is taken between the boosted output and the positive side of the DC input (in place of the negative input), introducing a fixed DC offset. Hence the output voltage is now the difference between the DC supply voltage and the boosted output voltage.

This apparently trivial change makes a considerable difference to the output terminal performance. The effective output voltage may now range from near zero to a value far exceeding the input voltage.

Since the inductor current is returned to the input, the input terminal current becomes discontinuous, dropping to zero when Q1 is "off." The output is positive with respect to the common positive input line, and the current in C1 remains discontinuous, as with the simple boost regulator.

1.5.7 Inverted (or Negative) Bootstrap Boost Regulator

Figure 4.1.16*d* and *e* shows this topology. As drawn in Fig. 4.1.16*d*, it is clearly the same as the previous positive bootstrap boost regulator, except that the supply polarity is reversed; this requires Q1 and D1 to be reversed. It has exactly the same properties as its positive bootstrap boost counterpart, but the output is now negative with respect to the common negative line.

It is interesting to note that a simple rearrangement of this circuit, Fig. 4.1.16*e* (if you trace through the electrical connections, you will find that they have not been changed), shows that this topology is what is normally referred to as the buck-boost topology.

1.5.8 Transformer-Coupled Boost Regulator (Flyback)

Figure 4.1.16*f* shows this topology. When this is compared with Fig. 4.1.16*a*, it can be seen by inspection to be a positive boost regulator. The difference is that the input is transformer-coupled. T1 must perform the functions of a transformer and a choke.

The output is fully isolated, and by its nature and by adjustment of the turns ratio, the output voltage may have a wide range of voltage both above and below the supply voltage, the basic limitation being that the secondary voltage (as reflected to the primary) must exceed the peak primary voltage. Under these conditions, this topology may be used for power factor correction. It is often referred to as a "flyback" converter.

1.5.9 Bootstrap Power Factor Correction Boost Regulator

Figure 4.1.16*g* shows a more complete arrangement of the positive bootstrap regulator of Fig. 4.1.16*c* as applied to a power factor correction application. This topology is very useful where the key boost property (the ability to control the input current over the complete range of the applied haversine) is required, but the output voltage is required to be less than the input voltage. It should be noticed that the boosted DC output voltage, although constant in amplitude, "rides" on top of the input haversine voltage.

The input haversine voltage is present in equal parts on both output lines with respect to the common line. Hence, this method is suitable only for a floating load which is not connected in any other way to the common line. A good example would be an electronic ballast. Using this topology, a ballast designed for 110 V DC could be power factor corrected and supplied from a 277-V ac supply.

1.6 BUCK REGULATORS

Buck regulators are included in this section because they are often used as output stages in power factor correction applications using combinations of converters.

Although they can also be used more directly for power factor correction, they are not ideal for this application, as they will not control the input current when the input voltage is less than the output voltage.

Figure 4.1.17 shows three basic topologies of typical buck regulators. The major disadvantage of buck regulators for power factor applications is that the input current cannot start to increase until the haversine input voltage exceeds the output voltage. This will cause some distortion at the voltage crossover points. Further, the input current is discontinuous (pulsing at high frequency), and so additional RFI input filtering would be required at the input. Hence, buck topologies are not normally preferred for power factor correction applications.

However, in some applications, the overall power factor can still be quite good and the requirements of IEC 1000-3-2 class D can be satisfied with this topology.

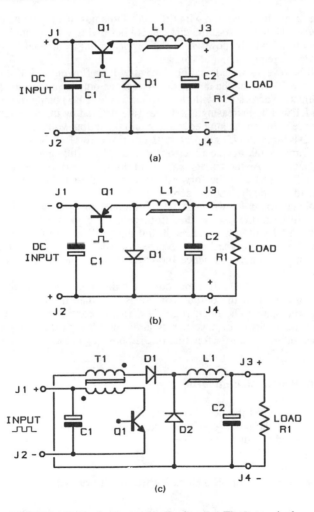

FIG. 4.1.17 Basic power topologies. *(a)* Three-terminal, nonisolated positive buck regulator. *(b)* Three-terminal, nonisolated negative buck regulator. *(c)* Isolated, four-terminal, transformer-coupled buck regulator; often referred to as forward converter.

For low-voltage outputs, when isolation is not required, and in transformer-coupled applications, the advantage of having a single power conversion stage may make the buck topology an attractive choice. Remember, tapings or overwindings on the choke or transformer will provide other semistabilized output voltages, which may also be isolated. Hence, the basic buck topologies are included in this review of power factor circuits.

1.6.1 Operating Principle of the Positive Buck Regulator

Figure 4.1.17a shows the topology of the positive buck regulator, in which the supply and output are positive and the common line is negative. It operates as follows:

Assume a starting condition with Q1 "on" and a positive input and output voltage established on J1 and J3. Assume that current is flowing via Q1 into L1. The voltage at the input to L1 will be positive, at the supply voltage potential (which by design must exceed the required output voltage).

D1 will be reverse-biased, and no current will be flowing in D1 at this time. The voltage across L1 will be the input voltage less the output voltage and will act in the forward direction across L1, with the left-hand side (LHS) being more positive. The current in L1 will be increasing at a linear rate defined by the inductance of L1 and the difference between the applied voltage and the output voltage.

When Q1 turns "off," current will continue to flow in L1 in the forward direction, rapidly taking the junction of the Q1 emitter and the top end of D1 (cathode) negative. As the voltage passes through zero, D1 will conduct to maintain the current flow in L1 in the forward direction and will clamp the voltage at the LHS of L1 to a diode drop below zero.

The voltage across L1 will now be the output voltage (across C2) plus a diode drop, but will be in the reverse direction, with the right-hand side of L1 positive. The current in L1 will now decrease during the "off" state of Q1. Since, for steady-state conditions, the forward and reverse volt-seconds across L1 must equate to zero, the output voltage and hence the current can be adjusted by controlling the duty ratio of Q1.

It should be noticed that the input current is discontinuous and the output current is continuous. Further, the output voltage is less than the input voltage, but the polarity is the same. The buck regulator, in the continuous- or discontinuous-conduction mode, does not have the dynamic problem of the right-half-plane-zero that is inherent in the transfer function of the boost regulator.

1.6.2 Negative Buck Regulator

Figure 4.1.17b shows the negative version of the buck regulator. It has the same characteristics as the positive version, except that the common line is now positive and the input and output are negative.

1.6.3 Transformer-Coupled Buck Regulator (Forward)

Figure 4.1.17c shows a transformer-coupled version of the buck regulator. It is clear by inspection that this is the same as the circuit shown in Fig. 4.1.17a, except that the square-wave power pulse from the switching transistor Q1 has been reflected through to the secondary of transformer T1. This circuit operates in the same way as that in Fig. 4.1.17a, except that the transformer provides isolation between input and output. The turns ratio provides for any combination of input and output voltages. The topology is often referred to as a single-transistor forward converter.

1.7 COMBINATIONS OF CONVERTERS

It has been shown that the boost-type power factor correction circuit has the potential to accept the line input voltage, after rectification (the haversine waveform), and maintain the shape and phase of the input current as a similar haversine, so that at the input to the bridge rectifier, there will be a sine wave current in phase with the voltage to simulate a resistive load and provide the required near-unity power factor at the input terminals.

As previously explained, the output voltage will be a stabilized DC which must exceed the maximum applied voltage at all times. Hence, for example, if the input voltage has a maximum rms value of 137 V, then the peak value would be 192 V; since the output voltage must exceed this value, 200 V DC may be chosen.

In the case of a simple boost regulator (as shown in Fig. 4.1.16a), there is a direct connection between the input and output terminals through L1 and D1, and it is not possible to provide short-circuit protection. Also, the output voltage is fixed (or at least it must be above a defined minimum).

In many cases, the fixed output voltage will not be exactly what the load requires. Further, it may be necessary to adjust the output voltage, provide short-circuit or overload protection, or provide isolation between input and output. Consequently, it is quite common to follow the PFC boost stage with a second power processing section. As a result, combinations of converters will often be found in many power factor correction applications; some examples follow.

It should be remembered that the properties of the boost topology will be carried through to any combination of converters which includes a continuous-conduction boost stage. Hence the right-half-plane-zero limitation will apply to the transfer function of the combination.

1.7.1 Boost-Buck Combination

Figure 4.1.18a shows a typical tandem combination of a positive boost regulator followed by a positive buck regulator.

In this combination, the negative line is common to both input and output. L1, D1, Q1, and C2 form the front-end power factor correction boost regulator. The high voltage developed across C2 is reduced to a lower, adjustable voltage at the output, C3, by a buck regulator stage, Q2, L2, D2, and C3. The buck stage provides voltage regulation, and Q2, by breaking the direct current path to the output, has the potential to provide current limiting.

A disadvantage with this combination is that the buck stage transistor Q2 is difficult to drive, as it does not share a common return with the first stage Q1. Although drive transformers may be used, they limit the range of duty ratio control.

1.7.2 Boost-Buck with Common Emitter

Figure 4.1.18b shows the combination of a positive boost regulator with a negative buck regulator. This arrangement is sometimes preferred because the two switching

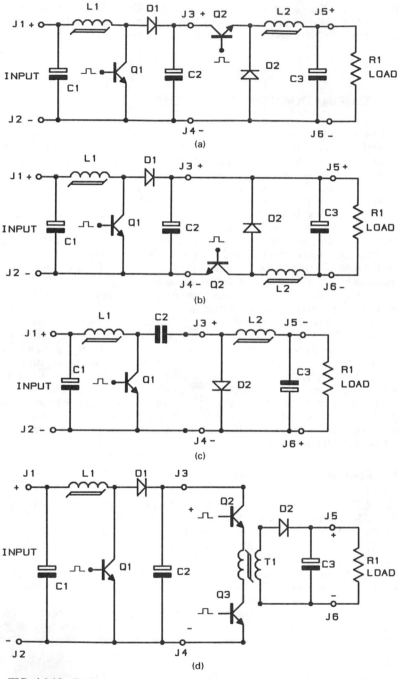

FIG. 4.1.18 Basic power topologies. (a) Nonisolated, three-terminal, noninverting boost-buck combination. (b) Nonisolated, four-terminal, noninverting boost-buck combination in which the switching devices share a common return line for easy driving. (c) Three-terminal, nonisolated, inverting Ćuk[10]-type boost-buck-derived regulator. (d) Isolated, four-terminal, transformer-coupled boost-buck converter.

devices share a common return line; this permits a single drive IC to provide the drive directly to both transistors, without the need for a drive transformer. This topology is used later in the applied engineering example.

1.7.3 Ćuk[10] Regulator

Figure 4.1.18c shows how the combination shown in Fig. 4.1.18a becomes the Ćuk regulator when the output diode D1 is replaced by a capacitor C2 and the switching device Q2 is removed. The combination remains a boost-buck combination.

1.7.4 Transformer-Isolated Combinations

Figure 4.1.18d shows a boost regulator, for power factor correction, followed by a two-transistor transformer-coupled forward converter of the buck type. In this example, by duty ratio control and selection of turns ratio, the output voltage is fully adjustable and is totally isolated from the input supply.

1.7.5 Conclusions on Topologies

Although the boost circuit of Fig. 4.1.16a has become the preferred industry standard for power factor correction in higher-power applications, there is a great deal of room for innovative design. If power factor correction can be integrated into the main control converter, then the number of power stages will be reduced, efficiency will be improved, and cost will be reduced.

It is useful to bear in mind that any arrangement which can maintain the input current sinusoidal and in phase over the required working range can be used. There are many possibilities: All the boost-derived and the various current-fed topologies have the potential to provide the required sine wave input current for integrated power factor correction applications. Only a few of these possibilities have been shown here.

1.8 INTEGRATED CIRCUITS FOR POWER FACTOR CONTROL

It has been shown that there are several linear control variables which must be processed by any control circuit used for power factor correction. It will also be shown later that compensation for input-voltage changes requires a more complex nonlinear adjustment. Further, to implement this compensation, the loop gain of the current control loop is best adjusted as the input voltage changes. This requires a variable-gain pulse-width modulator, which is difficult to implement with discrete components.

Fortunately, many integrated circuits are now available for this function. These ICs have been designed for various topologies and applications, and the designer must make the best choice for the intended application. Low-cost 8-pin ICs are normally designed for variable-frequency, complete energy transfer, boost-type topologies and are more suitable for lower power levels. Applications would include ballasts for fluorescent lamps up to, say, 150 W. Other topologies used at lower power levels include the buck-boost or bootstrap boost, where the output voltage is required to be less than the peak input voltage. The buck-type topology can be used for very low output voltages and the flyback boost for higher output-voltage applications.

The more complex 14- to 20-pin ICs are normally reserved for high-power, high-performance applications. Very often these will be used with continuous-conduction, average current or hysteretic, boost-type topologies. These combinations provide the best power factor, lowest harmonic distortion, and low-input RFI noise current.

Since many different IC designs are available, it would be helpful to examine some of the more basic key parameters which will be intrinsic to the design of most high-performance ICs for boost applications.

1.8.1 Key Requirements for a Boost PFC Control IC

To identify the key control requirements, we should start by reviewing in more detail the functional properties inherent in the power sections of a typical continuous-conduction boost regulator, and then we should consider how the boost regulator functions. Figure 4.1.19 shows the basic arrangement of the power elements in a typical three-terminal power factor correction boost circuit, together with a simple control block. The input to the unit is a sine wave from a normal line supply. The input to the buck regulator section is a unidirectional haversine from a full-wave rectifier bridge (BR1). We should notice the following fundamental functional parameters.

1.8.2 Key Parameters of a Continuous-Conduction Boost Regulator

1. Common Line. With reference to Fig. 4.1.19, after ac line rectification (BR1), the negative line is common to both input and output; this allows the drive to the power

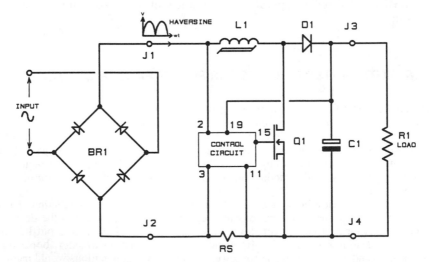

FIG. 4.1.19 Basic power topology for a nonisolated, power factor correction, positive boost regulator, showing the essential elements of the power and control sections.

switching device Q1 to be common to the negative line, eliminating the need for a drive transformer. For the same reasons, most control ICs will have a negative common terminal (pin 11 in this case).

2. Inductor Size. The positive rectifier output (a haversine waveform at 120 Hz) is fed to the input of a relatively large inductor (L1), which must ensure that the current at the input is relatively free from the high-frequency high-power switching currents found in Q1 and D1. Also, L1 must maintain continuous conduction for most of the haversine. These requirements limit the minimum value of the inductance.

Of course, the low-frequency 120-Hz haversine current must be free to flow through L1 with little impedance. This, together with the required current slew rate, limits the maximum value of the inductor L1.

The high-frequency noise rejection provided by L1 allows the input filter to more easily satisfy the agency RFI limits at the ac line input.

3. Current Sensing. Since the positive input current is continuous (for most of the haversine) and relatively noise free, the negative return line current will be the same. Hence, a resistive shunt R_S in the negative return line will provide a voltage analogue signal of the haversine current to the control circuit IC on pin 3. Because of the direction of current flow, this signal is negative with respect to the negative common line at pin 11.

4. Output Capacitor Size. The high-frequency discontinuous output current from D1, together with the 120-Hz modulation current, flows into a relatively large output capacitor C1, which is chosen to ensure that the output ripple voltage will be relatively low (typically in the range of 5 to 20% of V_{out}). To deal with the high- and low-frequency ripple currents, C1 will often consist of a combination of film and electrolytic capacitors.

The output capacitors are common to the negative supply line and return most of the ripple currents to this line. This, together with the load current and the remainder of the load ripple current, alternates with the ripple current from Q1, so that the sum of the currents in the negative return line is the same as the mean current in L1. Hence, the mean current in the return line is the same as the mean current in L1.

5. Small-Signal Transfer Function. As with all continuous-current boost topologies, in order to increase the output current the "on" period of Q1 is increased (the intention is to increase the inductor current). However, the immediate effect will be to decrease the "off" period of Q1, and hence the period when D1 conducts is also reduced. This will initially decrease the output current instead of increasing it. It may take several cycles for the current in L1 to increase sufficiently to increase the output current.

This is the basic dynamics of the noncompensatable right-half-plane-zero, and the small-signal loop gain must be reduced, at a relatively low frequency, to prevent current control-loop instability (see Part 3, Chap. 9).

1.8.3 Boost Regulator Function

With reference to Fig. 4.1.19, we can now examine the function of the boost regulator. In simple terms this is as follows:

Assume a starting condition such that the input voltage at L1 has some positive value, current is flowing in L1 in the forward direction (left to right), and the output voltage exceeds the input voltage.

When Q1 turns "on," the forward voltage across L1 is the instantaneous haversine voltage, and the forward current will increase in L1 during this period. Since L1 is relatively large, the current change in L1 during this single "on" period of Q1 will be relatively small.

When Q1 turns "off," induction in L1 will maintain the current flow in the forward direction, and the voltage on Q1 and the anode of D1 will quickly increase until D1 is forward-biased. At this point, L1 current flows via D1 into the output capacitor C1 and into the load.

Since by design, the output voltage always exceeds the input haversine voltage, the voltage across L1 is now reversed while D1 conducts (having a magnitude of $V_{out} - V_{in}$), and the current in L1 will decrease while Q1 is "off" and D1 conducts.

Hence, by adjusting the duty ratio of Q1, the current in L1 can be made to increase during the rising edge of the input haversine and decrease during the trailing edge, as required, to make the mean current waveform track the applied 120-Hz haversine voltage waveform. Essentially, this would appear to be a very simple function; however, consider the following requirements.

1.8.4 Key Control Requirements

1. *Modulation of Q1.*   From the preceding we can see that the prime requirement of the control IC is to control the switching action of Q1 to maintain a good haversine current waveform, which tracks the applied voltage haversine waveform, so as to simulate a purely resistive load.

2. *Output voltage control.*   For the boost process to be continuous, the output voltage on C1 must exceed the input haversine voltage at all times. Hence, a second requirement of the control IC is to maintain the output voltage constant and above $V_{peak(haversine)}$.

3. *Power control.*   If the output voltage is maintained constant and the load resistance changes, the output current and output power must also change. Hence, the rms input current must also change in order to keep input and output power equal, while still maintaining the haversine waveform shape. Thus, a third requirement of the control IC is to adjust the longer-term input rms current in response to load changes.

4. *Input voltage compensation.*   If we assume that the load resistance and output voltage remain constant for a period, then the output current and output power will also remain constant for the same period. However, if the input voltage changes during this period, then to maintain a constant input power, the rms input current must again change. Hence, the final requirement of the control IC is to adjust the mean haversine input current to compensate for any longer-term input voltage changes.

It turns out that this last requirement is the most difficult task, because the open-loop gain of the power stage changes, with changes in the input to output voltage differential. (This effect is covered more fully in Part 4, Sec. 1.9.10.)

In general, the control circuit must adjust the modulation of Q1 to compensate for any combination of the above variables in the longer term, while maintaining a good haversine current waveform in the shorter (per cycle) term. We can summarize these requirements as follows:

Key Requirements of the Control IC

1. The prime requirement is to ensure that the input current waveform follows a good, clean, harmonic-free haversine waveform, in phase with the applied haversine input voltage.

2. A second requirement is to maintain the output voltage constant *with variations in load current* and hence output power.

3. A third requirement is to adjust the rms input current to maintain the input power equal to the output power *for load changes.*

4. A fourth requirement is to adjust the rms input current to maintain the input power equal to the output power *for input voltage changes.*

In the boost topology, requirements (2) and (3) *are both satisfied* by maintaining the *output voltage constant*, so normally only three requirements, (1), (2), and (4), need to be considered.

In principle, all power factor control ICs of the boost type must satisfy the basic requirements outlined above. However, they may achieve these ends in many different ways.

Having established the basics, we can progress to considering the essential internal elements of a typical control IC for boost applications.

1.9 TYPICAL IC CONTROL SYSTEM

We will now consider how the previous functions may best be implemented in a typical boost-type power factor control IC. Although the following analysis is based upon the Micro Linear 4826,[88] most ICs used for this application will have similar basic elements, although the implementation may differ in detail.

A good understanding of the function of each element of the control IC is essential for the most effective implementation in a complete system. The designer has under his control much of the design of the analogue control circuit and the power sections, and must optimize each design for best overall performance.

1.9.1 Power Section

In the following example, it will be assumed that the control IC will be used in a power factor correction boost regulator with a power section and control block similar to those shown in Fig. 4.1.19.

To aid understanding, a more specific application will be considered. The input is to be 60 Hz with the universal input voltage range of 90 to 270 V rms via a line RFI filter. The line input is full-wave-rectified by BR1 to give an input haversine voltage at 120 Hz to L1. The output voltage will be 385 V DC, so that the output voltage will exceed the peak input voltage at all times. The load is simulated by a variable resistor R1.

1.9.2 IC Building Blocks

The control functions will be more easily understood by building up the internal control blocks in sequence, as follows:

Figure 4.1.20 shows the power section of Fig. 4.1.19, with additional detail provided for the control block. The boost power section is made up of L1, D1, C1, and Q1. A current-sensing resistor R_s is also shown in the negative return line.

Shown in the control block are the first two sections of the control circuit. The function of these parts is discussed next, with reference to Fig. 4.1.20.

1.9.3 Current Sensing

A low-value shunt resistor R_s in the common negative return line will provide a voltage analogue signal of the current on pin 3 (negative with respect to the common pin 11). This signal is proportional to the mean L1 inductor current, and hence the input current. In high-power applications, other low-loss current-sensing devices may be used, for example, Hall effect current transducers.

1.9.4 Pulse-Width Modulator (PWM)

The main switching FET, Q1, is driven by a pulse-width modulator, A2, at output pin 15. The inverting input to the PWM (pin 1 of A2) is a 5-V triangular waveform at a fixed frequency of 50 kHz, ranging from, say, 1 to 6 V. (Large signals are normally used to provide a good signal-to-noise ratio.)

The duty ratio of the square-wave drive to Q1 can be adjusted from zero to 100% by adjusting the DC signal voltage applied to the noninverting PWM input (pin 2 of A2); the signal voltage can range from <1 to >6 V. Thus, modulating the DC signal voltage can modulate the duty ratio of Q1 and hence the current in L1.

1.9.5 Current Error Amplifier (IEA)

Figure 4.1.20 also shows the previous stage, A1, which is the current error amplifier (IEA). This is a high-gain virtual-ground amplifier, with its inverting input (pin 1 of A1) connected to the common negative line (pin 11) and its output signal voltage connected to the PWM input. The noninverting input of the IEA (pin 2 of A1) is connected to the junction of two resistors R1 and R2, which function as described next.

1.9.6 Current Reference Signal

The input haversine voltage from the bridge rectifier BR1 is applied via pin 2 to a variable-current section of the IC, shown as the "gain modulator." At this stage, we can simulate the action of the gain modulator with an adjustable resistor R1. R1 converts the input haversine voltage to a haversine reference current, which is applied to the current summing point, SP, located at the junction of R1 and R2 and connected to the current error amplifier virtual-ground input at pin 2 of A1.

To maintain the summing-point voltage at virtual ground near zero (the voltage of pin 11), an equal and opposite negative current must flow down R2 to pin 3. To maintain this condition throughout the applied voltage haversine, the negative voltage on pin 3 (and hence the current) must track the positive voltage on pin 2 with a ratio determined by the ratio of resistance R2/R1.

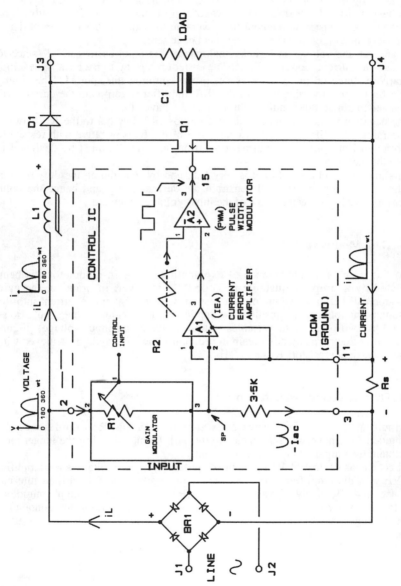

FIG. 4.1.20 Power factor correction boost regulator, showing the essential control elements of a typical fast inner-loop current-control stage, used to maintain a haversine current in L1.

1.9.7 Inner Fast Current Control Loop

In closed-loop operation, the duty ratio of Q1 will be adjusted on a fast cycle-by-cycle basis (as the haversine voltage waveform changes relatively slowly). The current will be adjusted throughout the haversine to maintain the IEA amplifier at balance and the voltage at the summing point near zero.

In general, the voltage on R_s is an analogue of the input current in L1 (since R_s completes the current loop from BR1). The current returning to BR1 via R_s develops a negative voltage on pin 3 with respect to the common line, pin 11. The aiming voltage for the noninverting input of the IEA amplifier is zero, and the current will be adjusted to obtain this condition throughout the haversine.

The negative voltage at pin 3 is dropped across R2, internal to the IC (between the summing point SP and pin 3). Hence pin 3 of the IC is negative with respect to the common line, pin 11, in normal operation, and the IC must be designed to tolerate this negative input.

This inner current control loop is very fast, and by modulating the duty ratio of Q1, the closed-loop control will constrain the current in R_s (and hence the input current) to closely follow the shape of the input voltage haversine.

1.9.8 Gain Modulator

Up to this point, for convenience of explanation, the gain modulator has been simulated by a simple adjustable resistor R1. However, in practice, the gain modulator will be an active circuit which will deliver a haversine current into the summing point not only as a result of the applied input haversine voltage, but also as required by two other input variables (output voltage and input voltage). Figure 4.1.21 shows diagrammatically (again in principle) how this could be achieved with two series variable resistors R1a and R1b.

1.9.9 Output Voltage Regulation

In principle, the effective series resistance of R1a and R1b (and hence the magnitude of the haversine reference current) will be modulated in the longer term to maintain the output voltage on C1 constant.

Direct linear control of R1a is applied by the high-gain voltage error amplifier A3 (VEA). This amplifier compares the output voltage on C1 with an internal reference, and effectively adjusts the value of R1a so as to adjust the magnitude of the reference haversine current (and hence the mean output current) to maintain the output voltage on C1 constant. By this means, the input current is adjusted to maintain the output voltage constant *with variations in load*.

This outer voltage control loop is made relatively slow compared with the inner current control loop.

1.9.10 Compensation for Input Voltage Changes

This variable has been considered last, as it is considerably more complex than the other variables and requires special attention.

If the longer-term input voltage changes, there will be a tendency for the output voltage to change. Although the slow outer voltage control loop, control amplifier

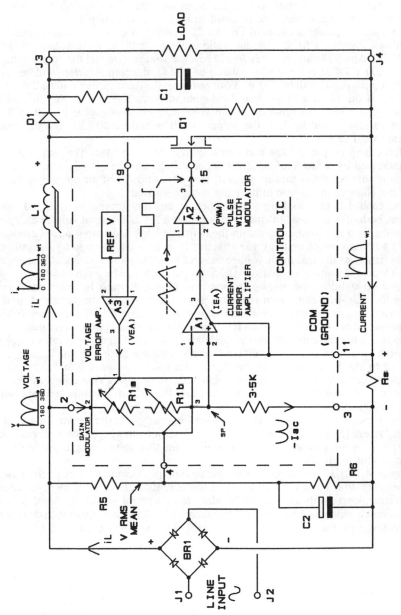

FIG. 4.1.21 Power factor correction boost regulator, showing the essential slow outer-loop control elements required to maintain the output voltage constant and compensate for input voltage changes.

(VEA), will compensate for this change in the longer term, simple linear control is not acceptable for this parameter.

To explain this, we must look at the open-loop transfer function of a typical boost regulator. This is best done by considering a specific example.

Consider the power sections of Fig. 4.1.21 working in open loop under fixed-DC conditions. Assume a fixed 50% duty ratio condition, with an input voltage of, say, 100 V DC. At a 50% duty ratio, there will be a 2:1 voltage gain and the output voltage will be 200 V DC. Let the load resistance be 200 Ω; the output current will be 1 A, and the output power will be 200 W. With zero loss, the input current will be 2 A.

Now, if the input voltage were to double to, say, 200 V with the duty ratio remaining at 50%, the open-loop output voltage would also double to 400 V. The output current would be 2 A, the output power would be 800 W, and the input current would be 4 A.

Hence both output voltage and output current have doubled. The output power has increased by a factor of 4, for an input voltage change of only a factor of 2. Hence it will be found that in general, the output power of an open-loop boost regulator changes in the ratio of the input voltage squared.

The control loop is looking at two inputs at this time, output voltage and input current, both of which tend to change by a factor of 2:1 as the input voltage changes. When the loop is closed, to maintain the output voltage and output power constant (using a linear control system), the effective loop gain will change by the ratio of V_{in}^2:1. Hence, as the rms input voltage changes over the industry standard range, 90 to 260 V ac, the unity-gain crossover frequency will change by a ratio of 8:1.

For loop stability, the unity-gain crossover frequency must be maintained well below the right-half-plane-zero frequency at all times. Hence, the large shift in the unity-gain crossover frequency would require a very low rolloff frequency in the compensation network. This would result in a very poor transient performance.

To compensate for this effect, the modulator gain is reduced by the ratio $1/V_{in}^2$ as the mean long-term rms input voltage changes. Hence a further input, $V_{rms(mean)}$, is applied to the gain modulator block to effectively increase R1b (reduce the modulator gain) for longer-term rms input voltage changes. This compensates for the change in open-loop gain so that the closed-loop gain remains constant with input voltage changes.

This removes the change in the crossover frequency, and the control loop can be much faster. (Since the gain modulator must not change its gain during the haversine, a very low pass filter is used on this mean rms input voltage compensation signal.)

Figure 4.1.21 shows in basic principle the complete control block of the power factor correction circuit. In addition to the essential elements shown, in a real application, loop compensation will be added to the control amplifiers. Also, various protection circuits will be found in a commercial IC. Typically these include output overvoltage protection, peak input current limiting, soft start, and brownout protection.

1.9.11 Brownout Protection

This protection parameter will be considered in more detail, as it is an essential requirement, not only to deal with below normal input voltage conditions (brownout), but even for safe turn-"on" and turn-"off" of the boost regulator.

It should be remembered that the boost topology in the closed loop will attempt to continue to supply a constant output voltage and power as the input voltage falls.

To maintain constant output power as the input voltage is reduced, the input current will increase in the same ratio. Hence, for each 50% decrease in input voltage, the input current will double. Clearly, if this progression were not limited, it would lead to destruction of the switching device at some low input voltage. Hence some form of input-current limiting or low-voltage shutdown is essential for reliable operation. This can be applied in many ways, and will not be covered in more detail here.

Now that the essential requirements of a PFC control IC have been established, it will be useful to look more closely at the design of a professional IC available for this application.

1.10 APPLIED DESIGN

In this section, we will consider the design steps required to implement a commercial 2200-W PFC variable-output DC power supply design, using one of the standard commercial control ICs for the control functions. The unit is required to provide a regulated output voltage, which can be adjusted over the range 50 to 400 V for an isolated resistive load. (Hence, in this application, the output need not be isolated from the input, and an isolation transformer is not required.)

1.10.1 System Specifications

The first step is to establish a realistic specification for the required application. As with most things in life, there is no free ride, and this is particularly true for power factor correction equipment. There will be a penalty to pay for extending the power range, output-voltage range, input-voltage range, and range of control. Hence the specification should not extend these parameters further than the essential needs. These factors all have an impact on size, weight, efficiency, cost, performance, and reliability. It is just not good engineering to specify more than the minimum requirements for the application, providing a realistic, but small, safety margin.

The following outline specification will be used to design this 2.2-kW supply:

1. Input voltage (rms values)
   a. Nominal: 227 V, 50/60 Hz
   b. Working range: −20% to +10% (220 to 305 V)
   c. Brownout voltage: 200 V
2. Output voltage (DC values, adjustable and nonisolated)
   a. Nominal: 300 V DC
   b. Working range: 50 to 400 V DC
3. Output current (DC amperes)
   a. Nominal: 6.6 A at 300 V output
   b. Maximum: 8 A
   c. Range: Constant-resistance load (hence current decreases with decreasing voltage)
4. Output power (true watts)
   a. Nominal: 2 kW
   b. Working range: 450 to 2200 W (open-circuit- and short-circuit-protected)
5. Output ripple
   a. 10% maximum
6. Power factor and harmonic content
   a. To satisfy IEC 1000-3-1 Class C over the power range from 450 W to 2 kW

7. Efficiency

    a. >95% at 2 kW

8. Agency requirements

    a. Safety: U/L, CSA, VDE
    b. EMI: FCC Class A

1.10.2 Specification Review

The following points are important with regard to the above specification, and should be considered in all applications.

1. Input Voltage. It will be shown that the brownout voltage (minimum working voltage) defines the maximum current stress on the power components; hence it should be as high as possible to minimize this stress.

The power supply will be designed to provide full output power down to the brownout voltage value (although some degradation in performance may be permitted between the brownout value and the low end of the normal working range).

In this example, at 2.2 kW output and 90% efficiency, the peak input current at the brownout voltage of 200 V will exceed 17 A rms. If the brownout voltage were lower, say 90 V, the current would exceed 38 A.

Although universal input voltage ranges (80 to 257 V) are often specified for PFC units, the high current penalty remains for all designs, and this tends to limit this type of unit to the lower-power applications.

At the other end of the scale, the peak input voltage sets the working voltage stress on the power components. Again, the input voltage specification should not be higher than necessary. Remember, in the boost circuit, the output voltage must exceed the peak input voltage (>435 V in this example). Some allowance should be provided for input voltage transients, and transient protection should be provided.

2. Output Voltage Control. In this application, the output is common to the input (line isolation is not essential for an isolated load). This opens up the possibility of eliminating a second power stage. However, a further critical requirement to allow the output to be taken directly from the boost section is that the output voltage must exceed the peak input voltage at all times.

In this application, a range of output voltages is required (50 to 400 V DC), and short-circuit protection is also required; hence a direct output from the boost section is not possible. For this application, an additional voltage control power stage is required, and a simple buck regulator will be used.

3. Output Current Limiting. An important property of the buck regulator is that the output power may remain constant at all output voltages. Hence, a 2.2-kW buck stage has the potential of providing up to 44 A at 50 V output. Only 8 A is required in this application, so some form of output current limiting would reduce the stress on the output power components at the lower output voltages.

4. Output Power. There are two important parameters here. The first is the range of input voltage and output voltage over which the power can reasonably be expected to remain at the maximum value. This is required in order to establish the maximum currents for the power components.

The second important parameter is the range of power over which the full harmonic specification is expected to apply. The performance of the PFC circuit tends to degrade as the load falls, because the feedback current signal is getting progressively smaller. Hence, the signal-to-noise ratio for the PFC stage degrades at low powers.

In this example, because of the constant-current nature of the load, the lowest power at 50 V will be 400 W, and full specification performance could be expected over the range 2.2 kW to 400 W.

5. Power Factor and Harmonic Content.   These limits depend on the end user's application and the country of use. In general, the more stringent IEC 1000-3-2 Class C requirements will apply to the harmonic content and power factor for the most stringent applications.

6. Output Ripple.   In this application, up to 10% high-frequency ripple is permitted (40 V at 400 V output), so only the minimum output filtering is required, reducing the size and cost of the output components.

7. Efficiency.   Clearly, this should be made as high as possible, and at a 2-kW power level, efficiency values exceeding 95% are possible with a nonisolated system of this type. (Values of 96% were measured in the finished unit.)

8. Agency Requirements.   The RFI limits are important to the initial design, as stringent Class B limits tend to require a larger input inductor for the boost regulator section, to reduce the switching frequency ripple. This high-frequency ripple component is difficult and expensive to eliminate with an input EMI filter.

The less stringent Class A limits (office and industrial applications) in general (at this time) do not limit emissions below 150 kHz. Hence when Class A limits apply, it is wise to choose a switching frequency which brings the second and possibly third harmonics below 150 kHz (50 kHz is used in this design).

With the specification well defined, we can, with confidence, select the power arrangement to best meet the needs. This completes the specification review.

1.10.3 Choice of Power Section Topology

The need for the highest power factor and lowest harmonic distortion at high power demands that the best PFC methods should be chosen, irrespective of cost and other factors.

A natural selection would be the positive boost topology for the power factor correction front end. Since the maximum input voltage is to be 305 V rms, the peak input at the crest of the rectified haversine would be 430 V, and the boost output must exceed this value. To provide a working margin, a regulated output voltage of 450 V is chosen for the boost stage; also, standard electrolytic capacitors are available for this voltage.

Further, since the output voltage is to be variable over the range 400 to 50 V DC, a second adjustable voltage power stage is clearly required. In our favor, the output need not be isolated from the input, and a transformer-type converter is *not* required.

This allows the use of a simple three-terminal buck regulator. This would make the output power stage very efficient, simple, and low in cost. Hence a good choice of topology would be a direct-coupled boost-buck combination as shown in Fig. 4.1.18*b*. A more complete version of this with a control block is shown in Fig. 4.1.22. The control block will drive both the boost and the buck sections directly, as shown, and contains the control IC as well as the additional components required to complete the control functions.

1.10.4 Basic Principles for Boost Section

With reference to Fig. 4.1.22, the boost power factor correction front end would operate as follows:

The 60-Hz ac line input is taken in via the RFI line filter to the rectifier bridge inputs J1 and J2. This sine wave is full-wave-rectified by BR1 to produce a 120-Hz haversine at the input to L1. The haversine would have a peak voltage of 430 V at 305 V rms line input.

This 120-Hz haversine is applied to L1 and the control circuit in the boost power factor correction section, which maintains the required haversine current waveform in L1 and delivers a fixed DC output of 450 V via D1 to the common intermediate capacitor C1. The 450-V DC voltage is then reduced to the required DC output by the following buck regulator stage.

1.10.5 Buck Section

The negative rectified line from BR1 is common to both input and output (except for a small voltage drop across the current sense resistor R_s and R_T). Hence, both power transistors can be directly driven by a drive signal from the control IC, without the need for a drive transformer, since the IC is also common to the negative line J4 at pin 11.

This is a major advantage, as we will find that the drive to the buck regulator section requires duty cycles of up to 90%, which are very difficult to obtain, except by direct drive methods. The design of the power stages will be covered in more detail later.

1.11 CHOICE OF CONTROL IC

Although many IC combinations by various manufacturers can be used, the IC chosen in this example will be the Micro Linear ML4826-1[88], because it is particularly suitable for this boost-buck combination. It provides the drive for both sections in a single IC. The full specification for this IC can be obtained from the manufacturer and should be studied. This, together with application notes 16, 33, and 34 from Micro Linear, will provide a more complete understanding of the application of this IC.

The following design review covers some of the more important parameters of interest in this design, and assumes that the reader will have or will obtain the data sheet for the ML4826-1 IC and the application notes, to which many of the following comments apply. This information may be obtained directly from Micro Linear Corporation, 2092 Concourse Drive, San Jose, CA 95131, or from a local

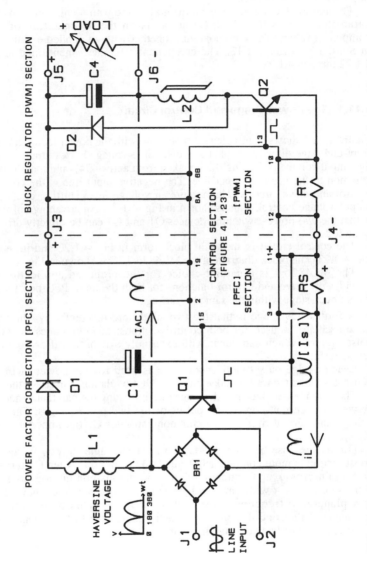

FIG. 4.1.22 Basic power and control parts of a power factor correction boost-buck combination providing regulated and variable DC outputs.

distributor. It may also be downloaded from Micro Linear's Web site, http://www.microlinear.com.

The author has developed additional nomograms for some of the transfer parameters of this IC, which further assist in the optimization of the external component values. These nomograms, together with an applied example, are shown in Secs. 1.11 through 1.13. The component references mentioned apply to Figs. 4.1.22 through 4.1.28.

1.11.1 The Power Stage and Control Circuit

Figure 4.1.22 shows a suitable power section, with the power factor correction boost front-end stage directly coupled to the variable-voltage buck output stage. Note that the output voltage is floating (being taken from across C4), and that the output is *not* common to the negative input line. The negative input line is the common for the control and drive circuits. This requires a voltage level shifting circuit, to provide output-voltage control. This is shown in Fig. 4.1.27. An advantage of this topology is that the two power switching devices Q1 and Q2 can be directly driven from the same IC.

The control circuit is shown in block form in Fig. 4.1.22 and in more detail in Fig. 4.1.23. The block diagram of the ML4826-1 IC is shown in Fig. 4.1.24.

The ML4826-1 is a natural choice for the proposed application because it provides the drive and control functions for both the input PFC boost stage and the output buck stage within the same IC package.

A further advantage is that the two stages are inherently synchronized, as the same oscillator is used for both control sections; this eliminates the possibility of noise "jitter," which can occur with externally synchronized or nonsynchronized arrangements.

Leading-edge pulse-width modulation is used for the power factor correction boost stage, and the trailing-edge pulse-width modulation is used for the buck stage.

This arrangement has the advantage of eliminating the gap that is normally found between the current conduction pulses of the two power stages. This reduces the voltage and current ripple in the common capacitor C1 (an important advantage at high power levels).

Further, because both power stages run at the same (lower) frequency and the boost section is more continuously loaded (Q2 turns "on" when Q1 turns "off"), the unity-gain crossover frequency of the continuous-conduction boost section is increased, giving a wider margin between the crossover frequency and the right-half-plane-zero frequency (see Part 3, Chap. 9). This permits the crossover frequency of the control loop to approach one-half the line frequency, providing a faster response and a wider stability margin.

1.11.2 Protection and Ancillary Features

Figures 4.1.23 and 4.1.24 show, in block schematic form, the basic elements of the control circuit and the internal arrangement of the ML4826-1 IC. The combination has the following ancillary features, which improve the overall reliability.

1. PFC Current Limit (Input Current Limit). Within the IC, a fast comparator (A1) monitors the current signal voltage developed across the external shunt resistor R_s as it appears between pin 3 and the common analogue ground, pin 11 on the IC. If this

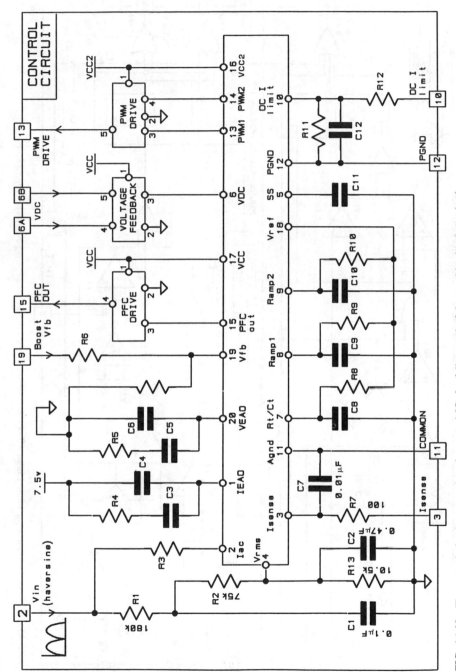

FIG. 4.1.23 The control circuit, external to the control IC, for full control of the combination boost-buck power factor correction converter shown in Fig. 4.1.22.

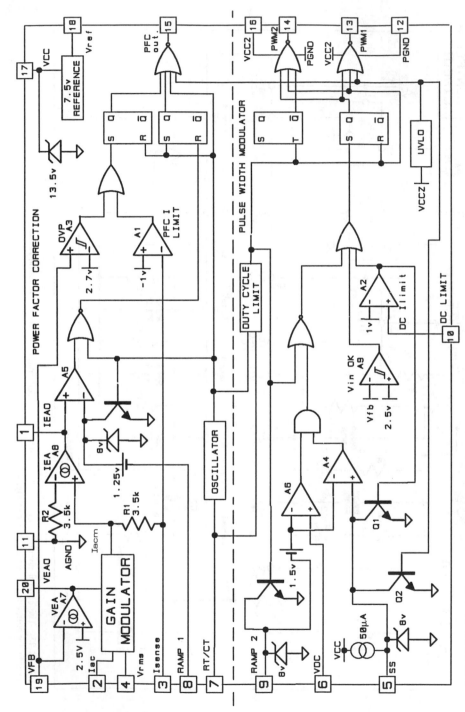

FIG. 4.1.24 Block diagram of the Micro Linear ML 4926-1 control IC used in Fig. 1.4.23. *(Courtesy of Micro Linear.)*

signal exceeds 1 V negative, the drive pulse to Q1 is terminated. This provides a pulse-by-pulse peak current limit on the input power switch to protect Q1. This limit may be activated for transient conditions outside of the normal control-loop response time.

2. Pulse-Width Modulator Limit (Output Current Limit). A fast comparator (A2) is provided to shut down the drive to the buck stage (Q2) if the "DC limit" signal on pin 10 exceeds 1 V. In this design example, the limiting signal will be developed from a small resistive shunt R_T in the source of external power device Q2, so as to provide pulse-by-pulse current limiting in Q2. At the same time, this also provides output DC current limiting. This is possible because when Q2 is conducting, its current is derived from L2, so that its mean level is an analogue of the DC output current.

After a current limit shutdown, Q1 in the IC discharges the soft-start capacitor C11 via pin 5, and soft-start action is initiated to restart the buck section after the overload is removed (see paragraph 5).

3. PFC Output Overvoltage Protection. A fast comparator (A3) with hysteresis is provided in the PFC control section of the IC to terminate the conduction of Q1 if the PFC output voltage on the external capacitor C1 exceeds the required preset voltage by more than 8%.

This protects the buck section from excessive voltage stress, which could occur if the load were suddenly removed. (When Q1 turns "off," the stored energy remaining in L1 is dumped into the output capacitor C1, causing only a few volts increase in output voltage, because C1 is large compared with L1.)

4. PFC Soft Start. Soft-start action is provided for the PFC section by taking the compensation capacitors C3 and C4 on pin 1 of the IC to the 7.5-V positive reference voltage on pin 18. (This is shown in Fig. 4.1.23.)

During start-up, the compensation capacitors (which are initially discharged) take pin 1 positive as the 7.5-V reference develops, giving a minimum pulse-width drive to Q1. As the compensation capacitors charge, the voltage on pin 1 drops and the pulse width is progressively increased to the required value, giving soft-start action to the PFC stage.

5. Buck Regulator Soft Start. A capacitor C11, connected externally to pin 5 of the IC, is charged at a constant current of 50 A from the IC during power-up. This provides soft-start action for Q2 and the buck section via modulator A4 in the IC.

The beginning of this soft-start action is delayed by Q2, which remains "on" until the IC supply voltage V_{cc} has been correctly established.

Further protection is provided by amplifier A9, which inhibits drive pulses to the PWM section (Q2) until the correct working voltage on external capacitor C1 has been established by the PFC boost section. This prevents the buck section from loading the input boost section until the correct working voltage has been established on C1.

Also, during an output overcurrent shutdown, the soft-start capacitor is discharged by Q1, invoking a soft-start recovery action (see paragraph 2).

6. Reference Voltage. A 7.5-V reference voltage is available at pin 18 and is used for internal and external control functions and bias.

7. Noise Immunity. The large haversine input voltage at the input of L1 is applied to pin 2 of the control circuit. R3 provides a current to pin 2 of the IC which follows the haversine shape.

Within the IC, this pin 2 current flows via the gain modulator to become the source for the much smaller haversine reference current signal I_{acm} at the output of the gain modulator on R1.

By this means, the large input haversine voltage is converted to a current by the high-value series resistor R3 connected from L1 to the gain modulator input at pin 2. (Two or three equal-value series resistors are normally used for R3 to reduce the voltage stress.) Using a current signal (rather than a small voltage signal, as found in some ICs) gives much better noise rejection to this critical parameter.

In addition, the amplitudes of the voltage ramps used by the pulse-width modulators A5 and A6 in both the PFC section and the buck drive section are large (typically 5 V), giving very good signal-to-noise ratios for these modulators.

Two common return (ground) pins are provided. Noise-sensitive analogue signals return to pin 11, and power drives return to pin 12. This allows the large peak currents of the drive outputs to be isolated from the control signals, further reducing noise problems.

Finally, the use of transconductance amplifiers, A7 and A8, provides further noise rejection, as shown next.

8. Loop Compensation. The voltage and current error amplifiers, A7 and A8, within the IC are transconductance types, which give an output current proportional to the input voltage differential. This allows the loop compensation components to be returned to the common line, rather than to the inverting input of the amplifier.

This makes the compensation components less noise-sensitive. Also, the voltage-divider resistors for the voltage error amplifier are no longer linked to the compensation capacitors, and are not part of the compensation network. Hence, they may have a much wider range of values, and adjusting the boost output voltage does not change the loop compensation parameters.

1.11.3 The ML4826-1 Control Sections

The internal organization of the IC is shown in Fig. 4.1.24. The IC is designed with two sections, a power factor control section (PFC) and a pulse-width modulator section (PWM). These are linked together by three key parameters.

The first is the working frequency; it can be seen that the same oscillator drives both sections to provide intrinsic synchronization between the PFC drive and the PWM drive.

The second is a link between the modulation methods, so that leading-edge modulation is used for one section and trailing-edge modulation is used for the other.

Finally, an undervoltage sensing circuit is provided (A9) which prevents the output buck stage from powering up until the input voltage is within the working range and the boost-stage output voltage is fully established. We will now look more closely at the two sections.

1.11.4 Power Factor Correction Section (PFC)

This section provides all the control functions for the boost power factor correction stage.

Gain Modulator.   A critical element of the design is the gain modulator. This section provides the modulated haversine reference current signal I_{acm} for the current error amplifier (IEA). The shape and amplitude of this current signal define the shape and amplitude of the input current, and hence the quality of the final power factor correction.

The gain modulator applies the internal haversine current reference signal I_{acm} to R1 and the current error amplifier, in the form of a modulated current drive. For best results, this signal must have exactly the same shape as the input voltage haversine. It is obtained from the current I_{AC} injected into pin 2 via an external resistor R3 connected to the input to L1.

By closed-loop control, the noninverting input of the virtual-ground current error amplifier will remain near zero, as the haversine current signal I_{acm} from the current modulator is summed to zero with a negative current feedback signal from the 3.5-kΩ resistor R1. The input to this resistor at pin 3 is the negative voltage developed across the external current shunt R_s as a result of the mean haversine current returning to BR1 in the negative supply line.

This balance is maintained by the current error amplifier, which acts on the pulse-width modulator A5 to adjust the duty ratio of Q1, and hence the mean current flowing in L1, such that the current waveform in R_s tracks the haversine reference signal in both shape and phase.

Longer-Term Current Amplitude Modulation.   Two much slower outer control loops adjust the effective resistance of the gain modulator, and hence the amplitude of the haversine current signal I_{acm}, and thus the mean rms supply current and boost output voltage. These variables are DC output voltage control and mean rms input voltage compensation.

1. *DC output voltage control.* A first signal voltage, from the boost output capacitor C3 via R6 and R14 to pin 19 and the voltage error amplifier (VEA), adjusts the effective resistance of the gain modulator, and hence modulates the amplitude of the haversine reference current signal I_{acm}. It does this in a linear way, so as to maintain the output voltage on C1 constant.

2. *Mean rms input voltage compensation.* A second input, proportional to the longer-term rms input voltage V_{rms}, is applied to pin 4. This signal is well filtered by R1, C1 and R2, C2 to give the long-term mean value of the haversine input voltage. Its function is to adjust the gain of the gain modulator to compensate for longer-term supply voltage changes.

   As explained in Part 4, Sec. 1.9.10, the change in the gain is required to follow an inverse square of the rms input voltage, to normalize the overall gain and take out the square-law dependency introduced by the boost power stage.

Brownout Current Limiting.   At and below the brownout voltage (1.2 V), the gain of the gain modulator is rapidly reduced, to prevent any further increase in input current and reduce the stress on Q1.

Figure 4.1.25 shows how the relative gain changes with the voltage applied to pin 4. The onset of brownout is just below 1.2 V at pin 4. At voltages above 1.2 V, the relative gain follows an inverse square law, and this is the normal working range.

During power-up or power-down and under brownout conditions, the supply voltage will fall below the minimum working voltage (220 V in this example). As the voltage falls, the boost action will tend to increase the input current to maintain the output voltage and power constant.

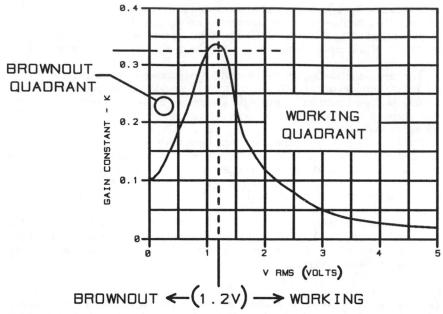

FIG. 4.1.25 Variation in gain of the "gain modulator" with mean V_{rms} input voltage change, showing normal square-law working quadrant and brownout power-limiting quadrant.

As the supply voltage decreases, the voltage on pin 4 will fall toward 1.2 V, with the relative gain increasing as $1/V_{in}^2$ as the voltage approaches the turnover point at 2.1 V.

At the brownout voltage (200 V in this example), the voltage on pin 4 will be exactly 1.2 V, as preset by the external divider resistors R1, R2, and R13.

It is intended that the current will have reached its maximum limiting value at this point. Therefore, as the voltage continues to fall and the voltage on pin 4 drops below 1.2 V, the relative gain decreases rapidly, as shown in Fig. 4.1.25, and you would expect the current to decrease.

However, there is a conflict at this point, because as the brownout action starts to reduce the gain of the modulator below 1.2 V, the output voltage starts to decrease.

Because of the large gain of the voltage error amplifier, if it were permitted sufficient gain margin, it would swamp the decrease in gain introduced by the gain modulator, and the current would continue to increase as the voltage falls. This could lead to failure of Q2 in a high-power system.

To prevent this action, it is necessary to correctly size several parameters, so that irrespective of the demands of the voltage error amplifier, the gain modulator will have "topped out" and cannot increase its output current I_{acm} any further at or near the brownout voltage point. To optimize this action, we need to look more closely at the interaction of the various inputs to the gain modulator.

1.12 POWER FACTOR CONTROL SECTION

Figure 4.1.24 shows the internal organization of the ML4826 control IC, with the top part showing the power factor control section. Figures 4.1.22 and 4.1.23 show the control components that are external to the IC. These component values must be optimized to provide the best performance.

The Micro Linear data sheets and application notes 16, 33, and 34 show how these component values may be calculated. It is expected that the reader will use this information, along with the following additional notes, during the design process. Hence, the following notes and nomograms are not complete and are only intended to assist in the calculation process provided in the application notes.

1.12.1 Shunt Resistor Value R_s

As we have seen, the output of the gain modulator within the IC is a haversine signal current, which is applied to the noninverting input of the virtual ground current error amplifier (IEA). At balance, this current is taken out via the internal 3.5-kΩ resistor R1, by the negative voltage signal applied to pin 3.

In turn, this voltage feedback signal is the negative voltage developed across the external current sense resistor R_s due to the return current flow in the negative line returning to BR1.

Under normal working conditions, the closed current control loop will adjust the current during the term of a haversine, to ensure that the current waveform in R_s matches the voltage haversine applied to L1. Hence, for best signal-to-noise ratio, the shunt resistance and hence voltage feedback signal on pin 3 should be as large as possible.

The absolute maximum value of R_s is defined by the peak load current and the maximum voltage allowed at pin 3 of the IC. The peak voltage permitted at pin 3 is 1 V, as this is the value defined by the PFC I limit comparator A1. However, prior to this value's being reached, the gain modulator (by design) will have "topped out"; that is, it will have reached its maximum output current.

The maximum gain modulator output current is specified as 200 μA, but it is not very well defined; it was measured as 250 μA on several ICs. This translates to a voltage on pin 3 of minus 0.875 V. (This is the voltage developed across the internal 3.5-kΩ resistor R1 at 250 μA.) As this is a critical parameter, the measured value of 250 μA and 0.875 V will be used in this engineering example to define the value of R_s and hence the maximum current limit value.

At maximum input power (2.2 kW) and minimum brownout voltage (200 V rms), the rms input current would be 11 A. The peak haversine current would be 15.55 A. This current must develop 0.875 V across the shunt resistor R_s, and the shunt value is easily calculated as follows:

$$I_{rms} = \frac{P_{max}}{V_{min(rms)}} = 11 \text{ A rms}$$

$$I_{peak} = \sqrt{2} \times I_{rms} = 15.55 \text{ A}$$

$$R_s = \frac{V_{pin3(max)}}{I_{peak}} = \frac{0.875}{15.55} = 0.056 \ \Omega$$

1.12.2 Establishing the Value of the I_{ac} Current-Setting Resistor R3

It is now necessary to set up the other inputs to the gain modulator so as to get its maximum output current at the brownout voltage of 200 V rms.

If the current limit occurs early (before the brownout voltage is reached), the output power may be less than intended, and the output power may be limited at low input voltages. If the current limit is set too late, the current in Q1 can become very high at low input voltages, and the stress on Q1 could be excessive. The correct conditions for the modulator inputs can be established as follows:

The value of the external resistor R3 (between the input of L1 and pin 2 of the IC) is defined by the brownout voltage and the peak value of the current required into pin 2, I_{ac}, at this brownout voltage.

The value of I_{ac} will be the current required to just bring the output of the gain modulator to its maximum limited output, 250 μA, at the brownout voltage, with the output of the voltage error amplifier (VEA) topped out at its maximum value, and the V_{rms} at pin 4 set at the gain modulator turnover point of 1.2 V. This information is not provided in the data sheets, and was obtained as follows:

The voltage control amplifier (VEA) was biased fully "on" so that it was topped out and was giving its maximum drive to the gain modulator. (In normal operation, when the parameter is correctly set up, this will occur as soon as the output voltage begins to fall, as a result of the input current limiting action.)

The voltage on pin 4 was then adjusted in the range 0 to 3 V, and the current into pin 2 of the IC (I_{ac}) was adjusted to just obtain the topped-out condition of the gain modulator. By this process, the transfer characteristic shown in Fig. 4.1.26 was developed.

We are now required to find the value of the input current I_{ac} that must be applied to the modulator at pin 2 so that it will just saturate the modulator and provide the required maximum output of 250 μA at the peak value of the brownout voltage (200 V rms).

At 200 V rms input, the peak haversine voltage applied to R3 will be 283 V.

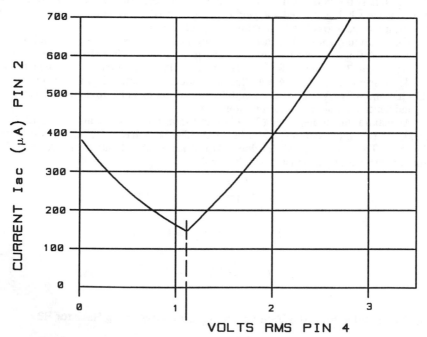

FIG. 4.1.26 I_{ac} current transfer characteristics of the gain modulator, for mean V_{rms} input voltage change.

From the transfer characteristic in Fig. 4.1.26, the peak I_{ac} at 1.2 V is 140 μA. To the first order, R3 can be calculated as follows:

$$R3 = \frac{V_{(brownout)peak}}{140\ \mu A} = \frac{283}{140\ \mu A} = 2\ M\Omega$$

Ideally, R3 would consist of two or more series resistors to reduce the voltage stress. These would be adjusted on test to optimize the value of I_{ac} so as to just get the maximum output from the gain modulator and define the maximum current limit exactly at the brownout voltage.

Figure 4.1.25 shows the variation in gain of the gain modulator. Notice that if I_{ac} is increased at the brownout value of 1.2 V, then the voltage amplifier will not be fully topped out at 1.2 V, and it will be necessary to reduce the voltage below 1.2 V to reduce the modulator gain and force the voltage amplifier to its topped-out limiting value.

Since the modulator gain drops quite rapidly at this point, it will be noticed from Fig. 4.1.26 that increasing I_{ac} to 160 μA reduces the brownout voltage by only 20% (with only a slight increase in maximum current). Since this provides a good working margin for drift and component tolerances, 160 μA will be used in this example. Hence R3 will be replaced by two series resistors of 876 kΩ each. (Two resistors are used to reduce the voltage stress.)

1.12.3 PFC Output Voltage Setting Resistors

The output voltage of the boost section is selected to exceed the peak supply voltage by at least 10 V; in this case, a voltage of 450 V is chosen.

Selection of the output voltage setting resistors is then quite straightforward, as the reference voltage for the voltage error amplifier (VEA) is 2.5 V. The network R6, R14 is chosen to provide 2.5 V at pin 19 with an output of 450 V DC on C1. Once again, two or more resistors should be used in series for R6 to reduce the voltage stress; 2 kΩ/V (0.5 mA) is suitable for this resistor network.

1.12.4 V_{rms} Network

This parameter is a little more difficult to set up. A simple resistor-divider network is used to set the voltage on pin 4 to 1.2 V. The large 120-Hz attenuation provided by capacitors C1 and C2 ensures that the voltage on pin 4 is the mean value of the applied haversine.

The choice of capacitors is a compromise. Too little attenuation at 120 Hz will give undesirable modulation of the gain at the haversine frequency. Too large an attenuation will give a large delay in the response to input voltage changes. The values shown for C1 and C2 have been found to be a good compromise with the resistor values shown. Additional information on this selection is shown in the Micro Linear application notes.

To prevent excessive current below the brownout voltage, the divider ratio is calculated at the mean brownout voltage (200 V) to give 1.2 V at pin 4. The mean value of the haversine voltage from BR1 applied to the network will be

$$0.9 \times V_{rms} = 0.9 \times 200 = 180\ V$$

The ratio of the divider network will be 180/2.1 = 85.7:1. Hence the resistor ratio is

$$\frac{R1 + R2}{R13} = 85.7{:}1$$

Once again, R1 may be two series resistors to reduce voltage stress.

1.12.5 Loop Compensation

The voltage error amplifier is compensated by C5, C6, and R5. Because the amplifier is a transconductance type, the compensation components are returned to the common return, rather than to the inverting input. This reduces noise pickup and removes the voltage-divider resistors R6 and R14 from the compensation loop.

Unity gain is chosen to occur well below the line frequency; the optimum calculations are shown in application note 33.

The current loop is much faster, and again a transconductance amplifier (IEA) is used. In this case, the compensation components C3, C4, and R4 are returned to the 7.5-V reference, to provide soft start to the PFC section. Application note 33 provides the information to optimize the current-loop compensation.

1.13   BUCK SECTION DRIVE STAGE

The internal organization of the ML4826 is shown in Fig. 4.1.24. The lower part of the control circuit (the pulse-width modulator stage) is intended to provide the drive to a push-pull transformer of an isolated DC/DC converter. For this reason, two drive outputs alternate between pins 13 and 14 (180° out of phase). Each output will have a maximum duty cycle range of 0 to 48% of the total period. The ML4826-1 produces an alternate drive pulse on pin 13 or pin 14 for each drive pulse on pin 15 (the PFC boost section drive).

However, in this application (Fig 4.1.22), the buck output stage has a single power switching FET, Q2. For the buck stage to obtain an output of 400 V on C4 from the 450-V input on C1 (the voltage on C1 is the output of the boost stage), the maximum effective duty ratio must go to 89%. This can be obtained by linking both IC outputs (pins 13 and 14) to a single drive stage for Q2. (This interface block is shown in Fig. 4.1.23 as the "PWM Drive" block.)

This parallel connection of pins 13 and 14 cannot be made directly, as each drive is active in both its high and low states and there would be a conflict between the two drive stages. A special drive buffer is used in this application to "or" the two drive signals and obtain an effective duty cycle of up to 89% for the buck regulator stage, while retaining an active turn-"off" and turn-"on" action for Q2.

1.13.1 PWM Drive Circuit

Figure 4.1.27 shows the drive circuit. Alternate rectangular, pulse-width-modulated pulses are applied to inputs J6 and J7 via pins 13 and 14 on the IC. Diodes D1 and D2 "or" these pulses via R4 to the input of the totem-pole driver transistors Q3 and Q4 and to the output J5.

At full pulse width, the 48% duty ratio pulses from pins 13 and 14 interleave to provide an effective 98% duty ratio at the output. This ratio is adjustable down to zero. D1 and D2 provide active turn-"on" of Q3, but are not active during turn-"off." Active turn-"off" is provided by Q1, Q2, and Q4, which turn "on" during the

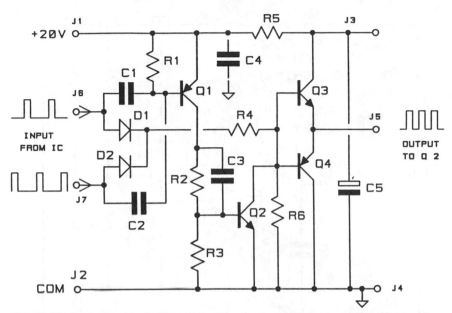

FIG. 4.1.27   Buck stage drive buffer, with "or" function to provide a wide range for the duty ratio.

trailing edge of the applied drive waveform to pull down on Q3 and Q4. R6 maintains a low condition for the remainder of the zero drive pulse condition.

This arrangement, together with the choice of the ML4826-1 IC, provides one power pulse to the buck transistor Q2 for each power pulse of the boost stage Q1. Pulse-width modulation is applied on the leading edge for Q1 and on the trailing edge for Q2.

1.13.2 Voltage Control Section

Reference to Fig. 4.1.22 shows that the output voltage is developed across C4, which is not directly referenced to the IC common line, pin 11. For correct voltage control, the IC requires an input to pin 6 with respect to pin 11 which is proportional to the differential voltage across C4.

Figure 4.1.23 shows the voltage feedback block, which provides a single level-shifted output to pin 6 of the control IC with respect to the common line, pin 11. This voltage is proportional to the output voltage of the buck section, as found across C4. Figure 4.1.28 shows how this is obtained. It contains an operational amplifier A1 and a voltage error amplifier A2. The two inputs to A1 are the differential voltage from C4 (the output voltage of the buck stage). The output of A1 is a control voltage with respect to the common line pin 2 which is proportional to the differential output voltage on C4, as applied to pins 5 and 4 of the block.

This control voltage is applied to the inverting input of the error amplifier A2 via R2. Simple pole zero loop compensation is provided by C1, R1, and R2. D1 becomes forward-biased and speeds up the loop in the event of a large transient increase in output voltage (as may occur when the load is suddenly reduced).

VR1 provides a variable reference to A1 and allows the output voltage to be adjusted. The output of A1 at pin 3 is a voltage with respect to the common pin 2, which may be applied directly to the PWM control input of the IC at pin 6.

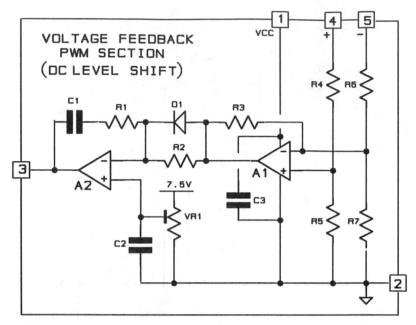

FIG. 4.1.28   Output voltage level-shifting circuit and voltage error amplifier stage with variable reference.

1.13.3   Setting Up the PWM Component Values

The control components external to the IC are shown in Fig. 4.1.23, and the control IC is shown in Fig. 4.1.24. Unlike the case with the PFC section, very little calculation is required to establish the optimum values of the pulse-width modulator components.

1. *Soft start.*   A capacitor (C11) on pin 5 sets the soft-start delay and ramp-up period of the PWM buck drive stage. The capacitor on this pin is charged from the IC at a constant current of 50 μA, and will provide a delay and then a progressively increasing pulse width as the capacitor charges above 2.7 V.

   The start of charge is delayed by Q1 and Q2 within the IC until both the supply voltage to the IC and the output voltage of the PFC boost stage have been correctly established. Full output is obtained when the voltage on pin 5 ramps up to near the peak value of the ramp on pin 9, plus a 1.5-V offset. The typical peak ramp voltage will be 5 V. Hence, C11 is selected to reach 6.5 V within the required soft-start time, when charged at a constant current of 50 μA.

2. *Current limit.*   A fast-acting, pulse-by-pulse peak-current limit is provided by comparator A2 within the IC. The limiting voltage on pin 10 is 1 V, and the external shunt R_T is chosen to give this voltage at the required peak current in Q2.

   In this topology, the peak current in Q2 is also the peak output current, which is to be 8 A. Hence R_T is set to 0.13 Ω. (A 10-W resistor is required.) Some noise rejection is provided by R12 and C12, and fine current adjustment is provided by R11.

3. *Ramp 2.*   The PWM modulator ramp is defined by R10 and C10. These values are developed as shown in the Micro Linear data sheet and application notes 16, 33, and 34.

1.14 POWER COMPONENTS

Figure 4.1.22 shows the major power components external to the control circuits. Of particular interest are the inductors L1 and L2 and the larger capacitors C1 and C4. Since there are no absolute equations for the selection of these components, it can be difficult to decide on the optimum values.

In practice, the choice depends on several factors related to required performance, application, stress, temperature, cooling, weight, size, and cost, and so the selection is. often a compromise. In general, as regards best performance, the larger these components are, the better. The following are some guidelines that should help the designer in this selection process.

1.14.1 L1 Inductor (Choke) Size

In this example, L1 is the power factor inductor, or more correctly choke, since it has a DC component. The main function of L1 is to limit the current flow from the low-impedance supply into Q1 when Q1 is turned "on", connecting L1 to the common line through Q1. Since this design is to be a continuous-condition mode, the choke is also required to maintain the input current flow nearly constant during a switching cycle; hence it will maintain a current flow into C1 via D1 when Q1 is "off."

Maximum Choke Size.   Clearly a very large inductance will meet the above requirements much better than a small inductance, so is there a maximum limit? Again, clearly there is, as the choke should not be so large as to impede the flow of the 120-Hz haversine current.

We can get an approximate limiting value for the inductance by calculating the value required to limit the input current at 120 Hz to the required full load current at the nominal input voltage without any other components, as follows:

Input voltage nominal = 277 V rms

Input power at 90% efficiency = 2.4 kW

Input current W/V = 8.8 A rms

Maximum reactance of L1 at 120 Hz = V_{rms}/I_{rms} = 31 Ω

Hence, the maximum inductance would be of the order of 40 mH. This would be a very large choke at 8.8 A, and it is clear that it is very unlikely that the maximum inductance would become a limiting factor in any real design.

Minimum Choke Size.   We now consider the factors limiting the minimum inductance Reducing the inductance will increase the switching frequency ripple current. Large ripple currents directly stress the switching components Q1 and D1 and the output capacitor C1. Less directly, they put additional demands on the line input RFI filter, since the ripple current flows via the input bridge rectifier BR1 to the filter. Hence, if L1 is small, a larger RFI filter would be required in order to give acceptable noise current at the line input. Any one or more of these may become the limiting factor on the minimum size of L1.

In particular, some conducted-mode RFI specifications apply strict limits on the low-frequency line noise down as far as 10 kHz. Since a small L1 will increase the requirements of the line filter, it is often more cost-effective to increase the inductance of L1, rather than use a larger line filter.

An arbitrary factor often used to define the inductance of L1 is to choose to limit the peak ripple current to some reasonable percentage of the peak loading current; typically ripple currents in the range of 5 to 20% will be used.

1.14.2 L1 Choke Design Parameters

For this design, we will work on the basis of a maximum of 15% ripple current so that we can calculate a median arbitrary value for L1.

Since the haversine input voltage to L1 changes throughout the cycle, the duty cycle also changes and the switching frequency ripple current will change throughout the haversine cycle. Maximum ripple current will occur when the haversine voltage is half the output voltage, at which point the duty cycle is 50%. L1 can then be calculated as follows:

The maximum input current occurs at full load and minimum input voltage. Hence

V_{in} minimum (V_{min}) = 220 V rms

Maximum input power (P_m) at 90% efficiency = 2400 W

Rms input current = P_m/V_{min} = 10.9 A rms   (4.1.7)

Peak choke current at 120 Hz = $\sqrt{2} \times I_{max}$ = 1.414 $\times$ 10.9 = 15.4 A   (4.1.8)

Hence

Peak-to-peak ripple current = 15% I_{peak} = 2.3 A   (4.1.9)

The output voltage on C1 is 450 V. Maximum ripple current will occur when $V_{haversine}$ is 1/2 V_{out}, or 225 V. At this point the duty ratio will be 50%, and at 50 kHz, Q1 will be "on" for 10 μs. There will be a near linear rate of change of current in L1 when Q1 is "on," as follows:

$$\frac{di}{dt} = \frac{2.3}{10 \ \mu s} = 230 \times 10^3 \ A/s$$

Since $L \ di/dt = e$ (the applied voltage of 225 V),

$$L = \frac{225}{230 \times 10^3} = 0.98 \ mH \qquad (4.1.10)$$

Hence, the value of L1 for 15% ripple current (2.3 A p-p) would be 1 mH. The value is arbitrary, and larger or smaller values may be chosen, depending on the tradeoff required between ripple current performance and the size, weight, and cost of the choke.

It should be noticed that L1 has a unidirectional 120-Hz haversine applied to it. The peak current in L1 is 15.4 A [Eq. (4.1.8)]. The change in the 120-Hz current at the peak of the haversine is relatively slow (compared with the switching frequency); hence, for the choke design, the peak 15.4 A can be considered a DC current. Therefore, the choke must not saturate at 15.4 A DC plus half the high-frequency ripple current component of 2.3 A [Eq. (4.1.9)], and a maximum current of 16.6 A DC would be used for the choke design.

The relatively small high-frequency ripple current leads to low switching losses in the choke core and favors the use of low-cost, low-permeability, iron powder cores for this application. If ferrite cores are preferred, then a suitable air gap must be used to prevent saturation of the core by the DC component.

1.14.3 Choke L1, Applied Design

In general, suitable design methods for L1 will be found in Part 3, Chaps. 1, 2, and 3. An example of one particular design for this application is also shown in Appendix 1.A.

L1 Design Parameters

Inductance: 1 mH [Eq. (4.1.10)]

Peak current: 16.6 A

Maximum mean current: 10.9 A rms

Ripple current: 2.3 A p-p (0.663 rms) [Eq. (4.1.9)]

Frequency: 50 kHz

Output voltage maximum: 450 V DC

Input voltage for maximum ripple current: 225 V

Duty ratio at 225 V: 50% = 10 μs

Copper Loss. The winding copper loss (I^2R loss) in this type of choke should be given special consideration. It is common practice to neglect the loss caused by the relatively small ripple current because the DC current component is typically much larger (in this example, 10.9 A rms compared to 0.663 A rms).

Since, in simple terms, the loss is proportional to $I_{rms}^2 R_c$, where R_c is the winding resistance, it would appear that the contribution from the 120-Hz component would be 126 R_c and the contribution from the 50-kHz component would be only 0.44 R_c × F_r, where F_r is the ratio of ac to DC resistance at 50 kHz, and thus it would appear reasonable to neglect the loss due to high-frequency ripple.

Further, it would also appear that the skin effect (which is much more applicable to the high-frequency ripple) could also be neglected. Hence, it is common practice to use a single solid wire for the choke winding.

However, calculations in Appendix 1.A show that this simplistic approach is not valid, because with a single solid wire, the ac/dc resistance ratio F_r can be very large, and so stranded wire should be used for best results in this example. (See Appendix 1.A.)

1.14.4 L2 Choke Size

L2 is the output buck regulator choke. Once again, the choice of inductance for L2 is arbitrary. Large values of inductance give low ripple currents, but are large and expensive and may limit the output current slew rate (load transient response).

Small values of inductance will give large ripple currents and will result in a change from continuous conduction to discontinuous conduction at a higher minimum current. Large ripple currents will stress Q2, D2, and C4 and result in larger output ripple voltages.

As with L1, L2 is often designed for a ripple current in the range of 5 to 20% of the peak-to-peak value. The maximum ripple current will occur when the output voltage is half the supply voltage, at which point the duty cycle is 50%.

In this example, the input voltage is 450 V. The maximum output current is to be 8 A. The frequency is the same (50 kHz). Hence, L2 may be calculated in a way similar to that shown in Sec. 1.14.1.

The buck stage input voltage (from C1) is 450 V constant. With V_{out} set at 225 V, Q2 will be "on" or 50% duty, or 10 μs in this example, and the voltage across L2 is 225 V for 10 μs. There will be a linear increase in current in L2 when Q2 is "on"; hence

$$\text{Maximum output current} = 8 \text{ A DC}$$
$$\text{Peak-to-peak ripple at } 10\% = 0.8 \text{ A p-p}$$
$$\frac{di}{dt} = \frac{0.8}{10 \text{ μs}}$$
$$L\frac{di}{dt} = 225 \quad \text{and} \quad L = 2.8 \text{ mH}$$

Once again, an iron powder core may be used for this choke. The design is covered more fully in Part 3, Chaps. 1, 2, and 3. The choke must support a DC current of 8.8 A minimum. The particular design approach shown in Appendix 1.A may be used, with the DC current reduced to 8.8 A and the inductance increased to 2.8 mH.

1.14.5 Capacitor Selection

C4. The output capacitor C4 is not very highly stressed, as the ripple current from L2 into C4 is a continuous triangular waveform at a maximum of 0.8 A p-p. However, the load may have large ripple currents, and if so, this must be considered in the selection of C4. Otherwise, a simple low-ESR electrolytic capacitor of a size that will give an acceptable output ripple voltage would be selected. The voltage rating should be at least as high as the supply voltage (in this case, 450 V).

C1. C1 is a quite different proposition. This capacitor must carry the 120-Hz haversine current and the switching frequency current. It should be noticed that the current from D1 into C1 is discontinuous at the switching frequency (when Q1 is "on," the current in D1 is zero).

The use of leading-edge modulation on Q1 and trailing-edge modulation on Q2 will reduce the ripple current in C1, because as Q1 turns "off" and D1 conducts current into C1, Q2 will turn "on" and divert current away from C1. The overall effect has been shown[89] to reduce the effective ripple current in C1 by up to 30%.

The waveform applied to C1 is a near-rectangular waveform at the switching frequency, which changes in amplitude throughout the applied haversine. Hence C1 must have a low impedance at the switching frequency and must be large enough to give an acceptably low ripple voltage at the 120-Hz haversine frequency (that is, it must have a low ESR and a high capacitance). At high power it is difficult to satisfy both requirements in a single component.

Figure 4.1.29 shows a method that can be used in high-power applications to reduce the stress on C1. C1 is split into one or more electrolytic capacitors C1a and C1b and one or more film capacitors C2a and C2b. The film capacitors are intended to take most of the high-frequency current and the electrolytic capacitors the low-frequency current.

The electrolytic capacitors will have large capacitance values (typically 470 μF or more in this application), and the film capacitors will be quite small. (A typical 3.3-μF 450-V metalized polyester film capacitor may have a ripple current rating of 15 A or so at 50 kHz.)

To encourage the high-frequency current to flow in the film capacitors, a resistance is placed in series with each electrolytic capacitor. In this example, an NTC (negative temperature coefficient resistor) is used. This also provides inrush protection, as explained in Sec. 1.14.6. High-frequency ripple currents are best minimized in the electrolytic capacitors, as they tend to flow in a small section of the foil near the terminals, causing local heating in the capacitor.

We can establish a suitable resistance value for the NTC by calculating the impedance of the film capacitors at the switching frequency. Assume two 3.3-μF film capacitors are used in parallel.

$$X_c = \frac{1}{2\pi fC} = 0.96 \ \Omega \text{ for each capacitor, to give a total of } <0.5 \ \Omega$$

If 2 Ω is chosen as the working resistance of each NTC, then two-thirds of the high-frequency ripple current will flow in the film capacitors.

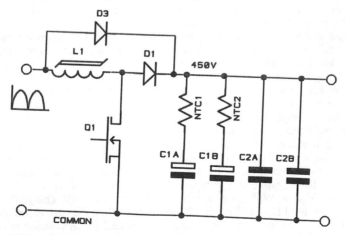

FIG. 4.1.29 Boost input stage with inrush limiting current bypass diode D3, and ripple current steering arrangements.

However, the NTCs will increase the low-frequency ripple voltage at the output of C1 as follows:

Each 470-μF capacitor will have an impedance at 120 Hz of 2.8 Ω, and the NTC will approximately double the ripple voltage. Since half the ripple current flows in each electrolytic capacitor, at the maximum input current of 15 A the peak-to-peak ripple voltage will be approximately 28 V (in a DC of 450 V), or about 6%. This is quite satisfactory for the intended application.

1.14.6 Inrush Current Limiting

Refer to the basic PFC stage, shown in Fig. 4.1.22. When first turned "on," the capacitor C1 is discharged, and current will flow via L1 and D1 into C1 to charge the capacitor. If the unit is turned "on" at the peak of the applied line voltage (430 V), the rate of change of current in L1 would be 287 A/ms, and clearly a very large current would flow before the half cycle is finished.

Since L1 will have been designed to saturate at some current above 15 A, it is likely that L1 will be saturated. If Q1 turns "on" when L1 is near saturation, the current will not be limited, and Q1 will fail.

Figure 4.1.29 shows a better arrangement, in which L1 is shunted by diode D3. Also, the large electrolytic capacitors have NTCs in series with them. Although the NTC is chosen to have a hot resistance of 2 Ω or less, when cold the resistance will be typically >50 Ω, and the inrush current will be limited to <20 A. With D3 shunting the inductor L1, the inrush current is diverted away from the inductor, preventing saturation. When C1 is fully charged and the boost circuit is active, D3 is reverse-biased and is out of the circuit.

In normal operation, the ripple current in the electrolytic capacitors will maintain heating of the NTCs and hence maintain their low resistance.

1.14.7 Low-Loss Snubber Circuit

Figure 4.1.30 shows a low-loss snubber circuit. The high voltage and fast switching of Q1 and Q2 tend to result in voltage spikes on these components during the high-current turn-"off" edge. These effects should be minimized by good layout, and can be further reduced by snubber circuits.

Simple snubber circuits are lossy and will dissipate transient energy in a snubber resistor. In high-power applications this energy may be quite large. Figure 4.1.30 shows a low-loss snubber circuit in which the recovered energy is used to drive a 24-V cooling fan and useful work is done.

The snubber works as follows: Consider Q1 during the turn-"off" edge. As the voltage on the Q1 drain rises, the previous drain current is diverted via C5 and D3 into the large storage capacitor C7. This reduces the rate of change of voltage on the Q1 drain and much reduces any tendency for voltage overshoot.

At the end of turn-"off," the voltage across C5 will be the voltage on C1 (450 V) minus the 24 V on C7. When Q1 turns "on," the left side of C5 goes to zero and the right side of C5 goes negative by 426 V, reverse-biasing D3 and bringing D5 into conduction. The negative voltage applied to L3 will cause current to flow in L3 in the direction shown. With Q1 now fully "on," this current will charge the right side of C5 positive until D3 conducts at +24 V, at which time the current will continue to charge C7. The time constants of L3 and C5 and the 24 V are chosen to ensure that the current in L3 has dropped to a low level before Q1 turns "on" again. This returns C5 and L3 to their original state for the next turn-"off" of Q1. The same action applies for Q2, with C6, D4, and D6. Z1 clamps C7 at 24 V, but the main loading is the 24-V fan. C5 and C6 are selected to just provide the required fan current.

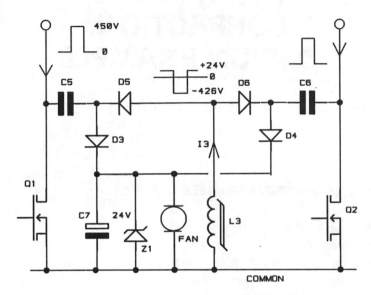

FIG. 4.1.30 Low-loss voltage snubber circuit with 24-V fan drive.

APPENDIX 1.A

BOOST CHOKE FOR POWER FACTOR CORRECTION: DESIGN EXAMPLE

This example is for the choke L1 in Part 4, Sec. 1.14.3.

1.A.1 BASIC DESIGN PARAMETERS FOR L1

Inductance: 1 mH [Eq (4.1.10)]
Peak current: 16.6 A
Maximum average current: 10.9 A rms at 120 Hz
Ripple current: 2.3 A p-p (0.663 A rms)
Switching frequency: 50 kHz
Output voltage maximum: 450 V DC
Input voltage for maximum ripple current: 225 V
Duty Ratio (at 225 V input, 450 V output) 50%: 10 μs

1.A.2 CHOICE OF CORE TYPE

For this design, we will consider an iron powder core material (rather than a ferrite or MPP) because the 50-kHz ripple current is quite small compared to the 120-Hz low-frequency current. This should allow the use of a low-permeability, higher-loss iron powder material, which would be much more cost-effective.

1.A.3 SIZE OF CORE

This can be established by an iterative design process (choose a core size and try an outline design), by selection based on experience, or by using a software program supplied by the manufacturer.

Micrometals provides suitable software programs for its cores free of charge. As an example, we will consider a Micrometals E305 core size. The data for this core will also be found in Micrometals' "Power Conversion & Line Filter Applications" data book.[90] Other manufacturers make similar cores.

1.A.4   CORE MATERIAL

To minimize the core loss and maintain a near-linear inductance over the wide 120-Hz haversine current range, the Micrometals #2 material will be considered. (A 120-Hz haversine is applied to the choke in a power factor boost-type regulator application.)

This #2 material has a low core loss and a low permeability ($\mu = 10$) and is unlikely to saturate with the high haversine current component, and so it is unlikely that a core gap will be required. The permeability of the #2 material remains linear to a DC magnetization force exceeding 150 oersteds (Oe). (See Micrometals' data book, page 1 and page 24, fig. 2.)

1.A.5   OUTLINE DESIGN

The chosen core is Micrometals E305-2, or similar, and the data book provides the following:

Core parameters:

A_I value = 75 nH

Magnetic path length (ℓ) = 18.5 cm

Core pole area (A_e) = 562 mm^2

Core volume (V) = 104 cm^3

Bobbin parameters:

Winding width (W) = 48 mm

Flange height (h) = 12 mm

Mean length turn * (MLT) = 15.5 cm

1.A.6   CALCULATE TURNS

The required inductance is 1 mH, and the first iteration for turns can be made using the published A_I value for the E305-2 core.

$$H = N^2 A_I \quad \text{and} \quad N = \frac{H}{A_I}$$

Therefore
$$N = \sqrt{\frac{1 \times 10^{-3}}{75 \times 10^{-9}}} = 115 \text{ turns}$$

---

*From Micrometals' "Iron Powder Cores" data book, issue 1, 1998, page 61.

1.A.7   CALCULATE MAGNETIZATION AND CHECK PERMEABILITY

The near-DC current flowing in this 115-turn winding near the peak of the applied haversine will set up a magnetization force H (oersteds) that will push the core toward saturation and may reduce the permeability. If this happens, then the new value A_{l2} will be used to recalculate the turns; this can be an iterative process.

Note:

$A_{l2} = A_l \times$ %perm (from "Power Conversion" data book, page 24, fig. 2)

Turns (N) = 115

Haversine current (I) = 120-Hz peak current (16.6 A)

Magnetic path length (ℓ) = 18.5 cm

$$H = \frac{0.4\,\pi\,NI}{\ell} = \frac{0.4\,\pi \times 115 \times 16.5}{18.5 \text{ cm}} = 129 \text{ Oe}$$

From the graph shown in the data book (fig. 2, page 24), the permeability remains near 100% at this magnetization level for the #2 material and the turns do not need to be adjusted. Saturation starts above 350 Oe, and a good saturation margin is provided for any overcurrent condition.

1.A.8   CALCULATE CORE LOSS

Establish ac flux density swing (B_{p-p}), gauss, at 50% duty cycle (½ voltage):

$$B = \frac{V \times t \times 10^4}{N \times A_e} = \frac{225 \times 10 \times 10^4}{115 \times 562} = 350 \text{ G}$$

where V = voltage across inductor when Q1 is "on" at ½ voltage
$\quad\ t$ = "on" time, μs (at 50% duty)
$\quad\ N$ = turns
$\quad\ A_e$ = area of pole, mm^2

Figure 1, page 29, of the data book, shows the core loss for the #2 material. Because the peak flux density shown on this nomogram is half the peak-to-peak value calculated above, the nomogram will be entered on the lower scale with 1/2 the p-p value.
To enter the nomogram, use

$\qquad B = 350/2 = 175 \text{ G}$
$\qquad f = 50 \text{ kHz}$ (the pulse repetition frequency)

From the nomogram, the loss is found to be 15 mW/cm^3.

Hence
$$\text{Total core loss} = \text{loss} \times \text{core volume (in cm}^3)$$
$$\text{Core loss} = 15 \times 10^{-3} \times 104 = 1.56 \text{ W}$$

1.A.9 COPPER LOSS

The copper loss consists of two parts. The first is simple DC current loss I^2R_c, where I is the rms input current (the rms value of the 120-Hz haversine current at nominal load) and R_c is the DC winding resistance.

The second part of the loss is more complicated. It is the loss due to the high-frequency ripple current, which works against the effective ac resistance of the winding. In chokes, this effect is often neglected, as it is assumed that the ac loss will be small because the 50-kHz ripple current is much less than the 120-Hz current.

However, the following calculations will show that the ac loss can be quite large if a single conductor is used for winding the choke (a common practice).

1.A.10 DC WINDING RESISTANCE R_C

The bobbin width is 48 mm, and a #13 AWG wire (1.95 mm) will have a current rating of 11.8 A (page 3.23). This wire size will give 23 turns per layer. With 5 layers, the buildup would be 9.75 mm plus insulation thickness. The bobbin height is 12 mm, and the winding will just fit. Assume that a single #13 AWG wire is used for the winding; if we use 23 turns per layer, 5 layers will give the required 115 turns.

1.A.10.1 Winding Length

The length of the winding will be 115 × MLT.

$$W_\ell = 115 \times 15.5 = 1782 \text{ cm}$$

1.A.10.2 Winding Resistance

The resistance of #13 wire at 60°C is 80 μΩ/cm. Hence, the winding resistance R_c at 60°C will be

$$R_c = 1782 \times 80 \times 10^{-6} = 0.143 \ \Omega$$

1.A.10.3 DC Winding Loss

$$\text{Copper loss} = I\text{rms}^2 \times R_c$$
$$= 10.9^2 \times 0.143 = 17 \text{ W}$$

1.A.11 EFFECTIVE AC WINDING RESISTANCE

Because of the skin effect at high frequency, the effective ac winding resistance R_{ac} is higher than the DC resistance for the high-frequency ripple current component. In general,

$$R_{ac} = R_c \times F_r$$

where F_r is the ratio of the effective high-frequency resistance to the low-frequency resistance R_c. Figure 3.4B.1 on page 3.100 shows the value of F_r with respect to a factor which is related to the geometry of the winding and the skin depth at the working frequency.

The factor is

$$\frac{h \sqrt{F_\ell}}{\delta} \quad \text{for rectangular wire}$$

and becomes

$$\frac{0.866 \, d \sqrt{F_\ell}}{\delta} \quad \text{for round wire}$$

where d = wire diameter, mm
δ = skin depth, mm
N_1 = turns per layer
W = winding width, mm
$F_\ell = 0.866 N_1 d / W$

From Fig. 3.4B.3, page 3.101, the skin depth at 60°C and 50 kHz would be 0.3 mm. Evaluating F_ℓ,

$$F_\ell = 0.866 \times N_1 \times d/W$$

$$= \frac{0.866 \times 23 \times 1.95}{48} = 0.81 \text{ (dimensionless)}$$

The factor $\dfrac{0.866 \, d \sqrt{F_\ell}}{\delta}$ may then be calculated:

$$\frac{0.866 \times 1.95 \times \sqrt{0.81}}{0.3} = 5.07$$

Entering Fig. 3.4B.1 with this value and 5 layers yields an F_r of about 80. Hence

$$R_{ac} = R_c \times F_r = 0.143 \times 80 = 11.5 \ \Omega$$

1.A.12 AC COPPER LOSS

The ripple current is a triangular waveform with a peak-to-peak amplitude of 2.3 A. The rms value of this is $0.577 \, I_{p\text{-}p} = 1.33$ A.

$$\text{Ac copper loss} = I_{rms}^2 \times R_{ac} = 1.76^2 \times 11.6 = 20 \text{ W}$$

It has been shown that if a solid conductor is used for winding the choke, the ac copper loss will exceed the DC loss and the temperature rise will be much greater than expected. Tests on similar designs have confirmed this effect.

Hence there is an advantage in using multistrand wire when winding high-frequency chokes, even when the ac ripple current is quite small. The requirement is not as stringent as in a transformer design, as the ripple current is small. In this example, using 13 strands of #24 AWG, with a current rating of 11.9 A, was found to be quite satisfactory.

1.A.13   TEMPERATURE RISE

This will depend on the total loss ($P_{core} + P_{copper}$), the surface area or volume, and the environment. For free air conditions, Fig. 3.3.5, page 3.52, or Fig. 3.16.10, page 3.219, provides the information to establish the expected temperature rise.

GLOSSARY OF POWER SUPPLY TERMS

Power supplies, and switchmode power supplies in particular, are the product of a relatively new and rapidly developing engineering discipline. As such, many of the specialized terms used to describe them are still developing, and have yet to become generally recognized and fully defined.

The following list, although by no means comprehensive, gives some of the more generally recognized terms, together with the author's definition, based on generally accepted industry usage at the time of printing. In some cases, the general meaning of a term may differ from its specific definition as applied to power supplies. In such cases, the "general meaning" (GM) is given first, followed by the "power supply specific definition" (PSS).

International standards and definitions for power supplies are yet to come, and may in due course change some of the examples given here.

Ambient Temperature. GM: Environmental temperature
   PSS: For convection-cooled supplies, the temperature of the air directly below the supply in a free flow position. For fan-cooled supplies, the air temperature at the fan inlet. For conduction-cooled units, the temperature of the heat exchanger interface.
Amplifier, Differential. An amplifier with two signal ports whose output voltage amplitude and polarity are proportional to the voltage difference and polarity between the voltages applied to the inverting and noninverting input ports.
Amplifier, Transconductance. An amplifier with two signal ports whose output current is proportional to the difference between the voltages applied to the input ports.
Ancillary Functions. All functions provided in a power supply which are not directly connected with the generation of stabilized outputs. In MTBF calculations, ancillary functions are sometimes disregarded in the calculation if a failure of an ancillary function is not to be considered a critical failure.
Automatic Crossover. PSS: This term is normally reserved for laboratory-grade variable power supplies which have both a constant-voltage and a constant-current characteristic. Such supplies have the ability for "automatic crossover" between the two operating modes in response to load variations or adjustment of the supply controls. The mode of operation is normally indicated by a front panel indicator light.
Auxiliary Outputs. All outputs of a multiple supply other than the main output. Auxiliary outputs usually have a lower power rating and limited performance.
Bandwidth. GM: This describes a system's frequency response, normally the difference in frequency between the upper and lower "half power" or 3-dB-down frequencies.
   PSS: When applied to power supplies, this normally refers to the frequency band over which output ripple and noise components are to be measured or specified.

Biphase Rectifier. A circuit which rectifies both half cycles of an ac input, utilizing a transformer with a center-tapped winding.

Bleeder Resistor. A resistor whose main purpose is to provide loading. It may be provided to discharge capacitors for safety reasons, or to preload output lines in switchmode power supplies, to prevent an excessive voltage rise when the external load is removed. Also called "dummy load" or "preloading resistor."

Bridge. GM: A circuit configuration, normally used for precision measurement, in which two potential dividers (resistive and/or reactive) are brought to balance to compare an unknown reactance with a precisely known value.

PSS: In power supplies, a bridge-type circuit configuration normally refers to the method of connecting switching devices, in bridge or half-bridge arrangements, for converter applications. Also, a method of connecting diodes to provide full-wave rectification from single-phase inputs.

Brownout. GM: A drop in supply voltage to a value below the minimum normally specified by the supply authority, but above zero.

Brownout Detection. PSS: Used with reference to the power failure warning circuit. It describes a warning circuit capable of recognizing the critical supply voltage at which an "impending failure" warning must be given, to ensure a specified "hold-up time" at full load. The critical voltage will normally be in the "brownout zone."

Brownout Zone. PSS: Normally defined as a voltage between the lowest value specified for full operation and the lowest value at which the power supply will continue to operate. The power supply is not normally expected to meet full specifications in the brownout zone.

Bus. GM: With reference to IEEE. − 488, this refers to the interconnecting control lines.

PSS: For a power supply, it is any interconnection used to control or communicate with a power supply or between supplies—for example, between P terminals used for forced current sharing, or in master-slave applications; also, remote programming and shutdown.

Busbar. GM: A heavy copper bar, used for interconnections of high current capability. Very often, a busbar will not be insulated and provides multiple load connections.

Carryover. PSS: Describes the ability of a switchmode power supply to continue to provide regulated outputs during a short input failure as a result of the energy stored in the supply (normally in the input capacitors). Also see "holdup time."

Centering. PSS: A term normally reserved for multiple-output supplies. It describes the unavoidable deviation from nominal output voltage of auxiliary and semiregulated outputs as a result of design or manufacturing limitations. (Usually an effect caused by the need for the number of transformer turns to be a complete integer.

Also, the action of adjustment of an output voltage to bring it to the center of a specified range.

Choke. An inductor specifically designed to carry a large DC current component. (To prevent saturation, chokes will often have gapped cores or cores of especially low permeability material.)

Choke, Nonlinear. A choke specifically designed to have a very nonlinear characteristic. A nonlinear choke will normally have high inductance when the DC current is low, and a reasonably constant, but low, inductance when the DC current is in the middle to upper end of the current range. (This nonlinearity is often induced by stepped air gaps in the core, or by using compound cores of different permeability materials.)

Often used to extend the lower end of the range of continuous operation in wide-range switchmode regulators.

Choke, Polarized. A choke whose core has been magnetically prebiased toward one end of the B/H characteristic. (Because a larger flux excursion is now available in the reverse direction, a choke of this nature can withstand a larger DC current component.) Magnetically "hard" (rare earth) magnets, positioned in the core gap, are normally used for the magnetic prebiasing.

Chokes, RFI. Chokes which are specifically designed to have a high self-resonant frequency, so as to provide maximum impedance at RF frequencies. Various space winding or wave winding techniques are used to minimize interwinding capacitance.

Choke, Swinging. A choke whose inductance is designed to increase significantly as the

DC current is reduced toward zero. The swinging choke has a more linear characteristic than the "nonlinear choke," the change in permeability being a function of the properties of the bulk core material.

Common-Mode Ripple and Noise. The components of ripple and noise voltages or currents, which exist between input or output lines and a defined ground plane

Compliance Current. PSS: A little-used term that describes the range of current (and hence resistance) over which a constant- (stabilized) voltage power supply will maintain a constant voltage. More common terms are "maximum current" or "current limit value." These terms assume a compliance current range from zero or 10% to the limit value.

Compliance Voltage. PSS: A term, normally reserved for constant-current supplies, describing the range of load developed voltages (and hence load resistance) over which the power supply is capable of maintaining constant current.

Complimentary Tracking. The interconnection of two supplies in such a way that the output voltage amplitude of one supply will follow (track) the output voltage of the other, but will be of opposite polarity.

Conditional Stability. PSS: An undesirable condition of quasi-stability in which the phase shift exceeds 180° at frequencies below crossover. (In conditionally stable power supplies, large-signal instability may occur when loads, temperatures, or other parameters are changed.)

Conducted-Mode EMI. That part of the unwanted interference energy that is conducted along the supply or output leads. Conducted-mode EMI is limited by national and international standards.

Constant-Current Limit. PSS: A method of overload protection in which the output current remains constant regardless of the load resistance in the protected overload range.

Constant-Current Supply. GM: Any high-impedance current source whose current is essentially constant regardless of the load resistance.
PSS: Describes a type of power supply in which the major controlled parameter is the output current. Such supplies will maintain the output current constant for a range of load resistance, normally from zero to some maximum value defined by the compliance voltage.

Constant-Voltage Supply. PSS: A power source in which the main controlled parameter is the output voltage.

Control ICs. PSS: Dedicated integrated circuits used for the control of power supplies, both switchmode and linear.

Convection Cooling. PSS: A method of air cooling the power supply which is dependent on the development of convection currents in a free air environment.

Converter. PSS: A general term for any switchmode power supply which converts a DC voltage at the input to a DC voltage at the output while providing galvanic (often transformer) isolation. Where regulation is not provided, the term "DC transformer" is more correctly used. Where input-to-output galvanic isolation is not provided, the term "switchmode regulator" is normally used.

Cross-Connected Load. PSS: A load which is connected between a positive and a negative output terminal in a bipolar (series) power supply connection, without reference to the common line. Cross-connected loads can cause lockout when foldback or reentrant current protection is used.

Crossover. PSS: The ability of a power supply to change its mode of operation between constant current and constant voltage in response to a load change or control adjustment.

Cross Regulation. PSS: The regulation effects measured on one output as a result of changes on other outputs. (Very often, in multiple-output supplies, the main output is fully regulated, and cross regulation would refer to the output voltage variations on the auxiliary outputs as a result of load changes on the main output.)

Crowbar Protection. PSS: A method of overvoltage protection in which the offending output is short-circuited in the event of an overvoltage stress. (The short-circuiting device will normally be an SCR.)

Current Foldback. PSS: A method of overload protection, sometimes referred to as "reentrant protection," in which the output current is reduced in the overload region as the load resistance moves toward zero. This method of protection can give lockout problems and is normally restricted to linear supplies, where it is required to prevent excessive dissipation in the regulator element.

Current-Mode Control. PSS: A switchmode power supply control technique in which a fast-acting control loop defines the maximum current in the switching element on a pulse-by-pulse basis.

In constant-voltage supplies, the fast current control loop is then adjusted by a slower voltage control loop to provide a constant output voltage. The two control loops thus form a voltage-controlled current source.

This method of control effectively eliminates the output filter inductor from the small-signal model, automatically improving the stability margin and small-signal dynamic performance. It also has the advantage of providing fast current limiting.

Cycling. PSS: A recovery mode during which a power supply will repeatedly attempt to restart after or during a stress condition (e.g., a short circuit, overload, overvoltage, etc.).

DC Current Transformer. A type of current transformer in which a DC primary current controls the pulsating output current. (Very useful for isolated low-loss current limiting in high-current supplies. See Sec. 14.8 in Part 3.

DC Transformer. A square-wave DC-to-DC converter which provides DC transformation and galvanic isolation without regulation. DC transformers often use simple self-oscillating push-pull topologies.

Derating. The reduction in some specified operating parameter as a result of some change in another parameter. For example, a reduction in a supply power rating as a result of an increase in ambient temperature.

Differential-Mode Ripple and Noise. That part of the input or output ripple and noise voltage which exists between two supply or output lines with respect to each other. In multiple-output units, the ripple and noise voltage between output lines and a common return line.

Direct-Off-Line Switcher. PSS: A switchmode power supply which provides isolated DC outputs from ac line inputs without using line frequency transformers.

Drift. A time-related variation in a defined parameter, normally caused by self-heating or aging effects.

Drop-Out Voltage. PSS: The input voltage below which the output can no longer be fully maintained.

Duty Cycle Control. PSS: A method of control in fixed-frequency switchmode supplies in which the "on" period of the power switch is adjusted to control the output.

Duty Ratio Control. PSS: A method of control in switchmode power supplies in which the ratio of the "on" period to the "off" period of the power switch is controlled to maintain a constant output. (Note: This differs from "duty cycle" control in that the total period need not be constant, giving a variable-frequency performance.)

Dynamic Load. GM: A load that is varying, an active load.

PSS: An electronic load which can be rapidly changed to test transient response. May also refer to an adjustable constant-current electronic load used for test purposes.

Efficiency. Ratio of output power to input power as a percentage. (Note: True power must be used, with due allowance for power factor. With capacitive input filters, often used for "off-line" SMPS, the power factor will be approximately 0.63.)

EMI. Electromagnetic Interference. Also referred to as RFI (radio-frequency interference). EMI levels, both radiated and conducted, are controlled by national and international standards.

ESR. Effective series resistance, a term normally applied to electrolytic capacitors. In the classical capacitor model, the effective resistance is normally shown as a pure resistance in series with an ideal capacitor. However, this model can be somewhat misleading, as ESR is, in itself, a frequency-dependent parameter.

Faraday (Electrostatic) Shield. An electrostatic shield, usually copper, placed between a source of high-voltage noise and a low-noise area. Typical examples would be electrostatic screens in transformers and the screens between switching elements and heat sinks.

FCC. Federal Communications Commission. A U.S. Federal regulatory body which, among other things, defines maximum permitted EMI conduction and radiation levels in the United States.

Filter. PSS: Normally refers to power level low-pass filters, intended to give nearly continuous DC output currents. Power filters differ from signal filters in that input and output impedances are not matched and are often variable. Because of the need for energy storage, power filters are large.

Flyback. The property of an inductor that enables it to reverse its terminal voltage when the conducted current is interrupted. The term is often applied to transformers and diodes which utilize this flyback voltage property.

Flyback Converter. PSS: A switching power supply in which the energy is transferred from the input to the output during the "off" state of the primary switching device, using the flyback property of the transformer.

Foldback Current Limiting. PSS: See "current foldback."

Forward Converter. PSS: A switching power supply in which the energy is transferred from the input to the output during the "on" state of the primary switching device. (Generally any converter in the buck family.)

Full-Wave Rectifier. A circuit that rectifies both half cycles of an ac input.

Gain, Closed-Loop. PSS: The response of a power supply control loop to a voltage inserted in series with the control amplifier input after negative feedback is applied.

Gain, Open-Loop. PSS: The ratio of output change to input change (normally voltage) of the control loop before feedback is applied. Often a calculated value only because of the difficulty of making open-loop measurements.

Gain Margin. The amount (in decibels) by which the closed-loop gain is less than unity when the total phase shift is 360°.

Galvanic Isolation. PSS: A term coined by power supply engineers to indicate that there is no direct DC link between two points (e.g., input and output) except by intended converter action; the points are "passively isolated."

Ground Loop. PSS: Normally refers to noise-generating current loops set up as a result of grounding the common output of the power supply at more than one place. (In switchmode power supplies, ground loops are normally minimized by grounding the common output of the supply at the output terminals. If the system demands other ground points, loop currents can be minimized by a common-mode inductor in the output leads.)

Heat Sink. PSS: A term generally used to describe a thermal shunt, normally metal, used to aid in the transfer of heat from a hot spot (often a semiconductor) to an "infinite heat sink" (often free air).

Heat Sink (Infinite). GM: A medium of such large thermal mass that it can absorb (sink) heat indefinitely with negligible change in temperature (e.g., free air).

Hiccup Mode. Normally applies to a mode of protection in which the supply repeatedly attempts to restart during an overload or short circuit condition. (See also "cycling.")

Holdup Time. PSS: The time period during which the output voltages of a power supply will be maintained within their regulation limits after removal of the input supply. (See "carryover.")

Inrush Current . PSS: The peak input current that flows into the power supply when it is first switched on.

Inverter. A device which provides an ac output from a DC supply.

Leakage Current. Current which flows between two points that are normally considered isolated. (For medical power supply applications, the ground return leakage current must be very low for safety reasons. For this application, the current returning to ground through any input filter component is considered part of the leakage current.)

Line Regulation. PSS: The change in the controlled output parameter as a result of an input line voltage change. Normally expressed as a percentage change of output for a full-range input voltage change.

Linear Regulator. PSS: A dissipative series or shunt regulator technique which gives continuous control of the regulated parameter. More commonly, these will be series transistor voltage regulators.

LISN. (Line impedance stabilizing network.) A line filter and impedance network of specified wideband design, used to divert system noise into the measurement equipment. Required for conducted-mode RFI measurements.

Load Effect. (See "load regulation.")

Load Regulation. The change in the controlled output parameter as a result of a defined load change on the measured output. Usually given as a percentage change in output for the defined load change

Lockout. PSS: A condition in which the power supply fails to establish its correct working condition on initial start-up or after a transient load. (This is often caused by incompatibility between the load-line characteristic and the shutdown protection characteristic of the power supply.)

Magnetic Amplifier. A wound magnetic device in which a control signal applied to one winding controls the flow of current in a second winding, with amplification provided by the turns ratio, the design of the magnetic path, and the magnetic properties of the core.

Main Output. PSS: In multiple-output supplies, the principal output, often the highest-current or highest-power output. The control loop is normally closed to the "main output," and therefore it will have the best overall performance.

Margining. A defined change in output voltage around the nominal value used to check the performance of the load circuit. Often used to describe the provision of such means in the power supply.

Master. PSS: The controlling power supply in a parallel "master-slave" configuration.

Master-Slave Operation. A method of interconnecting two or more units in such a way that a defined "master" unit controls the behavior of the remaining "slave" units. Normally power supplies must be specially designed for this mode of operation if required.

Modulator. PSS: An electronic control device which varies one parameter in response to the changes of another, used as part of the control loop to vary the output of a power supply. A typical example would be a pulse-width modulator or current-mode control modulator in a switchmode supply.

MTBF. Mean time between failures. Ideally, a statistical prediction of likely failure, given real-time life measurements. More often, a prediction based upon a statistical stress analysis, using a recognized standard. For power supplies the procedures and figures specified in MIL-HDBK 217E will often be used.

Nominal Value. The ideal, intended, or center value of the defined parameter.

Open-Frame Unit. PSS: A power supply which is not totally enclosed; normally a component part of the main enclosed system.

Output Impedance. PSS: The magnitude of the complex resistive and reactive components seen by looking into the power supply output terminals at frequencies other than DC. Normally measured by dividing the rms output voltage change by the forced rms load current change over the required range of frequency.

Note: At high frequency, the output impedance tends to the ESR of the output capacitors. At low frequency, the output impedance is a function of the control loop, and tends to the output resistance of the supply at DC.

Overshoot. A transient excursion of the controlled parameter outside the regulation limits, as a result of initial turn-on or turn-off, or a sudden load variation.

Overvoltage Protection Circuit (OVP). PSS: An independent circuit which will prevent excursion of the output voltage beyond a defined limit in the event of a control circuit failure.

Note: A design criterion for effective overvoltage protection requires that a single component failure must not result in an unprotected overvoltage condition.

Some overvoltage protection circuits (for example, crowbar protection) will provide a measure of overvoltage protection for externally induced overvoltage stress conditions.

Parallel Forced Current Sharing. A method of parallel operation in which power supplies communicate so as to share the load current equally without a designated master. Often used in conjunction with parallel redundant operation.

Parallel Operation. The connection of two or more supplies in parallel, to increase the current capability.

Parallel Master-Slave. The operation of two or more supplies in parallel, using a control bus, in such a way that a single designated master controls all output currents. Master-slave operation has the advantage that slave power supplies share the load equally.

Parallel Redundant Operation. A method of parallel operation, using output isolating diodes, in which the failure of a single supply will not result in a loss of power.

The total current capability of the parallel system must exceed the load requirements to a point where the failure of a single unit will not result in a system overload. Hence, the remaining supplies must have sufficient current margin to take up the extra load among them without a reduction in output voltage. Hence, under normal conditions, one or more of the supplies may be considered "redundant" in terms of the normal current requirements

PARD. PSS: (Periodic and random deviation.) The total wideband ripple and noise components at the output of a switchmode power supply. Usually expressed as a peak-to-peak noise voltage for a bandwidth from DC to 30 MHz.

Phase Margin. The difference between the closed-loop phase shift at unity gain crossover and a total loop phase shift of 360°.

Piggyback Regulator. A series arrangement of voltage reduction devices or regulators such that the voltage stress is shared. (One supply carries the other on its back, so to speak.)

Postregulator. PSS: A regulator positioned in the output side of a DC-to-DC converter. Postregulators are often applied to multiple-output supplies to provide additional regulation on auxiliary outputs. Postregulators may be linear or switching, and in high-power applications, magnetic amplifier or saturable reactor techniques are commonly used.

Power Factor. GM: The ratio of real power to apparent power in ac supply applications. PSS: In direct-off-line power supplies with capacitive input filters, current flows only at the peak of the applied sine wave. The product of rms input voltage and rms current gives the input VA (apparent power, not true power). The power factor of a capacitive input filter is typically 0.63 for single-phase and 0.9 for three-phase supplies; hence real power is approximately VA × 0.63 or VA × 0.9.
(To measure true input power accurately, a wattmeter is required. This may be a dynamometer or digital type, but it should have a bandwidth exceeding 1 kHz for reasonably accurate results.)

Power Failure Detection Circuit. PSS: A circuit which detects the onset of brownout, or input power failure, and provides a warning signal.

Power Failure Signal. A signal (often TTL compatible) provided by a power failure detection circuit. In switchmode supplies, the power failure signal should occur before the end of the holdup time, to give sufficient warning for housekeeping in the external load before the supply voltage falls below its regulation limits.

Power Supply. GM: A power source. Normally applied to the prime source, e.g., line supply, battery, or generator.
PSS: More often, a unit which conditions a source of unregulated power to provide a regulated output. Not necessarily a prime source of power.

Preregulator. PSS: A regulator that precedes the converter or DC transformer to provide input regulation. Very often the preregulator is controlled by feedback from one of the output lines, to provide overall regulation. This technique is very useful for multiple-output units.

Programming. The control of output parameters by remote external means (e.g., resistance, voltage, or digital signals).

Recovery Time. The time taken for the controlled output parameter to recover to a value within the regulation limits after a step change in load.

Reentrant Current Protection. See "current foldback."

Remote Programming. See "programming."

Remote Sensing. PSS: A provision for sensing the voltage at the load, rather than at the power supply output terminals. This method of connection provides compensation for voltage drops in the supply lines, so as to maintain the voltage at the load constant.

Resolution. The smallest increment of change that can be made in the controlled parameter.

Response Time. PSS: The time taken for the controlled output parameter to change from 10% to 90% of its programmed change in response to a step change in the programming signal.

Reverse Protection. PSS: Protection of the power supply output components and regulators by shunt- and series-connected diodes. Used to prevent voltage reversal across the output terminals or internal regulators. Also required for series or bipolar operation of constant-voltage supplies with cross-connected loads.

Saturable Reactor. PSS: An inductor designed and controlled in such a way that it will saturate after a defined volt-second stress. Saturable reactors are used as magnetic duty ratio controllers in converter outputs.

Self-Recovery.   The ability to automatically recover to normal operation after removal of a stress condition.

Sequencing.   The control of a multiple-output unit in such a way that output voltages are established or turned off in a defined order or sequence.

Series Regulator.   PSS: A regulator in which the active element is in series with the supply output. A term often applied to linear regulators.

Short-Circuit Protection.   PSS: A method of overload protection which will prevent failure of the supply if a short circuit is applied to the output. Some flyback switchmode units are designed so that they will withstand a short circuit, but may not survive long-term intermediate overload conditions.

Shunt Regulator.   A method of control in which the power controller is in parallel with the output terminals. Shunt regulators have the advantage of absorbing minimum power when the output is fully loaded. A shunt-regulated output can source or sink current and is useful for servomotor control during overrun conditions.

Slew Rate.   PSS: The maximum rate of change of the control circuit and power modulator in response to an overdrive condition.

SMPS.   PSS: Acronym for "switchmode power supply."

Snubber.   PSS: (1) A network used to reduce the stress on a switching component. (2) A network used to reduce the rate of change of voltage on diodes or switching devices to minimize RF components and dv/dt stress.

Soft Start.   PSS: A method of controlling the initial rate of increase in duty ratio in a switchmode power supply during the turn-on transient. (Used to reduce the stress on internal components and to prevent transformer saturation.)

Source Effect.   See "line regulation."

Squegging.   PSS: In switchmode power supplies, squegging refers to a nondamaging, bounded instability mode, characterized by a short period of switching action followed by a short quiet period. Squegging normally occurs under light loading conditions, where the pulse width approaches the storage time of the switching device, so that further progressive control is not possible.

Switching Regulator.   PSS: A switchmode DC-to-DC regulator in which input and output share a common line.

Thermal Protection.   PSS: A method of preventing failure of the power supply as a result of an overtemperature condition. (The power supply may be shut down should a critical component exceed a defined temperature.)

Topology.   PSS: A term referring to the general arrangement of parts to form a particular circuit form, e.g., boost topology.

UPS.   PSS: (Uninterruptible power supply.) A power supply that will continue to provide regulated outputs without interruption in the event of failure of the primary supply. (UPSs use more than one energy source; the backup may be DC from a battery or ac from generators or inverters. In most applications, standby operation is provided only for a limited period.)

Voltage Stabilizer.   PSS: A device used to maintain a constant voltage, for example, a zener diode. Voltage stabilizers are not themselves a source of energy.

REFERENCES

1. McLyman, Colonel Wm. T., *Transformer and Inductor Design Handbook*, Marcel Dekker, New York, 1978. ISBN 0-8247-6801-9.
2. McLyman, Colonel Wm. T., *Magnetic Core Selection for Transformers and Inductors*, Marcel Dekker, New York, 1982. ISBN 0-8247-1873-9
3. Kraus, John D., Ph.D., *Electromagnetics*, McGraw-Hill, New York, 1953.
4. Boll, Richard, *Soft Magnetic Materials*, Heydon & Sons, London, 1979. ISBN 0-85501-263-3. & ISBN 3-8009-1272-4.
5. Smith, Steve, *Magnetic Components*, Van Nostrand Reinhold, New York, 1985. ISBN 0-442-20397-7.
6. Grossner, Nathan R., *Transformers for Electronic Circuits*, McGraw-Hill, New York, 1983. ISBN 0-07-024979-2.
7. Lee, R., *Electronic Transformers and Circuits*, Wiley, New York, 1955.
8. Snelling, E. C., *Soft Ferrites—Properties and Applications*, Iliffe, London, 1969.
9. Middlebrook, R. D., and Ćuk, Slobodan, *Advances in Switch Power Conversion*, Vols. I and II, Teslaco, Calif., 1983.
10. Ćuk, Slobodan, and Middlebrook, R. D., *Advances in Switchmode Power Conversion*, Vol. III, Teslaco, Calif., 1983.
11. Landee, Davis, and Albrecht, *Electronic Designer's Handbook*, McGraw-Hill, New York, 1957.
12. The Royal Signals, *Handbook of Line Communications*, Her Majesty's Stationery Office, 1947.
13. Langford-Smith, F., *Radio Designer's Handbook*, Iliffe & Son, London, 1953.
14. Pressman, Abraham I., *Switching and Linear Power Supply, Power Converter Design*, Haydon, 1977. ISBN 0-8104-5847-0.
15. Dixon, Lloyd H., and Potel Raoji, *Unitrode Switching Regulated Power Supply Design Seminar Manual*, 1985.
16. Severns, Rudolph P., and Bloom, Gordon E., *Modern DC-to-DC Switchmode Power Converter Circuits*, Van Nostrand Reinhold, New York, 1985. ISBN 0-442-21396-4.
17. Hnatek, Eugene R., *Design of Solid-State Power Supplies*, 2d Ed., Van Nostrand Reinhold, New York, 1981. ISBN 0-442-23429-5.
18. Shepard, Jeffrey D., *Power Supplies*, Restin Publishing Company, 1984. ISBN 0-8359-5568-0.
19. Chryssis, George, *High Frequency Switching Power Supplies*, McGraw-Hill, New York, 1984. ISBN 0-07-010949-4.
20. KitSum, K., *Switchmode Power Conversion*, Marcel Dekker, New York, 1984. ISBN 0-8247-7234-2.

21. Oxner, Edwin S., *Power FETs and Their Applications*, Prentice-Hall, Englewood Cliffs, N.J., 1982.

22. Bode, H., *Network Analysis and Feedback Amplifier Design*, Van Nostrand, Princeton, N.J., 1945.

23. Geyger, W., *Nonlinear-Magnetic Control Devices*, Wiley, New York, 1964.

24. Tarter, Ralph E., *Principles of Solid-State Power Conversion*, Howard W. Sams, Indianapolis, 1985.

25. Hanna, C. R., "Design of Reactances and Transformers Which Carry Direct Current," *Trans. AIEE*, 1927.

26. Schade, O. H., *Proc. IRE*, July 1943.

27. Venable, D. H., and Foster, S. R., "Practical Techniques for Analyzing, Measuring and Stabilizing Feedback Control Loops in Switching Regulators and Converters," *Powercon*, 7, 1982.

28. Middlebrook, R. D., "Input Filter Considerations in Design and Application of Switching Regulators," *IEEE Industrial Applications Society Annual Meeting Record*, October 1976.

29. Middlebrook, R. D., "Design Techniques for Preventing Input Filter Oscillations in Switched-Mode Regulators," *Proc. Powercon*, 5, May 1978.

30. Ćuk, Slobodan, "Analysis of Integrated Magnetics to Eliminate Current Ripple in Switching Converters," *PCI Conference Proceedings*, April 1983.

31. Dowell, P. L., "Effects of Eddy Currents in Transformer Windings," *Proc. IEE*, 113(8), 1966.

32. Smith, C. H., and Rosen, M., "Amorphous Metal Reactor Cores for Switching Applications," *Proceedings, International PCI Conference*, Munich, September 1981.

33. Jansson, L., "A Survey of Converter Circuits for Switched-Mode Power Supplies," Mullard Technical Communications, Vol. 12, No. 119, July 1973.

34. "Switchers Pursue Linears Below 100 W," *Electronic Products*, September 1981.

35. Snigier, Paul., "Those Sneaky Switchers," *Electronic Products*, March 1980, and "Power Supply Selection Criteria," *Digital Design*, August 1981.

36. Boschert, Robert J., "Reducing Infant Mortality in Switches," *Electronic Products*, April 1981.

37. Shepard, Jeffrey D., "Switching Power Supplies: the FCC, VDE, and You," *Electronic Products*, March 1980.

38. Royer, G. H., "A Switching Transistor DC to AC Converter Having an Output Frequency Proportional to the DC Input Voltage," *AIEE*, July 1955.

39. Jensen, J., "An Improved Square Wave Oscillator Circuit," *IERE Trans. on Circuit Theory*, September 1957.

40. IEEE Std. 587-1980, "IEEE Guide for Surge Voltages in Low-Voltage AC Power Circuits," ANSI/IEEE C62-41-1980.

41. "Transformer Core Selection for SMPS," Mullard Technical Publication M81-0032, 1981.

42. "Radio Frequency Interference Suppression in Switched-Mode Power Supplies," Mullard Technical Note 30, 1975.

43. Owen, Greg, "Thermal Management Techniques Keep Semiconductors Cool," *Electronics*, Sept. 25, 1980.

44. Pearson, W. R., "Designing Optimum Snubber Circuits for the Transistor Bridge Configuration." *Proc. Powercon*, 9, 1982.

45. Severns, R., "A New Improved and Simplified Proportional Base Drive Circuit," Intersil.

46. Redl, Richard, and Sokal, Nathan O., "Optimizing Dynamic Behaviour with Input and Output Feed-forward and Current-mode Control," *Proc. Powercon*, 7, 1980.

47. Middlebrook, R.D., Hsu, Shi-Ping, Brown, Art, and Rensink, Lowman, "Modelling

and Analysis of Switching DC-DC Converters in Constant-Frequency Current-Programmed Mode," IEEE Power Electronics Specialists Conference, 1979.

48. Bloom, Gordon (Ed), and Severns, Rudy, "Magnetic Integration Methods for Transformers," in *Isolated Buck and Boost DC-DC Converters*, 1982.

49. Hetterscheid, W., "Base Circuit Design for High-Voltage Switching Transistors in Power Converters," Mullard Technical Note 6, 1974.

50. Gates, T. W., and Ballard, M. F., "Safe Operating Area for Power Transistors," Mullard Technical Communications, Vol. 13, No. 122, April 1974.

51. Dean-Venable, H., "The K Factor: A New Mathematical Tool for Stability Analysis and Synthesis," *Proc. Powercon*, 10, March 1983.

52. Dean-Venable, H., and Foster, Stephen R., "Practical Techniques for Analyzing, Measuring, and Stabilizing Feedback Control Loops in Switching Regulators and Converters," *Proc. Powercon*, 7, 1980.

53. Tuttle, Wayne H., "The Relationship of Output Impedance to Feedback Loop Parameters," *PCIM*, November 1986.

54. Dean-Venable, H., "Stability Analysis Made Simple," Venable Industries, Torrance, Calif., 1982.

55. Tuttle, Wayne H., "Relating Converter Transient Response to Feedback Loop Design," *Proc. Powercon*, 11, 1984.

56. Dean-Venable, H., "Optimum Feedback Amplifier Design Control Systems," *Proc. IECEC*, August 1986.

57. Tuttle, Wayne H., "Why Conditionally Stable Systems Do Not Oscillate," *Proc. PCI*, October 1985.

58. Jongsma, J., and Bracke, L. P. M., "Improved Method of Power-Coke Design," *Electronic Components and Applications,* vol. 4, no. 2, 1982.

59. Bracke, L. P. M., and Geerlings, F. C., "Switched-Mode Power Supply Magnetic Component Requirements," Philips Electronic Components and Materials, 1982.

60. Carsten, Bruce, "High Frequency Conductor Losses in Switchmode Magnetics," *PCIM*, November 1986.

61. Clarke, J. C., "The Design of Small Current Transformers," *Electrical Review,* January 1985.

62. Houldsworth, J. A., "Purpose-Designed Ferrite Toroids for Isolated Current Measurements in Power Electronic Equipment," Mullard Technical Publication M81-0026, 1981.

63. Cox, Jim, "Powdered Iron Cores and a New Graphical Aid to Choke Design," *Powerconversion International*, February 1980.

64. Cox, Jim, "Characteristics and Selection of Iron Powder Cores for Induction in Switchmode Converters," *Proc. Powercon*, 8, 1981.

65. Cattermole, Patrick A., "Optimizing Flyback Transformer Design." *Proc. Powercon*, 1979, PC 79-1-3.

66. Geerlings, F. C., and Bracke, L. P. M., "High-Frequency Ferrite Power Transformer and Choke Design, Part 1," *Electronic Components and Applications,* vol. 4, no. 2., 1982.

67. Jansson, L. E., "Power-handling Capability of Ferrite Transformers and Chokes for Switched-Mode Power Supplies," Mullard Technical Note 31, 1976.

68. Hirschmann, W., Macek, O., and Soylemez, A. I., "Switching Power Supplies 1 (General, Basic Circuits)," Siemens Application Note.

69. Ackermann, W., and Hirschmann, W., "Switching Power Supplies 2, (Components and Their Selection and Application Criteria)," Siemens Application Note.

70. Schaller, R., "Switching Power Supplies 3, (Radio Interference Suppression)," Siemens Application Note.

71. Macek, O., "Switching Power Supplies 4, (Basic Dimensioning)," Siemens Application Note.

72. Bulletin SFB, Buss Small Dimension Fuses, Bussmann Division, McGraw-Edison Co., Missouri.

73. Catalog #20, Littlefuse Circuit Protection Components, Littlefuse Tracor, Des Plaines, Ill.

74. Bulletin-B200, Brush HRC Current Limiting Fuses, Hawker Siddeley Electric Motors, Canada.

75. Bulletins PC-104E and PC109C, MPP and Iron Powder Cores, The Arnold Engineering Co., Marengo, Illinois.

76. Publication TP25-575, HCR Alloy, Telcon Metals Ltd., Sussex, England.

77. Catalog 4, Iron Powder Toridal Cores for EMI and Power Filters, Micrometals, Anaheim, Calif.

78. Bulletin 59–107, Soft Ferrites, Stackpole, St. Marys, Pa.

79. SOAR—The Basis for Reliable Power Circuit Design, Philips Product Information #68.

80. Bennett, Wilfred P., and Kurnbatovic, Robert A., "Power and Energy Limitations of Bipolar Transistors Imposed by Thermal-Mode and Current-Mode Second-Breakdown Mechanisms," *IEEE Transactions on Electron Devices*, vol. ED28, no. 10, October 1981.

81. Roark, D. "Base Drive Considerations in High Power Switching Transistors," TRW Applications Note #120, 1975.

82. Gates, T. W., and Ballard, M. F., "Safe Operating Area for Power Transistors," Mullard Technical Communications, vol. 13, no. 122, April 1974.

83. Williams, P. E., "Mathematical Theory of Rectifier Circuits with Capacitor-Input Filters," *Power Conversion International*, October 1982.

84. "Guide for Surge Voltages in Low-Voltage AC Power Circuits," IEC Publication 664, 1980.

85. KitSum, K., *PCIM*, February 1998.

86. Spangler, J., *Proc. Sixth Annual Applied Power Electronics Conf.*, Dallas, March 10–15,1991.

87. Neufeld, H., "Control IC for Near Unity Power Factor in SMPS," Cherry Semiconductor Corp., October 1989.

88. Micro Linear application notes 16 and 33.

89. Micro Linear application note 34.

89. Micrometals' "Power Conversion & Line Filter Applications" data book.

INDEX

ABOUT THE AUTHOR

KEITH BILLINGS, President of DKB Power Inc. and engineering design consultant, has over 35 years of experience in switch-mode power supply design. He is a Chartered Electronics Engineer and a full member of Great Britain's Institution of Electrical Engineers.